Sustainable Materials for Electrochemical Capacitors

AF794735

Scrivener Publishing
100 Cummings Center, Suite 541J
Beverly, MA 01915-6106

Publishers at Scrivener
Martin Scrivener (martin@scrivenerpublishing.com)
Phillip Carmical (pcarmical@scrivenerpublishing.com)

Sustainable Materials for Electrochemical Capacitors

Edited by

Inamuddin

Department of Applied Chemistry, Aligarh Muslim University, Aligarh, India

Tariq Altalhi

Department of Chemistry, College of Science, Taif University, Saudi Arabia

and

Sayed Mohammed Adnan

Department of Chemical Engineering, Aligarh Muslim University, Aligarh, India

Scrivener Publishing

WILEY

This edition first published 2023 by John Wiley & Sons, Inc., 111 River Street, Hoboken, NJ 07030, USA and Scrivener Publishing LLC, 100 Cummings Center, Suite 541J, Beverly, MA 01915, USA
© 2023 Scrivener Publishing LLC
For more information about Scrivener publications please visit www.scrivenerpublishing.com.

All rights reserved. No part of this publication may be reproduced, stored in a retrieval system, or transmitted, in any form or by any means, electronic, mechanical, photocopying, recording, or otherwise, except as permitted by law. Advice on how to obtain permission to reuse material from this title is available at http://www.wiley.com/go/permissions.

Wiley Global Headquarters
111 River Street, Hoboken, NJ 07030, USA

For details of our global editorial offices, customer services, and more information about Wiley products visit us at www.wiley.com.

Limit of Liability/Disclaimer of Warranty
While the publisher and authors have used their best efforts in preparing this work, they make no representations or warranties with respect to the accuracy or completeness of the contents of this work and specifically disclaim all warranties, including without limitation any implied warranties of merchant-ability or fitness for a particular purpose. No warranty may be created or extended by sales representatives, written sales materials, or promotional statements for this work. The fact that an organization, website, or product is referred to in this work as a citation and/or potential source of further information does not mean that the publisher and authors endorse the information or services the organization, website, or product may provide or recommendations it may make. This work is sold with the understanding that the publisher is not engaged in rendering professional services. The advice and strategies contained herein may not be suitable for your situation. You should consult with a specialist where appropriate. Neither the publisher nor authors shall be liable for any loss of profit or any other commercial damages, including but not limited to special, incidental, consequential, or other damages. Further, readers should be aware that websites listed in this work may have changed or disappeared between when this work was written and when it is read.

Library of Congress Cataloging-in-Publication Data

ISBN 978-1-394-16623-7

Cover images: Pixabay.Com
Cover design by Russell Richardson

Set in size of 11pt and Minion Pro by Manila Typesetting Company, Makati, Philippines

Printed in the USA

10 9 8 7 6 5 4 3 2 1

Contents

Preface

Electrochemical capacitors are ideal for a wide range of elevated operations due to the highly adjustable qualities of the materials employed in their production. Many regard them as effective alternatives to conventional capacitors and batteries due to their ability to enhance the system according to the anticipated power/energy characteristics. Electrochemical capacitors with considerable surface areas of their carbon components are desirable, because of their cost effectiveness, adjustable forms, frequent availability, and environmental friendliness. As a result of recent research, many unique approaches for employing bio-derived components in highly efficient electrochemical capacitors are under development.

This book summarizes the progress in the usage of ubiquitous environmentally sustainable materials.

Chapter 1 describes the eco-friendly electrode materials used in electrochemical supercapacitor devices. The techniques of synthesis, characterization, and electrochemical performance of carbon-based, metal oxide-based, and carbon/metal oxide composite electrode materials are explained. Moreover, this chapter highlights the pros, cons, and future perspectives of eco-friendly supercapacitor electrode materials.

Chapter 2 discusses the importance of supercapacitor devices among energy storage devices. This chapter examines various environmental waste-derived carbon nanomaterials and general preparation methods, activation processes, recent advancements, challenges, and future perspectives.

Chapter 3 discusses the fabrication of metal hydroxide through precipitation, consolidation, ion exchange, sonochemical, hydrothermal, and polyol synthesis. Primary topics include the applications in nanotechnology, electrochemistry, pharmaceutical, and wastewater treatment. Furthermore, examples of metal hydroxide are summarized, along with the advantages and drawbacks in humans and animals.

Chapter 4 describes the nature, classifications, and various Porous Organic Polymers (POP). These POPs' history, chemistry, inner classifications, and synthetic strategies are discussed. POPs possess the advantages of both porous materials and conventional polymers, and have tunable natural porous size, shape, and morphology, which can be expanded for application in material chemistry and biology.

Chapter 5 explains the classification of naturally occurring polymeric materials that are employed for hydrogel fabrication, along with the diverse properties and various synthesis techniques of hydrogels. This chapter closely focuses on the utilization of hydrogels in electrochemical supercapacitors.

Chapter 6 details the progress of ionic liquids as new green electrolytes for supercapacitors. Topics include the research progress and future directions of pure ionic liquids, functional ionic liquids, mixed systems of ionic liquids and ionic liquids, mixtures of ionic liquids, and organic solvents or salts as super-capacitor electrolytes.

Chapter 7 focuses on different kinds of sustainable binders used for the construction of electrochemical capacitors. Furthermore, it discusses the properties of binders, the mechanism involved in the binding process, as well as various classifications of binding materials. Issues related to the use of conventional binders are also presented.

Chapter 8 discusses the use of non-flammable fluorinated electrolytes for electrochemical capacitors and their drawbacks. The primary focus is to give an overview of alternative fluorine-free sustainable electrolyte materials, categorizing them, and comparing their performance to fluorinated electrolytes. In addition, additives that can improve the performance of sustainable fluorine-free electrolytes are briefly discussed.

Chapter 9 outlines various electrolytes suitable for high-performance electrochemical supercapacitors. Topics includes electrolyte classification, types, with examples and the supercapacitor characteristics that are influenced by the nature of electrolytes employed.

Chapter 10 describes how biodegradable electrolytes can eliminate the use of fossil fuels, and their classes corresponding to manufacturing methods. Also included are the properties of biodegradable electrolytes, the factors that influence their ionic conduction, and the application of biodegradable electrolytes in fuel cells, batteries, and supercapacitors.

Chapter 11 explores the importance of supercapattery devices among the energy storage devices, such as batteries and capacitors. The chapter encompasses their working mechanisms, along with the various electrode materials, such as metal-based, carbon-based, and polymer-based materials and their composites. Furthermore, it concerns the utilization, advantages, and hassles of supercapattery devices.

Chapter 12 explains the supercapacitors which are electrochemical energy storage devices that can cover the power/energy gap between conventional capacitors and batteries/fuel cells. The choice of electrode material and electrolyte directly influences its performance. Ceramic-based materials are very promising, and this chapter deals with the various advances and challenges of using ceramic-based materials as supercapacitor electrodes.

Chapter 13 discusses the many doping techniques that are utilized to create sustainable functional materials for next-generation supercapacitors, including N-, S-, B-, and P-doped and co-doped carbon materials. This chapter goes into great detail about the intriguing effects of heteroatom doping with carbon on supercapacitors. Furthermore, it provides a thorough assessment of the electrochemical performance of mono- or multi-doped carbon composites acting as superior electrodes in energy storage systems.

Chapter 14 discusses the rapid development of wearable supercapacitors in practical applications. Additionally, the synthetic processes for material fabrication is explained with important examples. Specific emphasis is placed upon integrated devices and their applications with supercapacitors as a power source.

Chapter 15 has covered the history and process of electrospinning. The electrospinning process's working parameters, like solution, processing, and ambient parameters, are thoroughly described. Moreover, the preparation, pore-formation, and modification processes of electrospun polymer nanofibers are extensively discussed. An emphasis is placed upon the most recent and important developments in electrospun nanofiber applications.

Chapter 16 reviews the use of polysaccharides-based electrolytes and electrodes for electrochemical applications. This covers the use of polysaccharides for electrolyte synthesis—from salt-in-polymer-electrolyte to polymer-in-salt-electrolyte range—and electrode synthesis to obtain hierarchical electrodes with the suitable distribution of micro and mesopores using external- and self-templates. Nitrogen/oxygen-doped electrode synthesis is also summarized.

Chapter 17 highlights advanced polymer inks for printable supercapacitors. The working principles of supercapacitors are discussed. Printing technologies for fabricating supercapacitors, such as screen printing, inkjet printing, and 3D printing are emphasized.

Chapter 18 describes the significance, tunable parameters, and different synthesis techniques of activated carbon from biomass for supercapacitor application. Moreover, it provides a detailed description of the constituents, applications, advantages, and limitations of a few major sources, like plant, animal, and microorganism-derived carbon as electrode material, as reported in recent literature.

Our thanks go to Wiley and Scrivener Publishing for their continuous support and guidance.

Inamuddin
Department of Applied Chemistry, Zakir Husain College of Engineering and Technology, Aligarh Muslim University, Aligarh, India
Tariq Altalhi
Department of Chemistry, College of Science, Taif University, Taif, Saudi Arabia
Sayed Mohammed Adnan
Department of Chemical Engineering, Zakir Husain College of Engineering and Technology, Aligarh Muslim University, Aligarh, India

1

Sustainable Materials for Electrochemical Supercapacitors: Eco Materials

R. Kumar and R. Thangappan*

Advanced Functional Materials for Energy Research Lab, Department of Energy Science & Technology, Periyar University, Salem, Tamil Nadu, India

Abstract
Electrochemical storage systems like secondary batteries and ultra-capacitors need continual performance development due to the long-term demand for rechargeable devices with a high specific energy and rapid charging. Supercapacitors are a recent development for electrochemical energy storage devices. Electrode materials have a lot of interesting features, like a wide active area, high electrical, chemical, and mechanical properties. That kind of makes this a great material for supercapacitor devices. The most prominent factor of electrode materials is that they are environment friendly and non-toxic. Thereby further, recent researchers are working to develop eco-friendly materials for supercapacitor electrode materials by using the bio waste and waste product extracts. In this chapter fully focused on eco-friendly electrode materials for supercapacitors application. Then, the electrode materials synthesis method, characterization, and electrochemical performance are discussed. The electrode materials were categorized into three parts: carbon-based, metal oxide-based, and carbon/metal oxide composite. Furthermore, the future perspectives of electrochemical energy storage devices are addressed.

Keywords: Carbon, metal oxide, composite, eco materials, supercapacitor

1.1 Introduction

In the present century, renewable energy is beginning to replace fossil fuels and non-renewable energy sources. When energy production is a necessity, as well as energy storage is very important. Now, people used to the two major types of electrochemical energy storage systems like battery and ultra-capacitor [1]. Figure 1.1 plots a comparison of energy and power density for different devices that store energy. The batteries have limited specifications of low charge–discharge time, restricted life, and large mass in the portable devices. Now, the modern technology of energy storage material is supercapacitor. It's have an enhanced high power, low weight, flexible, inexpensive and eco-friendly storage device compared to

**Corresponding author*: thangappan@periyaruniversity.ac.in

Inamuddin, Tariq Altalhi and Sayed Mohammed Adnan (eds.) Sustainable Materials for Electrochemical Capacitors, (1–18) © 2023 Scrivener Publishing LLC

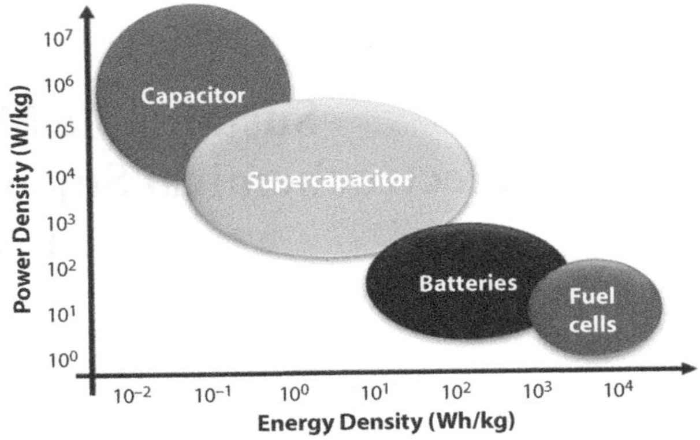

Figure 1.1 Energy versus power plot for various energy storage devices.

battery [2]. The supercapacitor as so called as an ultra-capacitor. According to the storage mechanism, it's classified those three types such as electrostatic double layer capacitance (non-faradic), electrochemical pseudocapacitance (faradic) and hybrid supercapacitor (faradic & non-faradic). In hybrid supercapacitor have electrostatic and electrochemical storage mechanism. Moreover, electrical double layer capacitance act on isolating charges at the conductive interface and pseudo capacitors store charges through reverse reduction and oxidation reactions at the electro active surface [3, 4]. EDLC has carbon related electrode materials included activated carbon, carbon nanotubes (CNT) and graphene [5–7]. Nevertheless, conductive polymers and transition metal oxides/hydroxides can be used as pseudo-capacitor positive terminal [8, 9]. Figure 1.2 depicted that different type's electrode material for the supercapacitor applications.

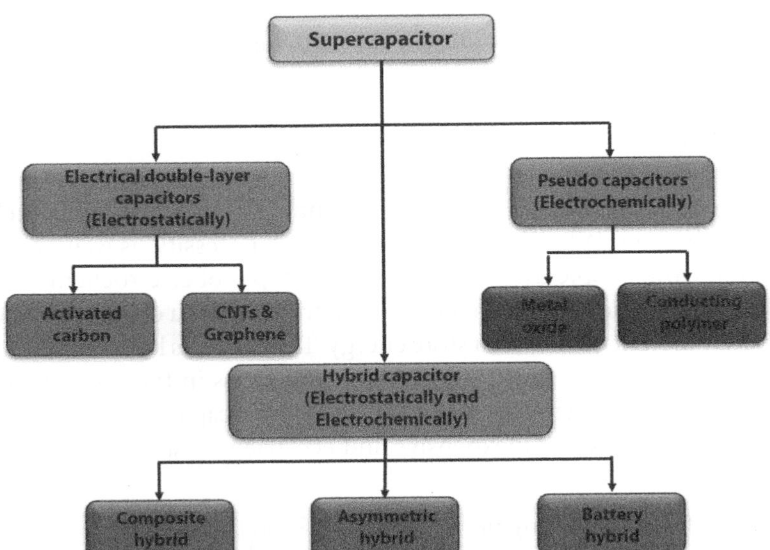

Figure 1.2 Various electrode materials for electrochemical supercapacitor.

An important function for the supercapacitor device has been played by the working electrode and the electrolyte. In particular, electrode materials must have vital features of low cost, high active surface area, high output power, and large duration [10]. The performance of active materials are depending upon the electron conductivity path way and structural of the electrode materials. Furthermore, the nanostructure of these materials was important in deciding electrochemical performance. Nanostructured electrode materials are divided into four different sorts of dimensionality, including 0D, 1D, 2D, and 3D [11, 12]. Recently, researchers have become interested in using biomass and green synthesis materials in supercapacitor electrode. The creation of a bio waste-based supercapacitor electrode material serves to supports in waste disposal by transforming waste into a valuable product of supercapacitor technology. Nevertheless, several harmful chemical agents or rigorous techniques have been utilized to prepare the materials for supercapacitors. Thus, green technology that uses electrochemical energy to reserve materials is evaluated from the perspectives of biomass sources and manufacture. Biomass has been processed in two distinct processes: pyrolysis and extract. Biomass-produced electrode materials are ideal for supercapacitor device because of their high active surface and rapid cycle durability [13].

In this chapter, recent advances on the synthesis methods, material structures, large contact surface, and electrochemical behavior of eco-friendly active electrode for supercapacitor application were discussed. Moreover, the electrochemical features of electrode materials including such specific capacity, cyclic stability, power, and energy density also reported. In this study, classified to three different types of environmental friendly supercapacitor anode and cathode like carbon based materials, metal oxides based material and both of them composite materials.

1.2 Eco-Carbon-Based Electrode Materials

Carbon is a vital component in the lives of human people as well as in the functioning of our current ecology and it is numerous in the environment. Since carbon related materials had a lot of effective surface area and electron conduction, when mostly used anode materials are graphene, carbon nanotubes, activated carbon and carbon aerogel [14]. Some of the researchers find to the waste material or green synthesis of electrode materials. Firstly, Kotchaphan Kanjana *et al.* eco-friendly synthesized using rubber seed shells, durian shells and palm petiole shells derived activated carbon in supercapacitor device. The activated carbon was made using an environmental friendly self-activation process with KOH as an alkali activator. The activated carbon derived from durian shell (DS) had the large capacitance of 178 Fg^{-1} and the highest cyclic stability tested to 4000 continuous charging and discharging cycles between the potentials of -0.2 to 0.2 V. This DS AC features a number of flakes on smooth surface, high porosity distribution, good electrochemical energy storage and charge carrier balance. Finally, it was concluded that the concentration of KOH had a significant impact on the porous morphology, more active area, crystalline phase, and graphitization mechanism, determining the supercapacitor performance of the produced electrode. There are several active agents included therein, although the KOH activator has gained significant interest due to its environmental friendliness as compared to other activating agents [15]. Moreover, Meenaketan Sethi *et al.* reported porous graphene has been synthesized that eco-friendly method of without using any toxic chemicals. A solvothermal method has been maintained at the common temperature for different time duration such as 16 to 32 hrs. The morphology structure of porous like graphene sheet is shown in Figure 1.3A (a, b). The porous graphene sheets (PG28) have a total active place of 420 m^2g^{-1} and obtained large specific capacity of 666 Fg^{-1} at 5 mVs^{-1} in positive potential window. After completing the 10,000 continuous GCD cycle, the PG 28 sample had 87% stability retention. Figure 1.3B (a–c) the resultant PG 28 electrode was tested for the symmetric supercapacitor application such as CV, GCD, and EIS. The symmetric device supports a high energy value of 26.3 Wh kg^{-1} and a peak power value of 6120 W kg^{-1}. Over the 5000 continuous cycles of galvanostatic charge-discharge, the PG 28 symmetrical supercapacitor electrodes preserved 93% of their initial capacitance value as shown in Figure 1.3B (d). The hierarchical porous structure PG 28 electrode material exhibited outstanding performance because it had a high active surface area and offered open ways but also a conductive path way for effective charge storage and ion movement [16]. Then Guoxiong Zhang *et al.* examined activated carbon from bamboo by carbonization and activation method. Various temperatures were kept during the carbonization process, with temperatures ranging from 700°C to 1000°C. Whereas prepared AC material at carbonized 900°C would have an effective highly

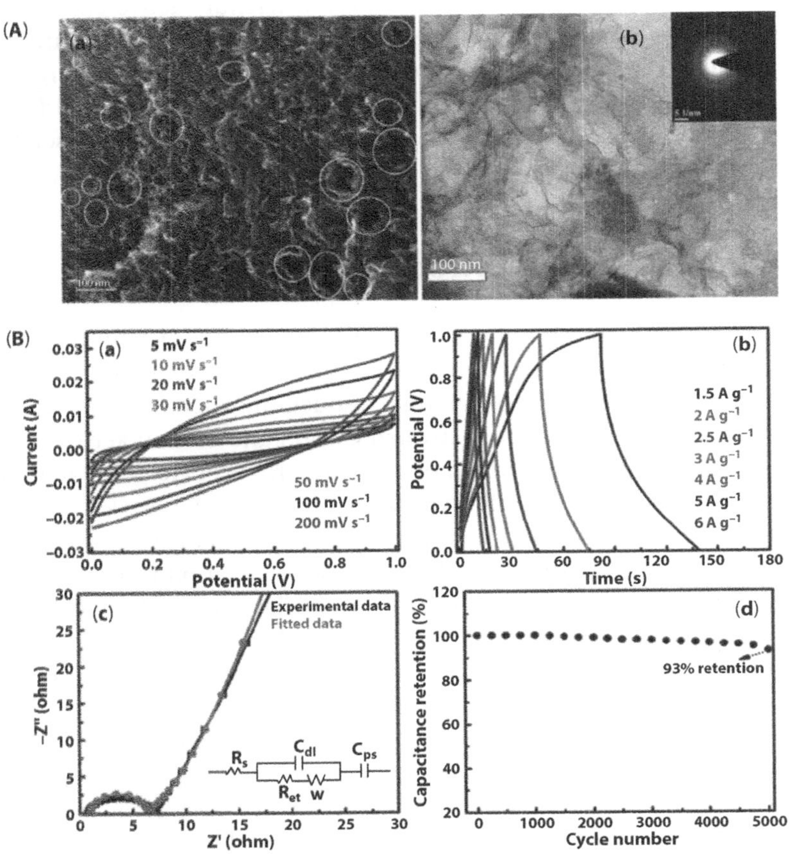

Figure 1.3 (A). (a, b) FESEM and TEM image of porous graphene sheet. (B). Electrochemical results of (a) CV, (b) GCD, (c) EIS and (d) cyclic stability for symmetrical supercapacitor fabricated from PG 28 electrode. Reproduced with permission [16]. Copyright 2019, Elsevier.

porous structure with a high outermost layer of 2221.1 $m^2 g^{-1}$, the excellent specific capacitive of 293 F g^{-1} at current range of 0.5 Ag $^{-1}$ and wonderful rate capability in 3 M KOH aqueous electrolyte solution. Assembled Symmetric device of the optimal electrode obtained an energy and power density (10.9 Wh/kg and 63 W/kg) and outstanding continuous cycle of 91.8% over 10000 cycle loops [17].

Divyashree A *et al.* produced that porous carbon nanosphere by pyrolysis method with three various types of coconut waste such as fiber (CF), leaves (CL) and stick (CS), respectively. The average particle size of the carbon spheres derived from coconut fiber approximately 20 nm and three different types of coconut waste derived carbon nanosphere SEM image shown in Figure 1.4 (a–c). Furthermore, the leaves and stick have a similar circular shape with particle sizes range between 30 to 60 nm. The pyrolysis used CF, CL, and CS coconut waste derived into carbon spheres had particle areas of 7 m^2/g, 4 m^2/g and 8 m^2/g, respectively. The synthesized wastes are tested in the electrochemical studies. Further the results show that the waste-derived electrode materials have high specific capacitance of 236, 116, and 208 Fg^{-1} from coconut fiber, leaves, and sticks, respectively [Figure 1.4 (d)]. Due to the fact that 2D flaks are coated with spheres, coconut fiber displayed a high capacitance and stability when compared to other waste [18]. Moreover, Xiao-Qiang Lin *et al.* fabricated to innovative, cost effective and environmental friendly electrode materials for supercapacitor application from Konjaku flour in way of precarbonization and KOH activation method. The precarbonized porous bio carbon was classified by different KOH activated agent concentrations, such as 3:1, 5:1, and 7:1. The surface study revealed that the surface of the 5:1 bio carbon approximately 1403 m^2g^{-1}. The 5:1 ratio of porous bio carbon displayed a maximum capacity of 216 F/g and a cyclic stability of 93.7 percent over 5000 cycles. Additionally, the symmetric device reached power/energy densities of 2.5 kW kg^{-1}/

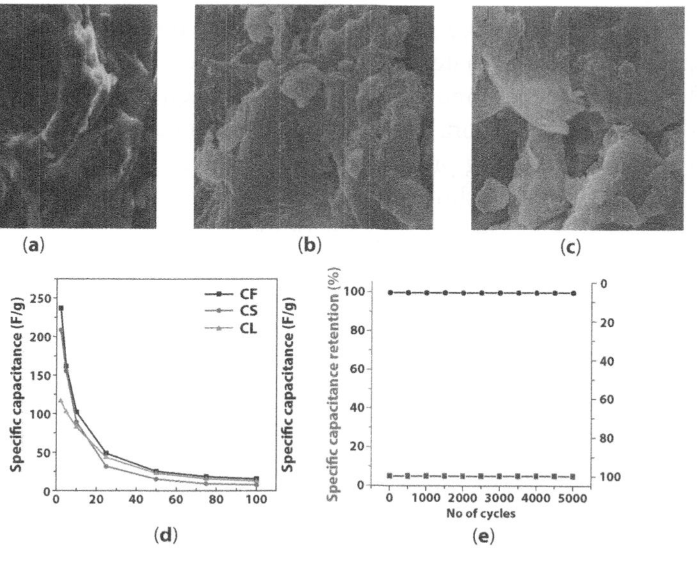

Figure 1.4 (a–c) scanning electron microscope images of coconut fiber (CF), coconut leaves (CL) and coconut stick (CS), respectively. (d) Specific capacitance of three different coconut wastes. (e) Capacitance retention of coconut fiber current density of 2.5 A/g for 5,000 cycles. Reproduced with permission [18]. Copyright 2016, Elsevier.

9.2 Wh kg^{-1} in 6 M KOH + 0.5 mM PPD electrolytes. Therefore, Konjaku flour derived porous bio carbon based electrode material was excellent supercapacitor device materials [19]. Then, Xiao-Li Su *et al.* used carbonization and KOH active process to synthesize loofah sponge-derived activated carbons. The activated carbons produced are categorized according to their amount of KOH activation. Now, porous structure SAC-4 electrode materials have been obtained an extremely large surface area of 2718 m^2g^{-1} according to the BET analysis. The prepared electrode sample has been tested for the electrochemical studies, compared to the different ratios of KOH activation electrodes; SAC-4 achieved a large capacitive of 309.6 Fg^{-1} at a current density of 1 Ag^{-1} in a three-electrode configuration with 6 M KOH electrolyte solution. With the 1 M Na_2SO_4 electrolyte, the symmetric device demonstrates power and energy output of 160 W/kg and 16.1 Wh/kg [20].

Sultan Ahmed *et al.* synthesized an activated carbon by using of *Butnea monsperma* flower pollens via thermal activation method and carbonized with $ZnCl_2$ activation agent. $ZnCl_2$ was added in various concentrations to the raw material of *Butnea monsperma* flower pollens. When compared to the other concentrations, the 1:2 ratio of AC produced the highest surface area of 1422.66 m^2g^{-1}. The activated carbon electrode of 1:2 showed a high capacitive of 130 Fg^{-1} and stability of more than 99 % in aqueous electrolyte for more than 10,000 cycles. Additionally, the large energy and power values of 19 kW/kg and 42 Wh/kg were determined [21]. Furthermore Yan Han *et al.* has been prepared the activated carbon from the Fish gill by using carbonization and KOH activation. At a mass ratio of 1:1, the pre-carbonized sample was combined with KOH and optimized for that device application. Fish gill derived Sheet like structures of activated carbon have high surface area due to found of oxygen and nitrogen functional groups. The AC nanosheets are exhibited the high specific capacitance of 334 F/g at 2 A/g and excellent cyclic stability in 6 M potassium hydroxide. Because of the synergistic effect of these properties on ion diffusion, transport and adsorption, and charge storage, the Fish gill derived activated carbon outperforms most bio-derived activated carbons. As prepared the eco-friendly material eligible to the supercapacitor electrode material [22]. Nannan Guo *et al.* synthesized environmental friendly material of tremella [Figure 1.5A (a)] derived activated carbon by using carbonization process and KOH activation. The activated carbon gets a high specific surface of 3760 $m^2\ g^{-1}$. After that KOH etching process, the activated carbons have a smooth surface and highly porous structure SEM shown in Figure 1.5A (b–f). By using the pyrolysis process the gas and unwanted particles are fully decomposed in the carbon surface. Compared to the different ratio of KOH activation, the ratio of 1:5 obtained the high capacitance value of 71 F g^{-1} at a current density 1 A g^{-1} in electrolyte solution of 6 mol of potassium hydroxide [Figure 1.5B (a–c)]. The symmetric supercapacitor was tested with three different electrolyte solutions: 6 mol potassium hydroxide (1V), 1 mol sodium sulfate (1.6 V), and neat 1-ethyl-3-methylimidazolium tetrafluoroborate (3V). The same anode and cathode terminal device displays a better energy and power value of 65.6 W h kg^{-1} and 19,700 W kg^{-1} [23].

The innovative bio char was created by extracting activated carbon from poultry litter waste. The poultry litters are then transformed to bio char via the pyrolysis process. Following that, the bio char was exposed to a chemical activation and carbonization procedure. Finally, the process obtained activated carbon with a 3000 m^2/g surface area. Daniele Pontirol *et al.* tested two different types of eco-friendly electrolyte in this two electrode configuration such as KOH and Na_2SO_4 respectively. The electrolyte of KOH and Na_2SO_4 reached the capacitance of 22 and 164 F/g [24]. There are several activating agents utilized

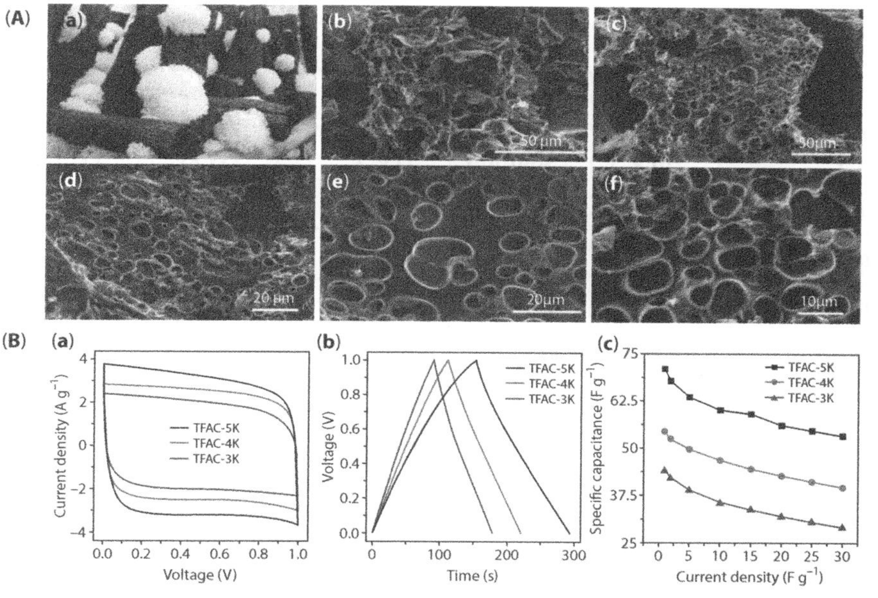

Figure 1.5 (A). (a) Image of tremella (b-f) SEM image of Biomass derived activated carbon. (B). (a–c) Electrochemical studies of activated carbon in 6 M KOH. Reproduced with permission [23]. Copyright 2017, Elsevier.

Table 1.1 Different types of bio waste derived carbon based electrode material in supercapacitor devices.

Source of electrode material	**Surface area (m^2g^{-1})**	**Ionic solution**	**Specific capacitance (Fg^{-1})**	**Power density ($W\ kg^{-1}$)**	**Energy density ($W\ h\ kg^{-1}$)**	**Ref.**
Bamboo	3000	6M KOH	300	-	-	[41]
European deciduous trees	614	1 M H_2SO_4	24	500	0.53	[42]
Tectona grandis leaf	514	1 M H_2SO_4	168	-	-	[11]
Sapindus trifoliatus nut shells	786	6M KOH	240.8	400	30	[43]
Hibiscus sabdariffa fruits	1720.46	2 M KOH	194.50	225	13.10	[44]
Willow catkins	645	6M KOH	340	-	-	[45]
Tree bark	1018	1 M Na_2SO_4	114	-	-	[46]
Finger grass flower	637.1	6 M KOH	315	6100	13.18	[47]

(*Continued*)

Table 1.1 Different types of bio waste derived carbon based electrode material in supercapacitor devices. (*Continued*)

Source of electrode material	Surface area (m^2g^{-1})	Ionic solution	Specific capacitance (Fg^{-1})	Power density ($W\,kg^{-1}$)	Energy density ($W\,h\,kg^{-1}$)	Ref.
Tea waste	1610	6 M KOH	332	-	-	[48]
Momordica charantia	1126	1 M H_2SO_4	186	6000	23	[49]
Mangosteen peel	2623	6 M KOH	357	401	17.28	[50]
Eucalyptus globulus seed	2388.38	1 M KOH	150	-	-	[51]
Potato peel	1911.5	1 M Na_2SO_4	323	800	45.5	[52]
Cotton	1508	6 M KOH	278	-	-	[53]
Jujube fruits	1135	6 M KOH	460	629	23.7	[54]
Albizia flowers	2757.63	6 M KOH	406	429	26.3	[55]
Solanum lycopersicum leaves	325.046	1 M H_2SO_4	345	61.34	43.13	[56]
Bean sprouts	397.15	1 M KOH	203.8	-	-	[57]

in the carbonization process, whereas Feiqiang Guo *et al.* produced activated carbon from peanut shell using a bimetallic active agent. The activators, in order, are the compounds with the formulas $ZnCl_2/MgCl_2$, $FeCl_3/MgCl_2$, and $FeCl_3/ZnCl_2$. Using BET surface analysis; the $FeCl_3/MgCl_2$ activation of carbon reveals a highly porous structure with a particle area of 1401.45 $m^2\,g^{-1}$. As prepared sample was achieved the specific capacitance (247.28 F/g at 1 A/g) and excellent cyclic loops. A symmetric supercapacitor with two activated carbon electrodes showed energy value of 32.7 Wh kg^{-1} and high power value of 588.3 W kg^{-1} [25]. The several types of carbon-based electrode materials obtained from biomass are reported in Table 1.1. According to the reviewed literatures above, biomass-derived carbon-based materials are suitable materials. Because, the porous shape that allows for ion movement and electrical interaction this positive findings suggest that activated biomass carbon compounds might be useful in supercapacitors.

1.3 Eco-Metal Oxide-Based Electrode Materials

Mostly, the behavior of ultra-capacitor is depends by the electrodes material composition. As a consequence, the scientists are researching highly active electrode for electrochemical energy storage. Commonly, the electrode materials should be low cost, environmental friendliness then high performance but some metal oxides have a highly toxic nature. The researchers have now contributed to the development of environmental friendly materials and green synthesis process. H.M. Abuzeid *et al.* synthesized α-MnO_2 from orange byproduct such like orange peel and orange juice using a green synthesis process. The MnO_2 was produced using a green approach that uses less solvents and organic substrates, which saves resources and costs, and also minimizes the toxicity of those substances. Disposable by-products were also used in the process. Orange juice and peel extract are used as a reducing agent to convert $KMnO_4$ to MnO_2 electrode material. Finally, to obtain orange peel extract, this has a greater surface area than orange juice extract. The performance of the cyclic voltammetry was evaluated throughout a potential range of -0.2 to 1.2 in 0.5 mol of sodium sulfate electrolyte. The orange peel extract produced MnO_2 reached the specific capacity of 139 F g^{-1} at the current value of 0.5 A g^{-1}. MnO_2 has been synthesized as an electrode material in a variety of ways, however this approach allows for the simple synthesis of MnO_2 for supercapacitor applications [26]. Likewise, Chae Eun Lee *et al.* developed on the pseudocapacitor properties of iron doped cerium oxide materials through the use of an environmentally friendly co-precipitation method. Then Iron particles are doped to the pure CeO_2 in the ratio of 1%, 5% and 10% respectively. Compared to the pure and Fe doped CeO_2, 10% of Fe doped CeO_2 exhibit the large capacitance value of 559 F g^{-1} at current range of 1 Ag^{-1}. The capacity of iron doped cerium oxide exceeds 4.6 times than the pure cerium oxide. For the reason of iron particles significantly increase the capacitance of CeO_2 electrode materials [27]. Whereas, G. Theophil Anand *et al.* generated ZnO nanoparticles using *Prunus dulcis* (almond gum) by green synthesis method. In this CV study, the possible window of 0 to 0.8 was examined using a variable scan rate. According to the pseudocapacitor mechanism in between of electrode and electrolyte, as a scan rate has been increased current also increased in CV analysis. So green synthesis approach of ZnO nanoparticle are best electrode material for supercapacitor applications [28]. Furthermore, H. E. Nsude *et al.* produced $CuFeS_2$ nanoparticles from touch me not (*Mimosa pudica*) leaves extract via a simple synthesis process with different annealed temperature. The four different temperatures are without annealed, 200°C, 250°C, and 300°C, respectively. As prepared electrode of $CuFeS_2$ nanoparticles at 250°C obtained the extreme capacitance of 501.4 F g^{-1}. By using a green synthesis approach, touch me not (Mimosa pudica) leaves extract have the best reducing agent for nanomaterial's production without polluting the environment [29]. Manab Kundu *et al.* also investigated NiO nanoparticles as electrode materials for supercapacitors. NiO was synthesized using *Hydrangea paniculata* flower extracts in a green synthesis process [Figure 1.6A (a)]. The average particle size of NiO nanoparticles was determined to be 33 nm in this TEM analysis [Figure 1.6A (b)], with an outer area of 78.472 m^2 g^{-1}. Figure 6B (a) depicts the CV curves of NiO nanoparticles at different scan rates in a applied voltage value of 0 to 0.5 V. Due to the high outer layer area of nickel oxide nanoparticles and the quick ion in rate of electronic diffusion during the faradic reduction and oxidation process, the anodic peaks shift positively while the cathodic peaks shift negatively in this CV curve. The galvanostatic charging/discharging curve was utilized to evaluate the capacitance of the NiO nanoparticle electrode material [Figure 1.6B (b)]. The green synthesis NiO nanoparticle electrode material obtained the specific capacitance of 752 Fg^{-1} at a current density of 2Ag^{-1}. Additionally, the NiO-NPs electrodes exhibit excellent continuous GCD up to 5000 cycles at 10 Ag^{-1}. Hydrangea paniculata flower extract was used as a nanoparticle conversion technology and achieved the best performance for supercapacitor studies [30].

Anacyclus pyrethrum plant extract were employed as a reducer in an involved to naturally synthesis RuO_2 nanoparticles and analyzed the electrochemical capability of RuO_2

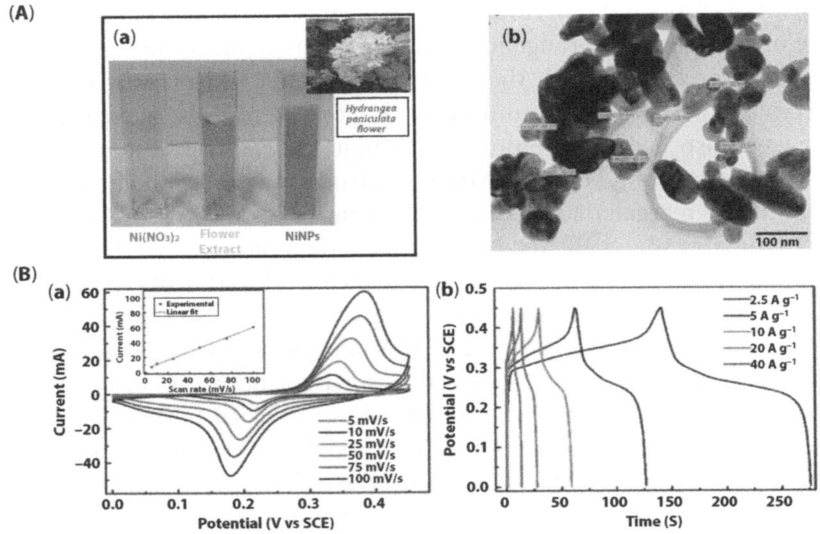

Figure 1.6 (A). (a) Hydrangea paniculata flower and extracts (b) TEM image of NiO nanoparticles. (B). (a) CV at various scan rate (b) GCD at various current value. Reproduced with permission [30]. Copyright 2021, Elsevier.

nanoparticles. Due to the fact that *Anacyclus pyrethrum* is an Indian herbal plant, it is readily accessible in markets. In TEM investigation, the average particle size of RuO_2 approximately 13 nm. RuO_2 active material coated on the surface of carbon conducting medium checked in a three electrode configuration, and the nanoparticles achieved an excellent capacitance rate of 209 F g^{-1} at a scan rate of 5 mV/s in 0.5 M of Na_2SO_4 ions conductive solution. Long-term cyclic loops of 98 % has been maintained in the electrode material over 1000 charge discharge loops [31]. E. Ismail *et al.* created a bio-synthesis technique for RuO_2 nanoparticles using *Aspalathus linearis* natural extract. This extract serves as the green synthesis method's redox and capping agent. The bio synthesis technologies are harmless and safe for the environment. The electrode setup was analyzed in the 0 to 0.5 V potential ranges. RuO_2 electrode material was deposited on the nickel foam surface, as well as the NiF/RuO_2 electrode attained highest specific capacitive value of 750 F g^{-1} at 10 Ag^{-1} and long-term cycling stability of 97% capacitance retained after 500 charge–discharge loops in 2 M potassium hydroxide. The porous nature of NiF/RuO_2 nanoparticles had outstanding performance, because the effective charge transfer and diffused ion in between of electrode and electrolyte surface [32]. Moreover, Irum Shaheen *et al.* examined the sol–gel fabrication of a ZnO–Co_3O_4 nanocomposite with *E. cognata* organic components. For the synthesis of nanomaterial, E. cognate acts as a binding agent and a reducer. The produced ZnO–Co_3O_4 nanocomposite exhibited significant homogeneity in particle arrangement with a porous structure. Preferably, the ZnO–Co_3O_4 nanocomposite should get a single particle size of 20 nm. The composite's specific capacitance was determined with cyclic voltammetry and found out to have been 165 F g^{-1}. The GCD graph was also used to compute an power and energy value of 7500 W/kg and 4.1 W h/kg. As an end, E. cognate is now one of the organic reducing agents used during fabrication of supercapacitor electrodes [33]. Furthermore, using a nontoxic aqueous phyto extract of guava leaves, Priyanka Lamba *et al.* produced NiO nanoparticles for use as supercapacitor electrodes. As a reducing and capping agent, guava leaf extract was added. The electrochemical properties of NiO nanoparticle electrodes were studied using 1, 5, 10, 15, and 20 mol of KOH ionic solution. The specific capacity of 85.31 Fg^{-1} in 5 M potassium hydroxide aqueous electrolyte at 2 mVs^{-1} demonstrates the supercapacitors strong capacitive performance. As a consequence, the performance of the electrode materials would be determined by the electrolyte solution [34]. Likewise, N. Matinise *et al.* used *Moringa oleifera* extract to create unique $ZnFe_2O_4$ nanocomposites. A green chemistry technique was used for the manufacture of $ZnFe_2O_4$ nanocomposite for electrochemical applications, which used Moringa Oleifera leaf extract as both a reducing and capping agent. The Glassy carbon electrode (GCE) supported $ZnFe_2O_4$ nanocomposites directly acted as the working electrode. Glassy carbon electrode (GCE) based $ZnFe_2O_4$ nanocomposites served as the working electrode directly. The CV performance of the GCE/$ZnFe_2O_4$ electrode was determined across a potential range of 0–0.6 V [35]. Pseudocapacitor electrodes have recently been employed in supercapacitors. Because the performance of synthesized metal oxide-based electrodes is higher than that of other electrode materials. Eco friendly metal oxides are achieved the remarkable performance for electrochemical supercapacitor and reduce the environmental effects. So the recent researchers are now developing the environmental friendly electrode materials for energy storage materials.

1.4 Eco-Carbon-Based Material/Metal Oxide Composite Electrode Materials

According to the aforementioned literatures, supercapacitor electrode materials based on carbon and metal oxide are created using waste materials and an environmentally acceptable synthesis approach. The researchers have indeed led to the advancement of a hybrid supercapacitor electrode material made of carbon-based materials and metal oxide composites. Because the hybrid composite electrode blends EDLC and Pseudocapacitance behaviors, it seems to have a high specific capacitance. C. Sasirekha *et al.* explored the eco-friendly preparation of ZnO/Carbon nanocomposite for supercapacitor electrode material in sol-gel method with sucrose as a capping agent. The ZnO nanoparticles are completely coated with carbon, which aids in effective charge transfer from the electrode/electrolyte contact. The CV and GCD of Pure Zink oxide and ZnO/Carbon composite terminals been studied at various scan rates and current densities in 1.0 M Na_2SO_4 at potentials ranging from 0 to 0.8V. According to the GCD curve, the ZnO/carbon composite does have a maximum capacity of 820 F/g at 1 A/g. The cyclic stability of composites attained 92 % capacitance retention after 400 loops. In a symmetric supercapacitor, the same two ZnO/carbon composite electrode materials demonstrate a capacitance of 92 F g^{-1} at current density of 2.5 A g^{-1} and an output energy of 32.78 W h kg^{-1}. For electrochemical devices, the ZnO/carbon composite was shown to have a high capacitance and a long lifespan [36]. Moreover, Y. N. Sudhakar *et al.* created rGO from Piper nigrum and CuO composite as anode for electrochemical capacitors. While maintaining the overall weight of the composite constant, several ratios of reduced graphene oxide–CuO nanocomposites were synthesized, namely 1:3 ratio, 1:2 ratio, 1:1 ratio, 2:1 ratio, 3:1 ratio, 4:1 ratio, 5:1 ratio, and 6:1 ratio. The morphological study of the reduced graphene oxide –CuO nanocomposite reveals that the CuO nano-leaves are well distributed over the rGO surface as leaves with equal sizes of around 100 nm. The electrode material was tested with several electrolyte solutions, including sodium sulfate, phosphoric acid, sulfuric acid, and (9:1) sulfuric acid: phosphoric acid. Comparing pure rGO and CuO electrodes, the 3:1 reduced graphene oxide/CuO nanocomposite had lower resistance and greater capacitive. For supercapacitors, the acid combination (9:1) was acceptable electrolyte. However, the symmetrical supercapacitor has a capacity of 137 F g^{-1}, and its cyclic stability is steady for charge–discharge cycles of up to 5,000 cycles [37].

Whereas, L. Jia *et al.* fabricated green synthesis technique of ultrafine porous carbon and MnO_2 nanowire as a anode and cathode material for asymmetric supercapacitors. Porous boron, nitrogen double doped carbon aerogel synthesized from the methyl cellulose. Importantly, this layered material minimized aggregation while increasing the effective area of contact, which resulted in exceptional electrochemical characteristics. The composite working electrode had the high specific capacitance of 338 F g^{-1} at a current density of 1 A g^{-1}. Under a broad voltage ranging from 0 to 1.8 V, the built ASC device demonstrated an ultra-energy density of 27.75 Wh kg^{-1} and an output power of 1.35 kW kg^{-1}. In this device demonstrated excellent cycle life, with just a 4.4 percent capacity reduction over 2000 cycles [38]. In particular, low cost, and eco-friendly single step synthesis process of NiO nanoparticles from *Opuntia ficus-indica* leaf juice for asymmetric pseudocapacitors has been developed by S.S. Gunasekaran *et al.* *Opuntia ficus-indica* is a polyphenolic plant that includes a variety of flavonoids and phenolic acids. The extract's phenolic components, flavonoids, lipids, amino acids and proteins function as a capping and reducing reagent during the NiO nanoparticle manufacturing process. Figure 1.7a an illustrated the production method for

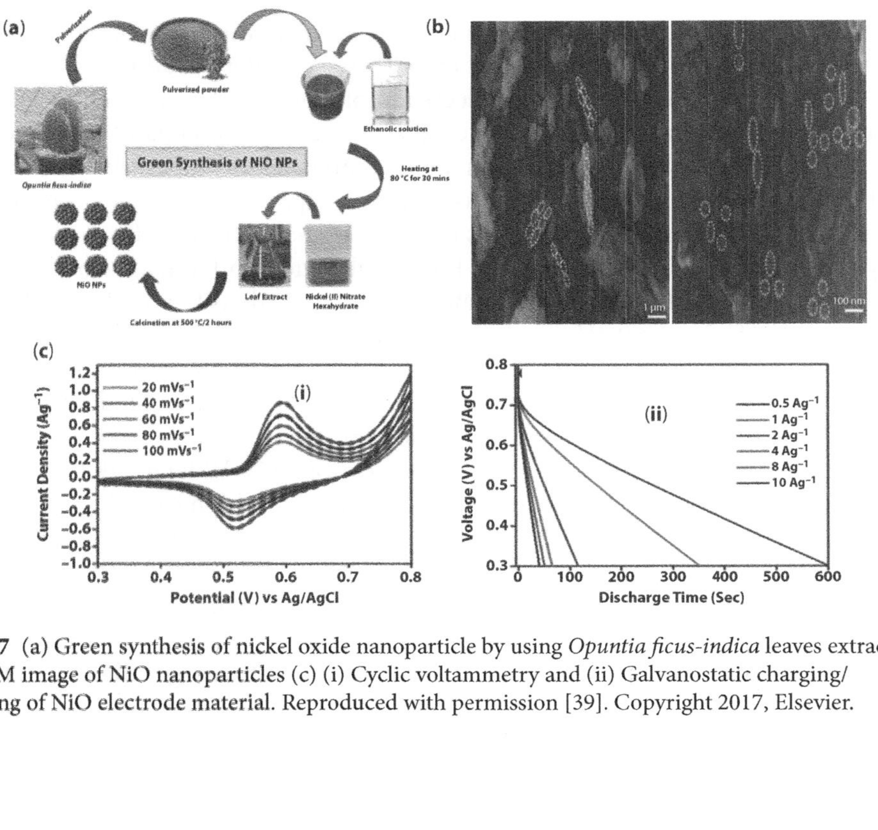

Figure 1.7 (a) Green synthesis of nickel oxide nanoparticle by using *Opuntia ficus-indica* leaves extract (b) FESEM image of NiO nanoparticles (c) (i) Cyclic voltammetry and (ii) Galvanostatic charging/discharging of NiO electrode material. Reproduced with permission [39]. Copyright 2017, Elsevier.

NiO nanoparticles, whereas Figure 1.7b illustrates the surface shape of manufactured NiO nanoparticles. NiO nanoparticles were synthesized and shown a large capacity of 644 Fg^{-1} at a current value of 0.5 Ag^{-1} as determined by the CV and GCD curves of electrochemical capacitor investigation [Figure 1.7c (i & ii)]. The assembled asymmetric supercapacitor device (NiO//AC) with carbon and as-prepared Nickel oxide nanoparticles as an anode and cathode terminals, demonstrated an exceptional energy value of 25 Wh kg^{-1} and a large power value of 40 W kg^{-1} with a capacitive retention of 96.5 % even over 10,000 cycles. Due to its various oxidation states, this NiO is suitable for high-efficiency electrochemical energy storage using a green manufacturing method based on leaf extracts [39]. Further, Sherief A. Al Kiey produced NiO/porous structured carbon composites using an easy and eco-friendly method based on banana peel waste materials. The NiO/pores carbon have a large outer area of 1187 m^2/g according to BET surface analysis. Since prepared samples have a similar specific surface area, ions are more accessible at the electrode/electrolyte contact. The NiO/pores carbon composite displays an extraordinary specific capacity of 811 F/g at 1 A/g, a ultra-rate capability of 84.55 % retention at 10 A/g, and excellent cycle stability over 1000 cycles. The activation of banana peels is a novel approach for preparing supercapacitor electrodes [40].

1.5 Conclusion

In summary, eco-friendly electrode materials for supercapacitor applications have advanced significantly in present century owing to their large surface area, exceptional stability, and low manufacturing cost. Total area of electrode materials has a significant impact on the specific capacitance of supercapacitor devices. Because the electrolyte may readily access the electrode's surface and ions diffuse across the electrode as a wholes. Formerly, several chemical synthesis-based supercapacitor electrodes would have been prohibitively expensive, toxic, and unstable. In this chapter, we discussed the three different types of electrode materials like carbon based materials, metal oxides based material, and both of them mixed materials in the way of green and eco- friendly method. For the first time, a carbon-based electrode manufactured from a broad variety of bio waste materials was disclosed, which not only solved the bio waste disposal problem but also made the whole process less expensive and low hazardous. Then, waste derived and green synthesis carbon delivered that high porous surface area so it's have an excellent specific capacitance and large durability. Green synthesis metal oxides based electrode materials also have outstanding supercapacitor performance. Then composite materials of carbon based and metal oxide electrode materials have extraordinary capacitance and long cyclic stability. As a conclusion, it is clear that more advancement are required to achieve high-performance energy storage systems using environmentally friendly technology, but there is reason to be optimistic about future development. So the research should focus on developing eco-friendly electrochemical capacitor devices with long-term durability and high capacity in a cost-effectiveness method and materials.

References

1. Wiston, B.R. and Ashok, M., Microwave-assisted synthesis of cobalt-manganese oxide for supercapacitor electrodes. *Mater. Sci. Semicond. Process.*, 103, 104607, 2019.
2. Gonzalez, A., Goikolea, E., Barrena, J.A., Mysyk, R., Review on supercapacitors: Technologies and materials. *Renewable Sustainable Energy Rev.*, 58, 1189, 2016.
3. Najib, S. and Erdem, E., Current progress achieved in novel materials for supercapacitor electrodes: Mini review. *Nanoscale Adv.*, 1, 2817–2827, 2019.
4. Mohamed Racik, K., Manikandan, A., Mahendiran, M., Prabakaran, P., Madhavan, J., Victor Antony Raj, M., Fabrication of manganese oxide decorated copper oxide (MnO2/CuO) nanocomposite electrodes for energy storage supercapacitor devices. *Phys. E Low-Dimens. Syst. Nanostruct.*, 119, 114033, 2020.
5. Zhang, L. and Zhao, X.S., Carbon-based materials as supercapacitor electrodes. *Chem. Soc. Rev.*, 38, 2520, 2009.
6. Dubey, R. and Guruviah, V., Review of carbon-based electrode materials for supercapacitor energy storage. *Ionics (Kiel)*, 25, 1419, 2019.
7. Wang, Y., Zhang, L., Hou, H., Xu, W., Duan, G., He, S., Liu, K., Jiang, S., Recent progress in carbon-based materials for supercapacitor electrodes: A review. *J. Mater. Sci.*, 56, 1732021, 2021.
8. Delbari, S.A., Ghadimi, L.S., Hadi, R., Farhoudian, S., Nedaei, M., Babapoor, A., Namini, A.S., Le, Q.V., Shokouhimehr, M., Asl, M.S., Mohammadi, M., Transition metal oxide-based electrode materials for flexible supercapacitors: A review. *J. Alloys Compd.*, 857, 158281, 2021.
9. Mohd Abdah, M.A.A., Azman, N.H.N., Kulandaivalu, S., Sulaiman, Y., Review of the use of transition-metal-oxide and conducting polymer-based fibres for high-performance supercapacitors. *Mater. Des.*, 186, 108199, 2020.
10. Sheng, S., Liu, W., Zhu, K., Cheng, K., Ye, K., Wang, G., Cao, D., Yan, J., Fe_3O_4 nanospheres *in situ* decorated graphene as high-performance anode for asymmetric supercapacitor with impressive energy density. *J. Colloid Interface Sci.*, 536, 235, 2019.
11. Taer, E., Melisa, M., Agustino, A., Taslim, R., Sinta Mustika, W., Apriwandi, A., Biomass-based activated carbon monolith from Tectona grandis leaf as supercapacitor electrode materials. *Energy Sources Part A Recovery Util. Environ. Eff.*, 2021.
12. Bi, Z., Kong, Q., Cao, Y., Sun, G., Su, F., We, X., Li, X., Ahmad, A., Xie, L., Chen, C.M., Biomass-derived porous carbon materials with different dimensions for supercapacitor electrodes: A review. *J. Mater. Chem. A*, 7, 16028, 2019.
13. Okhawilai, M. and Pattananuwat, P., Sustainable electroactive materials for energy storage. *Curr. Opin. Green Sustain. Chem.*, 28, 100431, 2021.
14. Sunaina, S., Chand, P., Joshi, A., Biomass derived carbon for supercapacitor applications: Review. *J. Energy Storage*, 39, 102646, 2021.
15. Kanjana, K., Harding, P., Kwamman, T., Kingkam, W., Chutimasakul, T., Biomass-derived activated carbons with extremely narrow pore size distribution via eco-friendly synthesis for supercapacitor application. *Biomass Bioenergy*, 153, 106206, 2021.
16. Sethi, M., Bantawal, H., Shenoy, U.S., Bhat, D.K., Eco-friendly synthesis of porous graphene and its utilization as high performance supercapacitor electrode material. *J. Alloys Compd.*, 799, 256, 2019.
17. Zhang, G., Chen, Y., Chen, Y., Guo, H., Activated biomass carbon made from bamboo as electrode material for supercapacitors. *Mater. Res. Bull.*, 102, 391, 2018.
18. Divyashree, A., Manaf, S.A.B.A., Yallappa, S., Chaitra, K., Kathyayini, N., Hegde, G., Low cost, high performance supercapacitor electrode using coconut wastes: Eco-friendly approach. *J. Energy Chem.*, 25, 880, 2016.

19. Lin, X.Q., Lu, Q.F., Li, Q., Wu, M., Liu, R., Fabrication of low-cost and ecofriendly porous biocarbon using konjaku flour as the raw material for high-performance supercapacitor application. *ACS Omega*, 3, 13283, 2018.
20. Su, X.L., Chen, J.R., Zheng, G.P., Yang, J.H., Guan, X.X., Liu, P., Zheng, X.C., Three-dimensional porous activated carbon derived from loofah sponge biomass for supercapacitor applications. *Appl. Surf. Sci.*, 436, 327, 2018.
21. Ahmed, S., Ahmed, A., Rafat, M., Investigation on activated carbon derived from biomass Butnea monosperma and its application as a high performance supercapacitor electrode. *J. Energy Storage*, 26, 100988, 2019.
22. Han, Y., Shen, N., Zhang, S., Li, D., Li, X., Fish gill-derived activated carbon for supercapacitor application. *J. Alloys Compd.*, 694, 636, 2017.
23. Guo, N., Li, M., Sun, X., Wang, F., Yang, R., Tremella derived ultrahigh specific surface area activated carbon for high performance supercapacitor. *Mater. Chem. Phys.*, 201, 399, 2017.
24. Pontiroli, D., Scaravonati, S., Magnani, G., Fornasini, L., Bersani, D., Bertoni, G., Milanese, C., Girella, A., Ridi, F., Verucchi, R., Mantovani, L., Malcevschi, A., Ricco, M., Super-activated biochar from poultry litter for high-performance supercapacitors. *Microporous Mesoporous Mater.*, 285, 161, 2019.
25. Guo, F., Jiang, X., Jia, X., Liang, S., Qian, L., Rao, Z., Synthesis of biomass carbon electrode materials by bimetallic activation for the application in supercapacitors. *J. Electroanal. Chem.*, 844, 105, 2019.
26. Abuzeid, H.M., Elsherif, S.A., AbdelGhany, N.A., Hashem, A.M., Facile, cost-effective and eco-friendly green synthesis method of MnO2 as storage electrode materials for supercapacitors. *J. Energy Storage*, 21, 156, 2019.
27. Lee, C.E., Choi, S.H., Kim, H.Y., Lee, S.S., Kim, S.K., An, K.-S., Enhanced pseudocapacitive performances of eco-friendly co-precipitated Fe-doped cerium oxide nanoparticles. *Ceram. Int.*, 47, 21988, 2021.
28. Theophil Anand, G., Renuka, D., Ramesh, R., Anandaraj, L., John Sundaram, S., Ramalingam, G., Maria Magdalane, C., Bashir, A.K.H., Maaza, M., Kaviyarasu, K., Green synthesis of ZnO nanoparticle using Prunus dulcis (Almond Gum) for antimicrobial and supercapacitor applications. *Surf. Interfaces*, 17, 100376, 2019.
29. Nsude, H.E., Nsude, K.U., Whyte, G.M., Obodo, R.M., Iroegbu, C., Maaza., M., Ezema, F.I., Green synthesis of CuFeS2 nanoparticles using mimosa leaves extract for photocatalysis and supercapacitor applications. *J. Nanopart. Res.*, 22, 352, 2020.
30. Kundu, M., Karunakaran, G., Kuznetsov, D., Green synthesis of NiO nanostructured materials using Hydrangea paniculata flower extracts and their efficient application as supercapacitor electrodes. *Powder Technol.*, 311, 132, 2017.
31. N.B., V.Y., Abdul Razack, S., Enhanced formation of ruthenium oxide nanoparticles through green synthesis for highly efficient supercapacitor applications. *Adv. Powder Technol.*, 31, 1001, 2020.
32. Ismail, E., Khamlich, S., Dhlamini, M., Maaza, M., Green biosynthesis of ruthenium oxide nanoparticles on nickel foam as electrode material for supercapacitor applications. *RSC Adv.*, 6, 86843, 2016.
33. Shaheen, I., Ahmad, K.S., Zequine, C., Gupta, R.K., Thomas, A.G., Malik, M.A., Green synthesis of ZnO-Co3O4 nanocomposite using facile foliar fuel and investigation of its electrochemical behaviour for supercapacitors. *New J. Chem.*, 44, 18281, 2020.
34. Lamba, P., Singh, P., Singh, P., Kumar, A., Singh, P., Bharti, Kumar, Y., Gupta, M., Bioinspired synthesis of nickel oxide nanoparticles as electrode material for supercapacitor applications. *Ionics (Kiel)*, 27, 5263, 2021.

35. Matinise, N., Kaviyarasu, K., Mongwaketsi, N., Khamlich, S., Kotsedi, L., Mayedwa, N., Maaza, M., Green synthesis of novel zinc iron oxide (ZnFe2O4) nanocomposite via Moringa Oleifera natural extract for electrochemical applications. *Appl. Surf. Sci.*, 446, 66, 2018.
36. Sasirekha, C., Arumugam, S., Muralidharan, G., Green synthesis of ZnO/carbon (ZnO/C) as an electrode material for symmetric supercapacitor devices. *Appl. Surf. Sci.*, 449, 521, 2018.
37. Sudhakar, Y.N., Hemant, H., Nitinkumar, S.S., Poornesh, P., Selvakumar, M., Green synthesis and electrochemical characterization of rGO–CuO nanocomposites for supercapacitor applications. *Ionics (Kiel)*, 23, 1267, 2017.
38. Jia, L., Shi, Y., Zhang, Q., Xu, X., Green synthesis of ultrafine methyl-cellulose-derived porous carbon/MnO2 nanowires for asymmetric supercapacitors and flexible pattern stamping. *Appl. Surf. Sci.*, 462, 923, 2018.
39. Gunasekaran, S.S., Gopalakrishnan, A., Subashchandrabose, R., Badhulika, S., Phytogenic generation of NiO nanoparticles as green-electrode material for high performance asymmetric supercapacitor applications. *J. Energy Storage*, 37, 102412, 2021.
40. Al Kiey, S.A. and Hasanin, M.S., Green and facile synthesis of nickel oxide-porous carbon composite as improved electrochemical electrodes for supercapacitor application from banana peel waste. *Environ. Sci. Pollut. Res.*, 28, 66888, 2021.
41. Yang, C.S., Jang, Y.S., Jeong, H.K., Bamboo-based activated carbon for supercapacitor applications. *Curr. Appl. Phys.*, 14, 1616, 2014.
42. Jain, A., Ghosh, M., Krajewski, M., Kurungot, S., Michalska, M., Biomass-derived activated carbon material from native European deciduous trees as an inexpensive and sustainable energy material for supercapacitor application. *J. Energy Storage*, 34, 102178, 2021.
43. Vinayagam, M., Suresh Babu, R., Sivasamy, A., de Barros, A.L.F., Biomass-derived porous activated carbon nanofibers from Sapindus trifoliatus nut shells for high-performance symmetric supercapacitor applications. *Carbon Lett.*, 31, 1133, 2021.
44. Hamouda, H.A., Cui, S., Dai, X., Xiao, L., Xie, X., Peng, H., Ma, G., Synthesis of porous carbon material based on biomass derived from hibiscus sabdariffa fruits as active electrodes for high-performance symmetric supercapacitors. *RSC Adv.*, 11, 354, 2020.
45. Wang, K., Zhao, N., Lei, S., Yan, R., Tian, X., Wang, J., Song, Y., Xu, D., Guo, Q., Liu, L., Promising biomass-based activated carbons derived from willow catkins for high performance supercapacitors. *Electrochim. Acta*, 166, 1, 2015.
46. Momodu, D., Madito, M., Barzegar, F., Bello, A., Khaleed, A., Olaniyan, O., Dangbegnon, J., Manyala, N., Activated carbon derived from tree bark biomass with promising material properties for supercapacitors. *J. Solid State Electrochem.*, 21, 859, 2017.
47. Senthil, R.A., Yang, V., Pan, J., Sun, Y., A green and economical approach to derive biomass porous carbon from freely available feather finger grass flower for advanced symmetric supercapacitors. *J. Energy Storage*, 35, 102287, 2021.
48. Khan, A., Senthil, R.A., Pan, J., Osman, S., Sun, Y., Shu, X., A new biomass derived rod-like porous carbon from tea-waste as inexpensive and sustainable energy material for advanced supercapacitor application. *Electrochim. Acta*, 335, 135588, 2020.
49. Aparna, M.L., Rao, G.R., Thomas, T., Momordica Charantia pericarp derived activated carbon with dual redox additive electrolyte for high energy density supercapacitor devices. *J. Energy Storage*, 48, 104048, 2022.
50. Yang, V., Senthil, R.A., Pan, J., Khan, A., Osman, S., Wang, L., Jiang, W., Sun, Y., Highly ordered hierarchical porous carbon derived from biomass waste mangosteen peel as superior cathode material for high performance supercapacitor. *J. Electroanal. Chem.*, 855, 113616, 2019.
51. Rajasekaran, S.J. and Raghavan, V., Facile synthesis of activated carbon derived from Eucalyptus globulus seed as efficient electrode material for supercapacitors. *Diam. Relat. Mater.*, 109, 108038, 2020.

52. Khalafallah, D., Quan, X., Ouyang, C., Zhi, M., Hong, Z., Heteroatoms doped porous carbon derived from waste potato peel for supercapacitors. *Renew. Energy*, 170, 60–71, 2021.
53. Jiang, W., Li, L., Pan, J., Senthil, R.A., Jin, X., Cai, J., Wang, J., Liu, X., Hollow-tubular porous carbon derived from cotton with high productivity for enhanced performance supercapacitor. *J. Power Sources*, 438, 226936, 2019.
54. Yang, V., Senthil, R.A., Pan, J., Rajesh Kumar, T., Sun, Y., Liu, X., Hierarchical porous carbon derived from jujube fruits as sustainable and ultrahigh capacitance material for advanced supercapacitors. *J. Colloid Interface Sci.*, 579, 347, 2020.
55. Wu, F., Gao, J., Zhai, X., Xie, M., Sun, Y., Kang, H., Tian, Q., Qiu, H., Hierarchical porous carbon microrods derived from albizia flowers for high performance supercapacitors. *Carbon N. Y.*, 147, 242, 2019.
56. Divya, P. and Rajalakshmi, R., Renewable low cost green functional mesoporous electrodes from Solanum lycopersicum leaves for supercapacitors. *J. Energy Storage*, 27, 101149, 2020.
57. Cao, X., Li, Z., Chen, H., Zhang, C., Zhang, Y., Gu, C., Xu, X., Li, Q., Synthesis of biomass porous carbon materials from bean sprouts for hydrogen evolution reaction electrocatalysis and supercapacitor electrode. *Int. J. Hydrogen Energy*, 46, 18887, 2021.

2

Solid Waste-Derived Carbon Materials for Electrochemical Capacitors

Shreeganesh Subraya Hegde* **and Badekai Ramachandra Bhat**†

Catalysis and Materials Chemistry Laboratory, Department of Chemistry, National Institute of Technology Karnataka, Surathkal, Mangalore, Karnataka, India

Abstract

An exponential increase in the atmospheric temperature, destructive environmental effects, and the unreplenishable nature of energy supplies have sparked universal attention to the need to find alternative energy sources, value-added chemical products, and fuel. Biomass-derived material has emerged as a fantastic alternative in this regard. These biomasses contain carbon as a skeletal backbone, allowing science and technology to function to a higher degree. Numerous naturally abundant environmental and agro-wastes have been researched to manufacture porous carbon for supercapacitor electrodes because of their ready availability, outstanding performance, and easy processing techniques. This chapter discusses various environmental waste-derived carbon nanomaterials and general preparation methods, activation processes, recent advancements, challenges, and future perspectives.

***Keywords*:** Agricultural waste, waste to energy, supercapacitor, sustainable materials, sustainable energy, biomass conversion

2.1 Introduction

Global population expansion has resulted in an enormous and unconventional increase in worldwide advancement. Consequently, associated needs are growing day by day, causing an explosive rise in creating various types of waste materials. An exponential increase in the atmospheric temperature, destructive environmental effects, and the unreplenishable nature of energy supplies have sparked universal attention to the need to find alternative energy sources, value-added chemical products, and fuel. Biomass-derived material has emerged as a fantastic alternative in this regard. These refer to organic materials originating from plants or animals often regarded as garbage. According to several latest scientific publications, 7–8 billion tons of solid garbage is created globally, posing significant health and environmental risk, thus showing a considerable challenge for the research world. When solid biomass and

**Corresponding author*: hegdeshreeganesh@gmail.com; ORCID: 0000-0003-2678-7481
†*Corresponding author*: ram@nitk.edu.in; ORCID: 0000-0002-8169-1055

Inamuddin, Tariq Altalhi and Sayed Mohammed Adnan (eds.) Sustainable Materials for Electrochemical Capacitors, (19–32) © 2023 Scrivener Publishing LLC

agricultural wastes are burned, several pollutants are discharged into the environment. The sort of biomass feedstock and its types have a significant impact on the nature of emissions. Numerous pollutants are routinely emitted into the atmosphere, including nitrogen oxide, carbon monoxide, and sulphur dioxide. Forest fires are caused by waste biomass, which can potentially raise global greenhouse gas emissions by 25%–30% every year [1, 2].

The correct handling of such a large volume of solid garbage is a primary environmental concern in and of itself [3]. To successfully address the issue, scientific progress necessitates the development of a novel technique capable of addressing and managing the use of such wastes through the application of sophisticated scientific technology. In this sense, nanotechnology and nanoscience have been significant blessings to the scientific society. They may concurrently serve as a lifeline for efficiently managing various universal duties, such as solid waste management and the global energy problem. As a result, waste handling solutions must be invented to fully address the challenge of solid waste management. Furthermore, the technological world has a tremendous challenge in developing a new approach for manufacturing ecologically responsible and economically friendly energy storage technologies to address the global energy issue. However, several research groups have published some unique research, such as effective solid waste treatment and an economically favorable synthesis route of carbon nano-structures (CNS) for diverse industrial uses, particularly in the energy sector.

Biological trash, abundant natural resources, or items that may be obtained for free or at a minimal cost are typically used to make biochar. More precisely, animal manure and agricultural by-products are frequently used as raw materials for biochar. These substances are produced in every nation but are rarely used and have disposal issues. Thus, they can be used as feedstocks to make biochar. For instance, trash or by-products from farming production, including corn, peanuts, sugarcane, oil palm, and other crop waste, are frequently used to make biochar. The use of biochar as a crucial tool for protecting the atmosphere has subsequently attracted a lot of interest. A blackened product called biochar is created when biomaterials are pyrolyzed in a limited or no oxygen condition. In other words, the carbon-containing substance known as biochar is abundant in minerals and carbon. According to earlier studies, the qualities and many uses of biochar have been linked to its carbon percentage. However, the elemental components of biochar also significantly affect its characteristics and, ultimately, its use in a variety of areas, as the latest scientific advancements have confirmed. Additionally, the elemental components of biochar significantly influence its shape, structure, and other characteristics, including pH, electrical properties, porous structure, and surface area per unit weight. The production of these carbons in large quantities is thought to be both inexpensive and environmentally friendly. Hence, activated carbon is frequently utilized for purposes such as soil remediation, energy production, greenhouse gas reduction, waste management, and enhancement of the soil's capacity to contain nutrients and retain water since it has many positive ecological attributes. It is generally known that the conversion of biomaterials into biochar lowers the atmospheric CO_2 concentration and generates alternative sources [4].

2.2 Solid Waste as a Source of CNS

Treatment of solid waste and the demand for renewable energy are two crucial research topics for adopting low-cost solid-waste utilization strategies. The development of low-cost, high-efficiency electrode materials is required to meet the rising demand for green and sustainable energy. Numerous naturally abundant environmental and agro-wastes (Figure 2.1 and Table 2.1) have been researched to manufacture porous carbon for electrode because of their readily available availability, outstanding performance, and easy processing techniques. The majority of solid waste comprises plastic waste, farm waste, municipal garbage, paper waste, polymeric, and technological waste. These waste materials contain carbon as a skeletal backbone, allowing science and technology to function to a higher degree. In contrast, based on the preparation technique, the activation method can modify the morphological and structural features of the material. The source and type of biomass are crucial

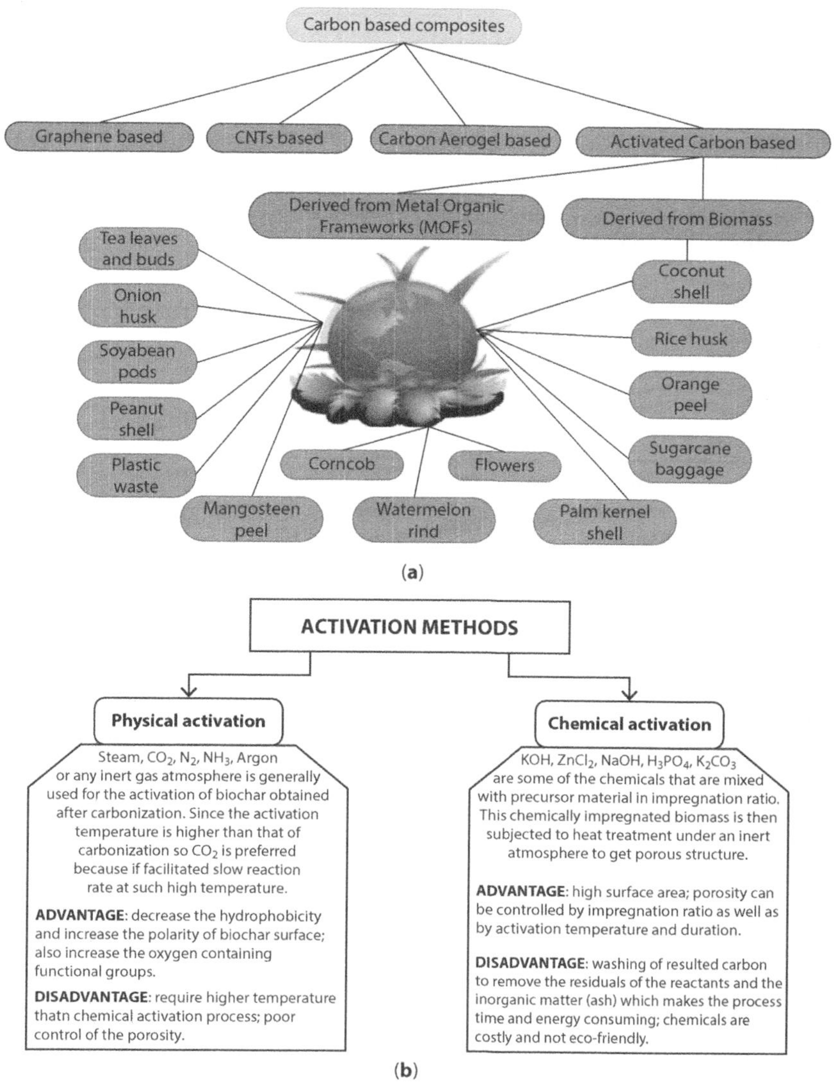

Figure 2.1 Illustration of (a) different carbon-based composites and solid-waste materials (b) Activation processes [46]. (Reprinted with permission from Elsevier [License Number 5292881309609]).

Table 2.1 Comparative view of various precursor materials and electrochemical performance.

Precursor	Activating agent	Electrolyte	Current density (Ag^{-1})	Specific capacitance (Fg^{-1})	Ref.
Sakura flower	KOH	6M KOH	0.2	265.8	[47]
Peanut shell	$ZnCl_2$	1M H_2SO_4	1.0	340	[48]
Rice husk	H_3PO_4	6M KOH	0.05	176	[49]
Potato waste	$ZnCl_2$	6M KOH	10	192	[50]
Tea leaves	KOH	2M KOH	1.0	330	[51]
Banana fibers	$ZnCl_2$	1M Na_2SO_4	0.02	74	[52]
Orange peel	KOH	1M H_2SO_4	1.0	460	[53]
Jujube fruit	NaOH	6M KOH	1.0	460	[54]
Soyabean pods	NaOH	1M Na_2SO_4	0.5	352.6	[55]
Cashew nut shell	KOH	1M Na_2SO_4	1.0	214	[40]
Watermelon rind	KOH	6M KOH	1.0	333.4	[56]
Mangosteen peels	NaOH	6M KOH	1.0	357	[57]
Onion husk	K_2CO_3	1M $TEABF_4$/AC	1.0	188	[15]
Sugarcane bagasse	KOH	1M H_2SO_4	1.0	298	[11]
Corncob	KOH	1M H_2SO_4	0.5	390	[6]
Tea waste buds	KOH	6M KOH	1.0	332	[58]

for producing carbon in a commercially sustainable way. Directly burning garbage is a frequent practice since developing nations lack an appropriate disposal technique, which raises atmospheric carbon dioxide concentrations and has a negative impact on both public health and the ecosystem. By transforming such leftovers into value-added products, thermochemical conversion via pyrolysis, gasification, and hydrothermal liquefaction provides an appropriate means of reducing the negative ecological effect of such wastes. As a result, agro-forest wastes can be viewed as an environmentally friendly resource for the creation and use of carbon.

The trash produced by a variety of industries includes significant amounts of carbon. These comprise agro-enterprises that create residues like potato peel, orange peel, oilseed cakes, etc., and industries or sawmills in the forest that produce sawdust. Each year, approximately 50 million tons of lignin is produced as a byproduct of the pulp and paper industry; less than 5% of this amount is economically used, and the remainder is burned as dry fuel. The manufacturing methods can be used based on the input. Depending on their accessibility and production in a worldwide context, many sources for the manufacture of carbon may be identified. The bulk of the trash, primarily in the form of lignocellulose, may be attributed to agricultural and forest waste. After the wood is harvested, forestry-related sectors produce debris such as stems, treetops, and branches, among other things, which is referred to as "forestry residual." Rice straw, wheat straw, rice husk, and maize stover are examples of the waste materials produced after crop harvesting that are referred to as "agricultural residues." It either serves as livestock feed or is just left to decompose naturally in the vast farming area. As a result, tobacco [5], corncob [6], corn silk [7], corn husk [8], rice husk [9], soybean residue [10], sugarcane bagasse [11], orange peel [12], lime peels [13], banana peels [14], onion husks [15], coconut coir pith [16], coconut fibers [17], coconut shell [18], jackfruit shell [19], longan shell [20], almond shells [20], soybean roots [21], coffee seed powder [22], bacterial cellulose [23], pine cones [24] and many more bio-waste-derived carbon materials are reported to show good performance as supercapacitor electrodes.

Additionally, sewage sludge produced during the municipal water treatment process in urban facilities can be used as a promising option for the creation of biochar. It is closely connected to the number of people. Wastewater sludge is often disposed of in landfills, fertilizers, and incinerators. These approaches, nevertheless, are risky and constrained in areas with little access to soil [25]. Compared to biochar made from biomass, sewage waste biochar generally has greater hydrogen to carbon ratio and higher levels of N, P, and K [26]. Additionally, neither the P nor the K content of the biochar would not be altered by the rise in pyrolysis temperature [27]. Using cattle dung as fertilizer is somewhat constrained by the rising levels of heavy metal ions in the ground. Such heavy metal ions are helped to get immobilized in the carbon framework by the thermal transformation of animal dung into carbon materials [28]. These carbon materials may be used in energy storage technologies such as supercapacitors since the existence of d-block metals can improve the biochar's capacity for charge storage [4].

2.3 Preparation and Activation Methods of Solid Waste-Derived CNS

Carbon nanomaterials are carbon-based nanoparticles with a wide range of benefits. So, increasing the yield of different CNS from environmental solid trash substances can be done by following various scientific and technological treatments. Physical vapor deposition (PVD), chemical vapor deposition (CVD), hydrothermal carbonization, torrefaction, organic oxidation-reduction techniques, and pyrolytic heating are the methods for converting carbonic trash into CNS. The properties of these materials are primarily dependent on particular surface interactions and the use of reagents [29]. On the other hand, heating temperature, residence time, atmosphere, pressure, and scan rate are the most critical factors that affect the layer development and fluctuations of the CNS in the pyrolysis reaction. The pyrolysis process is the thermal degradation of biomass in an inert environment to produce carbon materials at temperatures between 300°C and 900°C. The quantity and quality of biochar produced are affected by the nature of the raw substance and the adapted parameters or methods [30]. According to heating rate and temperature, pyrolysis is classified into two types: slow and fast pyrolysis. The slow process uses a lower heat input rate, usually 5–7°C/min, and a prolonged residence period (greater than one hour), and the rapid process follows a higher heat input rate, usually 100–200°C/min, and a shorter residence period (less than 10 s) [31]. Slow pyrolysis produces more charcoal, whereas rapid pyrolysis produces more bio-oil [32].

While using highly moist bio-waste material for biochar synthesis, it's common to dry it first, which is a time-consuming process. Hence, hydrothermal carbonization can be done to eliminate this stage to get biochar, and this process is also known as hydrochar (Figure 2.2). Hydrothermal transformation can be regulated by adjusting the operational parameters to get the appropriate ratio of solid, liquid, and gaseous outputs. The hydrothermal carbonization process is achieved at a lower pressure and temperature [33]. When the material is heated to a higher temperature and higher pressure, the char is reduced, and bio-oil production increases [34]. Hydrochar's physical and chemical properties differ extremely from those of biochar. For instance, the structure of the hydrochar is rich in alkyl groups, whereas the biochar's system is predominantly of aromatic groups [35]. Compared to biochar produced from a similar feedstock, hydrochar contains more oxygen moieties [36]. Changes in hydrothermal carbonization's solvent system could affect the active moieties on the superficial layer of biochar.

The biochar created by the aforementioned thermochemical procedures frequently needs physical or chemical activation in order to improve its surface properties, as well as other characteristics like specific surface area and the porous nature of the material that significantly enhance its charge storage properties. Hence, activating the waste-derived carbon material is a critical process for developing a porous surface structure and increasing the active area. Physical activation and chemical activation (Figure 2.1) are two methods for producing activated carbon materials from biomass [37]. Physical activation is the process of generating porous carbon structures from waste-derived carbon compounds by utilizing gaseous mediators including hot steam, oxygen, or carbon dioxide [38]. In contrast, chemical activation is connected with traditional chemical activating

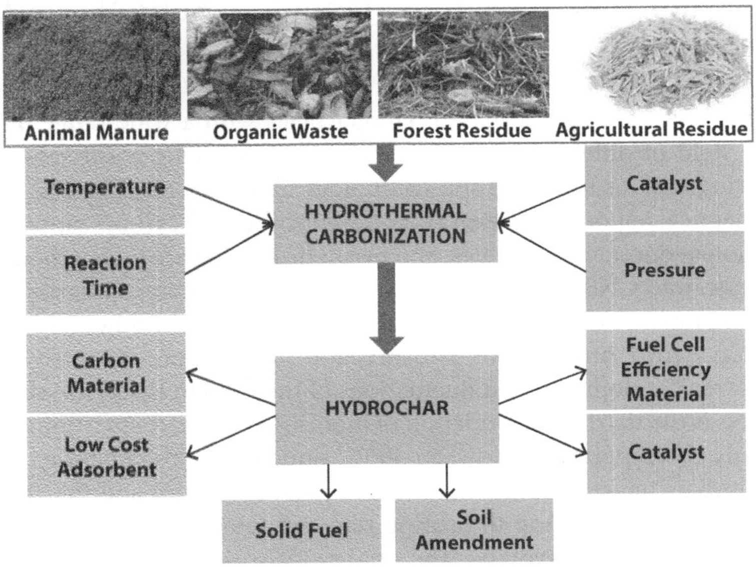

Figure 2.2 Schematic illustration of hydrochar preparation [34]. (Reprinted with permission from Elsevier [License Number 5294591067917]).

agents like acids, bases, or salts [39]. In physical activation, activation of the waste-derived carbon material is carried out after the carbonation step, whereas in chemical activation, both activation and carbonization of the materials can be done in a one move [40]. The synthesized carbon material is processed with an activating agent prior to carbonization. Potassium hydroxide, oxyacids, zinc chloride, and other activating chemical agents are commonly utilized. Chemical activation is more favorable because of the higher yield generation, lower cost, and the ability to generate porous structures at lower temperatures and in less time [41].

Despite its average surface area and improper pore size dispersion, immediate utilization of carbon for charge storage techniques is not advantageous. The carbon substance is activated in order to create porosity. Below, many kinds of porous substances made from biomass have been addressed according to their pore sizes. Activated carbon has also been doped with p-block elements such as nitrogen, phosphorus, or sulphur using inorganic or organic precursors to improve its electrochemical reactive regions and the electrode surface's hydrophilicity with electrolytic solutions. According to the processing conditions and feedstock, the existence of such heteroatoms may also be intrinsic to the feedstock, resulting in carbon with varying elemental components. Furthermore, the manufacturing circumstances can alter the functional group produced by the heteroatom substituents. According to reports, d-block metal oxides offer a lot of possibilities as an energy storage technology. Whereas the energy storage method for carbon is primarily electrostatic, the various oxidation states of ions aid in energy storage through electrochemical processes. Consequently, metal-doped carbon can provide a combination effect. Since the carbon substrate offers a greater surface area for the electrochemically effective substance, the redox processes brought on by the metal-incorporated carbon increase the capacitive performance.

2.4 Effect of Structural and Morphological Diversities on Electrochemical Performance

The electrochemical characteristics of CNS enormously vary with the structural features, such as specific surface area (SSA) and morphological behavior, which highly depend upon the structural framework of the solid waste precursor material. In connection with electrode materials, researchers are constantly looking for carbon electrodes that have a large active surface area, an optimal porous structure with heteroatom dopant insertion to improve the conductivity and capacitance without losing their stability [42]. Carbon compounds generated from biomass have sparked a lot of attention in energy storage materials because of their abundant supply and eco-friendly nature. The electrochemical properties of the produced CNS are significantly connected with morphological features such as graphitization or degree of carbonization, pore volume, pore size, and surface patterns. The morphological and structural characteristics of synthesized materials can be controlled and manipulated by doping heteroatoms and other substituents. The extent of carbonization affects the capacitance, power density, total resistance, power transmission, long cycle life, and energy density. The hydrophilicity of the synthesized material towards aqueous electrolyte depends on the extent of carbonization. This carbonization would result in better ion diffusion and ion transfer efficiency, which improves the function of the electrode material [43]. The lower resistance indicates a high level of carbonic activation or carbonization in the sample, implying a higher conductance in the specimen. However, when the degree of graphitization and carbonization progresses, the specific area decreases, resulting in distortions of the relative structure and pore volume [29].

2.5 Environmental Trash-Derived CNS in Electrochemical Capacitors

As shown in the previous investigations, the characteristics of environmental trash-produced CNS are greatly reliant on the biomass feedstock as well as experimentation circumstances. These CNS show excellent electrical conductance and have a large SSA, making them ideal materials for supercapacitor electrodes. A supercapacitor (SC) is an electrochemical energy storage system that can transfer charge to a rapid extent and supply a large amount of electric current in a short time. They are now often used in electric vehicles, UPS devices, data backup systems in information technology (IT) networks, day-to-day use gadgets, and many other applications. Figure 2.3 shows an overview of the developments in

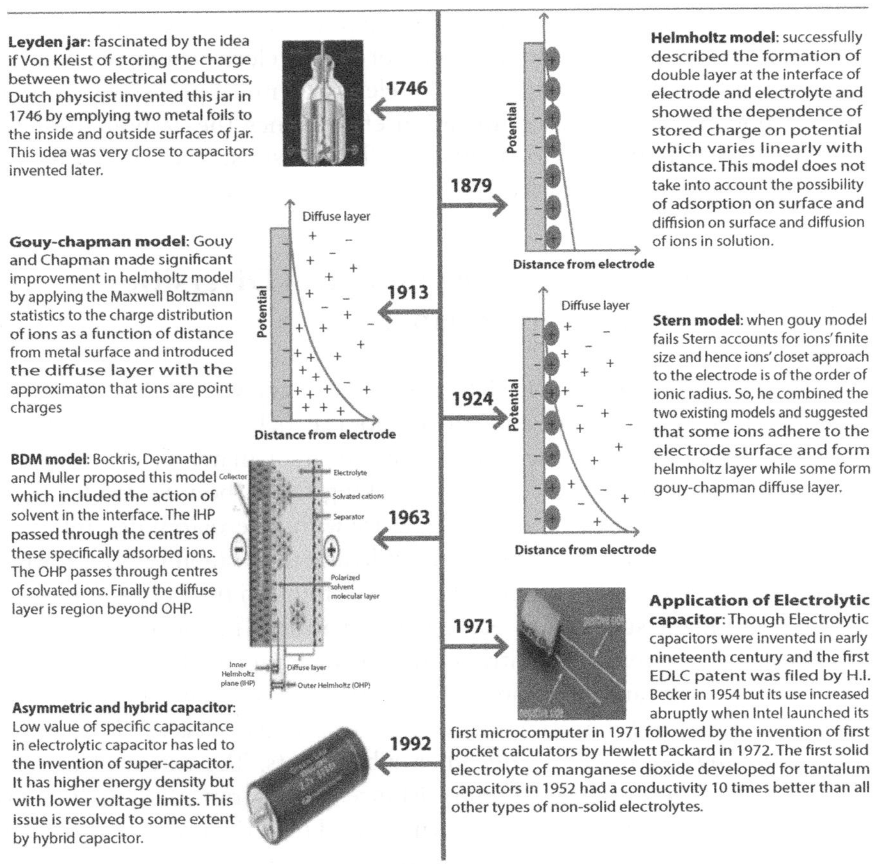

Figure 2.3 Illustration of sequential development in supercapacitor technology [46]. (Reprinted with permission from Elsevier [License Number 5292881309609]).

supercapacitor technology. The primary factors driving changes in electrical and chemical characteristics include differences in surface morphology, changes in the purity of CNS, chemical and thermal permanence, and chemical and structural characteristics. The conductance, capacitive, and storing qualities of CNS are influenced by their extensive surface area. The storing efficiency of SCs is mainly determined by the conductance behavior of electrode materials and the rate of ion diffusion. The electrochemical capacitor material research section is divided into two sections. The first consists of CNS, and the second consists of various metal-based materials and their composites, polymeric materials and their composites [29]. Table 2.1 shows the several types of solid waste-derived SCs materials as well as their electrochemical performances.

2.6 Challenges and Future Prospects

Converting biowaste to carbon material, followed by activation through a variety of chemical and physical techniques, produces various kinds of valuable carbon-rich materials. The quality as well as reliability of the generated biochar is a critical issue among the different sources of raw materials accessible. As a result, the uniformity of raw resources is critical to ensuring the applicability of CNS for a specific application. As a result, the selection of raw materials and the procedure for activation at a massive level must be thoroughly investigated. The appearance of inorganic substances and other contaminants in the waste that might be collected while the manufacture of carbon-rich materials. Hence, it must be cautiously examined because the existence of these substances can dramatically change the capacitive performance [44]. The exact regulation of the process, surface functional moieties, and morphological features remain a crucial problem to be tackled for substances that display both electrochemical and electrostatic charge storage behavior. Extraordinary substances like graphene, CNTs, CDs, and others can only be manufactured at a laboratory level. Therefore, it is vital to simplify the manufacture of these kinds of materials with a significant yield and commercial feasibility. The interactions of electrolytes have a considerable influence on the supercapacitor's charge storage performance [45]. Hence, the coupled impact of solid waste-derived carbon-rich electrode substances on different electrolyte materials has to be researched.

2.7 Conclusions

The global situation of diverse solid waste has reached a point where waste can only be controlled via efficient usage. On the other hand, one of nanotechnology's biggest successes is establishing a link between materials research and the social realm, which will address a major ecological concern and create harmony between technology and the social realm. Because environmental trash has high carbon content, it may be transformed into carbon nanomaterials. These CNS have unique incorporative capabilities and can be used in various industries, permitting the CNS and their composites to demonstrate their transitional qualities. However, more technical progress is still needed to mass-produce such a substance in a sustainable and ecologically friendly manner.

References

1. Balat, M., An overview of biofuels and policies in the European Union. *Energy Sources Part B Econ. Plan. Policy*, 2, 167, 2007.
2. Cook, J. and Beyea, J., Bioenergy in the United States: Progress and possibilities based on the final report for NREL subcontract no. ACD-5-15212-01 with the National Audubon Society. *Biomass Bioenergy*, 18, 441, 2000.
3. Ferrero, G.A., Fuertes, A.B., Sevilla, M., N-doped porous carbon capsules with tunable porosity for high-performance supercapacitors. *J. Mater. Chem. A*, 3, 2914, 2015.
4. Vijayaraghavan, K., The importance of mineral ingredients in biochar production, properties and applications. *Crit. Rev. Environ. Sci. Technol.*, 51, 113, 2021.
5. Chen, H., Guo, Y., Wang, F., Wang, G., Qi, P., Guo, X., Dai, B., Yu, F., An activated carbon derived from tobacco waste for use as a supercapacitor electrode material. *New Carbon Mater.*, 32, 592, 2017.
6. Karnan, M., Subramani, K., Srividhya, P.K., Sathish, M., Electrochemical studies on corncob derived activated porous carbon for supercapacitors application in aqueous and non-aqueous electrolytes. *Electrochim. Acta*, 228, 586, 2018.
7. Mitravinda, T., Nanaji, K., Anandan, S., Jyothirmayi, A., Chakravadhanula, V.S.K., Sharma, C.S., Rao, T.N., Facile synthesis of corn silk derived nanoporous carbon for an improved supercapacitor performance. *J. Electrochem. Soc.*, 165, A3369, 2018.
8. Usha Rani, M., Nanaji, K., Rao, T.N., Deshpande, A.S., Corn husk derived activated carbon with enhanced electrochemical performance for high-voltage supercapacitors. *J. Power Sources*, 471, 228387, 2020.
9. Jin, H., Wu, S., Li, T., Bai, Y., Wang, X., Zhang, H., Xu, H., Kong, C., Wang, H., Synthesis of porous carbon nano-onions derived from rice husk for high-performance supercapacitors. *Appl. Surf. Sci.*, 488, 593, 2019.
10. Ferrero, G.A., Fuertes, A.B., Sevilla, M., From soybean residue to advanced supercapacitors. *Sci. Rep.*, 5, 16618, 2015.
11. Wang, B., Wang, Y., Peng, Y., Wang, X., Wang, J., Zhao, J., 3-dimensional interconnected framework of N-doped porous carbon based on sugarcane bagasse for application in supercapacitors and lithium ion batteries. *J. Power Sources*, 390, 186, 2018.
12. Parveen, N., Al-Jaafari, A.I., Han, J.I., Robust cyclic stability and high-rate asymmetric supercapacitor based on orange peel-derived nitrogen-doped porous carbon and intercrossed interlinked urchin-like NiCo2O4@3DNF framework. *Electrochim. Acta*, 293, 84, 2019.
13. Ahirrao, D.J., Tambat, S., Pandit, A.B., Jha, N., Sweet-lime-peels-derived activated-carbon-based electrode for highly efficient supercapacitor and flow-through water desalination. *ChemistrySelect*, 4, 2610, 2019.
14. Liu, B., Zhang, L., Qi, P., Zhu, M., Wang, G., Ma, Y., Guo, X., Chen, H., Zhang, B., Zhao, Z., Dai, B., Yu, F., Nitrogen-doped banana peel–derived porous carbon foam as binder-free electrode for supercapacitors. *Nanomaterials*, 6, 18, 2016.
15. Wang, D., Liu, S., Fang, G., Geng, G., Ma, J., From trash to treasure: Direct transformation of onion husks into three-dimensional interconnected porous carbon frameworks for high-performance supercapacitors in organic electrolyte. *Electrochim. Acta*, 216, 405, 2016.
16. Sesuk, T., Tammawat, P., Jivaganont, P., Somton, K., Limthongkul, P., Kobsiriphat, W., Activated carbon derived from coconut coir pith as high performance supercapacitor electrode material. *J. Energy Storage*, 25, 100910, 2019.
17. Yin, L., Chen, Y., Li, D., Zhao, X., Hou, B., Cao, B., 3-Dimensional hierarchical porous activated carbon derived from coconut fibers with high-rate performance for symmetric supercapacitors. *Mater. Des.*, 111, 44, 2016.

18. Ashraf, C.M., Anilkumar, K.M., Jinisha, B., Manoj, M., Pradeep, V.S., Jayalekshmi, S., Acid washed, steam activated, coconut shell derived carbon for high power supercapacitor applications. *J. Electrochem. Soc.*, 165, A900, 2018.
19. Lee, K., Shabnam, L., Faisal, S.N., Hoang, V.C., Gomes, V.G., Aerogel from fruit biowaste produces ultracapacitors with high energy density and stability. *J. Energy Storage*, 27, 101152, 2020.
20. Yan, J., Nitrogen-doped oxygen-rich activated carbon derived from longan shell for supercapacitors. *Int. J. Electrochem. Sci.*, 15, 1982, 2020.
21. Guo, N., Li, M., Wang, Y., Sun, X., Wang, F., Yang, R., Soybean root-derived hierarchical porous carbon as electrode material for high-performance supercapacitors in ionic liquids. *ACS Appl. Mater. Interfaces*, 8, 33626, 2016.
22. Choi, J., Zequine, C., Bhoyate, S., Lin, W., Li, X., Kahol, P., Gupta, R., Waste coffee management: Deriving high-performance supercapacitors using nitrogen-doped coffee-derived carbon. *C*, 5, 44, 2019.
23. Ding, C., Liu, T., Yan, X., Huang, L., Ryu, S., Lan, J., Yu, J., Zhong, W.-H., Yang, X., An ultra-microporous carbon material boosting integrated capacitance for cellulose-based supercapacitors. *Nanomicro Lett.*, 12, 63, 2020.
24. Gadipelli, S., Howard, C.A., Guo, J., Skipper, N.T., Zhang, H., Shearing, P.R., Brett, D.J.L., Superior multifunctional activity of nanoporous carbons with widely tunable porosity: Enhanced storage capacities for carbon-dioxide, hydrogen, water, and electric charge. *Adv. Energy Mater.*, 10, 1903649, 2020.
25. Adhikari, S., Nam, H. *et al.*, Conversion of solid wastes to fuels and chemicals through pyrolysis, in: *Waste Biorefinery*, pp. 239–263, Elsevier, 2018.
26. Liu, Z., Singer, S., Tong, Y., Kimbell, L., Anderson, E., Hughes, M., Zitomer, D., McNamara, P., Characteristics and applications of biochars derived from wastewater solids. *Renewable Sustainable Energy Rev.*, 90, 650, 2018.
27. Yang, Y., Meehan, B., Shah, K., Surapaneni, A., Hughes, J., Fouché, L., Paz-Ferreiro, J., Physicochemical properties of biochars produced from biosolids in Victoria, Australia. *Int. J. Environ. Res. Public Health*, 15, 1459, 2018.
28. Li, F., Wu, X., Ji, W., Gui, X., Chen, Y., Zhao, J., Zhou, C., Ren, T., Effects of pyrolysis temperature on properties of swine manure biochar and its environmental risks of heavy metals. *J. Anal. Appl. Pyrolysis.*, 152, 104945, 2020.
29. Tatrari, G., Karakoti, M., Tewari, C., Pandey, S., Bohra, B.S., Dandapat, A., Sahoo, N.G., Solid waste-derived carbon nanomaterials for supercapacitor applications: A recent overview. *Mater. Adv.*, 2, 1454, 2021.
30. Wei, J., Tu, C., Yuan, G., Liu, Y., Bi, D., Xiao, L., Lu, J., Theng, B.K.G., Wang, H., Zhang, L., Zhang, X., Assessing the effect of pyrolysis temperature on the molecular properties and copper sorption capacity of a halophyte biochar. *Environ. Pollut.*, 251, 56, 2019.
31. Dhyani, V. and Bhaskar, T., A comprehensive review on the pyrolysis of lignocellulosic biomass. *Renewable Energy*, 129, 695, 2018.
32. Al Arni, S., Comparison of slow and fast pyrolysis for converting biomass into fuel. *Renewable Energy*, 124, 197, 2018.
33. Funke, A. and Ziegler, F., Hydrothermal carbonization of biomass: A summary and discussion of chemical mechanisms for process engineering. *Biofuels Bioprod. Biorefining*, 4, 160, 2010.
34. Nizamuddin, S., Baloch, H.A., Griffin, G.J., Mubarak, N.M., Bhutto, A.W., Abro, R., Mazari, S.A., Ali, B.S., An overview of effect of process parameters on hydrothermal carbonization of biomass. *Renewable Sustainable Energy Rev.*, 73, 1289, 2017.
35. Kumar, A., Saini, K., Bhaskar, T., Hydochar and biochar: Production, physicochemical properties and techno-economic analysis. *Bioresour. Technol.*, 310, 123442, 2020.

36. Liu, Z., Zhang, F.-S., Wu, J., Characterization and application of chars produced from pinewood pyrolysis and hydrothermal treatment. *Fuel*, 89, 510, 2010.
37. Chen, W., Yu, H., Lee, S.-Y., Wei, T., Li, J., Fan, Z., Nanocellulose: A promising nanomaterial for advanced electrochemical energy storage. *Chem. Soc. Rev.*, 47, 2837, 2018.
38. Dong, S., He, X., Zhang, H., Xie, X., Yu, M., Yu, C., Xiao, N., Qiu, J., Surface modification of biomass-derived hard carbon by grafting porous carbon nanosheets for high-performance supercapacitors. *J. Mater. Chem. A*, 6, 15954, 2018.
39. Huo, S., Zhao, Y., Zong, M., Liang, B., Zhang, X., Khan, I.U., Li, K., Enhanced supercapacitor and capacitive deionization boosted by constructing inherent N and P external defects in porous carbon framework with a hierarchical porosity. *Electrochim. Acta*, 353, 136523, 2020.
40. Merin, P., Jimmy Joy, P., Muralidharan, M.N., Veena Gopalan, E., Seema, A., Biomass-derived activated carbon for high-performance supercapacitor electrode applications. *Chem. Eng. Technol.*, 44, 844, 2021.
41. Su, X.-L., Chen, J.-R., Zheng, G.-P., Yang, J.-H., Guan, X.-X., Liu, P., Zheng, X.-C., Three-dimensional porous activated carbon derived from loofah sponge biomass for supercapacitor applications. *Appl. Surf. Sci.*, 436, 327, 2018.
42. Jiang, L., Sheng, L., Fan, Z., Biomass-derived carbon materials with structural diversities and their applications in energy storage. *Sci. China Mater.*, 61, 133, 2018.
43. Sun, L., Tian, C., Li, M., Meng, X., Wang, L., Wang, R., Yin, J., Fu, H., From coconut shell to porous graphene-like nanosheets for high-power supercapacitors. *J. Mater. Chem. A*, 1, 6462, 2013.
44. Duignan, T.T. and Zhao, X.S., Impurities limit the capacitance of carbon-based supercapacitors. *J. Phys. Chem. C*, 123, 4085, 2019.
45. Rawat, S., Mishra, R.K., Bhaskar, T., Biomass derived functional carbon materials for supercapacitor applications. *Chemosphere*, 286, 131961, 2022.
46. Saini, S., Chand, P., Joshi, A., Biomass derived carbon for supercapacitor applications: Review. *J. Energy Storage*, 39, 102646, 2021.
47. Ma, F., Ding, S., Ren, H., Liu, Y., Sakura-based activated carbon preparation and its performance in supercapacitor applications. *RSC Adv.*, 9, 2474, 2019.
48. Xiao, Z., Porous biomass carbon derived from peanut shells as electrode materials with enhanced electrochemical performance for supercapacitors. *Int. J. Electrochem. Sci.*, 13, 5370, 2018.
49. Wu, M., Li, L., Liu, J., Li, Y., Ai, P., Wu, W., Zheng, J., Template-free preparation of mesoporous carbon from rice husks for use in supercapacitors. *New Carbon Mater.*, 30, 471, 2015.
50. Ma, G., Yang, Q., Sun, K., Peng, H., Ran, F., Zhao, X., Lei, Z., Nitrogen-doped porous carbon derived from biomass waste for high-performance supercapacitor. *Bioresour. Technol.*, 197, 137, 2015.
51. Peng, C., Yan, X., Wang, R., Lang, J., Ou, Y., Xue, Q., Promising activated carbons derived from waste tea-leaves and their application in high performance supercapacitors electrodes. *Electrochim. Acta*, 87, 401, 2013.
52. Niu, J., Liang, J., Shao, R., Liu, M., Dou, M., Li, Z., Huang, Y., Wang, F., Tremella-like N,O-codoped hierarchically porous carbon nanosheets as high-performance anode materials for high energy and ultrafast Na-ion capacitors. *Nano Energy*, 41, 285, 2017.
53. Subramani, K., Sudhan, N., Karnan, M., Sathish, M., Orange peel derived activated carbon for fabrication of high-energy and high-rate supercapacitors. *ChemistrySelect*, 2, 11384, 2017.
54. Yang, V., Arumugam Senthil, R., Pan, J., Rajesh Kumar, T., Sun, Y., Liu, X., Hierarchical porous carbon derived from jujube fruits as sustainable and ultrahigh capacitance material for advanced supercapacitors. *J. Colloid Interface Sci.*, 579, 347, 2020.

55. Kong, X., Zhang, Y., Zhang, P., Song, X., Xu, H., Synthesis of natural nitrogen-rich soybean pod carbon with ion channels for low cost and large areal capacitance supercapacitor. *Appl. Surf. Sci.*, 516, 146162, 2016.
56. Mo, R.-J., Zhao, Y., Wu, M., Xiao, H.-M., Kuga, S., Huang, Y., Li, J.-P., Fu, S.-Y., Activated carbon from nitrogen rich watermelon rind for high-performance supercapacitors. *RSC Adv.*, 6, 59333, 2016.
57. Yang, V., Senthil, R.A., Pan, J., Khan, A., Osman, S., Wang, L., Jiang, W., Sun, Y., Highly ordered hierarchical porous carbon derived from biomass waste mangosteen peel as superior cathode material for high performance supercapacitor. *J. Electroanal. Chem.*, 855, 113616, 2019.
58. Khan, A., Senthil, R.A., Pan, J., Osman, S., Sun, Y., Shu, X., A new biomass derived rod-like porous carbon from tea-waste as inexpensive and sustainable energy material for advanced supercapacitor application. *Electrochim. Acta*, 335, 135588, 2020.

3

Metal Hydroxides

Rida Fatima[1,2], Sania Naseer[3], Muhammad Rehan Hasan Shah Gilani[1], Muhammad Aamir[3*] and Javeed Akhtar[3]

[1]Institute of Chemical Sciences, Bahauddin Zakariya University, Multan, Pakistan
[2]Department of Chemistry, Govt. College University Faisalabad, Sahiwal Campus, Sahiwal, Pakistan
[3]Materials Laboratory, Department of Chemistry, Mirpur University of Science and Technology (MUST), Mirpur (AJK), Pakistan

Abstract
Metal hydroxides, (MOHs) hydroxides of metals, having hydroxide anions as well as metallic cations are frequently sturdy foundations which completely ionize upon dissociation. Certain metal hydroxides are fragile electrolytes that disintegrate slightly in fluid arrangement only, and can be incorporated using various fabrication techniques such as solvothermal, sonochemical, ion-exchange, aqueous, polyol arrangement, and precipitation. Nanofibers made of MOHs have been intensively researched since the mid of 20th century when electron microscopy became accessible as a research instrument. MOHs can be practiced as a fire retardant and smoke suppressant in plastic by the endothermic reaction. They have applications in the field of pharmaceutical, as acid neutralizer in humans and animals, react with an overflow of corrosive and lower the harshness of the digestive process, which may allow the adverse effects of ulcers, acid reflux, or dyspepsia to be avoided. "The salt metal hydroxides of Aluminum, Magnesium as well as some of the alkali metals of Group 1A (Lithium, Sodium, Potassium, Rubidium, and Cesium) are the most grounded bases having strong stability and extensively soluble hydroxides. Calcium, Barium, and Strontium solid bases – all antacid earth metals – form dissolvable hydroxides, have low stability level as compared to hydroxides of salts. Except for Thallium hydroxide, other metal hydroxides such as magnesium, iron, bismuth, nickel, cobalt, and copper are only sparingly soluble in water but destroy acids. The hydroxides of beryllium, tin (divalent), lead, zinc, chromium (trivalent), aluminum, gold (trivalent) and other metals exhibit caustic as well as the other basic characteristics, i.e., they dissolve in aqueous combinations of either bases or acidic media."

Keywords: Metal hydroxides, MOH nanostructures, applications of MOH, MOH for supercapacitor, MOH for waste treatment

3.1 Introduction

Metal hydroxides are hydroxides of metals, consisting of anions and cations of hydroxide and metals respectively. Most often they are the strong bases, some of them ionize completely upon dissolution while certain MOHs dissolve partially in aqueous medium

**Corresponding author*: Amir.chem@must.edu.pk; aamirorg@gmail.com

Inamuddin, Tariq Altalhi and Sayed Mohammed Adnan (eds.) Sustainable Materials for Electrochemical Capacitors, (33–64) © 2023 Scrivener Publishing LLC

and termed as weak electrolytes. Under specific circumstances, MOHs might take on an extraordinary translucent morphological nanostructure. The incredibly smooth nanofibers, nanometres in distance across and a couple of micrometers long, structure precipitously in watery arrangements of specific progress metal particles (Cadmium ions, Copper ions, and Zinc ions). They normally have a huge amount of positively charged ions and exhibit a property that renders them exceptionally in aqueous media (H_2O), similar to natural polyelectrolytes. Different nano-composite structures are realistic from mixes of MOH nano strands and adversely charged compounds, and these filaments might be changed over into ultrathin unsupported films with an assortment of helpful applications [1, 2].

It is as yet muddled if MOH nanostrands ought to be named new materials. High goal electron magnifying instrument perceptions have shown that these nanostrands [3] display translucent constructions basically the same as those of the relating MOHs in the mass state [4]. Along these lines, in one sense, they are basically exceptionally dainty nanofibers [5] of MOHs. Notwithstanding, these nanostrands additionally have very high surface charge densities, similar to those of polyoxometalates [6]. Proof of this is given by essential organization investigation of cadmium hydroxide nanostrands, which demonstrates the equation $[Cd_{37}(OH)_{68}(H2O)_{nþ4}]_{6þ}$, actually intending that around one-6th of cadmium particles are emphatically charged [7]. This worth is near the negative charge thickness of an isopolymolybdate, which has an ellipsoidal group structure. Considering their surface properties and primary units, MOH nanostrands could likewise be ordered as direct polymers of cationic MOH groups [8]. It is essential to note, in any case, that the exceptionally decidedly charged design of these materials is balanced out in water, to such an extent that nanostrands eliminated from watery scattering lose a lot of surface charge by deprotonation, thus maybe portraying these constructions as extremely slight MOH nanofibers is more exact [9].

3.2 Method to Fabricate Metal Hydroxide

MOH can be synthesized by using various chemical strategies including solvothermal, sonochemical, precipitation, grafting, ion exchange method.

3.2.1 Precipitation Strategy

The precipitation system is a fundamental and advantageous strategy for organizing hydroxides metal [10] from drying watery salt game plans with the development of a foundation [11]. The status cycle's control is vexing, and the coordinated particles are truly absolute. Nonetheless, a grouping of hydroxides of metals with a glasslike and ill-defined nature could be consolidated using this method. Röntgen radiation diffraction and electronic microscope revealed that matter have penetrable plans as well as eccentricular and upsetting surfaces. Homogeneous substance material association is begotten with a diversity of engineered reactions in game plan [12].

3.2.2 Post-Uniting and Metal Cation Consolidation Strategy

Generally, it is feasible to support the synthetic soundness of the sorbents by crosslinking medicines or uniting responses [12]. Thusly, different researchers uncovered the mix techniques for hydroxides of metals using after-joining together or cation joining process. Different types of materials were cast-off by joining Chelating ligands on the mesoporous material's outer layer [13].

Chelating pragmatic gatherings are bound solidly with metals, which cause dispose of ionic species with a negative charge from H_2O. Different normal valuable gatherings affairs have been attached to the plans of mesoporous materials via after-mix joining or co-development strategy [14]. Regrettably, most updates to the phosphate adsorbent limit were accompanied by rising costs and waste stream generation [15]. Materials such as silica gel, biomass, and chitosan, are used as a polymer assistance for the game plan of sorbents with valuable get-togethers. Metal salt game plans were mixed with functioning materials to create metal adsorption districts, which were then washed with deionized water and isopropyl alcohol. A fluid plan of a negative ion polymer can be used to pre-treat the substrate, for instance, carboxy methyl cellulose. This methodology could accomplish benefits for growing phosphate adsorption limit [16].

3.2.3 Ion Exchange Method

The particle trade strategy is typically utilized when permeable sorbents have the capacity of cation trade. Particle trade happens between solvent particles, like potassium cation, sodium cation, calcium cation, and ammonium cation, and active catalytic materials, like lanthanum cation in Lanthanum nitrate hexahydrate [17, 18]. To achieve high action, the molecule shape might be constrained by selecting the pore size of the transporters. However, openings may not work well with metal particles or the strongly active metal colloidal particles. Besides, a few utilitarian parts joined in the permeable medium might let out of the sorbent into fluid arrangement [19].

3.2.4 Sonochemical Method

Sono compound technique is an easy course worked under surrounding environment. During ultrasonic irradiation, high frequency waves are generated through the arrangement, having caused the liquid media to alternate between maximum and minimum strain [20]. This has been widely assumed that ultrasonic lighting caused bubbling in a liquid medium, resulting in the formation, development, and implosive breakdown of air pockets. The breakdown of short-lived air pockets causes extreme nearby warming and high tension. These limited problem areas could generate temperature of around 5000°C and stress of more than 1800 kPa, as well as drive a variety of substance responses. We've been interested in the combination of hydroxide of metals nanostructured for a few years [21]. MOHs can be blended by means of sans surfactant response dissolvable H_2O. The light activated chemical activity of the nanostructured materials was also investigated.

3.2.5 Hydrothermal Method

Aqueous is a combination of two words: "hydro" for water as well as "warm" for warmth. Aqueous is a method of producing MOH and control with morphology structure at extreme temperatures [22]. Sir Roderick Murchison, a British geologist, pioneered this strategy (1792–1871) [23]. Under controlled conditions, the substance is strategy with a concurrence of watery dissolvable. The amalgamation in nucleation of precious stone where the temperature gradient is maintained in Teflon tube that is antecedent dissolvability ward to obtain the craving structure [24]. This is especially true for the admixture of solidified particles with varying shapes.

3.2.6 Polyol Synthesis

Polyol is an ingenious and simple strategy for the development for MOH, which allows for formation of crystals and progression control. Fragments with precise morphology are formed, with positive form, size, extend, and dispersion, as well as exceptional crystallinity [25]. Liquid polyol and reactants are deliberately changed and warmed at specific temperatures that the polyol stay unreactive.

Polyol limits the turn of events and joining of combined example. The resolvable metal originator frames a moderate throughout the reaction, which is also reduced to metal nanomaterial. The size of the particles can be well ordered by expanding the exterior cores for nucleation and expanding in temperatures for warm polyol breakdown [26].

3.3 Properties and Applications of MOHs

MOHs can be used in various fields including nanotechnology, flame retardants, adsorption, and water purification.

3.3.1 MOH Flame Retardants

Due to low rate of corrosion, less toxicity, decrease amount of smoke emission during combustion and other processing, MOH flame retardants are given preference over traditional halogenated based polymeric flame retardants [27]. Aluminum hydroxide $Al(OH)_3$, Aluminum tri-hydrate, Milk of magnesia and basic magnesium carbonate (BMC) are the MOH flame retardants. Water (H_2O) is generated upon heating of these hydroxides, as they decompose endothermically, which results in the decrease of temperature of substrate and reduces the heat flow. The decomposition temperature of Aluminum tri-hydrate ATH and basic manganese carbonate BMC ranging from 180° to 340 °C with a peak of approximately 320°–410°C. Above 300°C, MDH is the most endothermic hydroxide of metal in nature, BMC is the least one, while ATH is between the two [28]. However, to achieve the demanded degree of flame retardant material, an elevated level of MOHs is usually required, that results in the decrease of polymer matrix and poor mechanical properties [29]. Figure 3.1 shows fire retardant activity of $Mg(OH)_2$.

3.3.1.1 Alumina Tri-Hydrate (ATH) and Milk of Magnesia

Aluminum Hydroxide and Milk of Magnesia, is a corrosive and with the absence of halogen are fire retardant for various polymers or plastics [30]. The two hydroxides decay endothermically when warmed as per the responses:

$$2Al(OH)_3 \rightarrow Al_2O_3 + 3H_2O$$

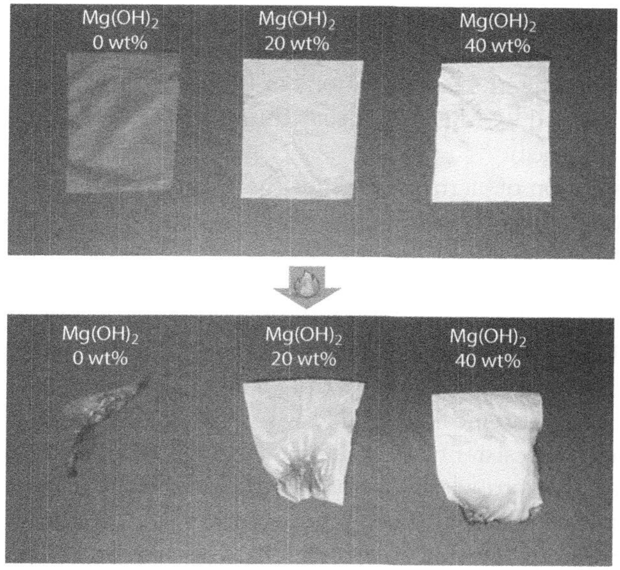

Figure 3.1 Representation of flame retardant test of $Mg(OH)_2$. Reproduced with permission from Ref. [112].

$$Mg(OH)_2 \rightarrow MgO + H_2O$$

The vaporous water stage is accepted to wrap the fire, in this way barring oxygen and weakening combustible gases. A heat protecting material, indistinguishable as compared to the capacity of scorch framed by fire retardant composed of phosphorous, may shape on the outer layer of the plastic that will directly interact with the flames of fire, diminishing the progression of potentially combustible deterioration items to the gas stage where burning occurs. The decay items in both responses are non-poisonous, and the mineral stages; particularly magnesium oxide is an antacid which reduces the chances of escape of acidic and damaging gases from plastic. The physical and artificial properties of $Mg(OH)_2$ and ATH. $Mg(OH)_2$ has a disintegration temperature that is 100°C higher than ATH, granting for a favorable handling degrees when compounding and expelling the flexible plastic. Furthermore, during the deterioration cycle, $Mg(OH)_2$ sock up most of energy [31, 32].

3.3.1.2 Utilization of $Mg(OH)_2$ as a Flame Resistance in Plastics

$Mg(OH)_2$ grades range from covered, micronized powders (for higher quality, greater stacking purposes) to uncoated $Mg(OH)_2$, which can be used as a direct replacement for ATH. Several studies were made to demonstrate the potential of $Mg(OH)_2$ as a fire resistant in plastics [33]. These have reasoned that $Mg(OH)_2$ is effective at reducing smoke emissions from plastic consumption. The following are synopses of a major relevant variables influencing the presentation of $Mg(OH)_2$ as a fire resistant material:

1) The endothermal breakdown of $Mg(OH)_2$, which begins at roughly 330 °C (versus around 230°C for ATH), reduces the temperature of the substrate, decelerating the rate of warm plastic degradation.

2) The entrance of water fume from $Mg(OH)_2$ decay lowers the fuel source existing in gaseous phase.
3) The typically elevated heat limitations of both $Mg(OH)_2$ and the deterioration items formed by $Mg(OH)_2$ decay diminish the nuclear power available to corrupt the plastic.
4) Such deterioration of elements extends the plastic's protection from the heat source through burn development.

The $Mg(OH)_2$ in this concentrate fundamentally brings down the general degree of smoke delivered. Moreover, the utilization of $Mg(OH)_2$ creates an impressive setback for the beginning of smoke development and notably eases back the pace of smoke discharge [34, 35].

A fire resistant should not degrade the actual qualities of the plastic in order to be useful in enhanced plastics. A typical adaptable wire Polyvinyl chloride plan employed a 30 PHR stacking of each fire resistant and resulted with a plastic. Plastics with greater loading requirement, such as poly-propylene and poly-amides, typically necessitate the use of specialist $Mg(OH)_2$ grades with unsaturated fat coatings or specified real characteristics. The unique properties of these substances consider high with practically zero debasement of the actual properties of the plastic. A new report was attempted to assess how much work expected to intensify metal hydrates into polypropylene. It is prominent that ATH expected either a Stearate or Silane treatment to be compounded as effectively as did the uncoated $Mg(OH)_2$ tests. Further decreases in the energy necessities for $Mg(OH)_2$ were acknowledged with either Silane or Stearate surface medicines [36]. The extra enhancements in processibility for specific polymers have been exhibited with metallocene catalyzed polymers which might reduce or kill the requirement for unique surface medicines of metal hydrate additive [37].

$Mg(OH)_2$ acts as a fire resistant and smoke suppressant in plastic by absorbing heat from the plastic as it degrades into MgO and H_2O. The aqueous fume produced reduces the fuel supply to the fire. By insulating plastic against heat and creating burn, decayed items keep potentially flammable gases from reaching the fire. Expanding regulation and concerns about the usage and reproducibility of halogenated flame reluctant make $Mg(OH)_2$ more desirable to plastics manufacturers. $Mg(OH)_2$ provides fire reluctant and smoky concealment from a corrosive, non-halogen chemical with low harmfulness ratings [38]. Much of the time, there is no need to make a trade-off between actual characteristics and fire retardancy of the plastic when using a valid grade of $Mg(OH)_2$. Regulation may be the prevailing principal incentive for determining $Mg(OH)_2$ instead of halogenated fire retardants in high-stacking situations where the use of extremely good grade or covered $Mg(OH)_2$ is directed [39].

3.3.2 MOHs Sludge Can Be Used as Latest Adsorbent

In the present naturally cognizant and profoundly directed assembling world, numerous printers like to utilize aqueous-based ink in their cycle. Aqueous-based inks, in which water accounts for at least 75% of the liquid element of the ink, have now been created as a substitute for dissolvable-based inks, which have now been deemed environmentally dangerous. The flexographic printing business, which manufactures paper and ridged bundling, mostly uses aqueous based inks and generates a significant amount of wastewater during washing [40]. This business has been studied because of the substantial consumption of aqueous based inks and extra water emission. Despite the fact that VOC contamination has decreased, the wastewater problem may have worsened. Over 5 percent of total of the inks used in wiping flexographic printing presses and plates was estimated to wind up in the wastewater [41]. This waste of time effluent is extremely turbid and tainted by inks including scattered colors, tars, and other added substance elements, and it should be processed before release. Flocculation is the typical treatment method for printing wastewater is fl which involves the use of conventional flocculants such as cationic polyelectrolytes and aluminum or ferric salts in acidic conditions. Shades, the primary colorant in aqueous based inks, are small nanoparticles that have been absorbed by scattering gum and have a negatively charged, creating a repulsive force and preventing clumping and settle down. This enhances the reliability of water-based inks while also addressing color build-up issues in manufacturing and printing [5]. Because the scattering steadiness of colloidal particles in the stream is due to the unfavorable electrostatic powers, the balance of this charge is critical in order to activate floc and precipitate for treating wastewater. Decidedly charged muck waste, on the other hand, can be used as an adsorbent to undermine the fine colloidal colors and adsorb a few surviving colors in a clever way. MOH muck is a dry byproduct of treating wastewater in the electro-plating industry. It is supplied through the precipitate of heavy metal particles in wastewater, such as Cr, Cu, Ni, Al, and Zn, using Ca or NaOH, and is used as a lessening specialist to minimize the conversion of Chromium (VI) to Cr (III). When an antacid is applied, along with other insoluble MOH, in a framework pH range of 7 to 10, $Fe(HO)_2$ is framed as well. Use of waste materials from different businesses would be of advantage, while perhaps not totally wiping out the expense of wastewater treatment [42]. The goal of this research was to assess the feasibility of shading expulsion from aqueous based ink wastewater by MOH muck in order to assist flexographic printers in lowering the cost of treating wastewater and reusing electroplating trash prior removal. However, all coagulation process ooze carrying heavy metals or mineral colors should be disposed of in a solid waste in accordance with Thai regulations [43].

3.3.3 Metal Hydroxide MOH Nanostructures

Nanotechnology has spread throughout the world, and intricate morphologies that were once incompatible with the imaginative mentality, such as nanoflowers and nanoribbons, are now consistently manufactured and represented. As compared to bulk materials, synthesis of nanostructured thin films has grabbed the attention in the recent decades due to the potential optical and electronic properties. Among all of the materials in the nature, nanostructures stand out enough to be noticed because of the properties based on their arrangement as well as on their size, shape, construction, stage, and size spread distribution [44]. Nanoparticles can be more or less reactive depending upon their morphology as the crystallographic planes are present on the exposed surface of the nanoparticles [28]. In the recent researches, they put together different MOH nanoparticles based on their reactivity, and discussed about their applications to make wall compositions (frescoes), paper, stone, and wood [45].

MOHs aren't very water-soluble, but they could be useful for things like social legacy protection, material, earthenware, and stain industry. Several methods for fabricating MOHs nano - structured substances have been demonstrated in recent years.

Nanofibers made of MOHs have been intensively researched since the 1940, when electron microscopy became accessible as a research instrument. Turkevich and Hillier reviewed previous electron microscopy research of metal oxide nanofibers in 1949 [46]. "It was previously reported that V_2O_5 nanofibers obtained by treating NH_4VO_3 with nitric acid had a very small, lace-like architecture with a width of 2.7 nanometer. This structure formed rapidly in fluid form after the polycondensation of vanadic acid, using a mechanism similar to that found during the formation of polyoxometalates. The early change metals, such as V, Mo, and W, structure a progression of anionic bunches, and the polyoxomolybdates exist in a variety of intricate nanomorphologies, including haggles. These nano-sized clusters have been shown to congregate into inorganic nanofibers using a straight natural polymer as a pattern. Polleux et al. used nanofibers that were very fine and transparent. Despite this, these nanofibers with a width of 1.3 nanometers quickly framed group like structures." [47, 48].

3.3.4 MOHs for Supercapacitor Electrode Materials

Based on charge storage strategy, supercapacitors are categorized into two classes, double layer capacitor and pseudo-capacitor. First type is elaborated by the diffusion of static charges and accumulation of these charges [49] on the interface of electrode and electrolyte. Former type is initiated by the Faradaic reaction [50] on the surface of electrode. Various metal hydroxides MOHs including hydroxide of nickel and cobalt have been considered as emerging materials for the fabrication of electrode in electrochemical capacitors as shown in Figure 3.2 [51]. For the material of positive electrode used in supercapacitor, mixed-MOHs compounds i.e. Nickel-Cobalt, Manganese-Cobalt, Nickel-Aluminum [52] are preferred over single compound as they enhanced the performance of electrode. Retention time shown by the metal hydroxides MOHs is much more as compared to crystalline material.

3.3.5 Drugs or Pharmaceutical Applications

$Al(OH)_3$ is used as an acid neutralizer in humans and animals under the nonexclusive name "algeldrate" (mostly felines or canines). It is preferred over other options, for example,

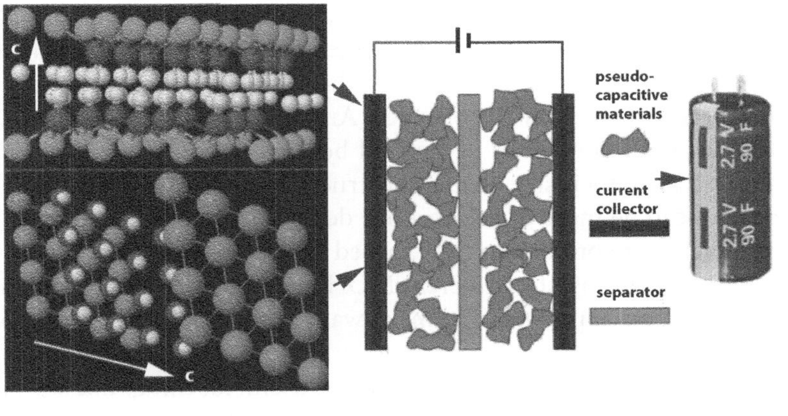

Figure 3.2 Representation of use of MOHs as electrode material in electrochemical capacitors. Reproduce with permission from Ref. [113].

sodium bicarbonate, since $Al(OH)_3$, being insoluble, does not raise the pH of the stomach over 7 and so does not cause the stomach to release excessive acid. "Alu-Cap, Aludrox, Gaviscon, and Pepsamar" are examples of brand names [53]. It reacts with an overflow of acid in the stomach, lowering the harshness of the digestive process, which may allow the adverse effects of ulcers, acid reflux, or dyspepsia to be avoided. Aluminum particles in these products hinder smooth muscle cell withdrawals in the digestive tract, delaying muscular action and increasing the time it takes for faces to travel through the colon. Whatever items are created to minimize such affects by taking into account equivalent convergences of $Mg(OH)_2$ or $MgCO_3$, which have balancing effects [54].

This chemical is also used to treat hyper-phosphatemia (high K or P levels in the blood) in people and animals suffering from renal failure [55]. Normally, the kidneys filter a lot of phosphate out of the blood, but renal failure can cause phosphorous to accumulate. When consumed, aluminum salt binds to phosphorous in the digestive tracts, reducing the amount of phosphorous that may be absorbed [56]. Encouraged aluminum hydroxide is incorporated as an adjuvant in certain immunizations (for example Bacillus anthracis antibody). Brenntag Biosector's Alhydrogel is a well-known brand of aluminum hydroxide adjuvant. Because it assimilates protein efficiently, it also balances out antibodies by stopping the proteins in the immunization from inciting or adhering to the compartment dividers throughout capacity. $Al(OH)_3$ is occasionally referred to as "alum," a name of a few sulphates [57, 58].

Immunization definitions containing aluminum hydroxide animate the resistant framework by initiating the arrival of uric acid, an immunological peril signal. This strongly attracts some types of monocyte, which differentiate into dendritic cells. Antigen Presenting Cells recognize the antigen, transport it to lymph nodes, and activate B- and T-lymphocytes [58]. It seems to add to enlistment of a decent Th2 reaction, so is valuable for vaccinating against microorganisms that are obstructed by antibodies. Notwithstanding, It has little power to animate cell (T helper cells) insusceptible reactions, which is important for protection against a variety of pathogens, and it is ineffective when the antigen is contain peptide bonds [59].

3.3.5.1 $Ca(OH)_2$ Used in Dental Practice

$Ca(OH)_2$ has been used for a number of purposes since it was first introduced into dentistry in the early twentieth century. The chemical has a high pH in its natural state, and its use in dentistry is mostly due to its mineralization and anti-bacterial induction property. Before looking into the precise effects of $Ca(OH)_2$, it is useful to discuss mineralization ideas inside mesenchymal tissues [60]. It has been well understood that an epitactic instrument operates by nurturing collagenous tissue beneath the surface of the skin. Only a few forms of collagen, such as dentine and sharpen, mineralize along these lines. The contact is most likely caused by charged bunches colliding with nearby macromolecules, resulting in epitactic foci. These focuses demand the presence of a nucleation location of which hydroxyapatite gem formation can take place [61]. There are several theories on who initiated the encounter. Some specialists believe the seed is chondroitin sulphate, while others feel it is a mineralization inhibitor. A vitamin D ward protein capable of restricting Ca, phospholipids and phosphoproteins has also been proposed as mineralization initiators [62, 63]. The ability to halt the interaction is equally important to the acceptance of mineralization.

One well-being factor could be the presence of substances such as pyrophosphate particles in the blood and tissue, which work as inhibitors. This activity is very less whenever pyrophosphates are processed by pyro phosphatase on the site mineralization [64].

Pyrophosphatase belongs to the soluble phosphatase family, which could explain why these enzymes are always found in mineralizing tissues. The vast majority of mineralized tissue investigation has concentrated on the usual mineralized tissues of the body. Regardless, other organs, such as the aorta, can cause dystrophic calcification, and type 1 collagen is also linked to this association. The dental mash is also susceptible to dystrophic calcification, which can be circular or dispersed in form and ruin the entire root channel architecture. $Ca(OH)_2$ started the mineralization process. $Ca(OH)_2$ seems to have a novel ability to trigger mineralization, even in tissues that have not been mineralized. Rasmussen and Mjor (1971) couldn't affirm that $Ca(OH)_2$ incited mature bone development under these circumstances, however they saw that when the material was put in direct contact with tissue, it started arrangement of sinewy tissue, with a periodic arrangement with young bones. These researchers also discovered that no critical responses occurred when $Ca(OH)_2$ was separated from have tissue using millipore channels, indicating that the material had no effect once it had diffused a certain distance. As calcium in $Ca(OH)_2$ is not consolidated in mineralized tissues, these tissues get all of the minerals from the dental mesh via blood flow. According to these beliefs, $Ca(OH)_2$ works as a starter rather than a substrate for fix. The device used to kick-start the reparative cycle with $Ca(OH)_2$ is a confusing mess. An rise in pH generated by free hydroxyl particles has been proposed to initiate or contribute to mineralization. Tolerating the hydroxyl particle as the only initiator of the interaction is difficult because other extremely antacid mixes, such as $Ba(OH)_2$ and $Ca_3(PO_4)$, fail to mineralize. $Ca(OH)_2$, on the other hand, acts as a nearby buffer against the acidic responses induced by the provocative cycle. A soluble pH may also kill the lactic corrosive excreted through osteoclasts, thereby preventing additional calcified tissue annihilation. Despite the fact that $Ca(OH)_2$ is not fused in the mineralized fix, it appears to be involved in the cycle's beginning [65]. Figure 3.3 depicts the events after application of $Ca(OH)_2$.

Osteogenic and mitogenic properties in a material are due to high pH and permeability of calcium and hydroxyl ions which changes the enzymatic pathway and eventually mineralization. The elevated pH may also stimulate antacid phosphatase activity, which is thought to be involved in the creation of hard tissue. The optimal pH for basic phosphatase activity is 10.2, which is the alkalinity produced by several $Ca(OH)_2$ arrangements [66]. High calcium levels enhance the activity of calcium pyrophosphatase during mineralization. If the regular self-limiting proteins (pyrophosphates) fail to operate, mineralization can continue uninterrupted [67]. A decrease in tight porousness due to increases in calcium particle number could lower serum flow inside the dental mash, reducing the clustering of the inhibitor pyrophosphate particle. This would have been consistent with the previously mentioned increase in calcium-subordinate pyrophosphatase levels, which results in unregulated mineralization of the mash tissue. This accounts for the increased rate of mineralized channels seen during pulpotomy and direct mash coating [68]. Instead of how much reparative dentine was framed over time, decreased blood supply to imperative tissues controls unregulated mesh mineralization because mesh mineralization was initiated by lower blood supply to pulpal veins due to parasympathetic hindrance loss [69].

It could cause cell respiratory reduction, which would end in neurotic calcification of the mash and possible annihilation of the root waterway. Surprisingly, in certain

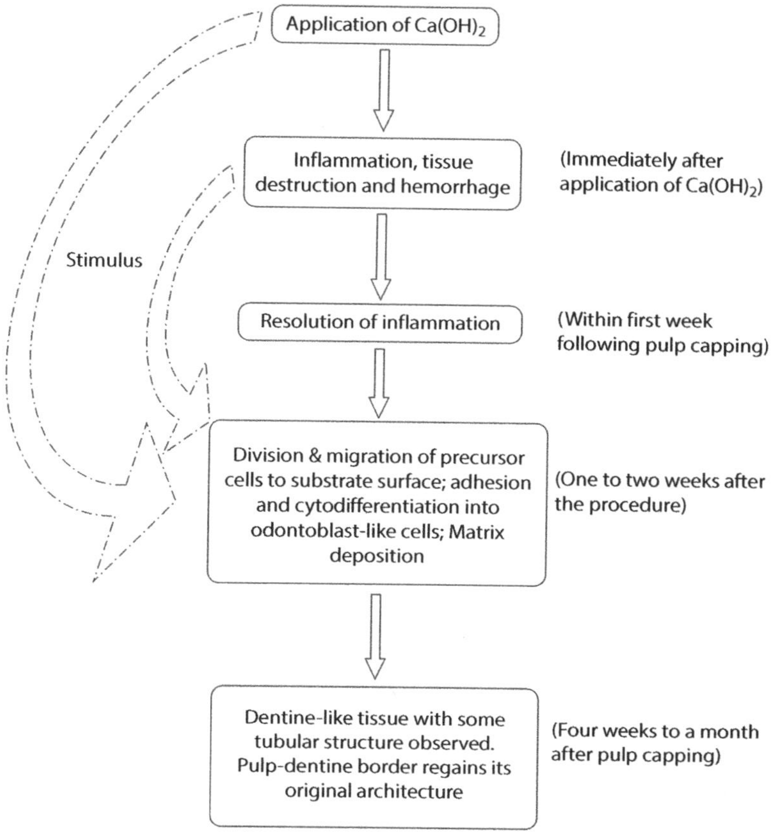

Figure 3.3 Representation of observed events following the application of $Ca(OH)_2$. Reproduced with permission from Ref. [114].

circumstances, $Ca(OH)_2$ appears to accelerate resorption. According to findings, if broken teeth are replanted and quickly filled with $Ca(OH)_2$, they may experience external substitution resorption (ankylosis) of the foundations. Resorption is less likely to occur if the root filling arrangement is deferred for a long period of time [70]. When $Ca(OH)_2$ was added, it was instantly set. The substance spread via the apical foramen, causing cementocytes to die. Inward resorption of deciduous teeth may happen after pulpotomies. This happens commonly near the junction of the coronal and radicular mashes [71]. Despite the fact that the elevated pH of $Ca(OH)_2$ may cause more damage to previously injured tissue, inward resorption is currently assumed to be aided by bacterial illness in the dentinal tubules, which mistakenly transport necrotic and vital parts of the mash. The dentine spectrum after the addition of $Ca(OH)_2$ to a critical mash, a mineralized border, also known as a 'dentine span,' is generally created (pulpotomy) [61]. Odontoblasts and connective tissue cells appear to be the source of this maintenance substance. It is possible that the impediment will not be completed entirely. Some versatile strategies regarding pH of the precursor material has been seen for the development of a dentine span to treat the tooth. A necrotic zone, for example, is framed contiguous the material as a result of a high pH material, and the dentine span then forms between this layer and the fundamental essential mash [72]. Finally, the necrotic tissue savages and vanishes, leaving a space between the covering layer and the expansion. The necrotic zone is generated as a result of a lower pH substance, such as Dycal, but it is resorbed before the arrangement of the dentine span, which is then framed directly against the covering material. High pH dentine spans are histologically indistinguishable from lower pH dentine spans, but are easier to discern through radiation exposure as there is separation between the scaffold and the $Ca(OH)_2$ [73]. It has been stated that $Ca(OH)_2$ is not the ideal substance to encourage dentine span arrangement. When $Ca(OH)_2$ was compared to a 50/50 mix of tricalcium phosphate and lease hydroxyapatite, the latter provided a thicker scaffold and faster growth of typical rounded dentine. Various substances, as zinc oxide-eugenol concrete and silicate concrete, were explored; however, despite not producing pupal aggravation (due to the absence of microbial contamination), dentine span arrangement was commonly feeble or nonexistent. Microscopic organisms are wiped out [74].

(1) As a dressing for puip. $Ca(OH)_2$'s antibacterial effects may account for some of its warming qualities. According to findings on gnotobiotic rodents, when a heavily exposed teeth in microbe-free organisms were treated, healing occurred regardless of the covering material employed. This suggests that under typical conditions, mending occurs due to $Ca(OH)_2$'s antibacterial activity rather of any influence it may have on mineralization. $Ca(OH)_2$'s bactericidal activities are assumed to be associated to pH and correlate with $Ca(OH)_2$'s ability to diffuse from the set material. $Ca(OH)_2$ has a lower warming potential that it shows when applied as a root trench dressing to treat continually aroused periapical tissue. This could be due to the prepared accessibility of solid veins in the periapical tissue, as opposed to their lack of passing on tissues in the mash's encased environment. It has been discovered that $Ca(OH)_2$ kills just the tiny organisms on the mash's outside layer, not those that have invaded necrotic tissue. As a result, the material influences the mending of an ignited mash, and its application has all the earmarks of being shown for the treatment of sound or marginally defiled pulps where microorganisms have not invaded the more profound part.

(2) There seems to be some uncertainty about the viability of $Ca(OH)_2$ as an intra-trench dressing when compared to other medications. In any event, the variable reactivity of distinct microbes to disinfection chemicals can explain a significant portion of this sensitivity. Any substance applied as a root waterway treatment should be assessed only as an antibacterial specialist, operating independently of tissue guard devices. Hence as opposed to its technique of activity in mineralization, $Ca(OH)_2$ includes a vague bactericidal activity inside the confines of the root waterway, soluble bases overall altering cell layers and protein structure. Although most small living beings are destroyed around pH 9.5, a couple can survive to pH 11 or higher. The major concern isn't how microorganisms are eliminated, but how vital tissues could be shielded from the toxicity of $Ca(OH)_2$. The removal of debris from primary tissues by a putrefaction zone appeared to avoid significant tissue damage. Ability of $Ca(OH)_2$ to degrade necrotic material – its efficacy is comparable to that of sodium hypochlorite, however it is less potent. Nevertheless, its prolonged stay

in the root trench, where it continually makes a therapeutic difference, should more than compensated [75, 76].

3.3.6 Removal of Toxins from the Water

Water is indispensable unit for the endurance of life in the universe however expansion in innovation and industrialization for enormous scope harms' the perfect water sources [77]. "Turbidity (physical), shading (physical), taste (physical), temperature (physical), chlorides (compound), fluorides (substance), iron and manganese (compound), lead and copper (synthetic), nitrate (synthetic), sodium (substance), sulphate (substance), zinc (synthetic), biochemical oxygen interest (synthetic), solids (physical), pH (compound), substance oxygen interest (substance) (physical). Surfactants (physical), require contaminations (substance), unstable natural mixes (synthetic), and hazardous metal and non-metal particles will also be depicted (compound)" [78].

An immense number of debasements including surfactants, pesticides, weighty metals and colors are entering the water on customary premise whose fixation is expanding step by step. These foreign substances taint water and make it risky for human, creatures and marine life.

Aluminum salts are used as flocculants in water filtering because of newly accelerated aluminum hydroxide structural gels. With time, this gel takes shape. Aluminum hydroxide gels can be dried (for example, using water-miscible non-watery solvents like ethanol) to form an undefined aluminum hydroxide powder that dissolves quickly in acids. Warming converts it to alumina, which are used as desiccants, adsorbents in gas cleaning, and impetus boosters [79]. Figure 3.4 shows the conventional method for heavy metal removal from wastewater.

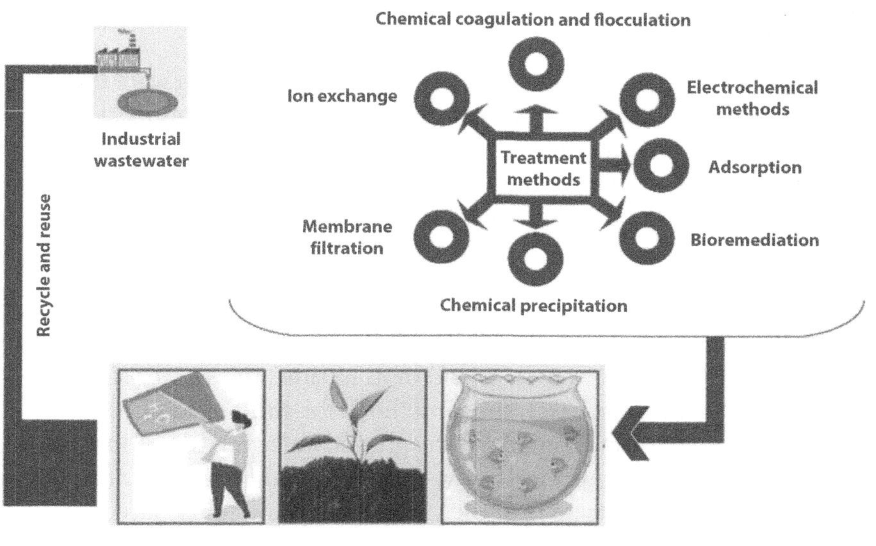

Figure 3.4 Conventional method for heavy metal removal from wastewater. Reproduced with permission from Ref. [115].

3.3.6.1 *Water's Physical and Chemical Characteristics*

TURBIDITY: Turbidity is a photometric measurement of how well suspended matter in water retains or loses the sun's brilliant light energy. The Jackson turbidimeter was the first device to detect turbidity.

COLOR: When the radiation of different wavelengths (visible light), collides with the retina of our eye, automatically the color is detected by naked eye. In this case, organic and inorganic materials that have been rotting for a long time make this idea come to light. This sense is created by materials degraded from vegetation and inorganic matter [80], which add color to water. This color may be offensive not because of health concerns, but because of its appearance. The United States Environmental Protection Agency (USEPA) establishes maximum pollutant levels to meet these standards (MCL). Color's secondary MCL is fifteen color unit [81]. This tone might not be good for your health, but it might be bad for style. Normal colors make water look yellow-caramel, so people usually think of pee as having this color. The platinum in potassium chloroplatinate is used to measure how much color there is (K_2PtCl_6). It takes 1 mg of Pt for every liter of K_2PtCl_6 to get one unit of shade. The big part of the SDWA (Safe Drinking Water Act) is the setting of rules. This declaration needs the foundation of important rules that protect the general well-being of people and the foundation of secondary rules that deal with things like taste, appearance, and color. To meet these requirements, the U.S. Ecological Protection Agency (USEPA) has set a maximum level of toxins (MCL). To add shading, the auxiliary MCL is made up of 15 units.

TASTE: The ability of the taste buds to make sense of things is referred to as taste. There should be no taste at all when drinking water is used for anything. Taste (or smell) isn't set in stone by figuring out the volume of the example A (in milliliters) and weakening it with a quantity of water (in milliliters) that doesn't taste or smell very much at all at a volume of 200 milliliters. In this way, the unit of taste (or smell) can then be spread out as far as a certain number.

SMELL: Odor is the sense that the olfactory nerves pick up on. If you want the water in your home to taste good, there should not be any smell to it. In small amounts, it won't hurt your health, but it will make the water taste bad. Having copper in the wrong places can make your stomach and digestive system hurt, also causes Wilson's illness. CPVC (chlorinated polyvinyl chloride) pipes are made of PVC (polyvinyl chloride). They can be used instead of copper pipes in the home.

NITRATE: Nitrate can cause blue babies, when we talk about nitrogen, the MCL is 10 mg/L. Before strict rules were put in place, sludge from wastewater treatment plants was spread on land and covered in ditches as a way to get rid of it. In time, the slop breaks down, and the nitrates are framed. So, in some cases, these methods for getting rid of nitrates have led to the nitrates getting into the dirt and making the groundwater very dirty. Indeed, these skills are still very good even today. Administrative offices should be happy with these practices if they figure out how much nitrate is in the groundwater and make sure it doesn't get to a level that isn't OK.

SODIUM: People who have heart, kidney, or circulatory diseases may be affected by the amount of sodium in their drinking water. It could make the blood pressure of people who aren't strong rise. The sodium content of ordinary table salt, which people use to season their food, is high. Because it makes up a large portion of ocean water, it would suffice in water supplies that have been contaminated by the sea.

SULFATE: One of the most important anions in water is the sulphate particle. When it is present in large concentrations in drinking water, it has a therapeutic or diuretic impact on people. 250 mg/L is the amount of SMCL that it has in a gallon.

ZINC: Zinc isn't thought of as bad for your health, but it can make drinking water taste bad. In grams per liter, it has SMCL that is 5 mg/L [82].

3.3.6.2 Types of Wastewater

There are two types of wastewater: clean and non-sterile. Non-sterile wastewaters are typically contemporary wastewaters. Sterile wastewaters (also known as clean sewage) are wastewaters that have not been contaminated with human waste. Clean wastewaters produced in homes are called homegrown wastewaters (homegrown sewages) [83]. Modern wastewaters will be wastewaters created during the time spent assembling. Accordingly, in light of the fact that a heap of assembling processes is utilized, a horde of modern wastewaters are additionally delivered. Sterile wastewaters generated in ventures could be referred to as modern clean wastewaters. Invasion and influx may also be introduced to these wastewaters. Sewers are used to transport wastewater. Several coincidental streams merge with them as they flow. Invasion, for example, refers to water entering sewers through breaches, defective connections, and sewer vents. This water is usually sourced from groundwater and isn't supposed to be disposed of in the sewer system. Another unintentional stream is inflow, which enters through openings that have already been purposely or unknowingly designed for its arrival. Continuous, instantaneous, or postponed inflow are the three main types of inflow [84]. Constant inflows enter the sewer framework indefinitely. Examples of these are releases from basement and establishment depletes that are consistently exposed to high groundwater levels, cooling water and depletes from humid places, and springs. Inflows that cause a rapid rise in stream volume in the sewer are known as direct inflows. Rooftop pioneers, sewer vent covers, and yard channels are probable sources of these. Delayed inflows are all those portions of the rain that don't reach the sewer right away and take many days to drain fully. Water drained from cellar basements after heavy rains and slow movements of water from ponded areas into sewage vent holes are examples of this waste. Invasion influx refers to the combination of penetration and inflows. The wastewaters mentioned above come from a variety of sources. Clean wastewater can come from a variety of sources, including private, commercial, institutional, and sporting facilities. Penetration inflow is definitely caused by precipitation and groundwater, however current wastewaters are caused by assembly enterprises. The amounts of these wastewaters really different sources are shifted and, once in a while, one piece of the writing would report an incentive for an amount of a boundary that contentions on data of the amount of a similar boundary detailed in one more part of the writing. For instance, numerous originators frequently expect that how much wastewater delivered is equivalent to how much water drank, including the stipend for penetration inflow, despite the fact that one source indicates that 60 to 130% of water polished off ends up as wastewater, another estimate indicates that 60 to 85% ends up as wastewater. Therefore, amounts gave beneath ought not to be utilized as outright insights, however just as guides. For more exact qualities, real information ought to be utilized [85].

3.3.6.3 *Treatment Techniques of Wastewater*

Normal treatment strategies for inorganic poisons include:
"Synthetic compounds are used in balancing and precipitation to adjust the pH of wastewater and combine with contaminants to create an insoluble compound (precipitant) that is removed from the wastewater through settling and, occasionally, filtration. Most heavy metals, phosphates, and sulphides are frequently eliminated through this cycle (Figure 3.4). Iron salts (ferrous and ferric chloride and sulphates); aluminum salts (aluminum sulphate, polyaluminum chloride, and sodium aluminate); lime; sodium bicarbonate and sodium carbonate (soft drink trash); sodium hydroxide; and sulphide salts are among the synthetics used (e.g., ferrous sulphide). Container and pilot testing are the greatest ways to determine which synthetic will be the most effective" [86]. Iron and aluminum salts remove phosphate and metals in a stoichiometric manner: higher toxin concentrations necessitate higher iron and aluminum fractions to obtain the best possible outcomes. Metals and phosphates, on the other hand, are generally discharged via hydroxide precipitation (the mechanism for lime and sodium hydroxide). Because the reaction occurs at pH 7.5–10.6, the pH of the solution should be altered on a regular basis. Furthermore, when processing phosphates, hydroxide frequently produces more solids than iron and aluminum salts. In the majority of balance precipitation systems, the material is delivered to the wastewater via synthetic metering syphons or feeders. Substance expansion can be controlled mechanically or naturally based on the wastewater stream, pH, or another interaction boundary. Before sedimentation, synthetic precipitate often necessitates a glitter mix and flocculation procedure. The following solids are then removed for further processing [84].

The exact synthetic measurements and ideal cycle execution rely on various elements, for example,

- The pH of the wastewater,
- Other competing particles,
- The alkalinity and buffering limit of the wastewater,
- The amount of natural material present, and
- The flocculation and blending force, as well as the detainment time.

Thus, container and pilot testing are unequivocally suggested prior to choosing a substance. Container examination on a regular basis can also help with dosing and execution of Hydroxide Precipitation Coagulation duties.

3.3.6.4 *Metal Hydroxide for Treatment of Wastewater*

Precipitation is commonly used to remove heavy metals from wastewater, for instance, MOH, followed by flocculation of the metal particles to create larger, heavier flocs that are then isolated from the water by sedimentation or buoyancy. Typically, metal fixation levels in solution can be reduced to less than 1 mg/L with this method, which has proven to be dependable as well as cost-effective and highly specialized. Generally, heavy metals are broken up under corrosive conditions and accelerated under soluble conditions, thus it is necessary to manage pH. To raise the pH and produce hydroxide particles, (NaOH; burning pop), [$Ca(OH)_2$; lime], or [$Mg(OH)_2$; hydrated Mg(OH)2] are commonly added to wastewater [87]. A significant number of hydroxide particles are expected to assure the completion of the precipitation reaction. The heavy metal particles react with hydroxide particles to form powerful MOH particles. The pH at which a metal is least soluble is referred to as its metal explicit (i.e., likely to accelerate). The bend's "knee" is the location where the metal is least soluble. Metal solvency is influenced by a variety of factors such as the presence of chelating experts, surfactants, various particles, temperature, and so on. The significant test in metals precipitation is the presence of various metals in numerous metal-completing wastewaters. Plan engineers should determine what pH value accelerates and removes the most metals. In addition, the overall concentration of metals in wastewater may fluctuate over time, causing the ideal pH to shift. In order to determine whether metals will accelerate at a pH near the cutoff threshold and whether profluent fixation will satisfy administrative criteria, container testing may be necessary on a daily or hourly basis. The tiny hydroxide particles should be coagulated into flocs once the MOHs have been accelerated. Heavier flocs will settle faster in clarifier. Hydroxide salts are typically too fine to settle quickly without the addition of a binder (polymer) to the wastewater before entering the clarifier. Excess MOH particles will be removed after explanation (sedimentation) [88]. To meet administrative cutoff points, wastewater pH needs to be changed before release. In compound expansion, it is frequently overlooked that the nature of Synthetics, specifically Acids, is one of the major points of contention. When conditions are severe enough, heavy metals can be found in acids and soluble bases. The reagent provider must acquire and certify the metal substance. n the coagulation of Iron and Aluminum Salt Precipitation, synthetic compounds with very low levels of heavy metals are available. Ferrous and ferric chloride. The iron and aluminum salts ferrous and ferric chloride, ferrous and ferric sulphate, aluminum sulphate (alum), and polyaluminum chloride are commonly used to remove inorganic toxins. pH-dependent applications can also use aluminum or iron hydroxides. The suitable iron or aluminum salt and the proper quantity are established by pilot testing or seat scale testing, as with other material precipitation processes [89].

3.4 Examples of Metal Hydroxide

3.4.1 Calcium Hydroxide $Ca(OH)_2$

The primary compound used to make $Ca(OH)_2$ was modern powdered calcium oxide ("unadulterated" range, "Reachem", Russia). CaO was heated in air for 4 hours and 30 minutes at 657 °C to remove the presence of $CaCO_3$ formed during long-term CaO preservation $CaCO_3$ degrades into CaO and CO_2 as a result. Allow the powder to cool to room temperature (27 °C). Following that, the suspension of CaO was immersed in refined water and placed on the inward bulk of the barrel-shaped quartz reactor. A 0.03 cm $Ca(OH)_2$ sheet covered 70 cm^2 of surface area. $Ca(OH)_2$ was used in place of calcium oxide. A supporting syphon removes water from the reactor as it works with the transition of $Ca(OH)_2$ to calcium carbonate in the air. A thick layer of $Ca(OH)_2$ was maintained on the reactor divider in the absence of calcium carbonate. After that, the reactor was placed in a high vacuum framework for testing. The reactor can be started up to a pressure of 10-7 mmHg [66].

3.4.1.1 *Utilizations of $Ca(OH)_2$ in Dental Detailing of $Ca(OH)_2$ (Glues)*

Glues containing $Ca(OH)_2$ are classed as either setting or non-setting. The former is employed as pit covering or sub-coating, as well as root trench sealers, while the latter is used in root channel treatment [61].

3.4.1.2 *Materials for Setting the Therapeutic Effect*

The beneficial characteristics of setting $Ca(OH)_2$ compounds are linked to specific pH. The number of unattached Ca and -OH particles that remain after the material is cured determines the final alternative [68], and it follows that particle departure from the set material will result in a drop in its mass. The hydrophobic concept of the material is one component that contributes to the accessibility of the hydroxyl particle. The more hydrophobic the last choice, the more unlikely dispersion is to happen [90], Hydrex, for example, was more hydrophobic than Dycal because it contained a paraffin dissolvable that prevented water from mixing into the set product [78].

An further factor to consider in the disintegration of $Ca(OH)_2$ is the impact of tiny organisms on the set material, which is related to microleakage. Microscopic organisms may come into contact with $Ca(OH)_2$. By converting the material to calcium carbonate, this could lower the pH of the material, explaining why earlier Dycai configurations seemed to disappear under efforts to rebuild them [68]. The accessibility of their hydroxyl particles can be used to rate the setting $Ca(OH)_2$ glues. The glue is to be chosen from close to the first spot of this list if antimicrobial action is to be expected (as in direct mashed covering), but if covering only is to be provided, the $Ca(OH)_2$ glue should be picked from close to the bottom of the list since it is more averse [72]. The rationale for considering $Ca(OH)_2$ in such nontherapeutic materials, when its substantial antibacterial impact isn't required, is its ability to respond quickly in the setting system. One inherent challenge in attempts to include unique setting type items as indicated by their synthesis is a lack of firmness in their plan [74]. Several of Dycal's constituent rate structures have been altered significantly since the model was first introduced. As can be observed, the difficulty for the producer is to strike a stability between a substance that is enough solvent to apply a restorative impact, however isn't dissolvable to the point where it breaks up away, so vitiating its optimum usage as a mechanical coating. Components of $Ca(OH)_2$ containing compounds that are used for setting. There are two main types of setting systems [91].

In order to create the two glue framework, calcium and zinc particles react with a salicylate chelating specialist, which is accelerated by water. The use of $Ca(OH)_2$ as a sealant for root canals has also been proposed [90].

To give adequate time for the gutta-percha root filling to be thick, the setting period is increased in these arrangements, most likely by substituting $Ca(OH)_2$ with calcium oxide. Analar $Ca(OH)_2$ can be employed as a vehicle either dry or with purified water. Clinically, a major drawback is formation of slurry which would be difficult to manage in the root channel by puddling; however, on the other hand, it can be blended with thick glue to deposit into the root trench with a combination transporter and packed with root waterway pluggers. Products of the highest quality Use methyl cellulose as a carrier with different quantities of water to prevent this issue. As a consequence, homogeneous glues of different consistency are made with outstanding care properties. Several admixtures have been suggested, including adjacent sedative arrangement, camphorated monochlorphenol (CMCP), beechwood creosote, Ledermix, and radiopacifiers. With the possible exception of the last, when barium sulphate is mixed in with $Ca(OH)_2$ in a 1:8 ratio, these combinations are currently undesired on the grounds that the added chemicals may accomplish very little and may unfavorably impact the beneficial effects of the $Ca(OH)_2$ [92].

3.4.1.3 *Covering of Pits*

$Ca(OH)_2$ setting glues are now widely used as coating materials. Their evident advantages, independent of their healing properties, are as follows:

(i) Due to the accelerating influence of moisture in the encompassing quality of the oral pit and from within the dentinal tubules, those have a rapid set in the cavity.
(ii) They don't hinder Bis-GMA saps' setting reaction, thus they're the best coating material for composite gum material.
(iii) The underlying configuration of the material in narrow sections is typically regarded to be sufficiently resistant to withstand the applied buildup pressures, which are necessary in any case for machine cut mixed amalgams.
(iv) Light-relieved pitches have been shown to be biocompatible and will not harm the pulp. However, due of the impacts of heat generated during setting, undesirable consequences are expected, especially in deep depressions around the mash. The actions of microscopic creatures and microleakage are most likely to be responsible for the disappearance of previous forms of these materials under restorative attempts. This method involves removing severely decayed dentine and covering the exposed fragile dentine with either $Ca(OH)_2$ or zinc oxide-eugenol concrete. Both materials are bactericidal, which helps to regulate but not totally destroy any remaining organisms. $Ca(OH)_2$ may provide a slight benefit in experiencing the same thing since it can also initiate the healing of a moment pulpal opening that was previously neglected. There's also evidence that $Ca(OH)_2$ boosts mineralization in the pit's dentine. This component may explain the detailed reduction in porousness (*in vitro*) of both dentine and the smear layer when $Ca(OH)_2$ is applied. However, it is projected that the usage of zinc oxide eugenol concrete will result in a similar rise in mineralization [76].

3.4.2 Magnesium Hydroxide $Mg(OH)_2$

The $Mg(OH)_2$ brucite valuable stone was used to prepare for the $Mg(OH)_2$ test. "The Geological Museum of the Institute of Geology and Mineralogy, Siberian Branch of the Russian Academy of Sciences, Russia, donated the brucite gem from the Tuva shop The brucite precious stone was ground in a corundum mortar in the air at room temperature under the $Mg(OH)_2$ ready test [32]. The acquired powder has a specific surface area of

2 m^2 per gram. By mixing, a suspension was prepared in refined water from the powdered $Mg(OH)_2$ test and supported on the inner mass of the tube-shaped quartz reactor. The reactor was similar to the $Ca(OH)_2$ reactor. The adsorption layer was then dried in the air. The $Mg(OH)_2$ reactor was connected to a high vacuum unit that included a gas supply framework, a Pirani check, and an APDM-1 mass spectrometer (made in SU). The quantum yield of photoadsorption (photodesorption) was calculated as the ratio of photoadsorbed (photodesorbed) atoms to quanta passing through the reactor's front facing (straightforward) divider" [33].

3.4.3 Copper Hydroxide

Schweizer's reagent, which is copper hydroxide in a smelling salts solution, has the fascinating ability to break down cellulose [34]. Because of this property, it is used to make rayon, a cellulose fiber. It's also used in the aquarium business because of its capacity to kill external parasites in fish, including as accidents, marine itch, Brooklynellosis, and marine velvet, without harming the fish. Although various water-soluble copper combinations can be effective in this task, they typically result in substantial fish mortality [36, 93]. Copper(II) hydroxide is also used as a burned colorant on occasion [37].

Copper(II) hydroxides have been combined with plastic paint to produce a product designed to regulate root growth in clipped plants. Auxiliary and horizontal roots thrive and spread, resulting in dense and robust underground root formation [38].

3.4.4 Graphene Hydroxide

Because of its extraordinary features, such as tremendous synthetic reliability, elevate conduction, and incredible mechanical qualities, graphene has received a lot of attention. Graphene's high unambiguous surface region makes it useful in a variety of applications, including polymer nanocomposites, energy capacity materials, and electrocatalysis. Unfortunately, graphene lacks substance beneficial groupings on its surface, limiting its dispersibility in the framework and applicability in many applications [94]. GO (Graphene oxide), as an oxidized derivative of graphene, is thought to be a promising nano-sized filler for the coming age of useful materials, as it has the amazing properties of graphene as well as having various compound useful gatherings [95]. Because of its several oxygen-containing functional groups, such as carbonyl, hydroxyl, epoxide, and carboxyl, GO has excellent dissolvability in solvents and provides numerous opportunities for the production of GO-based crossover nanocomposites. Recently, directing polymers have stood out in a variety of industries due to their profoundly - formed polymeric chains, ecological strength and electrochemical qualities, simple doping/dedoping science, and unusual leading component. Directing polymers, like as polyaniline (PANI), are appealing materials for high-performance energy devices, quick Faradaic response, and union efficacy due to their outstanding electrical conductivity and great electrochemical mobility. However, it has a poor cycle life because volume development and shrinkage can occur during the charge-release process, causing mechanical corruption of the terminals, weakening of the electrochemical exhibition, and limiting its usage in supercapacitors. Electrochemical mix or *in situ* polymerization procedures are used to improve interfacial interactions and work on electrochemical properties [96, 97].

3.4.5 Nickel Hydroxides

It has been a long time since significant advancements in nickel hydroxide research have been achieved, and as a result, we now know much more than previously. It is clear that these are vital elements based on the vast number of composing results that can be found by conducting a thorough search on the Internet for 'nickel hydroxide.' Regardless, this massive collection of writing can quickly become confusing and perplexing, particularly for new students or examiners who are just beginning to deal with nickel hydroxides and who may not yet be aware with all of the test strategies available. Furthermore, nickel hydroxide materials have been linked to a variety of applications in the fields of research, actual science, and strategic planning. $Ni(OH)_2$ has been utilized as a cathode material for battery propellers since the early twentieth century, according to the National Academy of Sciences [98]. Nickel hydroxides are enormous surface layer fragments generated electrochemically or by disintegration of nickel metal and nickel-based compounds. During the last decade of the twentieth century, assessments of nickel hydroxides were primarily based on these two districts. A direct arrangement was offered last year to sort out the point-by-point approach to acting of the electrochemical oxidation and reduction of nickel hydroxides to nickel(III) oxyhydroxide and the subsequent decreasing back to nickel(II) hydroxide, which had been proposed earlier in the decade [99]. Their arrangement comprises twice as much nickel hydroxide, $Ni(OH)_2$, and twice as much oxidized material, NiO_2H. Nickel-based batteries, including nickel cadmium and nickel metal hydride progressions, as well as other varieties, are now commercially accessible. McBreen recently examined the use of nickel hydroxides in modern battery improvements [100]. $Ni(OH)_2$, on the other hand, are not exclusively used in research labs studying batteries and usage processes. For the time being, these materials are linked to a number of practical applications, such as photocatalysis, electrocatalysis and electrosynthesis, supercapacitors, electrochromic devices, and electrochemical sensors, to mention a few. Numerous extended stretches of study have resulted in a portrayal of nickel hydroxides that is significantly more complete than the unambiguous two-stage model that was previously demonstrated. In terms of the current status and representation of nickel hydroxide materials, there have been several advancements.

There are numerous point by point methodologies of coordinating nickel hydroxide materials available today, including newer technologies like as microwave-aided association and sonochemistry, which are both promising from a genuine standpoint. All of this adds up to a thrilling and considerably interesting piece of work with an astonishingly rich historical background. In any event, this further implies that seeing all of the nuances of its advances and properties might be a difficult task in and of itself. Furthermore, creating a suitable method for synthesizing nickel hydroxide with certain chemical or physical properties on a reasonable scale may be a difficult task to say the least [101].

3.4.6 Aluminum Hydroxide

An overview of aluminum increases is offered, including those used in current contexts such as medications, food additives, beauty care products, and other home items. As with most metals, most aluminum compounds are solids with high dissolving centers. Given the fact that the aluminum (III)-cation (Al^{3+}) has a strong preference for the hydroxide molecule, the pH of aluminum salts is regulated by the presence or absence of hydrogen ions. In the same way that Mg^{2+} and Ca^{2+} particles seek out for complexing experts with oxygen-particle supporter objections, for example, carboxylate and phosphate get-togethers, recollecting for natural systems, Al^{3+} particles look out for complexing experts with oxygen-particle supporter objections. Different crystallographic structures of aluminum oxides, hydroxides, and oxyhydroxides exist, and each has a various surface property due to its different surface properties. The only exceptions to the regulation of the "European Economic Union Council Directive 67/1548" are aluminum powder and sodium aluminum fluoride (cryolite), with aluminum phosphide and sodium aluminum fluoride (cryolite) providing as instances of blends in which the anion makes them responsive. Finally, the subsequent logical processes devised for the investigation of manufactured speciation in solids and plans, as well as quantitative assessment, were used in the confirmation of aluminum and the unique verification of its multiple designs [102].

"Aluminum hydroxide is used to make additional aluminum compounds such as calcined aluminas, aluminum sulphate, polyaluminium chloride, aluminum chloride, zeolites, sodium aluminate, initiated alumina, and aluminum nitrate" [103].

3.4.6.1 *Sources of Human Exposure in the Environment*

Aluminum and its alloys make up around 8percent of the Earth's surface; aluminum can be found in silicates, cryolite, and bauxite rock, among other places. The majority of the reworking of aluminum in the environment is addressed by normal cycles. Accumulation of aluminum from common sources occurs as a result of acidic precipitation. The most common anthropogenic manifestations of aluminum compounds related with current cycles occur in the atmosphere. Aluminum is found in water and food for a variety of reasons [104].

Bauxite is a naturally occurring mineral that is required for the manufacturing of aluminum. Bauxite processing yields alumina, which may be electrolytically reduced to extract aluminum metal; scrap aluminum can also be recycled. The bauxite mine's aluminum hydroxide is conveyed. In 2004, it was claimed that vital aluminum was delivered in 41 countries, with China, Russia, Canada, and the United States being the biggest producers. In that year, the total yearly output of fundamental aluminum, alumina, and aluminum hydroxide was around 30 million tons, 63 million tons, and 5 million tons, respectively. According to the Aluminum Recovery Institute, more than 7 million tons of aluminum are frequently recovered from recycled old components [105].

Transportation, building and improvement, packaging, and electrical items are only a few of the most important commercial applications for aluminum metal and its alloys. One of the metal's fastest-growing applications is transportation. Aluminum powders are utilized in a variety of products, including cosmetics, fuel additives, explosives, and power generation. Aluminum oxides are employed in the production of paper, electrical separators and forces, lights, counterfeit pearls, combinations, and heat-resistant strands, as well as in the creation of abrasives, refractories, and ceramics. Aluminum hydroxide is a chemical compound that is widely found in medications and personal care products. In the food industry, aluminum is used to increase the concentration of additives, fillers, colorants, and anti-developing subject matter experts, as well as emulsifiers and baking powders; soy-based baby kid condition can contain aluminum. Ordinary aluminum minerals, particularly bentonite and zeolite, are utilized in a variety of sectors, including water treatment, sugar refining, food processing, and paper manufacturing. Aluminum has not been depicted as having a disease-causing nature; nonetheless, the "International Agency for Research on Cancer (IARC)" has identified "aluminum formation" as a human cancer caused by malignant development (for extra explanation, assuming no one minds, see effects on Humans, effects from Occupational Exposure, Cancer). There are limits on the quantity of aluminum buildup and aluminum oxide that can be introduced in a few countries. The establishment of restrictions on confirmation in food selections and drinking water has been established for non-word-related conditions; the decision to continue depends on sharp or utilitarian examinations rather than prosperity examinations [106, 107]. Aluminum hydroxide features, including physical characteristics and chemical characteristic, as well as analysis tools.

3.4.6.2 *Natural Levels and Exposure to the Environment and Humans*

Because aluminum is of crustal origin, Surface soils at uncontaminated locations serve as an important source of soluble aluminum compounds in surface water, as well as aluminum-containing particles in accumulation and enclosing air sprayers. Aluminum is classified as having a crustal origin. As one could imagine, the last alternative is abundant in high quantities in air testing conducted in agricultural organizations and when road dust is present. Aluminum has been shown to be influenced by environmental ageing in a variety of environments, including on land and at sea. The presence of aluminum and its different designs (species) in source water and after treatment with potassium alum as a shaggy during drinking water purification is of attention to researchers.

Workers in the aluminum creation and client projects, as well as aluminum welders, may experience critical openings in the metal or probable combinations of the metals with which they operate. As a result of the scarcity of word-related openings and the widespread utilization of aluminum-containing stomach settling experts and supported cerebral pain medications, food is the primary source of aluminum affirmation in the world [108, 109].

3.4.6.3 *Kinetics and Metabolism in Humans*

Using aluminum as a tracer using gas pedal mass spectrometry, researchers have been able to conduct safe analyses of aluminum toxicokinetics in humans while administering genuine openness meaningful doses. The oral aluminum bioavailability from word-related inward breath opening has been calculated to be less than 2%, whereas the oral aluminum availability from water is between 0.1% and 0.4%. The presence of citrate, an acidic pH, and uremia improve oral aluminum bioavailability, while silicon-containing compounds may decrease it. The bioavailability of oral aluminum is likewise inversely related to the presence or absence of iron in the body. When ingested orally, the bioavailability of aluminum in water is greater than that of aluminum hydroxide or sucralfate. When considerable amounts of aluminum hydroxide are eaten, oral aluminum bioavailability is believed to be less than 0.1%. It has been suggested that persons with "Alzheimer's disease (AD) and Down syndrome" have their oral aluminum retention increased. When daily aluminum consumption and urine disposal are considered, oral aluminum bioavailability from the eating regimen is expected to be between 0.1 and 0.3 percent.

The results of a few of trials including a regulated dietary regimen and tea are reliable in support of this estimation. The aluminum concentrations in steady state serum and whole blood are comparable. Transferrin (Tf) is responsible for around 90% of plasma aluminum, citrate is responsible for 7 to 8%, and phosphate and hydroxide are responsible for less than one percent. It is generally agreed that the normal plasma aluminum focus is between 1 and 2 g/L. Ordinary tissue aluminum fixations are more noticeable in the lungs (because to the entanglement of particles from the environment) than in bone or delicate tissues, but they are also noticeable in the lungs. According to the percentage of aluminum body difficulty in each of the organs, bone, lung, muscle, liver, and mind account for around 60, 25, 10, 3, and 1 percent of total aluminum body trouble. Fixations are reported to be higher in patients with uremia, and even higher in those with dialysis encephalopathy.

3.4.6.4 Animals

Aspiratory testimony of fly debris was identified in 2 to 12% of cases in creature research and was found to be inversely associated to molecule size in the opposite way. According to all available data, oral aluminum bioavailability from water is less than 3%. Oral aluminum bioavailability from food is different from oral aluminum bioavailability from water, according to the limited evidence available. Citric acid and, to a lesser extent, other carboxylic acids, enhanced solubility of the aluminum species, acidic pH, uremia, greater fraction of dissolvable aluminum species, and maybe fluoride, all boost the bioavailability of aluminum when consumed in high amounts. When administered orally, silicon-containing compounds have been demonstrated to impair the bioavailability of aluminum. Iron, calcium, and sodium levels are all related to oral aluminum bioavailability, and vice versa. Aluminum absorption from the gastrointestinal tract (GI) bears all of the hallmarks of occurring mostly in the distal digestive tract, according to the literature. Some mechanisms of gastrointestinal aluminum ingestion are supported by evidence, including salt transport processes, collaboration with calcium take-up, and paracellular dispersion, to name a few.

The aluminum penetrates the skin at an exceedingly shallow depth. Aluminum may have the ability to access the cerebrum from the nasal cavity in a direct manner, avoiding the fundamental flow, although there is no evidence to support this claim at the present time. The minimum retention time is followed by the minimum amount of drinking water. Consider bioavailability, which is the portion of a substance that is actually absorbed into the circulatory system. When it comes to persons who are not exposed to their jobs, food is once again the most important take-up point to consider. The increasing use of stomach settling specialists, padded migraine medications, and other therapeutic courses of action would presumably result in a significant increase in the take-up source, regardless of whether or not they were revealed to be effective.

Individuals' collaboration with calcium uptake and paracellular dissemination are important factors. Aluminum penetrates the skin at a very short depth. Although there is no solid evidence to support this method, it is probable that aluminum has the potential to enter the cerebrum from the nasal cavity in an immediate manner, skipping the fundamental flow. According to the researchers, the assimilation of aluminum from intramuscularly infused aluminum hydroxide and aluminum phosphate adjuvants is critical and may finally be accomplished. With ageing, the concentration of aluminum in tissue increases [110, 111].

3.5 Conclusions

The conclusions of this chapter demonstrate the routes of metal hydroxide synthesis by facile strategies as well as applications in various discipline of research. Metal hydroxide are comprehensively used not only in pharmaceutical industries for the purpose of diagnose and cure but also in pollution removal from the water bodies. Pollution removal from the real life water systems can be done by the process of adsorption using metal hydroxide. Copious implementations of metal hydroxide in the domain of nanotechnology, an emerging science, have been extensively studied. In the sphere of Electrochemistry, simple metal hydroxide and mixed metal hydroxide (having more than one metal) are utilized in supercapacitors, the former is preferred due to its enhanced conduction activity.

References

1. Ding, J., Fan, T., Shen, K., Li, Y., Electrochemical synthesis of amorphous metal hydroxide microarrays with rich defects from MOFs for efficient electrocatalytic water oxidation. *Appl. Catal. B: Environ.*, **292**, 120174, 2021.
2. Song, Q., Xu, D., Wang, W.D., Fang, J., Sun, X., Li, F., Li, B., Lou, J., Zhu, H., Dong, Z., Ru clusters confined in Hydrogen-bonded organic frameworks for homogeneous catalytic hydrogenation of N-heterocyclic compounds with heterogeneous recyclability. *J. Catal.*, **406**, 19–27, 2022.
3. Bo, X., Zan, L., Dastafkan, K., Zhao, C., The nature of synergistic effects in transition metal oxides/*in-situ* intermediate-hydroxides for enhanced oxygen evolution reaction. *Curr. Opin. Electrochem.*, 34, 100987, 2022.
4. Guo, J., Zhou, Z., Sun, D., Zeng, L., Ming, Q., Xing, Y., Huang, Z., Zhao, X., Recovering double-metal hydroxides precipitate from desalination process of saline wastewater as conditioner for excess sludge dewatering. *Chem. Eng. J.*, 434, 134787, 2022.
5. Wang, X., Li, L., Su, R., Zhao, Y., Gao, S., Guo, W., Luan, Z., Liang, G., Xi, H., Hierarchically porous metal hydroxide/metal–organic framework composite nanoarchitectures as broad-spectrum adsorbents for toxic chemical filtration. *J. Colloid Interface Sci.*, **606**, 272–285, 2022.
6. Elnagar, M.M., Hermann, J.M., Jacob, T., Kibeler, L., Tailoring the electrode surface structure by cathodic corrosion in alkali metal hydroxide solution: Nanostructuring and faceting of Au. *Curr. Opin. Electrochem.*, **27**, 100696, 2021.
7. Wan, L. and Wang, P., Recent progress on self-supported two-dimensional transition metal hydroxides nanosheets for electrochemical energy storage and conversion. *Int. J. Hydrog. Energy*, **46**, 12, 8356–8376, 2021.
8. Loh, A., Li, X., Taiwo, O.O., Tariq, F., Brandon, N.P., Wang, P., Xu, K., Wang, B., Development of Ni–Fe based ternary metal hydroxides as highly efficient oxygen evolution catalysts in AEM water electrolysis for hydrogen production. *Int. J. Hydrog. Energy*, **45**, 46, 24232–24247, 2020.
9. Chen, F., Zhou, W., Jai, L., Liu, X., Sasaki, T., Ma, R., Transition-metal hydroxide nanosheets with peculiar double-layer structures as efficient electrocatalysts. *Chem. Catal.*, 2, 4, 867–882, 2022.
10. Shiea, M., Querio, A., Buffo, A., Boccardo, G., Marchisio, D., CFD-PBE modelling of continuous Ni-Mn-Co hydroxide co-precipitation for Li-ion batteries. *Chem. Eng. Res. Des.*, **177**, 461–472, 2022.
11. Ma, X.-R., Wei, X.-Y., Dang, R., Guo, W., Ren, M.-J., Kang, Y.-H., Li, X., Gao, Y., Li, Y.-J., Bai, J.-J., Enhanced photoactivity for the methyl orange on heterostructure LaCO3OH/Ni–Fe

layered double hydroxides prepared from complexing agent-assisted homogeneous precipitation techniques. *J. Phys. Chem. Solids*, 110696, 2022.
12. Tan, W., Zhang, L., Gu, F., Bowman, S., Wang, P., Li, Y., Zhang, Y., The heavy metal adsorption and plant cultivation performance of grafting modified plant medium made with recycled fibers. *J. Cleaner Prod.*, **329**, 129788, 2021.
13. Kang, J., Hwang, M., Seong, K-d., Lyu, L., Ko, D., Piao, Y., Three-dimensional nanocomposite of graphene/MWCNT hydrogel grafted with Ni–Co hydroxide nanorods as high-performance electrode for asymmetric supercapacitor. *Electrochim. Acta*, **346**, 136258, 2020.
14. Monyake, K.C. and Alagha, L., Enhanced separation of base metal sulfides in flotation systems using chitosan-grafted-polyacrylamides. *Sep. Purif. Technol.*, **281**, 119818, 2022.
15. Li, Y., Yan, Z., Xiao, J., Yue, T., DNA-conjugated layered double hydroxides penetrating into a plasma membrane: Layer size, thickness and DNA grafting density matter. *NanoImpact*, **18**, 100222, 2020.
16. Masikhwa, T., Momodu, D., Oyedotun, K., Mirghni, A., Ndiaye, N., Manyala, N., Asymmetric supercapacitor based on cobalt hydroxide carbonate/GF composite and a carbonized conductive polymer grafted with iron (C-FP). *J. Alloys Compd.*, **769**, 376–386, 2018.
17. Wang, X., Song, H., Ma, S., Li, M., He, G., Xie, M., Guo, X., Template ion-exchange synthesis of Co-Ni composite hydroxides nanosheets for supercapacitor with unprecedented rate capability. *Chem. Eng. J.*, **432**, 134319, 2022.
18. Moss, R.A. and Gong, P.K., Proton exchange and chemoselectivity in metal cation and hydroxide ion hydrolyses of phosphonoacetate diesters. *Tetrahedron Lett.*, **44**, 42, 7845–7848, 2003.
19. Fang, D., Huang, L., Fan, J., Xiao, H., Wu, G., Wang, Y., Zeng, Z., Shen, F., Deng, S., Ji, F., New insights into the arrangement pattern of layered double hydroxide nanosheets and their ion-exchange behavior with phosphate. *Chem. Eng. J.*, 441, 136057, 2022.
20. Numan, A., Duraismay, N., Omar, F.S., Gopi, D., Ramesh, K., Ramesh, S., Sonochemical synthesis of nanostructured nickel hydroxide as an electrode material for improved electrochemical energy storage application. *Prog. Nat. Sci.: Mater. Int.*, **27**, 4, 416–423, 2017.
21. Lee, E., Park, A.-H., Park, H.-U., Kwon, Y.-U., Facile sonochemical synthesis of amorphous NiFe-(oxy) hydroxide nanoparticles as superior electrocatalysts for oxygen evolution reaction. *Ultrason. Sonochem.*, **40**, 552–557, 2018.
22. Shakir, I., Almutairi, Z., Shar, S.S., Hydrothermally prepared binary metal hydroxide electrode with a rational design and an innovative structure for electrochemical capacitors. *Ceram. Int.*, **48**, 4, 4424–4432, 2022.
23. Zhang, H., Li, H., Guo, Y., Li, G., Fan, C., Wang, K., Chen, J., Mild hydrothermal preparation of a layered metal hydroxide salt with microtube/rod morphology. *Particuology*, **8**, 3, 192–197, 2010.
24. Niu, Z., Li, G., He, D., Fu, X., Sun, W., Yue, T., Resource-recycling and energy-saving innovation for iron removal in hydrometallurgy: Crystal transformation of ferric hydroxide precipitates by hydrothermal treatment. *J. Hazard. Mater.*, **416**, 125972, 2021.
25. Saroj, S.K., Pal, S., Nagarajan, R., Polyol intercalation in copper substituted zinc hydroxide acetate and evaluation of its adsorptive role towards Congo red dye. *Appl. Clay Sci.*, **185**, 105411, 2020.
26. Gómez-Fernández, S., Gunther, M., Corcuera, M.A., Eciza, A., Impact of the combined use of layered double hydroxides, lignin and phosphorous polyol on the fire behavior of flexible polyurethane foams. *Ind. Crops Prod.*, **125**, 346–359, 2018.
27. Wang, R., Chen, Y., Liu, Y., Ma, M., Hou, Y., Chen, X., Ma, Y., Huang, W., Synthesis of sugar gourd-like metal organic framework-derived hollow nanocages nickel molybdate@ cobalt-nickel layered double hydroxide for flame retardant polyurea. *J. Colloid Interface Sci.*, 616, 234–245, 2022.

28. Oualha, M.A., Omri, N., Oualha, R., Nouioui, M.A., Abderrabba, M., Amdouni, N., Laoutid, F., Development of metal hydroxide nanoparticles from eggshell waste and seawater and their application as flame retardants for ethylene-vinyl acetate copolymer (EVA). *Int. J. Biol. Macromol.*, **128**, 994–1001, 2019.
29. Yeon, D., Yunju, L., Myung-Hyun, R., Yong, M.L., New flame-retardant composite separators based on metal hydroxides for lithium-ion batteries. *Electrochim. Acta*, **157**, 282–289, 2015.
30. Hong, N., Zhan, J., Wang, X., Stec, A.A., Hull, T.R., Ge, H., Xing, W., Song, L., Hu, Y., Enhanced mechanical, thermal and flame retardant properties by combining graphene nanosheets and metal hydroxide nanorods for Acrylonitrile–Butadiene–Styrene copolymer composite. *Compos. A: Appl. Sci. Manuf.*, **64**, 203–210, 2014.
31. Li, P., Dang, L., Li, Y., Lan, S., Zhu, D., Enhanced flame-retardant and mechanical properties of epoxy resin by combination with layered double hydroxide, Mg2B2O5 whisker, and dodecyl dihydrogen phosphate. *Mater. Des.*, 217, 110608, 2022.
32. Xu, Z.-s., Yan, L., Chen, L., Synergistic flame retardant effects between aluminum hydroxide and halogen-free flame retardants in high density polyethylene composites. *Proc. Eng.*, **135**, 631–636, 2016.
33. Lu, Y., Wu, C., Xu, S., Mechanical, thermal and flame retardant properties of magnesium hydroxide filled poly (vinyl chloride) composites: The effect of filler shape. *Compos. A: Appl. Sci. Manuf.*, **113**, 1–11, 2018.
34. Shen, H. and Liu, Y., One-step synthesis of hydrophobic magnesium hydroxide nanoparticles and their application in flame-retardant polypropylene composites. *Chin. J. Chem. Eng.*, **26**, 10, 2199–2205, 2018.
35. Liang, J.-Z., Tensile and flexural properties of polypropylene composites filled with highly effective flame retardant magnesium hydroxide. *Polym. Test.*, **60**, 110–116, 2017.
36. Yang, Y., Niu, M., Li, J., Xue, B., Dai, J., Preparation of carbon microspheres coated magnesium hydroxide and its application in polyethylene terephthalate as flame retardant. *Polym. Degrad. Stab.*, **134**, 1–9, 2016.
37. Liu, L., Hu, J., Zhuo, J., Jiao, C., Chen, X., Li, S., Synergistic flame retardant effects between hollow glass microspheres and magnesium hydroxide in ethylene-vinyl acetate composites. *Polym. Degrad. Stab.*, **104**, 87–94, 2014.
38. Liu, J., Yu, Z., Chang, H., Zhang, Y., Shi, Y., Luo, J., Pan, B., Lu, C., Thermal degradation behavior and fire performance of halogen-free flame-retardant high impact polystyrene containing magnesium hydroxide and microencapsulated red phosphorus. *Polym. Degrad. Stab.*, **103**, 83–95, 2014.
39. Liu, S.-P., Flame retardant and mechanical properties of polyethylene/magnesium hydroxide/montmorillonite nanocomposites. *J. Ind. Eng. Chem.*, **20**, 4, 2401–2408, 2014.
40. Cheng, Z., Yang, J., Li, L., Chen, Y., Wang, X., Flocculation inspired combination of layered double hydroxides and fulvic acid to form a novel composite adsorbent for the simultaneous adsorption of anionic dye and heavy metals. *J. Colloid Interface Sci.*, 289, 120741, 2022.
41. Bi, R., Yin, D., Lei, B., Chen, F., Zhang, R., Li, W., Mercaptocarboxylic acid intercalated MgAl layered double hydroxide adsorbents for removal of heavy metal ions and recycling of spent adsorbents for photocatalytic degradation of organic dyes. *Sep. Purif. Technol.*, **289**, 120741, 2022.
42. Wang, J., Zhang, L., Zhang, T., Du, T., li, T., Yue, T., Li, Z., Wang, J., Selective removal of heavy metal ions in aqueous solutions by sulfide-selector intercalated layered double hydroxide adsorbent. *J. Mater. Sci. Technol.*, **35**, 9, 1809–1816, 2019.
43. Li, L., Qi, G., Wang, B., Yue, D., Wang, Y., Sato, T., Fulvic acid anchored layered double hydroxides: A multifunctional composite adsorbent for the removal of anionic dye and toxic metal. *J. Hazard. Mater.*, **343**, 19–28, 2018.

44. Jiang, Q., Wang, L., Zhao, W., Xu, X., Li, Z., Li, Y., Zhou, T., Hu, J., Carbon dots decorated on the ultrafine metal sulfide nanoparticles implanted hollow layered double hydroxides nanocages as new-type anodes for potassium-ion batteries. *Chem. Eng. J.*, **433**, 133539, 2022.
45. Beigi, F., Mousavi, M.S.S., Manteghi, F., Kolahdouz, M., Doped nafion-layered double hydroxide nanoparticles as a modified ionic polymer metal composite sheet for a high-responsive humidity sensor. *Appl. Clay Sci.*, **166**, 131–136, 2018.
46. Li, Z., Liu, J., Zhao, Y., Shi, R., Waterhouse, G., Wang, Y., Wu, L.-Z., tumg, C.-H., Zhang, T., Photothermal hydrocarbon synthesis using alumina-supported cobalt metal nanoparticle catalysts derived from layered-double-hydroxide nanosheets. *Nano Energy*, **60**, 467–475, 2019.
47. Rasouli, N., Movahedi, M., Doudi, M., Synthesis and characterization of inorganic mixed metal oxide nanoparticles derived from Zn–Al layered double hydroxide and their antibacterial activity. *Surf. Interfaces*, **6**, 110–115, 2017.
48. Alinejad, B., Mahmoodi, K., Ahmadi, K., A new route to mass production of metal hydroxide/oxide hydroxide nanoparticles. *Mater. Chem. Phys.*, **118**, 2-3, 473–476, 2009.
49. Zhu, Y., Huang, C., Li, C., Fan, M., Shu, K., Chen, H.C., Strong synergetic electrochemistry between transition metals of α phase Ni– Co– Mn hydroxide contributed superior performance for hybrid supercapacitors. *J. Power Sources*, **412**, 559–567, 2019.
50. Shi, C., Du, Y., Guo, L., Yang, J., Wang, Y., Construction of interconnected NiCo layered double hydroxides/metal-organic frameworks hybrid nanosheets for high-performance supercapacitor. *J. Energy Storage*, **48**, 103961, 2022.
51. Ma, Q., liu, H., An, S., Han, X., Cui, J., Zhang, Y., He, W., Layered double metal hydroxide coated nickel oxide embedded carbon fiber to form open petal-shaped nanosheet arrays as electrode materials for high-performance supercapacitors. *J. Energy Storage*, **44**, 103455, 2021.
52. Yang, G., Takei, T., Zhu, Y., Iranmasesh, E., Liu, B., Li, Z., Wang, J., Hirala, P., Amaratunga, G.A., Fontaine, O., Constructing an efficient conductive network with carbon-based additives in metal hydroxide electrode for high-performance hybrid supercapacitor. *Electrochim. Acta*, **397**, 139242, 2021.
53. Mallakpour, S., Azadi, E., Hussain, C.M., Recent advancements in synthesis and drug delivery utilization of polysaccharides-based nanocomposites: The important role of nanoparticles and layered double hydroxides. *Int. J. Biol. Macromol.*, **193**, 183–204, 2021.
54. Asiabi, H., Yamini, Y., Alipour, M., Hamsayei, M., Hosseinkhani, S., Synthesis and characterization of a novel biocompatible pseudo-hexagonal NaCa-layered double metal hydroxides for smart pH-responsive drug release of dacarbazine and enhanced anticancer activity in malignant melanoma. *Mater. Sci. Eng.: C*, **97**, 96–102, 2019.
55. Ranjbar, E., Namazi, H., Pooresmaeil, M., Carboxymethyl starch encapsulated 5-FU and DOX co-loaded layered double hydroxide for evaluation of its in vitro performance as a drug delivery agent. *Int. J. Biol. Macromol.*, 201, 193–202, 2022.
56. Szabados, M., Gacsi, A., Gulyas, Y., Konya, Z., Kukovecz, A., Csanyi, E., Palinko, I., Sipos, P., Conventional or mechanochemically-aided intercalation of diclofenac and naproxen anions into the interlamellar space of CaFe-layered double hydroxides and their application as dermal drug delivery systems. *Appl. Clay Sci.*, **212**, 106233, 2021.
57. Pšenička, M., Škoda, J., Pospíšil, M., Structural arrangement and properties of layered double hydroxide drug nanocarrier intercalated by sulindac and mefenamic acid solved by molecular simulation methods. *Appl. Clay Sci.*, **189**, 105560, 2020.
58. Kumari, P., Pal, B., Das, R.K., Superior adsorptive removal of eco-toxic drug diclofenac sodium by Zn–Al LDH· xBi2O3 layer double hydroxide composites. *Appl. Clay Sci.*, **208**, 106119, 2021.
59. Zhang, Q. *et al.*, A biocompatible bismuth based metal-organic framework as efficient light-sensitive drug carrier. *J. Colloid Interface Sci.*, 617, 578–584, 2022.

60. Ma, B., Luo, Y., Zhou, L., Shao, Z., Liang, R., Fu, J., Wang, J., Zang, J., Hu, Y., Wang, L., The influence of calcium hydroxide on the performance of MK-based geopolymer. *Constr. Build. Mater.*, **329**, 127224, 2022.
61. Alventosa, K.M. and White, C.E., The effects of calcium hydroxide and activator chemistry on alkali-activated metakaolin pastes. *Cem. Concr. Res.*, **145**, 106453, 2021.
62. Khan, S.R., Jamil, S., Janjua, M.R.S.A., Radiation assisted synthesis of dumb bell shaped calcium hydroxide nanostructures from egg shells and study of its thermal and catalytic applications. *Chem. Phys. Lett.*, **710**, 45–53, 2018.
63. Zamanian, A., Yasaei, M., Ghaffari, M., Mozafari, M., Calcium hydroxide-modified zinc polycarboxylate dental cements. *Ceram. Int.*, **39**, 8, 9525–9532, 2013.
64. Akashi, A., Matsuya, Y., Nagasawa, H., Stability constants of Ca-alkyl salicylate complexes and their effect on the dissolution of calcium hydroxide dental cements. *Biomaterials*, **13**, 11, 771–774, 1992.
65. Jin, H., Li, Y., Wang, Q., Dong, M., Yang, M., Chen, W., Wang, S., Zhang, H., Zheng, S., Cao, C.Y., A strontium and amorphous calcium phosphate dipped premixed injectable calcium silicate-based ceramic for dental root canal sealing. *Ceram. Int.*, **47**, 23, 33738–33750, 2021.
66. Shenbagam, V.K. and Chaunsali, P., Influence of calcium hydroxide and calcium sulfate on early-age properties of non-expansive calcium sulfoaluminate belite cement. *Cem. Concr. Compos.*, **128**, 104444, 2022.
67. Murad, M. and Duggal, M., Intra-canal antibiotic pastes may promote thicker dentin walls and calcium hydroxide may promote more apical closure. *J. Evid. -Based Dent. Pract.*, 22, 2, 101728, 2022.
68. Wang, J., Yin, J., Kong, X., Influences of PCE superplasticizers with varied architectures on the formation and morphology of calcium hydroxide crystals. *Cem. Concr. Res.*, **152**, 106670, 2022.
69. Afkhami, F., Rostami, G., Batebi, S., Bahador, A., Residual antibacterial effects of a mixture of silver nanoparticles/calcium hydroxide and other root canal medicaments against Enterococcus faecalis. *J. Dent. Sci.*, 17, 3, 1260–1265, 2021.
70. Lei, J., Law, W.W., Yang, E.-H., Effect of calcium hydroxide on the alkali-silica reaction of alkali-activated slag mortars activated by sodium hydroxide. *Constr. Build. Mater.*, **272**, 121868, 2021.
71. Antony, D., Yadav, R., Kalimuthu, R., Kumuthan, M.S., Phyto-complexation of galactomannan-stabilized calcium hydroxide and selenium-calcium hydroxide nanocomposite to enhance the seed-priming effect in Vigna radiata. *Int. J. Biol. Macromol.*, **194**, 933–944, 2022.
72. Pokorná, D., Cuřínová, P., Pola, J., Calcium hydroxide effect in degradation of aqueous naphthalene: Nucleophilic substitution of hydrogen at the C (sp2)–H bond. *Polycycl. Aromat. Compd.*, 41, 841–850, 2019.
73. de Azevedo Basto, P., de Lima, V.E., de Melo Neto, A.A., Effect of curing temperature in the relative decrease peak intensity of calcium hydroxide pastes for assessing pozzolanicity of supplementary cementitious materials. *Constr. Build. Mater.*, **325**, 126767, 2022.
74. Yamanaka, S., Hirano, S., Uwai, K., Tokuratku, K., Design of calcium hydroxide–based granules for livestock sanitation. *Case Stud. Chem. Environ. Eng.*, **2**, 100005, 2020.
75. Camerini, R., Poggi, G., Ridi, F., Bagloini, P., The kinetic of calcium silicate hydrate formation from silica and calcium hydroxide nanoparticles. *J. Colloid Interface Sci.*, **605**, 33–43, 2022.
76. Liang, Y., Mechanical and fracture properties of calcium silicate hydrate and calcium hydroxide composite from reactive molecular dynamics simulations. *Chem. Phys. Lett.*, **761**, 138117, 2020.
77. Gautam, R.K., Singh, A.K., Tiwari, I., Nanoscale layered double hydroxide modified hybrid nanomaterials for wastewater treatment: A review. *J. Mol. Liq.*, 350, 118505, 2022.

78. Rahman, M.T., Kameda, T., Kumagai, S., Yoshika, T., A novel method to delaminate nitrate-intercalated MgAl layered double hydroxides in water and application in heavy metals removal from wastewater. *Chemosphere*, **203**, 281–290, 2018.
79. Ye, X., Shi, X., Xu, H., Feng, Y., Jin, B., Duan, P., Enhanced catalytic activity of layered double hydroxides via in-situ reconstruction for conversion of glucose/food waste to methyl lactate in biorefinery. *Sci. Total Environ.*, 829, 154540, 2022.
80. Feng, X., Long, R., Wang, L., Liu, C., Bai, Z., Liu, X., A review on heavy metal ions adsorption from water by layered double hydroxide and its composites. *Sep. Purif. Technol.*, **284**, 120099, 2022.
81. Sam, J., Kirankumar, P., Sanath, K., Prathish, K., Development of saleable chloride free iron oxide from hazardous waste in titanium industries via layered double hydroxide formation. *J. Environ. Manage.*, **290**, 112566, 2021.
82. Frau, F., Atzori, R., Ardau, C., Medas, D., Podda, F., Dore, E., Idini, A., Tempesta, G., Agrosi, G., A two-step pH control method to remove divalent metals from near-neutral mining and metallurgical waste drainages by inducing the formation of layered double hydroxide. *J. Environ. Manage.*, **271**, 111043, 2020.
83. Foruzin, L.J., Rezvani, Z., Habibi, B., New ternary-component layered double hydroxide as a low-cost and efficient electrocatalyst for water oxidation: NiCaFe-LDH from eggshell bio-waste. *Appl. Clay Sci.*, **188**, 105511, 2020.
84. Higgins, P., Siddiqui, S.H., Kumar, R., Design of novel metal hydroxide bio-nanocomposite (CBCS@ LDH) for the scavenging of ciprofloxacin (CPN) drug from its aqueous solution: Kinetic, isotherm, thermodynamic and reusability studies. *Curr. Res. Green Sustain. Chem.*, 5, 100298, 2022.
85. Sadeghalvad, B., Azadmehr, A., Hezarkhani, A., Sulfate decontamination from groundwater by metal layered double hydroxides functionalized high phosphorus iron ore waste as a new green adsorbent: Experimental and modeling. *Ecol. Eng.*, **106**, 219–230, 2017.
86. Sohrabi, H., Khataee, A., Ghasemzadeh, S., Majidi, M.R., Orooji, Y., Layer double hydroxides (LDHs)-based electrochemical and optical sensing assessments for quantification and identification of heavy metals in water and environment samples: A review of status and prospects. *Trends Environ. Anal. Chem.*, **31**, e00139, 2021.
87. Licona-Aguilar, Á., Torres-Huerta, A., Dominguez-Crespo, M., Palma-Ramirez, D., Conde-Barajas, E., Negrete-Rodríguez, M.X.L., Rodríguez-Salazar, A.E., García-Zaleta, D.S., Reutilization of waste biomass from sugarcane bagasse and orange peel to obtain carbon foams: Applications in the metal ions removal. *Sci. Total Environ.*, 831, 154883, 2022.
88. Dong, Y., Ahang, M., Xie, C., Effect of reaction conditions on agglomeration of aluminum hydroxide in the recovery of waste aluminum-catalyst. **248**, 116978, 2020.
89. Wang, T., Zhang, D., Fang, K., Zhu, W., Peng, Q., Xie, Z., Enhanced nitrate removal by physical activation and Mg/Al layered double hydroxide modified biochar derived from wood waste: Adsorption characteristics and mechanisms. *J. Environ. Chem. Eng.*, **9**, 4, 105184, 2021.
90. Kumar, P., Malhotra, B., Phalswal, P., Khanna, P., Salim, A., Singhal, R., Mukhophadhyay, A.K., Effect of reaction rate on the properties of chemically synthesized calcium hydroxide nanoparticles. *Mater. Today: Proc.*, **28**, 2305–2310, 2020.
91. Kumar, P., Kumari, J., Kumar, L., Salim, A., Singhal, R., Mukhopadhayay, A.K., Joshi, R., Influence of chemical synthesis process on the properties of calcium hydroxide nanoparticles. *Mater. Today: Proc.*, 60, 153–159, 2021.
92. Bahman, S., Sara, G., Somayeh, H., Parvin, T., Kalhori, K.M., Mona, S., Reza, F., Combined effects of calcium hydroxide and photobiomodulation therapy on apexogenesis of immature permanent teeth in dogs. *J. Photochem. Photobiol. B: Biol.*, **207**, 111867, 2020.

93. Tang, H., Zhou, X.-B., Liu, X.-L., Effect of magnesium hydroxide on the flame retardant properties of unsaturated polyester resin. *Proc. Eng.*, **52**, 336–341, 2013.
94. Zhao, M., Xue, S., Xiao, H., Gao, J., Cheng, X., Jing, Y., Jia, J., Wu, H., Facile *in-situ* electrochemical fabrication of highly efficient nickel hydroxide-iron hydroxide/graphene hybrid for oxygen evolution reaction. *Int. J. Hydrog. Energy*, **47**, 25, 12547–12558, 2022.
95. Shen, G., Zhang, L., Gu, Z., Zheng, Z., Liu, Y., Tan, G., Jie, X., Zinc aluminum-layered double hydroxide (LDH)-graphene oxide (GO) lubricating and corrosion-resistant composite coating on the surface of magnesium alloy. *Surf. Coat. Technol.*, 437, 128354, 2022.
96. Shi, M., Sun, X., Bai, Q., Zhang, Y., Yu, S., Liu, M., Wang, L., Sui, N., William, W.Y., Graphdiyne/graphene heterostructure supported NiFe layered double hydroxides for oxygen evolution reaction. *Colloids Surf. A Physicochem. Eng. Asp.*, 637, 128217, 2022.
97. Yang, Y.J., Chen, S., Jiang, C., Yang, P., Wang, N., Chang, Y., Hierarchical web-like NiFeMn ternary hydroxides microstructure assembled on reduced graphene oxide for binder-free supercapacitor electrode. *Diam. Relat. Mater.*, 125, 109009, 2022.
98. Xue, X., Zhong, J., Liu, J., Hou, Z., Wu, X., Li, S., Yu, M., Hydrolysis of metal-organic framework towards three-dimensional nickel cobalt-layered double hydroxide for high performance supercapacitors. *J. Energy Storage*, **31**, 101649, 2020.
99. Chang, J., Zang, S., Liang, W., Wu, D., Lian, Z., Xu, F., Jiang, K., Gao, Z., Enhanced faradic activity by construction of pn junction within reduced graphene oxide@ cobalt nickel sulfide@ nickle cobalt layered double hydroxide composite electrode for charge storage in hybrid supercapacitor. *J. Colloid Interface Sci.*, **590**, 114–124, 2021.
100. Hu, J., Liang, Y.Q., Wu, S.L., Li, Z.Y., Shi, C.S., Luo, S.Y., Sun, H.J., Zhu, S.L., Cui, Z.D., Hierarchical nickle-iron layered double hydroxide composite electrocatalyst for efficient oxygen evolution reaction. *Mater. Today Nano*, **17**, 100150, 2022.
101. Roushani, M., Hosseni, H., Hajinia, Z., Rahmati, Z., Rationally designed of hollow nitrogen doped carbon nanotubes double shelled with hierarchical nickel hydroxide nanosheet as a high performance surface substrate for cortisol aptasensing. *Electrochim. Acta*, **388**, 138608, 2021.
102. Ullah, S., Shamraiz, U., Badshah, A., Raza, B., Farooqi, A.S., Tofil, H.M., Zeb, M.A., Alfantazi, A., Sulfur enriched cobalt-based layered double hydroxides for oxygen evolution reactions. *Int. J. Hydrogen Energy*, 47, 72, 30799–30809, 2021.
103. Gorla, G., Civati, F., Maurich, C., Burnello, V., Paganica, R., Rampazzi, L., Giussani, B., New strategies to quantify aluminum hydroxide in powder coatings by thermogravimetric analysis and ATR-FT-MIR spectroscopy coupled with chemometrics. *Microchem. J.*, **172**, 107005, 2022.
104. Tiffo, E., Belibi, P., Mbah, J., Thamer, A., Pougnong, T., Baenla, J., Effect of various amounts of aluminium oxy-hydroxide coupled with thermal treatment on the performance of alkali-activated metakaolin and volcanic scoria. *Sci. Afr.*, **14**, e01015, 2021.
105. Rakshita, M., Babu, A., Jayanthi, K., Bathula, S., Kumar, K., Haranath, D., Studies on contact angle measurements in superoleophobic aluminum hydroxide nanoflakes. *Mater. Lett.*, **315**, 131938, 2022.
106. Kooijman, S., Vrieling, H., Verhangen, L., deRidder, De Haan, A., Van Riet, E., Hecck, A., Kresten, G., Pennings, J., Aluminum hydroxide and aluminum phosphate adjuvants elicit a different innate immune response. *J. Pharm. Sci.*, 111, 4, 982–990, 2022.
107. Zhang, T., Jiang, H., Shang, S., Zhang, K., Gao, W., Synthesis of aluminum hydroxide/zinc borate composite inhibitor and its inhibition effect on aluminum dust explosion. *Chem. Eng. Sci.*, **248**, 117204, 2022.
108. Zhang, B., Huang, C., Liu, L., Zhang, H., Chen, X., Zhao, Q., Experimental investigations on synergistic inhibition of aluminum hydroxide and aerosil on ignition sensitivity of Niacin dust. *J. Taiwan Inst. Chem. Eng.*, **122**, 23–31, 2021.

109. Maach, N., Georgin, J.F., Berger, S., Pommay, J., Chemical mechanisms and kinetic modeling of calcium aluminate cements hydration in diluted systems: Role of aluminium hydroxide formation. *Cem. Concr. Res.*, **143**, 106380, 2021.
110. Yin, H., Dai, H., Liang, G., Inhibition evaluation of magnesium hydroxide, aluminum hydroxide, and hydrotalcite on the flame propagation of coal dust. *Process Saf. Environ. Prot.*, **157**, 443–457, 2022.
111. L'Haridon-Quaireau, S., Laot, M., Colas, K., Kapusta, B., Delpech, S., Gosset, D., Effects of temperature and pH on uniform and pitting corrosion of aluminium alloy 6061-T6 and characterisation of the hydroxide layers. *J. Alloys Compd.*, **833**, 155146, 2020.
112. Kim, S., Han, T., Jeong, J., Ryou, M., Lee, Y.M., A flame-retardant composite polymer electrolyte for lithium-ion polymer batteries. *Electrochim. Acta*, **241**, 553–559, 2017.
113. Cheng, J., Zhang, J., Liu, F.J.R.A., Recent development of metal hydroxides as electrode material of electrochemical capacitors. *RSC Adv.*, **4**, 73, 38893–38917, 2014.
114. Sangwan, P., Sangwan, A., Duhan, J., Rohilla, A., Tertiary dentinogenesis with calcium hydroxide: A review of proposed mechanisms. *Int. Endod. J.*, **46**, 1, 3–19, 2013.
115. Babeker, T.M.A. and Chen, Q.J.C.P.R., Heavy metal removal from wastewater by adsorption with hydrochar derived from biomass: Current applications and research trends. *Curr. Pollut. Rep.*, **7**, 1, 54–71, 2021.

4

Porous Organic Polymers: Genres, Chemistry, Synthetic Strategies, and Diversified Applications

V. Renuga

Department of Chemistry, National College (Autonomous), Accredited A+ Grade, Tiruchirappall, Tamil Nadu, India

Abstract
Synthetic materials mostly at the molecular level exhibit porosity up to three-dimensional framework fashion. In particular, metal-organic porous frameworks have captivated a major attraction in this direction. Conversely, discrete molecules in the solid-state tend to pack in a manner to leave a small vacant space as much as possible, which leads the way to non-porous materials. This kind of perpetual assessment makes the present-day developments of porous natured molecules endured in two morphological states. They exist either as crystalline or as amorphous, and are further categorized either as molecules possess porosity in intrinsic nature (permanent covalent cavities) or in extrinsic nature (packed incompetently). The present review focuses on the nature of Porous Organic Polymers (POP) and discuss their classifications widely, based on their pore size, morphology, and pore volumes. Further, this review encircles the major types of POPs such as covalent organic frameworks (COFs), covalent triazine frameworks (CTFs), hyper-cross-linked polymers (HCPs), conjugated micro porous polymers (CMPs), porous aromatic frameworks (PAFs), extrinsic and intrinsic porous molecules and porous organic cages, and discussed their relative properties, synthetic strategy, characterization techniques adopted to confirm their formations and various applications rendered by them in detail.

***Keywords*:** Porous framework, discrete molecules, porosity, covalent organic framework, extrinsic porous molecules, intrinsic porous molecules, porous organic cages, covalent triazine frameworks

4.1 Introduction

In recent times, the discovery and development of sustainable materials' research have increased to retrieve anthropogenic resources, because, these resources were depleted and impart negative effects on other resources (e.g. the biosphere). To restore the long-term sustainability of human society and all living creatures, the development of sustainable chemicals and materials like modern or cutting-edge adsorbents and catalysts (e.g. monoliths with grades or ordered pore frameworks) in porous structures were used for millennia to lead to a more sustainable society. Hence, the incorporation of porous materials in all walks

Email: renugachem@nct.ac.in

Inamuddin, Tariq Altalhi and Sayed Mohammed Adnan (eds.) Sustainable Materials for Electrochemical Capacitors, (65–132) © 2023 Scrivener Publishing LLC

of life in the present-day scenario, including filtration, purification, cooling, detergents, water treatment, biogas upgrading and to capture carbon dioxide as catalysts were used to restore a sustainable society.

In the current times, the discovery and development of porous materials research had boomed enormously. Based on their structure, structural elements, size, properties, and mode of synthesis, porous materials are classified in different ways.

The word "porous" is defined as those materials permanently having voids and are interconnected to each other. Porous materials can permeate liquids apart from gases. Nature itself has designed materials as porous are biological tissues, charcoal, dried plant husks and terracotta. For many, modern-day practical applications, a number of new and advanced porous functional materials have emerged.

According to IUPAC [1] nomenclature, porous solids are defined by their "pore volume" (Vp), which corresponds to "volume covered by the pores" which are dependent on the methods of measurement, have to be mentioned". The term "porosity" is measured as a relationship between the total pore volume Vp and the apparent volume V of any system (except inter-particle voids).

The "pore size" is directly related to the interspace between two opposing walls of the pore. The arrangements of pore size on the lattice are directly related to the fraction of dAp/drp or dVp/drp as a function of rp. Here Vp, rp, and Ap are volumes, the radius of the pores, and the wall area. Here, the radius is assumed as the pores are cylindrical. Similarly, the radius of the pores' rp can be replaced by their width.

To a great extent, porous materials are classified based on morphological arrangements of atoms as amorphous and crystalline. Polymers with an amorphous morphology have their atoms loosely held – the chains are arrayed randomly throughout the material and are present in short-range order. Whereas polymers with crystalline morphology have the atoms arranged in an orderly manner and are present in a comprehensive order. The schematic representations of crystalline and amorphous forms are shown in Figure 4.1.

Following states of matter, organic materials with porous nature are further organized as porous liquids and porous solids [2].

Porous solids include hyper-cross-linked polymers (HCPs) [3–7], polymers of intrinsic microporosity (PIMs) [8], covalent organic frameworks (COFs) [9–11], conjugated micro porous polymers (CMPs) [12–14], covalent triazine frameworks (CTFs) [15–17], porous aromatic frameworks (PAFs) [18–29], extrinsic and intrinsic porous molecules [20], and porous organic cages [21]. Porous liquids are substance with varying compositions and extend their classifications as porous cages, discrete molecular cages and colloidal particles. The porous liquids with the above characters stabilize the electronic charges of the bulk molecules. The Graphical

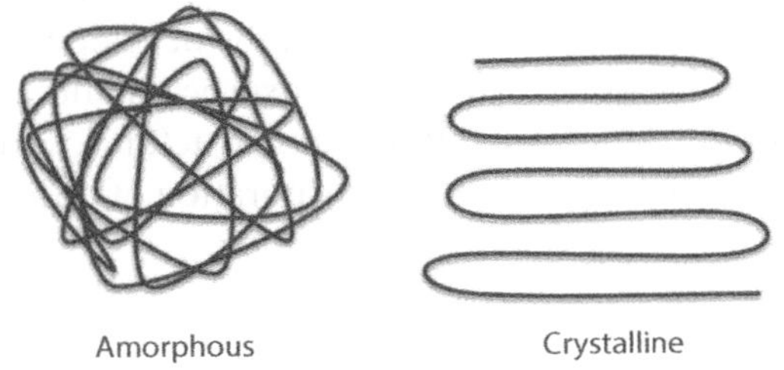

Figure 4.1 The schematic representations of amorphous and crystalline forms.

representation of the organization of porous materials is established based on the structural and chemical compositions of their skeletal units [22] are shown in Figure 4.2. Further, the classifications of organic porous polymers based on their synthesis method are shown in Figure 4.3.

Among the above polymers HCPs, PIMs and COFs exist in crystalline frameworks, whereas CMPs, CTFs, and PAFs exist in amorphous framework. Further the pores' position in a lattice of porous molecules distinguishes them as extrinsic and intrinsic porous molecules have existed as shown in Figure 4.4. It states that intrinsic porosity arrives that voids are present internally, that is, within a molecule. As a result, these intrinsic porous molecules appear as discrete molecules. For example, molecules that possess concave-shape (calixarenes [23], as well as 'porous organic cages' [24]) generally possess cavities in the premade design and also have pore windows, which are typical examples of intrinsic porous molecules.

In contrast, extrinsic porous molecules are molecules in which pores are present in between molecules due to the incompetent arrangements of molecular subunits. The classification of porous molecules as either extrinsic or intrinsic in crystalline organic molecules is shown in Figure 4.5.

According to International Science Council, porous substances are classified based on their dimensions as macro porous (d > 50 nm) [25] meso porous (2 nm < d < 50 nm)

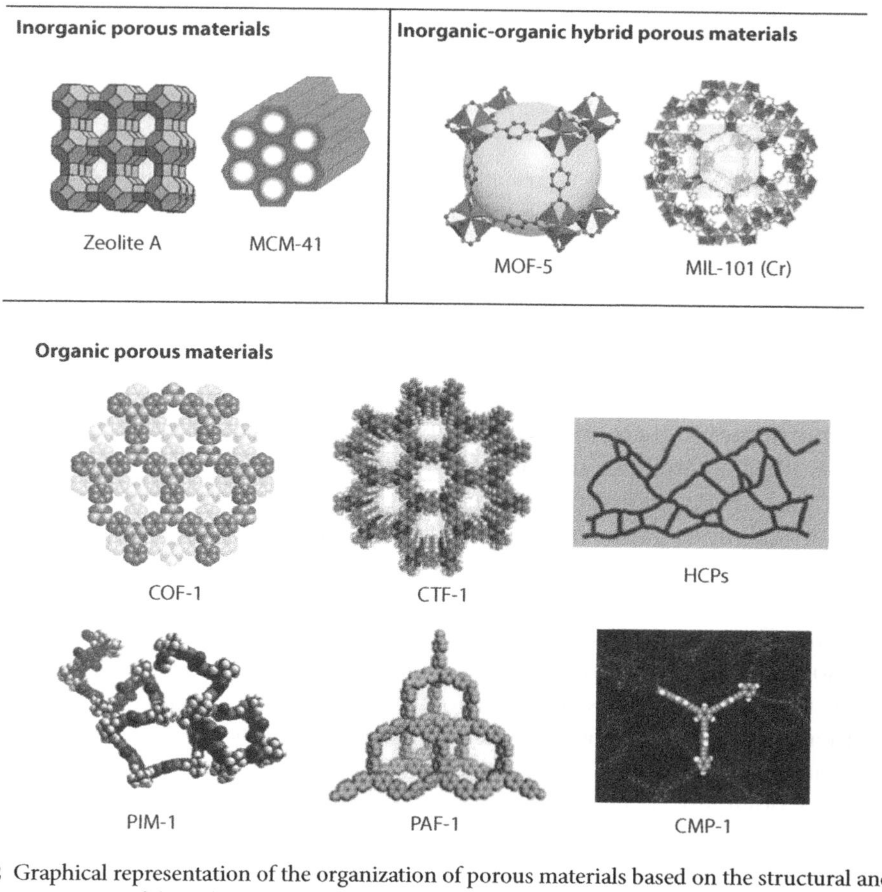

Figure 4.2 Graphical representation of the organization of porous materials based on the structural and chemical compositions of their skeletal units. (Adopted from Ref. [22] with permission).

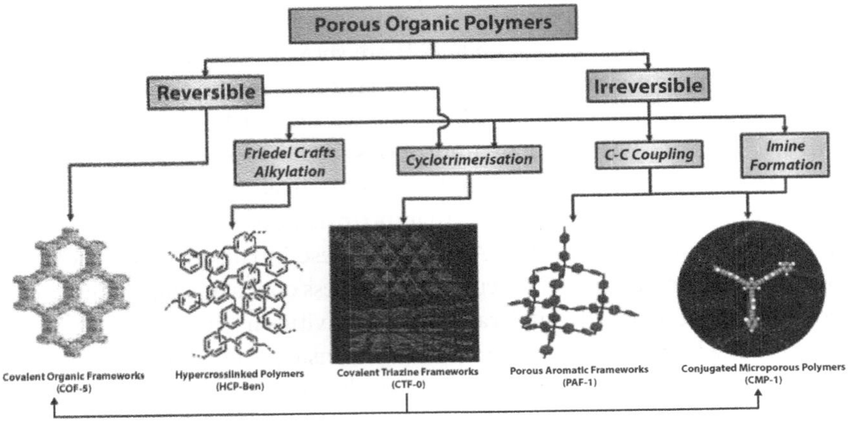

Figure 4.3 Classifications of organic porous polymers based on their synthesis method [2].

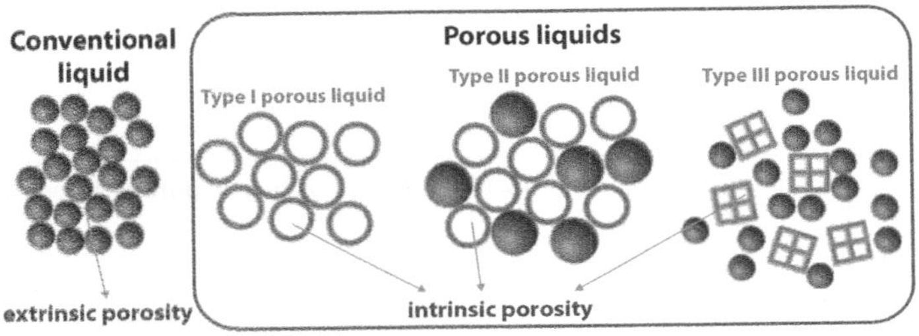

Figure 4.4 Intrinsic and extrinsic porosity present in substance possesses porous nature [24].

and micro porous (d < 2 nm). The physical and chemical reactions between the system of interest and porous materials are mainly depending on both size and surface of the porous materials.

The porous materials' surface leads to physical and chemical interactions between the porous materials and the system of interest more than that of their pore size. This surface chemistry contributes to their influence on pore geometries, morphologies, and the functional groups attached on the periphery of the surface of the porous materials [26, 27].

Based on the chemical constituents present in these porous materials, they are further classified as inorganic, organic, metal-organic frameworks, and conjugated micro porous materials.

For example, De Jong and co-workers' [28] tailored one of the most important inorganic materials Zeolite in micro-, meso- and macroporosity. They adopted different synthetic methodology to incorporate additional porosity over Zeolite materials to access its applications in catalysis, sensors and many other fields. The metal-organic frameworks are another emerging In-organic porous material. They are built by organic linkers with metal-containing nodes [29].

The organic polymers are classified as porous organic cages (POCs-discrete porous molecules) and porous organic polymers (rigid and directional). During POC formation, the molecular cages are formed initially and in a separate step, it is assembled in solid-state [30-32].

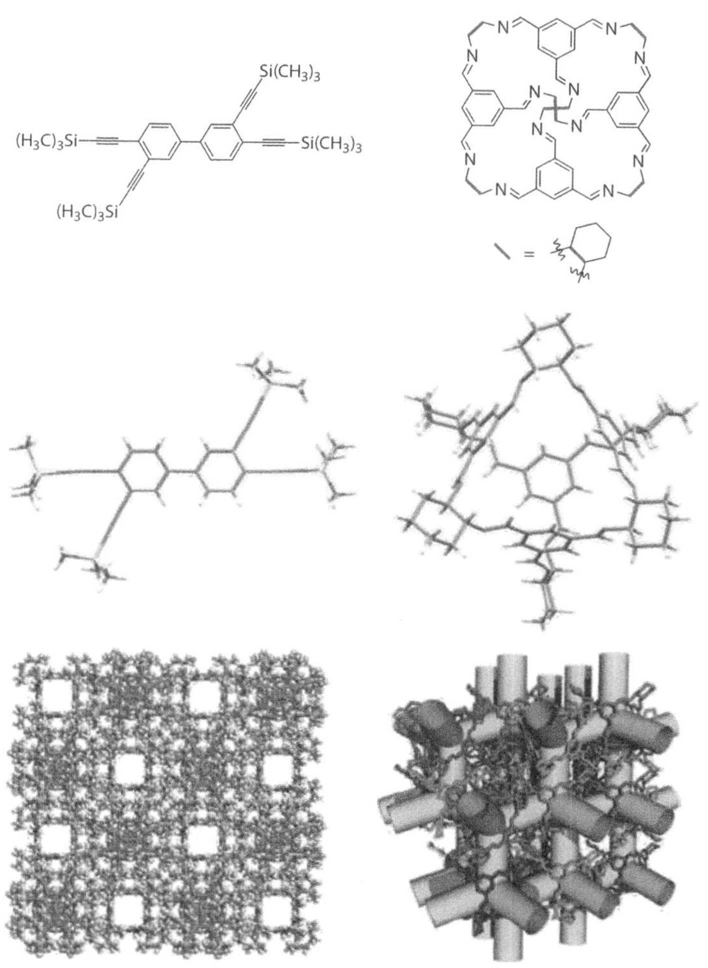

Figure 4.5 Molecular structure (top), the crystal structure of the individual molecule (middle), and crystal packing (bottom) for a tetrahedral imine cage. Figure reproduced with permission from Ref. 22, ©2009 Wiley.

The perceptual view of Richard M. Barrer and Vivien H. Shanson during 1976, initiates the importance of a small organic molecule (Dianin's compound) and derives their historical developments as shown in Figure 4.6. Figure 4.5 depicts the reminiscent of a porous solid [33-35] in detail. These compounds' absorbing capability of gases in their pure, solid form through the surface of the porous and crystalline lattice, has been considered mimics of zeolites and, they are named "organic Zeolites".

Later Barrer and Shanson suggest that the framework of these host structures (organic Zeolites) is less rigid compared to Zeolites and helps to pass-through the guest molecules into the holes. To avoid such circumstance in porous materials, Barbour describes, there are a number of small non-covalent interactions that takes place on the surface (Hydrogen bonds & Van der waals) rather than strong bonds (ionic, covalent or co-ordinate) defining the porosity with extended frameworks and was named as 'porosity without pores' [36].

In those days crystal engineers emulate porosity as a function of molecular scale. According to them, porous materials are having a skeletal framework with an infinite

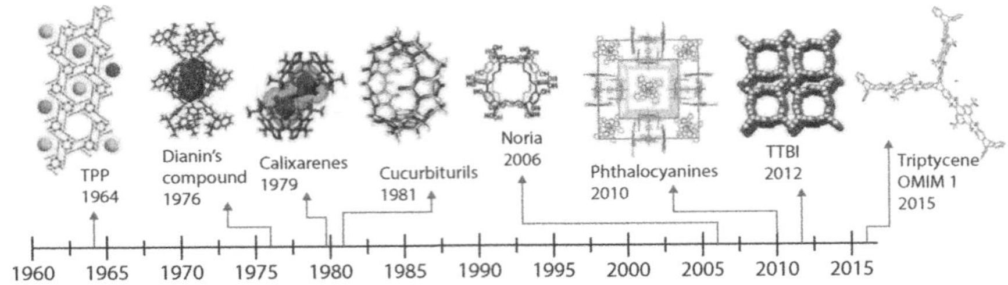

Figure 4.6 The historical development of porous molecules, from left to right: tris(o-phenylenedioxy) cyclophosphazene [20, 37, 38], Dianin's compound [33, 35, 39], calixarenes [40–47], cucurbiturils [48, 49], Noria [50, 51], phthalocyanines [52], triptycenetrisbenzimidazolone (TTBI) [53], and triptycene-based OMIM [54].

number of channels of the diameter of about 3A °. In metal-organic frameworks, there are an infinite number of such kinds of frameworks that exists and they are interconnected. The arrangements of such kinds of frameworks are joined firmly and this arrangement lead to a variety of applications in most the industries. The presence of huge number of tetrahedral atoms on the lattice are directly related to the pore size in Zeolites' on the lattice, these windows help to separate two foreign molecules either through Zeolite (or MOF) or through the "molecular sieving effect", when its pore size falls within the size range of those two molecules [55].

4.2 Family of Porous Organic Materials

The following segment sorts the various kinds of porous organic materials and their characteristic features in detail. The porous organic materials considered for the present discussions are porous solids including covalent organic frameworks (COFs), covalent triazine frameworks (CTFs), hyper-cross-linked polymers (HCPs), conjugated microporous polymers (CMPs), porous aromatic frameworks (PAFs), polymers of intrinsic microporosity (PIMs), extrinsic and intrinsic porous molecules, and porous organic cages.

4.2.1 Covalent Organic Frameworks (COFs)

4.2.1.1 Historical Development of Covalent Organic Frameworks COFs

Omar M. Yaghi (currently at UC Berkeley) and Adrien P Côté [9] were the first at the University of Michigan to introduce and synthesize Covalent Organic Frameworks (COFs) followed by a series of 2D COFs in the year 2005. They construct these covalent organic frameworks (COFs) mainly from light elements such as H, B, C, N, and O. These COFs building blocks are functionalized to form feathery substances for photonic, gas storage and catalytic applications. Côté *et al.* [9] first proposed the synthetic methodology by adopting condensation reactions between hexahydroxytriphenylene [$C1_8H_6(OH)_6$] and phenyl di-boronic acid {$C_6H_4[B(OH)_2]_2$} to synthesize COFs. In their laboratory, they first prepared crystalline COF-1 [$(C_3H_2BO)_6.(C_9H_{12})_1$] and $C_9H_4BO_2$ as COF-5 [9]. Their XRD analysis revealed that COF-1 has a space group of P63/mmc with staggered and COF-5 has space group of P6/mmm with eclipsed arrangements. The firm porous frameworks of crystal structures were formed through covalent bonds containing B,C and O atoms with pore sizes ranging from 7 to 27 A°. The uniqueness of the prepared COF-1 and COF-5 is, their ability to withstand till 600°C. COF-1 & COF-5 are formed as discrete molecules by boronic acids' precursors through condensation reactions as shown in Figure 4.7. In recent

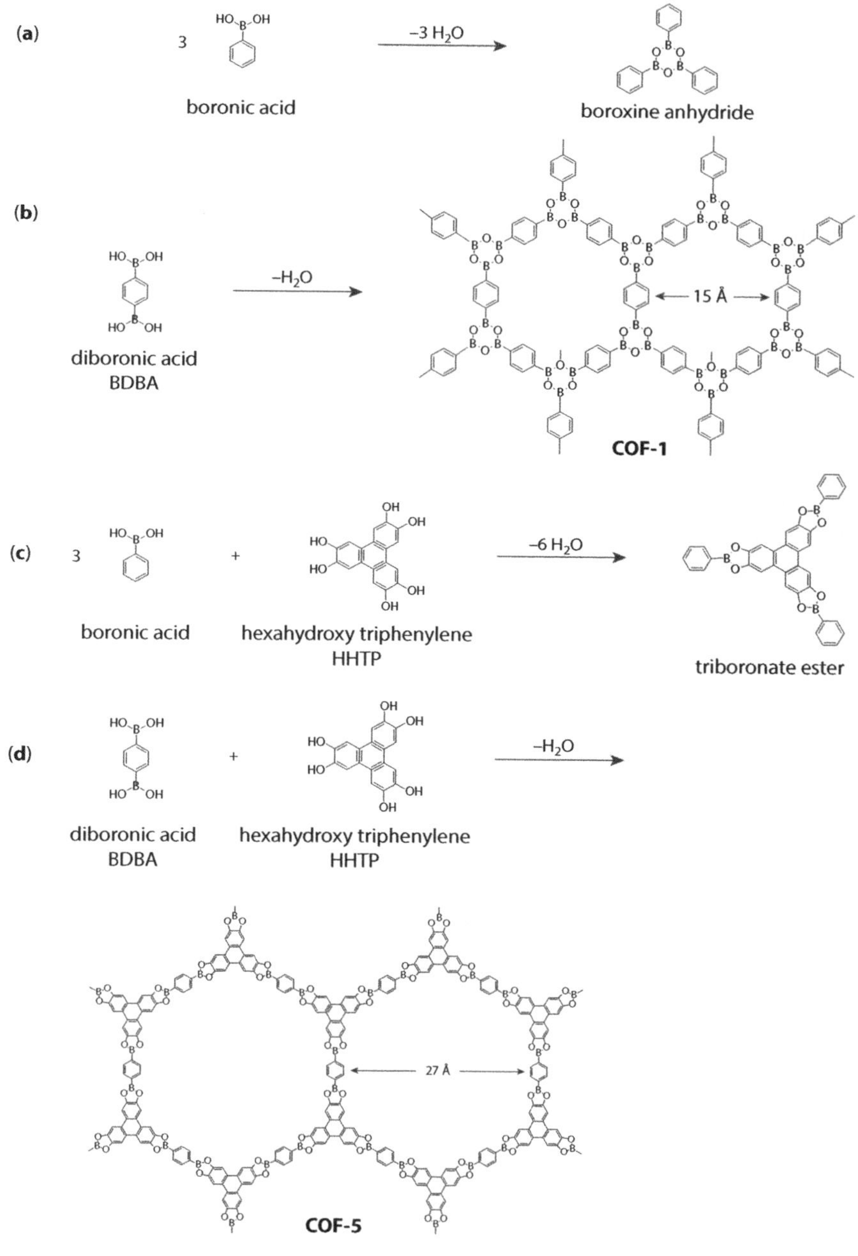

Figure 4.7 (a to d) The formation of individual molecules and extended COFs through condensation reactions between HHTP & boric acid [9].

times attempts had been made to synthesize COFs with compact arrangements of polymers in a linear fashion having crystalline structures with one-dimensional (1D) crystalline structures [56–61]. The enchanting properties of these COFs such as structural versatility, huge surface area with high chemical stability, and tuneable surface area lead to discussion in more detail in the following sections.

4.2.1.2 *Chemistry of Covalent Organic Frameworks (COFs)*

Among the porous organic material series, Covalent Organic networks (COFs) is the dynamic, versatile, and very important members. These COFs are constructed with covalent bonds as their building blocks and follow the principles of reticular chemistry. These reticular approaches was first introduced by Yaghi and co-workers [9] and follow variety of dynamic covalent chemistry (DCC) concepts, along with imination, triazine, hydrazine, and azine formation [11, 62, 63], which leads to high profile COFs.

4.2.1.3 *Classifications of COFs*

Based on the presence of the basic elements present in COFs skeleton (Figure 4.8a) it is classified as

1. Boron-containing COFs
2. Triazine-based COFs (CTFs)
3. Imide-linked COFs
4. Imine-based COFs

4.2.1.3.1 Boron-Containing COFs

In boron-containing COFs, boron plays a key component either as a building block through carbon–carbon cross-coupling reactions, or as a linker via the condensation reactions as shown in Figure 4.8b to construct the COF framework.

The precursors used to construct boron-based building blocks are triarylboranes [64–67], Polymers, boron-dipyrromethene (BODIPY) derivatives [68, 69] and tetraphenylborate [70–73]. For to construct COFs as boron-based linkers, heterocyclic linkers as precursors are involved in the multiple B–O or B–N covalent bonds [10, 74].

Yaghi and co-workers were the first ones to prepare boron frameworks of COFs as COF-1 and COF-5 through condensation between phenyl diboric acid and six-membered B_3O_3 rings through the solvothermal process to construct boronate ester-linked two-dimensional hexagonal and boroxine frameworks (Figure 4.6) [9]. containing COFs as COF-1 and COF-5 via condensation reactions of phenyl diboronic acids with planar six-membered B3O3 rings through solvothermal conditions to establish the bond between boroxine and boronate ester-linked two-dimensional hexagonal frameworks (Figure 4.6) [9]. The formed COF possess (BET exterior space of 711 m^2 g^{-1}) enlarged graphitic layers with 2D porous structure (pore size 0.7 nm). Based on the synthetic strategy adopted to synthesize these boron-containing COFs, they were further divided into two types as i) single framework unit through self-condensation (Figure 4.9a) [10, 75], ii) co-condensation of more than two building frameworks (Figure 4.9c).

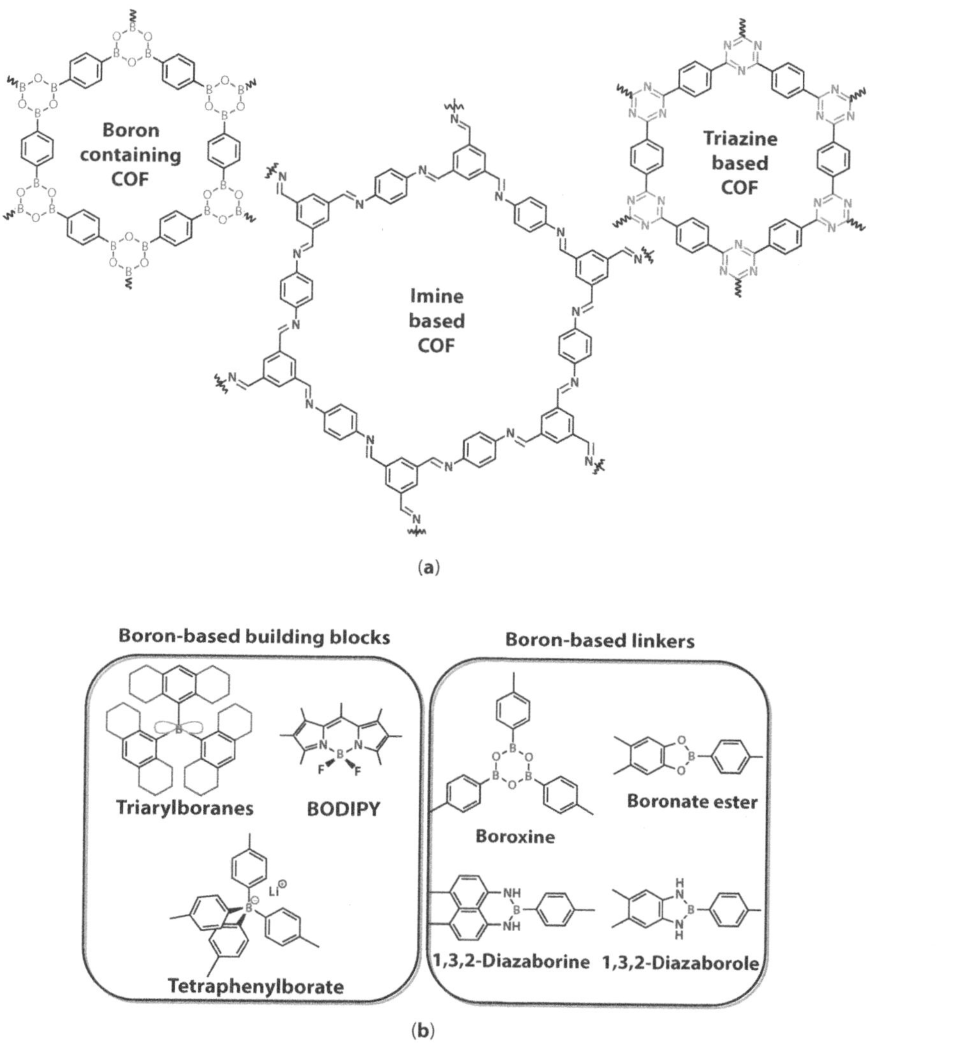

Figure 4.8 (a) Structures of diverse COFs. (b) The construction of building blocks of boron-containing COF with boron as the main frame work (left) & boron as linger (right).

However, the main drawback of these boron-containing COFs is, that they could not withstand atmospheric air or in water [76, 77] even for one hour. But literature survey reveals that COF-202 is the only COFs with boron that can withstand moist air even after 24 hours [78] and maintain its highest porosity with high crystalline nature.

4.2.1.3.2 Triazine-Based COFs (CTFs)

In 2008, Thomas and co-workers (Scheme 4.1) [15] first discovered CTF (CTF-1) with nitrile skeletal part of cyclotrimerization in the existence of ZnCl2 at 400 °C. The severe reaction conditions adopted for synthesizing these CTFs, limited their monomers to withstand for a long time. Hence it is hard to perceive significant monomers to construct CTFs.

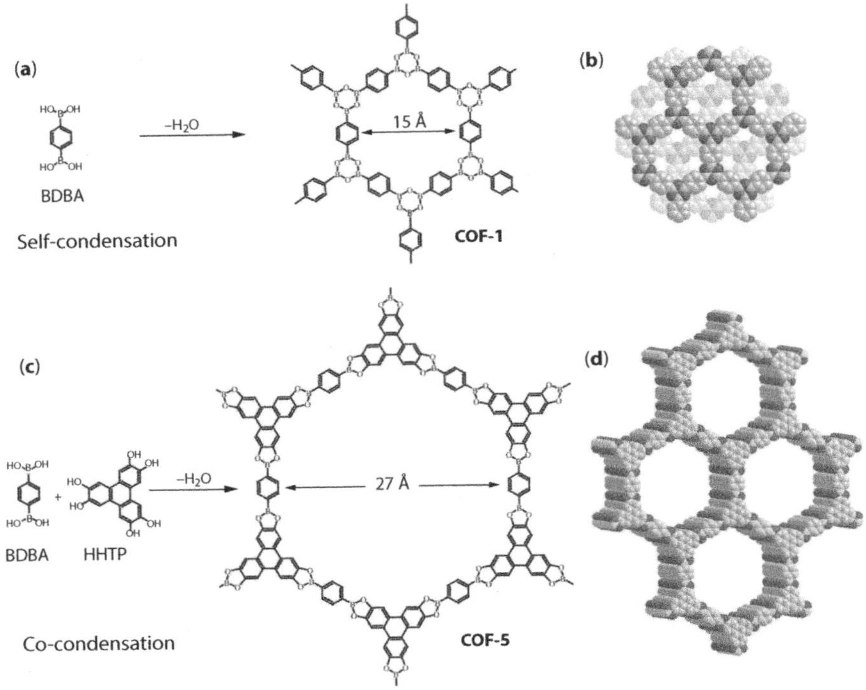

Figure 4.9 (a) The formation of COF-I through the process of self-condensation of BDBA and the corresponding crystalline structure (b). (c) Synthesize of COF5 by Co-condensation reaction between BDBA & the formed crystalline structure (d) 9 is shown [10, 75]. (Adapted from Ref. [10, 75] with permission from AAAS).

Apart from that, most of the CTFs lack maintenance of the ordered structure, which leads to poor crystallinity and porosity [79, 80]. In recent times, Tan's group [81] suggested a new low-temperature polycondensation reaction between aldehydes and amidines by adopting a gentle reaction environment to form CTFs as shown in Scheme 4.1.

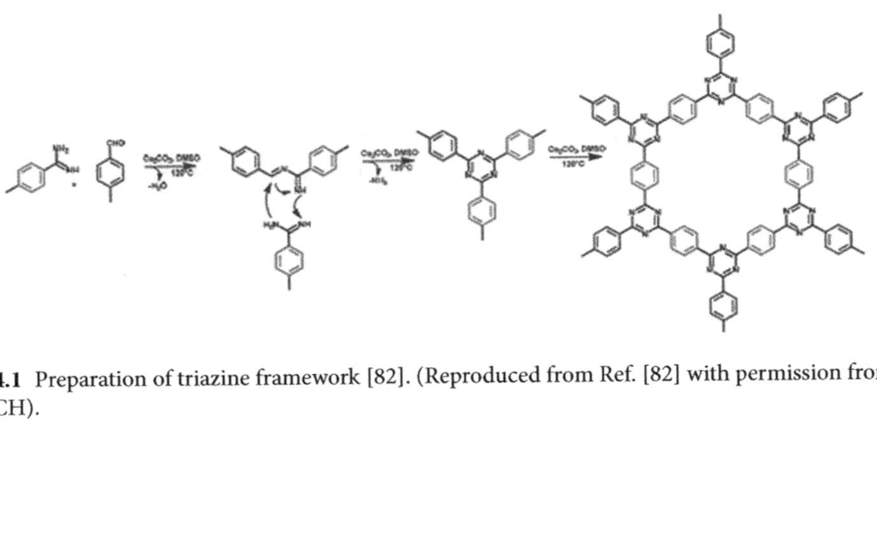

Scheme 4.1 Preparation of triazine framework [82]. (Reproduced from Ref. [82] with permission from Wiley-VCH).

4.2.1.3.3 Imide-Linked COFs

Yan's group [78] during 2014 first reported imide-linked COFs with different kinds of polyimide (Scheme 4.2) having a large surface area with a greater loading capacity of dye molecules for drug delivery [83] processes. The formed PI-COF can withstand up to 520°C later, Yang *et al.* synthesized advanced COFs of imide linked framework for fluorescence sensing purposes [84].

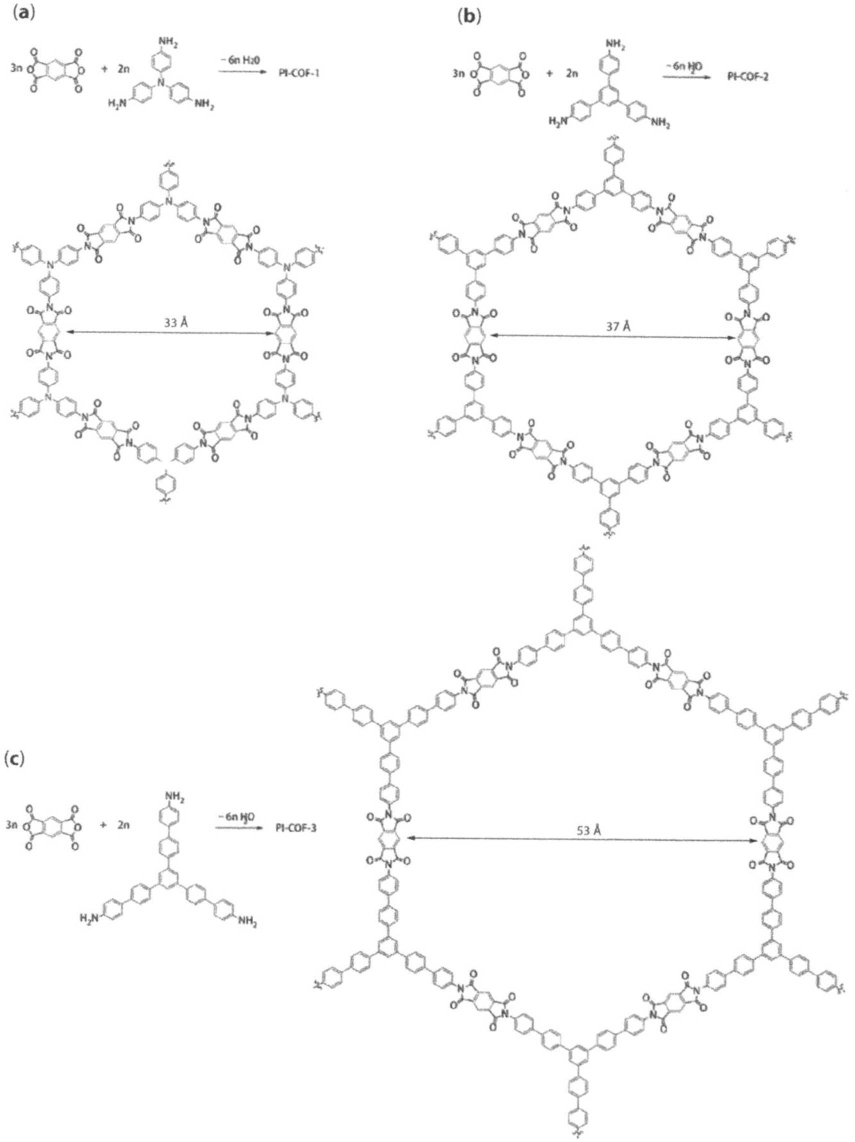

Scheme 4.2 The condensation reactions of Pyromellitic dianhydride (PMDA) with various reagents give 2D crystalline polyimide COF and termed PI-COF-1 (a) (triamine trisIJ4-aminophenyl)amine (TAPA), PI-COF-2 (b) (triamine 1,3,5-trisIJ4-aminophenyl)benzene (TAPB) and PI-COF-3 (c) (triamine 1,3,5-trisij4-aminoIJ1, 1-biphenyl-4-yl)benzene (TABPB) [78]. (Adapted from Ref. [78] with permission from Springer Nature).

4.2.1.3.4 Imine-Based COFs

Yaghi and co-workers [85, 86] 2009, first discovered the imine-based COFs by the reaction between amine [87] and aldehyde. These imine-based COFs fall into two categories of bond formation, such as i) "Schiff base" type of bond with –C = N-group between aldehyde and amine through co-condensation (Scheme 4.3) [87]. ii) "hydrazine" type of bond formed between aldehydes and hydrazides (Scheme 4.3b) [88]. Recent times, advanced COFs were prepared by squaraine-linked COF [89–91], which has broader applications in many fields.

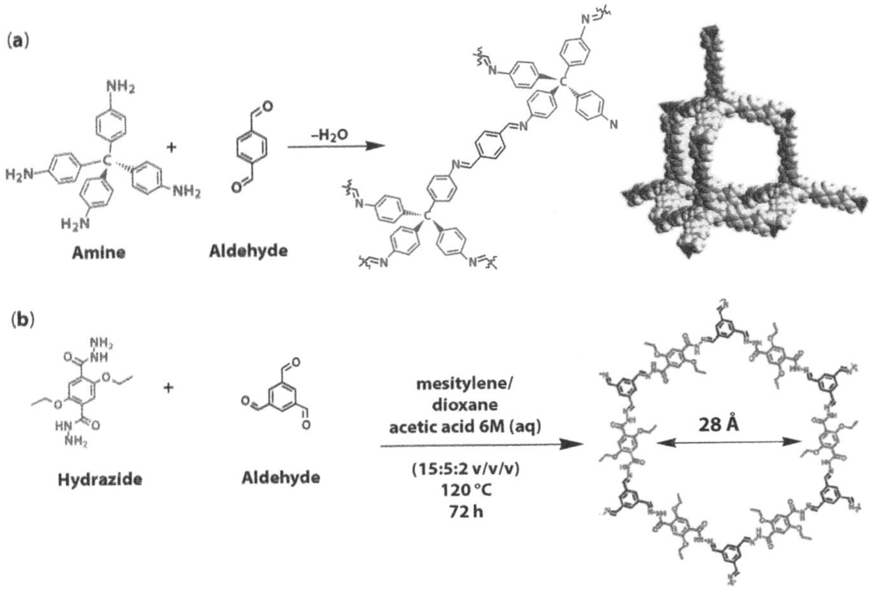

Scheme 4.3 (a) "Schiff base" synthesize of COF by co-condensation of aldehyde and amine. (b) "hydrazine" based synthesis of COF by co-condensation of aldehyde and hydrazide. (Adapted from Ref. [87] and [88] with permission from the American Chemical Society).

4.2.1.4 Synthetic Strategy Adopted for COFs Formation

By understanding the parameters (regulating thermal equilibrium) required to form highly ordered and crystalline COFs is a prerequisite to build covalent bond formation in COFs [87, 88]. These COFs can be designed according to our desire like porosity; surface functionalization; and structural regularity by careful understanding of the above-mentioned parameters. There are number of methods available for the successful formation of COFs [11, 92–97] and a few are solvothermal, ionothermal, microwave, sonochemical, mechanochemical, and light-promoted methods.

4.2.1.4.1 Solvothermal Synthesis

Solvothermal methods are one of the best methods to synthesize COFs in perfect crystalline form. In this method initially, the pyrex tube is degassed, it was then loaded with monomers containing suitable solvents. After sealing the pyrex tube, it was heated to a higher temperature in an autoclave at a constant period of time. The obtained precipitates were collected, washed and dried to produce powdered COFs. Yaghi *et al.* group synthesized the first two COFs (COF-1 & COF-5) by using the solvothermal method (Scheme 4.4) [9]. They assembled COF-1 as hexagonal sheet like networks by 1,4-benzenediboronic acid (BDBA) through a self-molecular dehydration process. Similarly, the condensation between 2,3,6,7,10,11-hexahydroxytriphenylene (HHTP) and BDBA generates COF-5 as enlarged layer-like structure with coplanar arrangements. This reaction is assisted by dioxane-based solvents, such as Mesitylene-dioxane, which helps to expand the nucleation of COF, by controlling the diffusion of monomers into the solution.

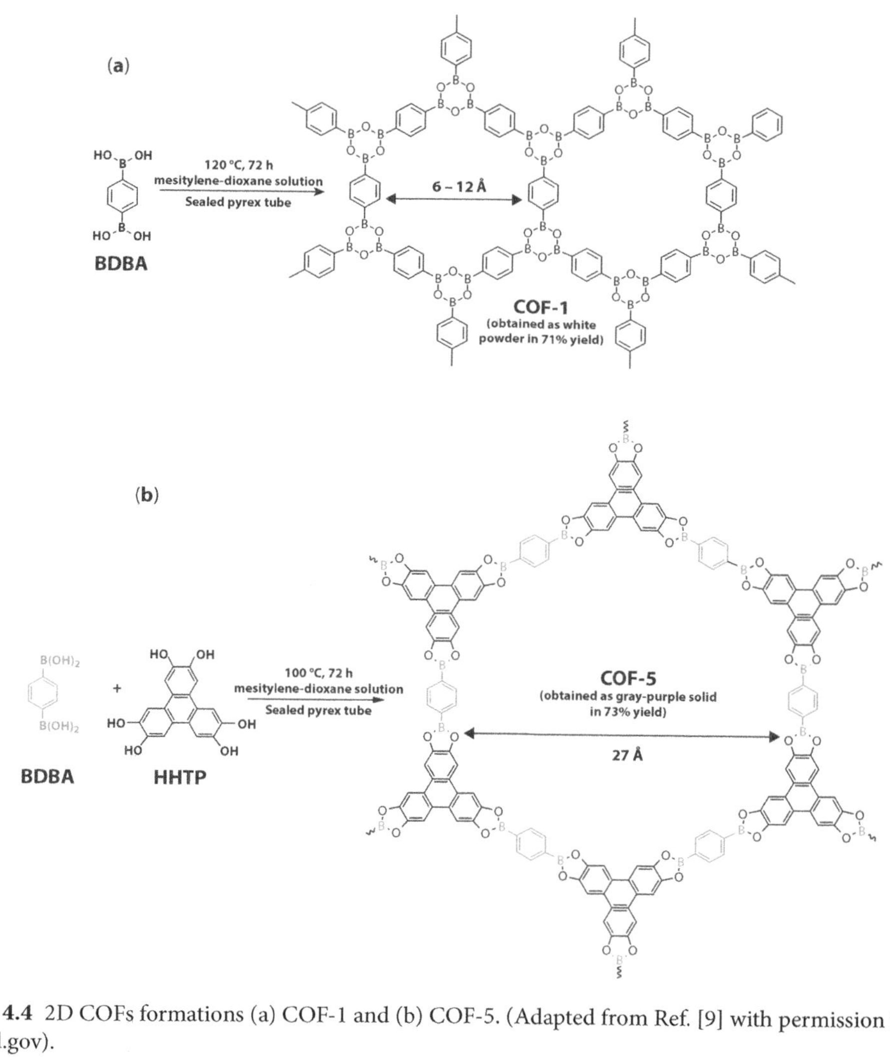

Scheme 4.4 2D COFs formations (a) COF-1 and (b) COF-5. (Adapted from Ref. [9] with permission from the PubMed.gov).

4.2.1.4.2 Microwave (MW) Synthesis

The Solvothermal methodology was generally preferred to synthesize COFs. Due to the harsh conditions adopted in this method, some uncustomary methods, such as microwave-assisted synthesize [98–100] were adopted in later days due to their shorter reaction times and ability to

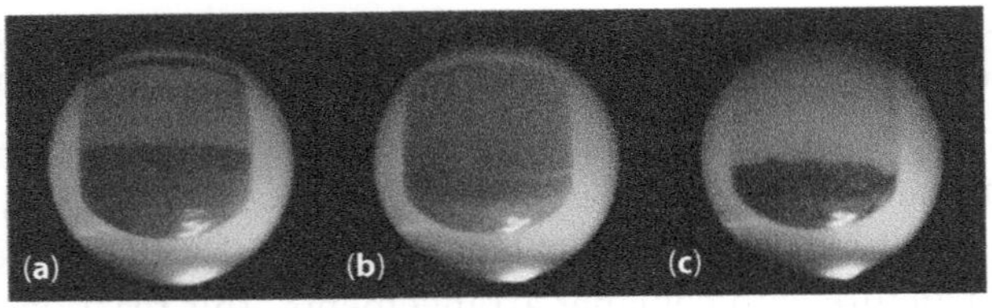

Figure 4.10 Representation of COF-5 formation: (a) appearance of COF-5 (grey) along with impurities (purple). (b) through MW operation the impurities are eliminated. (c) refined COF-5 [98]. (Adapted from Ref. [98] with Copyright 2009 American Chemical Society).

monitor the phase behavior the temperature, and pressure [101]. Banerjee *et al.* [102] assessed that the chemical stability of COFs can be obtained through an environmentally benign protocol (by using Microwave method) for the extensible synthesis of COFs in acidified water [103]. While synthesizing COFs with water as a solvent, generally exhibit unusual chemical stability than their counterparts (either mechanochemical or solvothermal method). The corresponding Brunaeur–Emmett–Teller surface areas [S BET(microwave) = 2019 m^2 g^{-1}] are considerably higher compared with the solvothermal products [S BET (solvothermal) = 1590 m^2 g^{-1}].

2D COF-5 and 3D COF-102 were synthesized by Cooper *et al.* [98] under MW irradiation at a power of 200 W by stirring at 100 °C for 20 min. It is 200 times faster than other methods. While synthesizing COF-5 by this MW method, the initially formed grey colored product had some impurities (purple color) as shown in Figure 4.10a. These unreacted purple-colored impurities were separated using acetone. The complete formation of COF-5 by the MW method is sketched in Figure 4.10.

4.2.1.4.3 Sonochemical Synthesis

Sonochemical methodology of synthesizing COFs gain greater attention as another root for most of the existing techniques, as it is proceed very fast and economical too [104, 105]. Ahn *et al.* [106] was the first person to introduce synthesized COF-1 and COF-5 by utilizing this strategy between mesitylene, BDBA, and 1,4-dioxane within an hour under the sonication method. By adopting this method it is possible to mix sparingly soluble BDBA with the help of mixed solvents using the cavitation phenomenon. When COF-1 was synthesized by adopting this strategy, its size reduces nearly 400 times compared to the standard method of preparation, with a higher surface area of 2122 m^2 g^{-1}. Furthermore, COF-5 was made by the reaction between HHTP and BDBA in a similar solvent system. Shim *et al.* followed the same sonochemical strategy to prepare COF-5 as core/shell architecture between graphene and carbon nanotube surface.

4.2.1.4.4 Mechanochemical Synthesis

The present-day synthetic chemists adopt a mechanochemical strategy, using a simple pestle and mortar for manual grinding. The advantage of this method is, that it is very fast and carries the reaction without solvent at room temperature.

Banerjee *et al.* [102], synthesized COFs through the MC method by adopting Schiff base condensations using mortar and pestle. The formation of COFs was confirmed by changing the color of the reactants during grinding and which can be visualized by the naked eye as shown in Figure 4.11. The formed COFs exhibit graphene-like layered structures, which

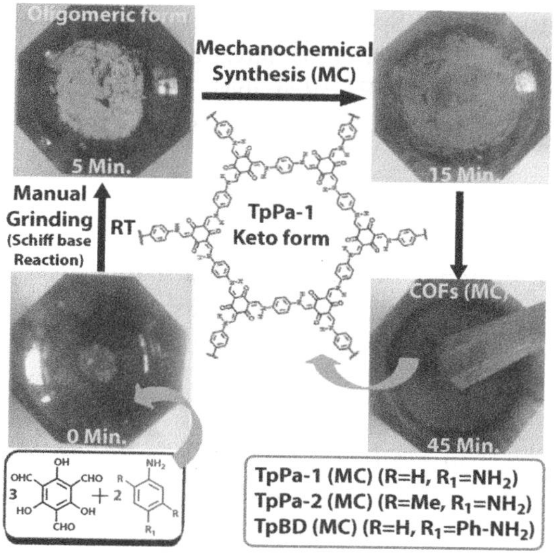

Figure 4.11 TpPa-1, TpPa-2, and TpBD MC syntheses using pestle and mortar [102]. (Reproduced from Ref. [102] with permission from the American Chemical Society).

are of greater industrial use. Wang *et al.* [107] adopted this strategy, by condensing Tp and 4,4′-azodianiline to prepare COF TpAzo in a minimum period (20 min). The obtained COF possess higher porosity and thermal stability with layered morphology.

4.2.1.5 *Characterization COF*

Generally, the synthesized COF materials were confirmed by recording their IR, XPS, XRD, SEM, and AFM.

4.2.1.5.1 Infrared Spectral Analysis of COF

The main elements present in COFs are C, H, O, N & B and are connected to each through C–N, C=N, C=O, B–O bonds in a molecule [108]. The infrared (IR) spectrum is one of the efficient tools to analyze the functional groups along with elements present in a particular material. For example, the first prepared COF-1 was confirmed by the presence of boron-containing rings in COF-1 [9], through IR study as shown in Figure 4.12.

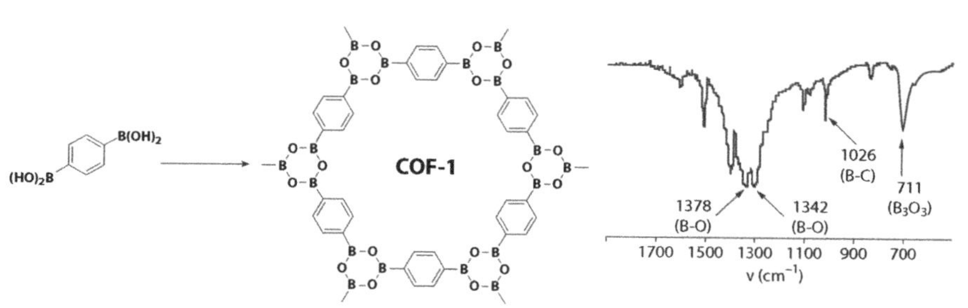

Figure 4.12 Condensation reactions of boronic acids used to produce COF-1 and the FT-IR spectra of COF-1. (Reproduced from Ref. [108] with permission from the Chemical communications: Chem comm).

4.2.1.5.2 X-Ray Photoelectron Spectroscopic Analysis of COF

XPS is an important and most helpful tool to analyze and determine the connection existing between monomers present in COFs. FT-IR analysis helps to confirm the elements present in the prepared COFs (Figure 4.13) and the same was confirmed by XPS analysis [108].

4.2.1.5.3 XRD Analysis of COF

The COFs possess a distinctive crystalline structure in contrast to the existing porous organic substances. The presence of the crystalline nature of COFs was confirmed by recording their XRD. For example, COF-10's establishment was ensured after cycling of ammonia adsorption is confirmed by XRD [109] and the corresponding XRD pattern is shown in Figure 4.14.

4.2.1.5.4 Scanning Electron Microscope (SEM)

To confirm the surface morphology, phase purity, and synthetic linearity of COF materials [110], SEM analysis is very helpful. For example, COF-102 and COF-103 SEM analysis depict spherical morphology and COF-105 exhibits platelet arrangements with lots of granular arrangements. In the same way, COF-108 shows a deformed spherical morphology.

4.2.1.5.5 Atomic Force Microscope (AFM)

2D patterned phthalocyanine COF-oriented thin films were grown by Colson *et al.* [111] during 2015 and become important and formed on graphene single layer. These layered arrangements were confirmed by AFM and the outcomes disclosed that the thickness of

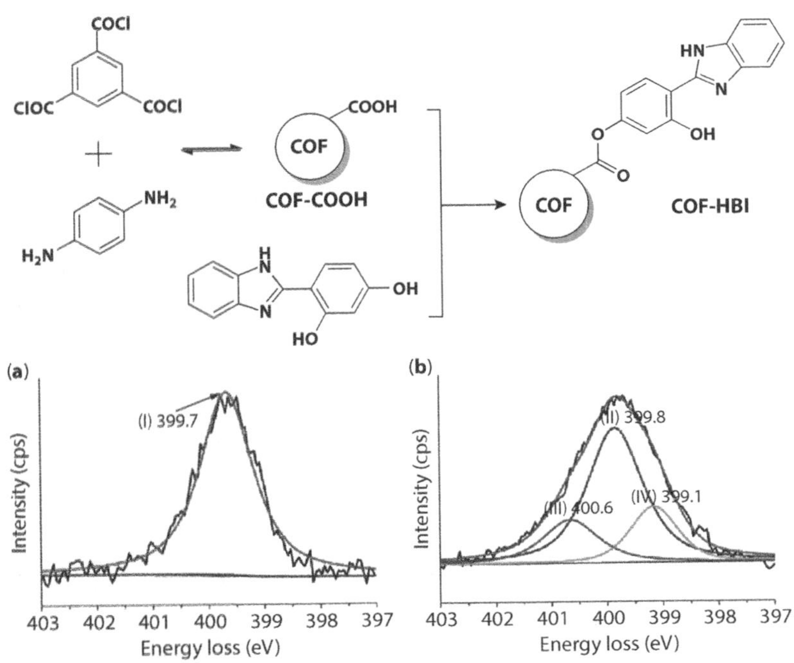

Figure 4.13 COF-COOH and COF-HBI synthesis and the corresponding N1s XPS spectra. (Reproduced from Ref. [108] with permission from the Chemical communications).

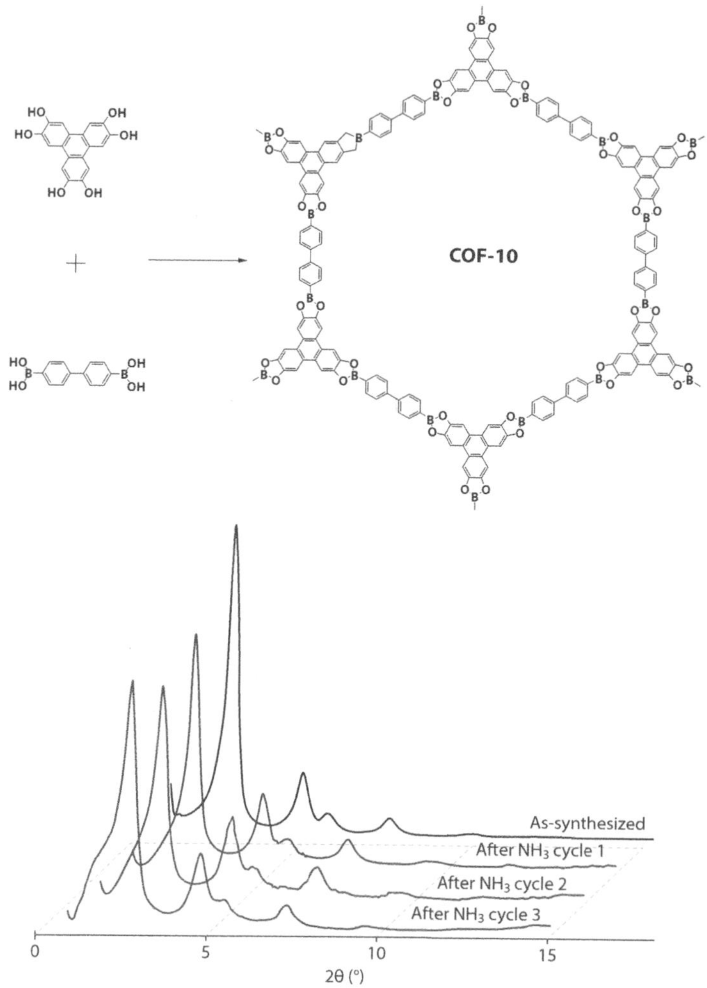

Figure 4.14 COF-10 synthetic strategy and the corresponding XRD patterns which follow adsorbing NH_3 for individual cycles. (Reproduced from Ref. [109] with permission from the Nat. Chem).

the COF was 134 ± 6 nm and is evident that COF possesses squares that are surrounded by wafers and their phase images were observed in the same region.

4.2.1.6 Applications of COF

COFs' inherent potentials such as, well-ordered channel arrangements, higher stability, larger surface area, and multi-dimensional properties make these COFs outstanding candidates for a plethora of applications. COFs act as promising materials in most industrial applications such as optoelectronics, small molecule adsorption, sensing, and drug delivery [112].

4.2.1.6.1 Gas Storage and Separation

The persistent nature of the porous structure present in COFs makes these COFs a superior material to separate and store gases including hydrogen, carbon dioxide, and methane

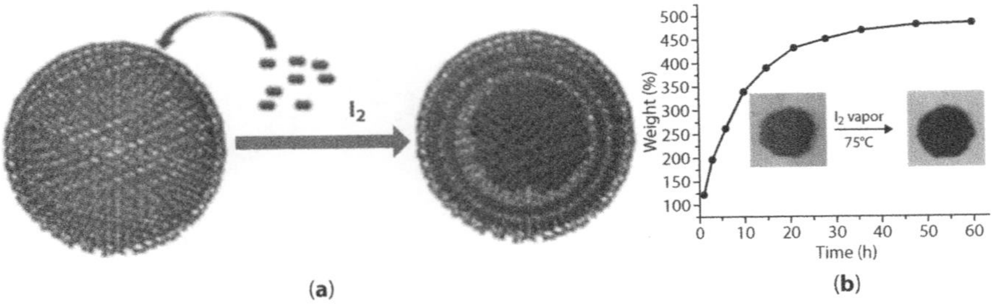

Figure 4.15 (a) The formed COF captured iodine (b) Gravimetric method of iodine assimilation for time (75°C) (Inset: images of COF in the absence and in the presence of iodine vapor). (Reproduced from Ref. [117] with permission from the Chemical communications).

[100, 113–115]. For example, Gascon's group made a membrane of mixed-matrix of Matrimid and azine-linked COFs and are used to remove CO2 from an equimolar mixture of CO_2 and CH_4 [116]. In the same way, Zhao's group formed a heteropore COF by condensing 4,4'-bis(bis(4-formylphenyl)amino)-[1,1':4',1"-terphenyl]-2',5'-dicarbaldehyde and 1,4-diaminobenzene with desirable iodine capture capacity (Figure 4.15) [117].

4.2.1.6.2 Catalysis

COFs' catalytic nature is directly related to the existence of the number of reactive sites within the pores of COFs. Eclipsed layered-sheet structure of COF-LZU1 was prepared by Ding *et al.* in 2011 [118], and it serves as an outstanding catalyst during the catalyzing processes of the Suzuki–Miyaura coupling reaction.

4.2.1.6.3 CO_2 Capture

The unique nature of COFs, as they possess uniform pore size, good stability and easy pore functionalization made these COFs a good CO_2 capturing material. COFs capture CO_2 under a pressure of 1 bar at temperatures of 273 and 298 K. To undergo effective and maximum CO_2 capture, COFs in 3D forms such as 3D COFs were utilized, which serve a scaffold and adsorb the maximum amount of CO_2 in a limited time interval. Yaghi *et al.* have synthesized outstanding carbon dioxide capturing 3D COF-102 and COF-103 in the range of 1200 and 1190 mg/g at 50 bar and 298 K [75] with a higher surface area of 3620 and 3530 m^2/g.

4.2.1.6.4 Sensor

COFs in various forms serves as excellent sensing agent. Jiang and co-workers prepared an azine-linked COFs a highly luminescent [119] material from 1,3,6,8-tetrakis (4-formylphenyl)pyrene and hydrazine. Further, thioether-based fluorescent COF (COF-LZU8) [120, 121] was synthesized by Wang's group to remove Hg^{2+}ions (Figure 4.16) [122]. Liu's group constructs luminescent COF-JLU3 through hydrazine hydrate and 1,3,5-tris (3′-tert-butyl-4′-hydroxy-5′-formylphenyl) benzene to detect Cu^{2+} [123] ions.

The pictorial representation of applications of COFs is shown in Figure 4.17.

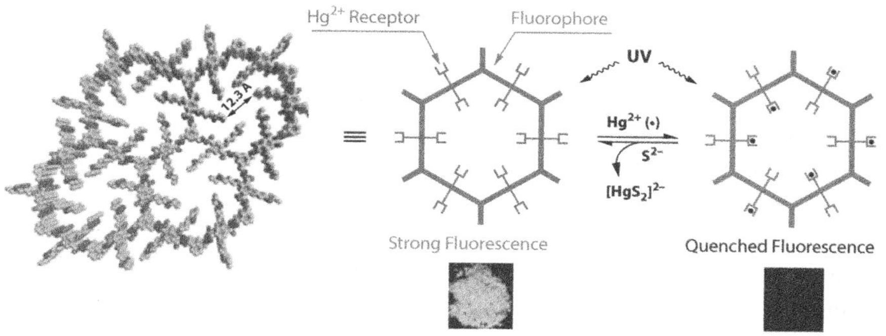

Figure 4.16 Utilization of the formed COF-LZU8 for both detection and removal of Hg^{2+} ions [123].

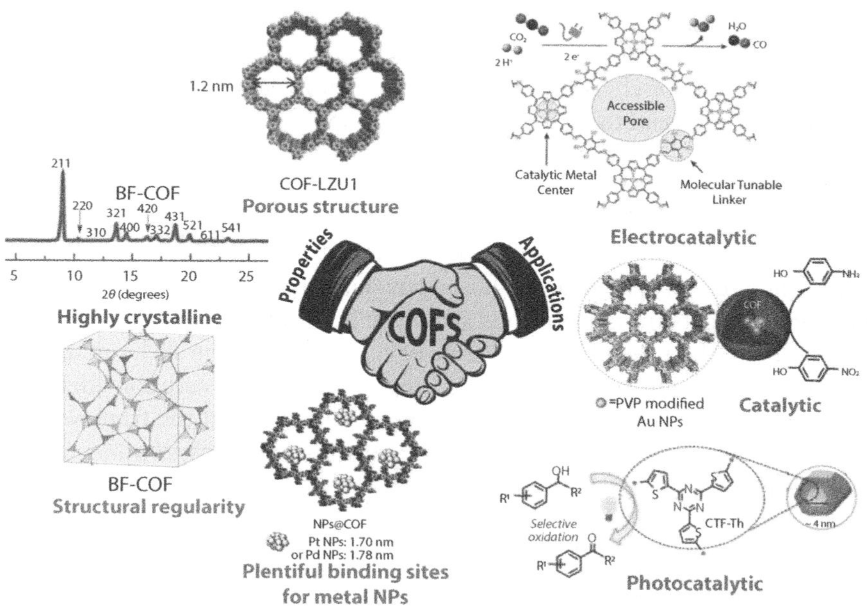

Figure 4.17 The pictorial representation of applications of COFs [123].

4.2.2 Covalent Triazine Frameworks (CTF)

4.2.2.1 Historical Development of CTF

Kuhn *et al.* in 2008 is the first person who discovered covalent triazine frameworks (CTFs) as a distinct subclass of crystalline COFs. The reaction methodology adopted to frame the CTFs are the reversible reaction of aromatic nitrile trimerization to form extended porous organic frameworks [15]. Most of the CTFs exist in amorphous form, due to the harsh synthesized methodology like ionothermal conditions were adopted to synthesize CTFs, in which $ZnCl_2$ acts as solvent as well as the catalyst. The presence of triazine-based porous frameworks as a skeleton in CTFs helps to prolong the exceptional thermal and chemical stability, both in acidic and basic mediums.

4.2.2.2 *Chemistry of CTFs*

The unique features of CTFs are their higher surface areas, tuneable pore sizes and adjustable building units. The presence of triazine-based porous frameworks, helps to undergo these CTFs in several possible executions, including gas storage, molecular separation, heterogeneous catalysis, etc. [94, 122–124]. The amenable chemical structures with designable pore walls and their inherent porous nature makes these CTFs an effective heterogeneous catalyst. Hence, these CTFs can lower activation energy for most of the reactions [125].

Apart from triazine linkage, there are some strong covalent bonds also formed between the lightweight elements present in CTFs such as C–C, B–N, C–N, B–O and C–O. These are formed either through the presence of metals or through radical polymerization of monomers [9, 92, 126] by irreversible chemical condensation reactions.

The persistent nature of porosity helps to hold high surface area, and the ability to withstand higher thermal and chemical stability along with easy functionalization of these CTFs. Adopting post-synthetic methodology, made it easy to incorporate these diverse guest molecules with a wide range of chemical functionalities into the frameworks and considered CTFs as a promising material to bridge the gap between heterogeneous and homogeneous catalysts.

CTFs are the only framework existing with higher nitrogen content compared to other frameworks and can be synthesized more economically using readily available chemical precursors [127–129]. Figure 4.18 reveals the above facts and represents some of the basic and most prominent features of Triazine-based covalent organic polymers (COPs) in detail.

The COPs which exists in microporous structure have limited ability to act as an effective catalyst, owing to their small diffusion rate of reactants into the skeleton. Hence, COPs with large pore volumes along with uniform arrangements of pores are considered desirable platforms for heterogeneous catalysis. The presence of nitrogen in the framework of CTFs along with their ordered meso porous arrangements of pores helps to transport any bulky substance under the catalytic conditions [130].

The main driving force behind CTFs as a heterogeneous catalyst is, that they hold the special structural features as indicated previously and these facts are considered to frame

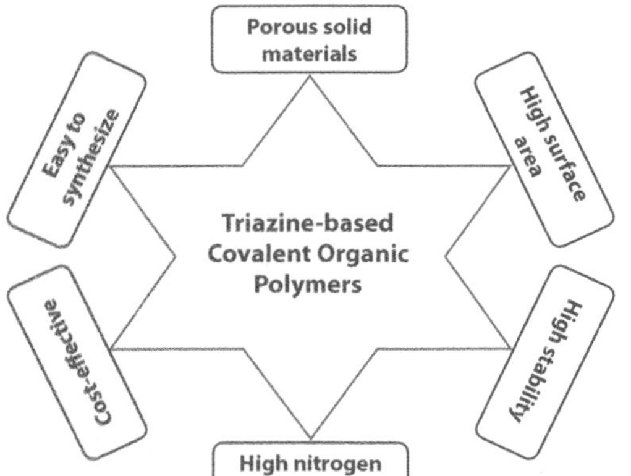

Figure 4.18 The most important and basic features of Triazine-based covalent organic polymers.

the structure and to prepare established triazine-COPs having numerous functionalities with high surface area porous network.

To design the targeted synthesis of triazine-based COPs, the monomers as a precursor should possess higher porosity along with suitable structure and chemical functionality to attain the desirable property.

4.2.2.3 Synthesize of CTFs

4.2.2.3.1 Trimerization Reaction

The synthesis of highly stable polymeric covalent triazine frameworks (CTFs) was first introduced by Thomas *et al.* They proposed an ionothermal trimerization method using the precursor containing carbonitrile as the main functional group and $ZnCl_2$ in the molten state, which acts as a catalyst as well as solvent between 400 and 600°C (Scheme 4.5) [15]. During the reaction, CTFs formed ultra-strong covalent bonds (aromatic C=N) from triazine linkage [10]. The formation of crystalline CTFs are followed reversible and polymerization processes and are optimized to adopt the strategy of trimerization reaction in ionothermal reaction.

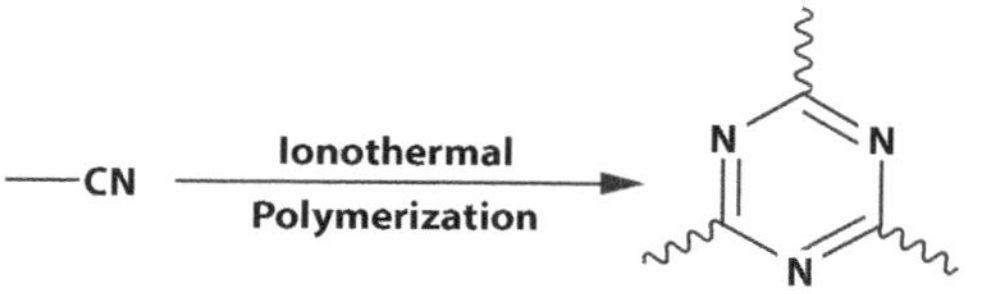

Scheme 4.5 Trimerization of carbonitriles.

The triazine-based CTFs with desired functionality along with desired surface area and pore size can be architecture engineered by selecting appropriate monomers. CTFs with different building blocks such as aromatic and heteroaromatic nitriles (Figure 4.19) having pores with periodic and symmetric connectivity have appeared as a consequence of an ionothermal reaction [17, 131–145].

The high stability of CTFs in adverse conditions includes strong acidic, basic and most organic solvents including acetone, $CHCl_3$, tetrahydrofuran (THF), dichloromethane (DCM) [15, 17, 131] and dimethyl sulfoxide (DMSO) may be utilized to prepare crystalline CTFs holding a huge number of nitrogen functionalized skeletons. The obtained CTFs furnish space for metals and reactant substrates as a guest molecules.

4.2.2.3.2 Schiff-Base Reaction

Mullen *et al.* 2009, found that CTFs can be synthesized by simple Schiff-base reaction by the one-pot poly condensations reactions between melamines and different aldehydes [146–151] in DMSO in an inert atmosphere at 180°C (72 h) (Scheme 4.6). The formed melamine-based CTFs were stable in all adverse conditions and is soluble in organic and aqueous solvents such as THF, acetone, DMSO, DCM and MeOH. CTFs with >40 wt% nitrogen content were also prepared using melamine as a starting compound in solvent-free conditions.

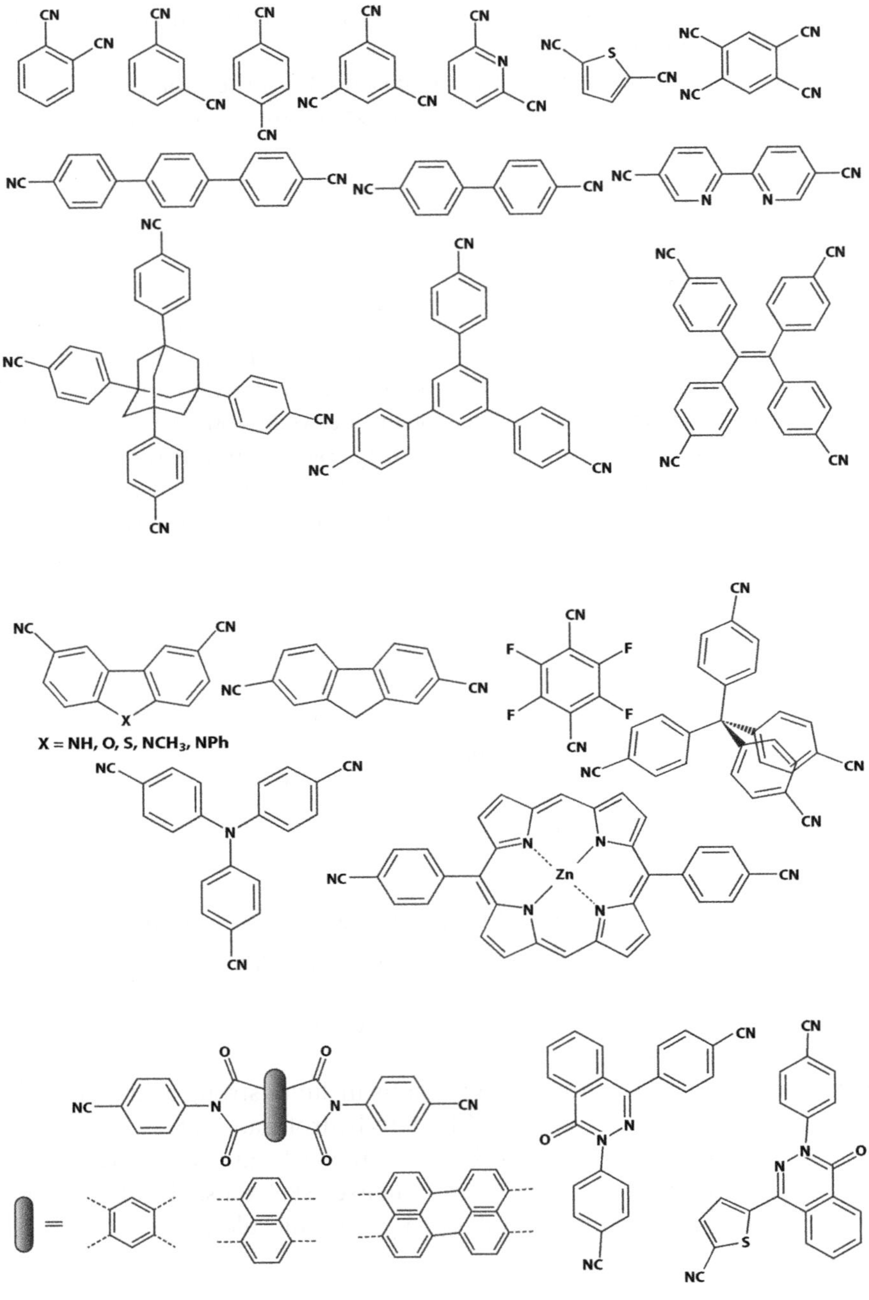

Figure 4.19 Carbonitrile monomers for trimerization reactions with distinct structures. (Adapted from Ref. [15] with Copyright ©2008 WILEY-VCH Verlag GmbH & Co. KGaA, Weinheim).

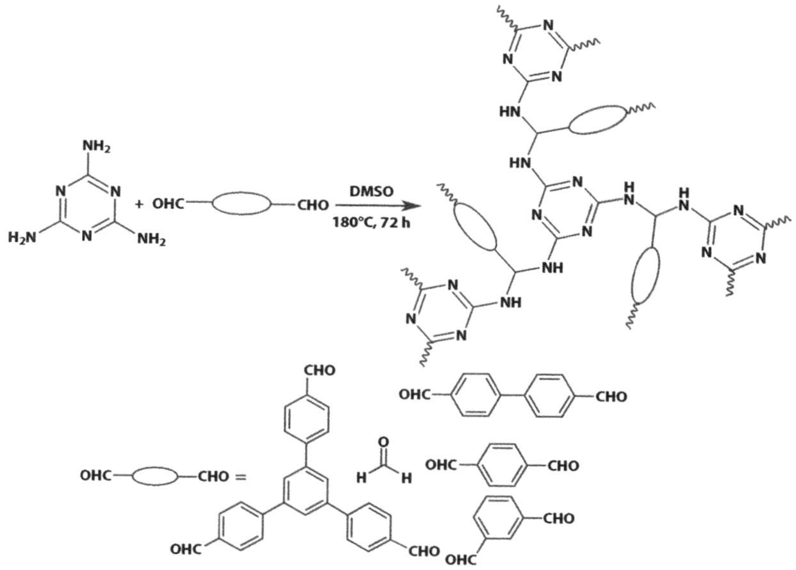

Scheme 4.6 Synthesis of melamine-based POFs. (Adopted from Ref. [146–151] with permission).

Pitchumani *et al.* [152] synthesized imine or hydrozone-linked CTFs through Schiff-base reaction by condensing aldehydes (1,3,5-tris-(4-formyl-phenyl)triazine (TFPT)) containing a triazine unit with different types of amines (phenylenediamine or terephthalohydrazide) as shown in Scheme 4.7.

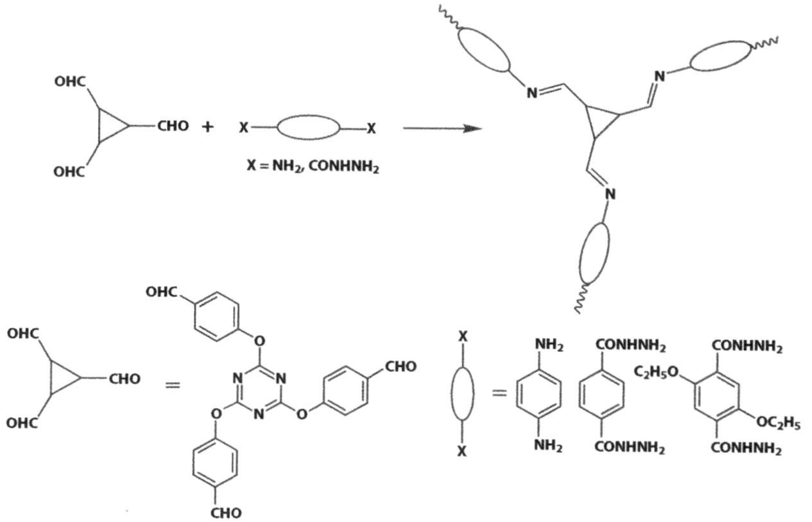

Scheme 4.7 Hydrazone-based CTPs or imine preparation. (Adopted permission from 2014 WILEY-VCH Verlag GmbH & Co. KGaA, Weinheim).

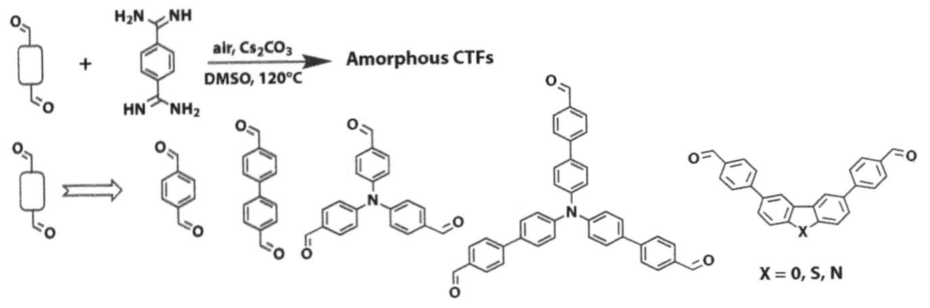

Figure 4.20 The polycondensation method of preparation of amidine as amorphous CTFs. (Adapted from Ref. [153] with permission).

Recently Tan *et al.* [81, 153] prepared CTFs by adopting Schiff base formation methodology by condensing aldehyde and an amidine dihydrochloride (Figure 4.20), followed by Michael addition. The above reactions utilize DMSO (solvent) and cesium carbonate (Cs_2CO_3-base) to prepare CTF entirely different method from other methods of preparation of CTF.

4.2.2.3.3 Nucleophilic Substitution Reaction

A triazine-based CTFs were prepared by the nucleophilic substitution reactions of triazine unit-built cyanuric chloride (CC). To adopt various applications of COPs, nucleophilic compounds with different types of COPs with varying porous arrangements were used (Scheme 4.8) and the formed materials were confirmed through recording their FT-IR, TGA and N2 isotherm analyses [154–158].

Yavuz *et al.* employed melamine and 1,3,5-benzenetricarbonyl trichloride to adopt nucleophilic substitution reactions at room temperature under various types of solvents such as 1,4-dioxane and DMAc/NMP) to form amide based COPs as shown in Scheme 4.9 [159].

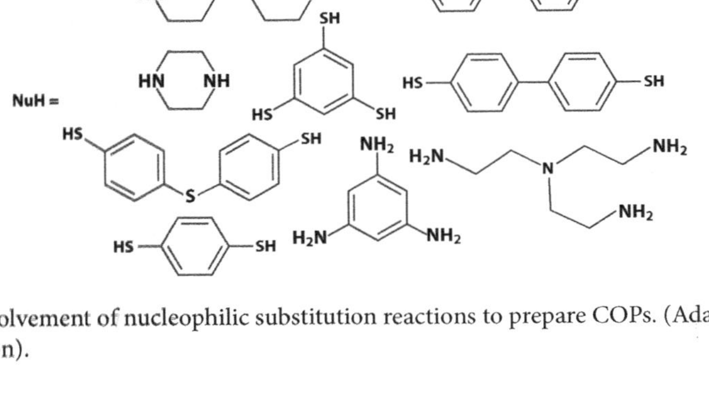

Scheme 4.8 The involvement of nucleophilic substitution reactions to prepare COPs. (Adapted from Ref. [154] with permission).

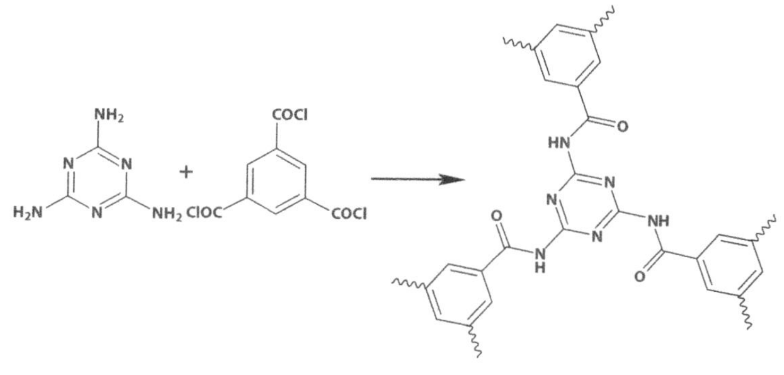

Scheme 4.9 Amide-based COPs preparation. (Adapted from Ref. [159] with permission).

4.2.2.3.4 Polycondensation Synthetic Route to Synthesize CTF

By adopting direct condensation methods of synthesizing triazine-based CTFs framework, the formed CTFs exhibit high specific surface area with stability. The reaction happens at 400°C in a degassed flame-sealed ampule and is catalyzed by P_2O_5 Scheme 4.10) [160] through condensation of aromatic amides. The CTFs formed by this method possess high crystallinity. CTFs are also formed either through Schiff base condensation reaction or through Michael addition. For example, CTFs with higher crystalline nature can be achieved in the presence of cesium carbonate (Cs_2CO_3) through polycondensation of amidines and aldehydes at 120°C (Scheme 4.11) [81].

An advanced in situ condensation approach has also been adopted to synthesize CTFs using dimethyl sulfoxide (DMSO) solvent [161] with alcohol into the aldehyde monomer through a slow oxidation process (shown in Scheme 4.1).

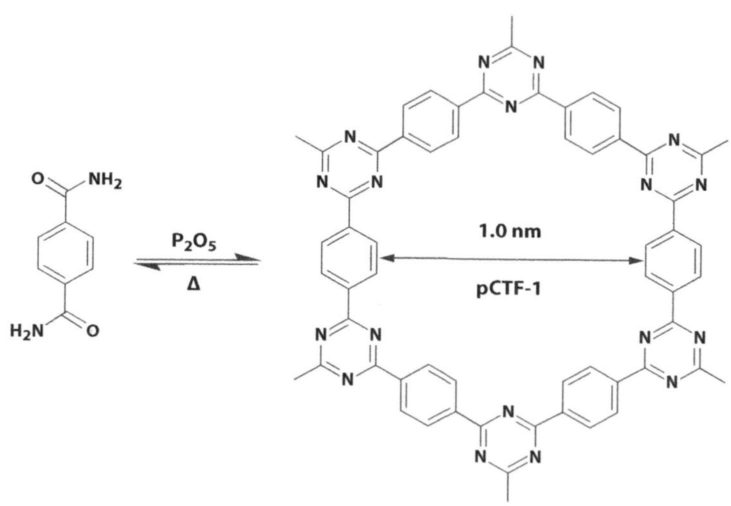

Scheme 4.10 The formation of discrete pCTF-1 using P_2O_5-catalyst through direct condensation of terephthamide (TA). (Adapted from Ref. [161] with permission).

Lotsch *et al.* synthesized micro porous CTF in an ordered manner using bi-pyridine with the high surface area of 1100 m^2g^{-1}. Here, bipyridine served as a scaffold unit and helps to coordinate Co, Ni, Pt and Pd by the impregnation method (Scheme 4.11) [43]. These metals are attached to the periphery of the surface of CTFs.

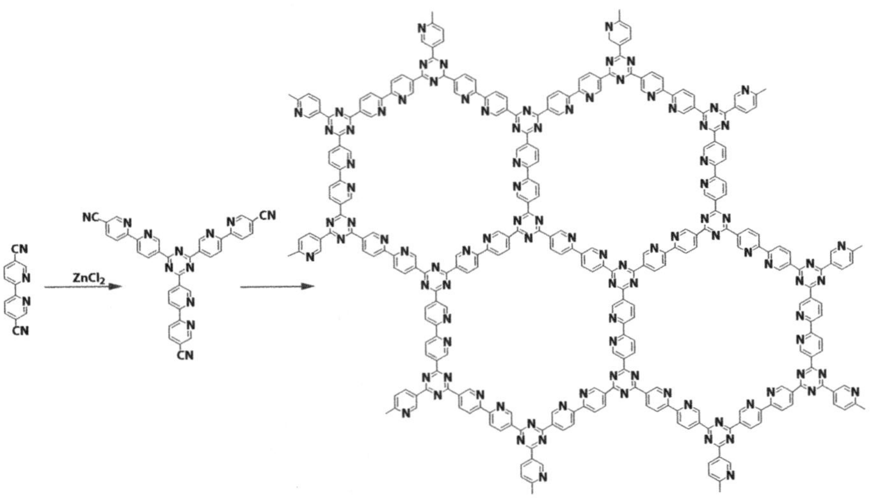

Scheme 4.11 Preparation of CTF with bipyridine unit. (Adapted from Ref. [43] with permission).

4.2.2.4 *Characterizations of CTFs*

The structure of CTFs can be confirmed by recording their PXRD [17, 162] or SXRD. Based on the monomer involved in the reaction, the CTFs can be formed either in amorphous or crystalline form. The presence of functional groups in the monomer which is used to synthesize CTFs was confirmed by recording their FTIR spectra by analyzing their vibrational stretching frequencies. The triazine link formed in CTFs units was confirmed by the presence of typical bands around ~1510 and 1350 cm^{-1}. CTFs when obtained through thermal methods were confirmed by the absence of stretching frequencies at 2260–2222 cm^{-1} for –CN group. To govern the level of carbonization that occur during the processes of synthesis, their C/N ratios were compared between the formed CTFs and the conceptual one, which helps to analyze the amount of nitrogen deprived in the CTFs synthesis process.

By recording NMR, it provides information about the chemical composition present in CTFs by exhibiting peaks correspond to the nature of functional groups present in the molecules. The porous properties of the formed CTFs were confirmed by nitrogen/Argon sorption measurements. Using this method, the distribution of pore size and volume of the pores can be found. The presence of various morphological arrangements of CTFs can be analyzed with the help of SEM& TEM. CTF's inner surface (TEM) and outer surface (SEM) provide statistical details of surface composition. The bonding nature and elemental composition of the formed CTFs were confirmed by recording their XPS. Thermo gravimetric analysis (TGA) is helpful to find the stability of CTFs at elevated temperatures and also the adverse reaction conditions (acidic and basic) stability.

4.2.2.5 *Applications of CTF*

4.2.2.5.1 Adsorption and Separation

The presence of triazine ring in the CTFs framework along with their high BET surface area and porosity offers CTFs, as an effective source for the separation of gases, metal ions and organic pollutants.

4.2.2.5.2 CO_2 Capture and Separation

The micro pores and ultra-micro porous nature of CTFs can make major contributions to the capturing and storage of CO_2, since the size of CO_2 (0.33 nm) matches with a pore size of CTFs [163]. Janiak *et al.* suggest that by adopting the ionothermal method to synthesize CTFs, their CO_2 capturing ability is considerably increased compared to other methods [138]. The amount of CO_2 captured by CTFs is decided by the reaction condition, such as the same precursors with different quantities of catalysts and variations in temperatures used in the reaction influence the quantity of CO_2 uptake by CTFs [131, 140]. Further, when increasing branched arm length [134] in the building block of CTFs decreases the CO^2 absorbing capacity. Further, Jian's groups suggested that when increasing the number of cyano groups in the CTFs skeleton for example, phthalazinone-based CTFs enhance the BET – surface area 1845 m^2 g^{-1},which leads to CO_2 absorption up to 17.1 wt% (273 k, 1 bar) [164]. Janiak and co-workers found one interesting concept related to CTFs, which is that by increasing nitriles on the framework of CTFs like adamantane-based CTFs, their heats of adsorption of CO_2 are also correspondingly increased and falls in the range of 25–28 KJ mol^{-1} at zero loading. The above adsorbing capacity is eventually higher than that of the CO_2 adsorption enthalpy of any activated carbons [188].

4.2.2.5.3 Electrocatalysis for Energy Storage and Conversion

The presence of the large number of nitrogen on the building blocks of CTFs may activate the CTFs as promising electro-catalysts, but it reduces their conducting ability. To retrieve this problem Nakanishi *et al.* adopted a strategy to increase the conductivity of CTFs, by incorporating carbon nanoparticles during the synthesis of CTFs (Figure 4.21) [165]. CTFs produced using the above method are static in methanol oxidation and act as a favorable cathode material in methanol fuel cells. The copper and carbon nanoparticles incorporated CTFs exhibit better electrocatalysts [166] properties, due to their oxygen reduction reaction (ORR). Kamiya *et al.* found that the above said prepared electrode material's onset potential is high (810 mV) compared to copper-based electro catalysts in neutral conditions.

Metal-free electro catalysts for ORR of CTFs were constructed by Hao *et al.* by bottom-up method (Figure 4.22) [167]. The fusion of hetero atoms like B or F on the CTF motif enhances their activity. For, B enhances the selectivity and F enhances the activity.

4.2.2.5.4 Heterogeneous Catalysis

The presence of high BET surface area and porosity of CTF, qualifies them as an optimistic candidate for heterogeneous catalysis and catalysts support for most of the reactions [167–169]. The high chemical and thermal stability even under harsh conditions make them remain stable under the cycle performance of these CTFs.

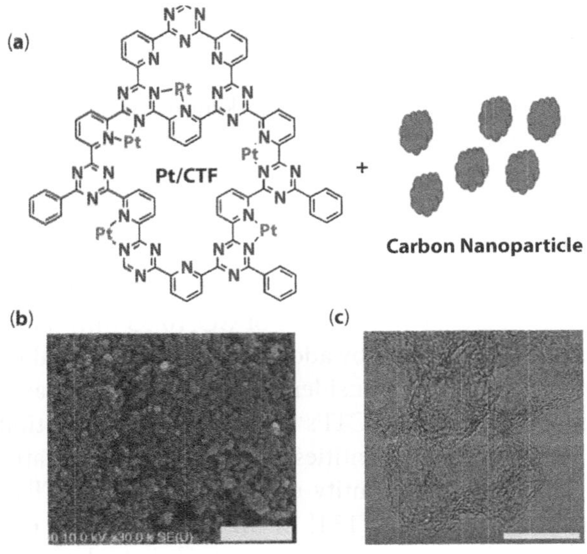

Figure 4.21 (a) Pt–CTF–CP (carbon particles) structure (b) and (c) SEM & TEM images of Pt–CTF–CP. (Adapted from Ref. [30] with permission).

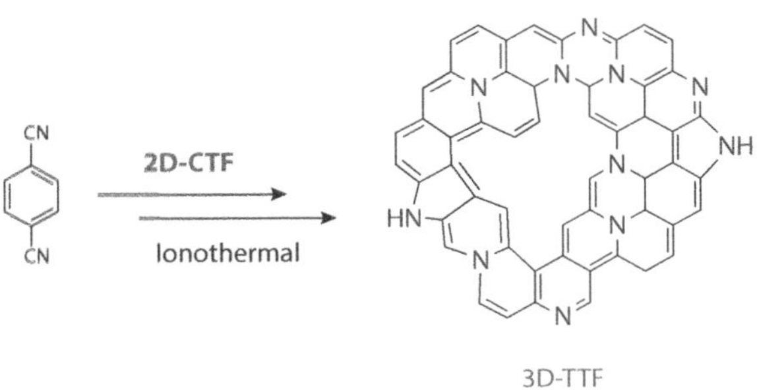

Figure 4.22 Bottom-up strategy to prepare 3D TTF (thermalized triazine-based framework) catalysts for oxygen reduction reaction by trimerization of terephthalonitrile at high temperatures. (Adapted from Ref. [167] with permission).

The catalytic performance of CTFs was further confirmed by Gascon co-workers [170] by comparing powder CTFs with monoliths coated thin CTFs and established that the monolith-supported CTFs exhibit better performance.

4.2.2.5.5 Supercapacitors and Batteries

CTF's high surface area, in addition to its nitrogen–rich structure serves as a perfect platform for super capacitors. Zhi *et al.* [171, 172] performed CTFs based super capacitors' energy storage capacity by selecting 2D-CTFs with micro porous structures and explored the correlation of nitrogen content and micro pores with super capacitance performance and revealed that hetero atom is not the only element to enhance super capacitor performance but also the micro porous structure of CTFs also enhance the supercapacitor behavior. The outline of the general applications of CTFs is shown in Figure 4.23.

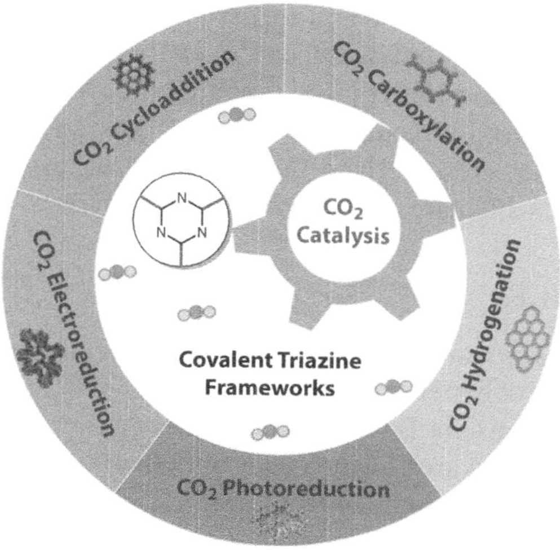

Figure 4.23 The outline of the general applications of CTFs.

4.2.3 Hyper-Cross-Linked Polymers (HCPs)

4.2.3.1 Historical Development

Hyper-cross-linked polymers (HCPs) is the one of the oldest and first invented porous polymers [173], which was established during the 1970s by Tsyurupa and Davankov. Hyper-cross-linked polymers (HCPs) possess more extensive cross linkers than traditional cross-linked polystyrene. The uniqueness of HCPs is their tenability, densely packed network, amorphous structure, and low cost.

The history of cross-linked polystyrene networks [174] was initially evolved and developed after innumerable years of research and development. Over time many researchers synthesized a number of networking polymers. The HCPs were one among the synthesized polymers in porous nature and were synthesized either by post-crosslinking using polystyrene [175] in the inflamed or by condensation of small building blocks. In the early days the prepared HCPs were used in adsorption and separation analysis.

HCPs were produced by knitting non-functional aromatic compounds (i.e., without distinctively, polymerizable groups) with external cross-linkers using iron (III) chloride as catalyst [176]. The huge array of aromatic compounds was considered external cross-linkers and act as monomeric materials and provide considerable scope for the design of HCPs.

4.2.3.2 Chemistry of HCPs

Compared to most of the existing porous polymers, HCPs are micro porous and exhibit enduring pores with high surface area [177–180] and larger volumes. The simple Friedel–Crafts chemistry is adapted to synthesize a highly dense cross-linked amorphous network of HCPs. The length of the monomer used to synthesize HCPs, plays a major role in designing the porous properties of HCPs. Han *et al.* concluded that, to get a high BET specific surface area [181–183] of HCPs, they architected 2D and 3D frameworks with non-planar rigid conformation and dendritic building blocks.

The microporous HCPs in flexible chain form, whose chain length decides the porous properties of HCPs, were synthesized by Qiao *et al.* [184] using carbazole as the precursor.

During 1930s, Staudinger and Heuer [185] prepared the Divinyl benzene (DVB)-styrene network as the first hyper cross-linked polystyrene by adapting a simple synthetic strategy. Later in 1945 D'Alelio *et al.* [186] derived the formation of ion-exchange resins through DVB-styrene as matrices, which are cited as gel-type polystyrene.

4.2.3.3 *Synthesis of HCPs*

The bis-chloromethyl groups functionalized aromatic molecules are considered as the main precursors in the synthesis of hyper-cross-linked polymers (HCPs). These precursors under optimized reaction conditions via Friedel-Crafts alkylation [6] leads a sequential order of HCPs with high a BET surface area of the order of 600–1904 $m^2 g^{-1}$ and form Davankov-type resins [187, 188] which are obtained from porous polymers.

The presence of high BET specific surface [189] area and requirement of mild reaction conditions [190] along with the inclusion of non-precious materials as catalyst [191], consider that HCPs as a widespread material in most of the industrial applications [177, 192, 193].

The general methods adapted to synthesize HCPs are solvent knitting methods [194], Scholl coupling reaction [195], the knitting method with FDA [196] functional group reaction [190, 197] and so on. Among these methods, the knitting method with FDA as an external cross-linker was first used as a cross-linker to knit aromatic building blocks [198]. The other most important monomers along with FDA used to prepare HCPs that provide outstanding porous properties are triptycenes [199] triphenylphosphine [192] benzimidazole [200], carbazole [201], 1,3,5-triphenylbenzene [200], naphthol-based monomers [194] etc.

4.2.3.3.1 Synthetic Strategy Adopted to Synthesize HCPs

Generally the formation of hyper cross linked polymers involved the post-crosslinking of polystyrene (PS) chains in the swollen state with the help of external cross-linkers. Scheme 4.12 exhibits the distinct steps involved in the course of the post-crosslinking reaction of PS. These external cross-linkers help to bridge the adjacent aromatic residues and formed a spatially non-planar cycle. The presence of numerous firm spacers present in the polymer chain holds higher order of crosslinking with low packing density and having very small pores (mean pore diameter = 2–4 nm) [202].

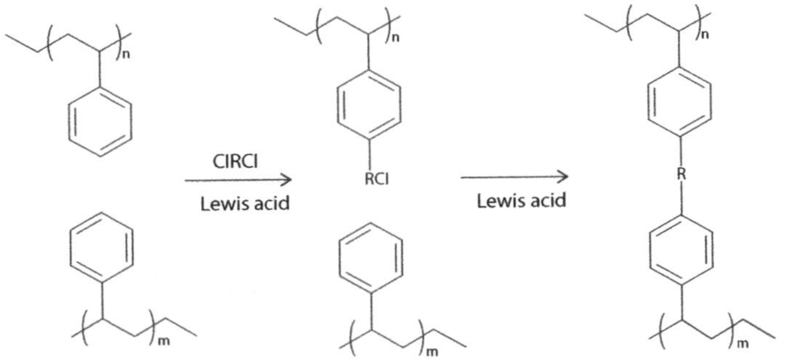

Scheme 4.12 Hyper cross linked polymer formation of PS via an external cross linker. (Adapted from Ref. [202] with permission).

Jerábek [203] prepared similar types of hyper cross-linked resins using internal electrophiles. A reactive co-monomer, vinylbenzyl chloride (VBC) along with pendent chloro methyl as a swellable monomer offered as a source of internal electrophiles in hyper cross-linking reactions. Adaptation of this strategy had better command over the physical format of formed HCPs. Scheme 4.13 represents the hyper cross-linked reaction of VBC-DVB via internal electrophile.

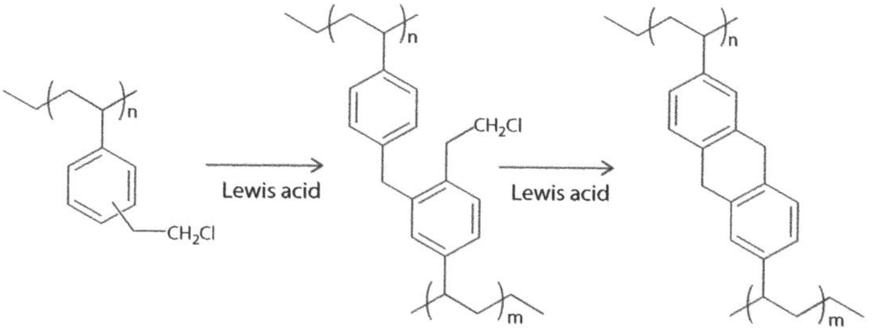

Scheme 4.13 Hyper cross-linked reaction of VBC-DVB via internal electrophile. (Adapted from Ref. [203] with permission).

4.2.3.3.2 Synthesis of a Distinct Class of Hyper Cross-Linked Polymers (HCPs)

HCPs were generally prepared through the method of post-crosslinking of polystyrenic chains/structures in an expanded/swollen state using external cross linkers. The adjacent aromatic residues present in the systems were bridged with the help of this external crosslinkers, through short, rigid (hyper)crosslinkers, and they are often installed through F-C reactions.

4.2.3.3.3 Classifications of First- and Second-Generation Frameworks of Polystyrene (Macro Porous Polystyrene and Gel-Type)

Copolymerization reactions between bulk styrene with 5%–8% of DVB [204, 205] and homogeneous gels formed hyper cross linked polystyrene polymers. Polystyrenes in linear, cross-linked (0.3%–2% DVB) and highly cross-linked (up to 30% DVB) macroporous frameworks were considered effective precursors to synthesizing HCPs [206, 207].

These precursors indulge hyper cross-linking frameworks in HCPs, along with the macro pores and micro pores formed during post cross-linking processes [188, 207, 208].

However, the cross-linking densities achieved by this method were considerably reduced, in contrast to the HCPs formed from gently cross-linked precursors due to steric constraints [206].

4.2.3.3.4 The Second-Generation Polystyrene Networks

McBurney in 1952. [209] first introduced the second-generation polystyrene networks – the macroporous polymers. Compared to the first-generation co-polymers, these second-generation copolymers hold permanent porosity and it was achieved in the presence of monomer-miscible porogen and between the reactions of DVB (6%–12%) and styrene. To retain the above structure [1] higher percentage of DVB (compared with the first-generation) is employed during the synthetic processes. The formed heterogeneous frameworks exhibit a whole pore volume of about 1 mL/g and surface area of 20–300 m^2/g. These pores are formed channels in between the zones of pores, and it can act as the best adsorbents for larger size molecules [204]. Though they are named "macro porous" (pore size >50 nm) gels [204], their approximate pore diameter falls about 20nm.

The second-generation networks named as Amberlite™ with different pore sizes [210] are commercialized by Haas and Rohm (now a part of Dow Chemical, Philadelphia, PA, USA).

4.2.3.3.5 Hyper Cross-Linked Polystyrene (Third Generation of Polystyrene Networks)

During the1970s, Davankov and Tsyurupa [208] were the first to invent the third generation of polystyrene networks known as "Davankov-type resins". They adopted a simple Friedel-Crafts type crosslinks methodology to prepare these third-generation polystyrene networks of homogeneous nature with a high degree of crosslinking (>40%).

The third generation hyper cross-linked polystyrene was prepared by stirring either linear or cross-linked precursor in a thermodynamically appropriate solvent (usually 1,2-dichloroethane) until it gets to expand. During this process free volumes between polymer chains are produced. To generate cross-linkers, Lewis acid catalysts such as $SnCl_4$ and $FeCl_3$ were added and this locked the polymer chain in the swollen state. To produce a microporous hyper cross-linked polystyrene network, the solvent used in the above steps was removed in a vacuum. Figure 4.24 describes the Schematic representation of hyper cross linked polystyrene framework formed by the post-crosslinking method.

Davankov and co-workers with the help of Friedel–Crafts alkylation synthesized high aromatic density hyper cross-linked polystyrene systems with the help of crosslinking agents (external), such as p-xylylene dichloride (XDC) [188, 209, 210], 1,4-bis-(chloromethyl)-diphenyl (CMDP),tris chloro methyl compound 1,3,5-tris-(chloromethyl)-mesitylene (CMM) and 4-bis-(p-chloromethylphenyl)-butane (DPB) [211], to crosslink polystyrene chains (Figure 4.25).

The effective crosslinking medium considered to crosslink polymer chains and small aromatic building blocks is formaldehyde dimethyl acetal (FDA) [211]. The precursors

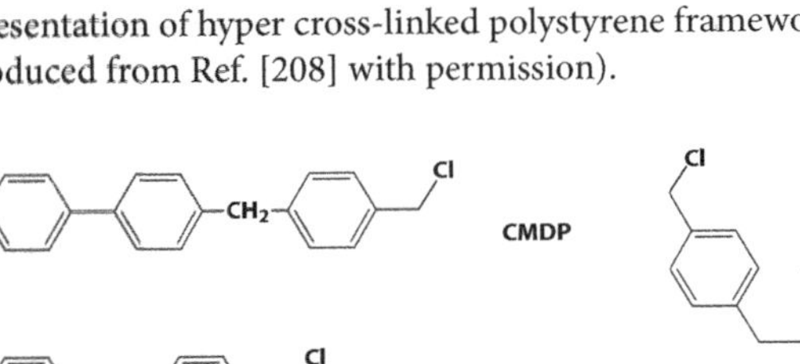

Figure 4.24 Schematic representation of hyper cross-linked polystyrene framework formed by post-crosslinking method. (Reproduced from Ref. [208] with permission).

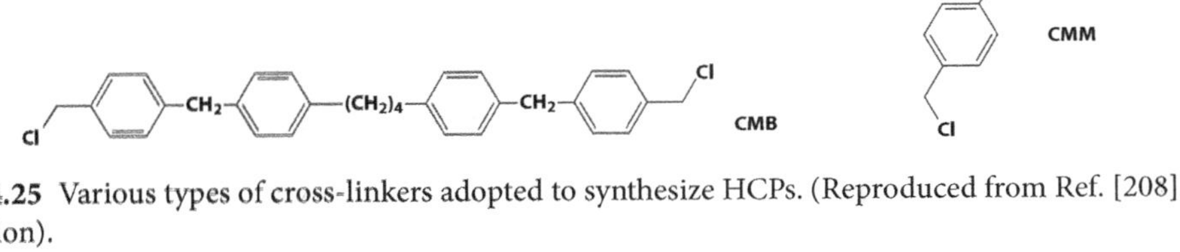

Figure 4.25 Various types of cross-linkers adopted to synthesize HCPs. (Reproduced from Ref. [208] with permission).

generally used to process these HCP formations are linear polystyrene and lightly cross-linked polystyrenes (0.3%–2% DVB) [206, 207]. For hyper cross-linking of HCPs along with the macro pores and micropores from post cross-linking were obtained from highly cross-linked [188, 208] macro porous polystyrenes (up to 30% DVB) as precursors. The precursors involved in the formation of highly cross linked HCPs, due to its bulkiness [206], it made hyper cross-linking HCPs a less denser material compared to the linear HCPs.

The third generation of polystyrene networks of HCPs involves the usage of Chloromethylated gel-type styrene-divinylbenzene resins (5% DVB) [203] (as internal electrophiles) as predecessors and the reactions were proceeded by post-cross linked manner via Friedel-Crafts reactions. The formed framework exhibited surface area up to about 1000 m^2/g. The rigid nature of chloromethyl groups restricts the formation of methylene bridges in the network due to the absence of the neighboring phenyl rings to the chloromethyl groups [212, 213].

4.2.3.3.6 Other (Divinyl Benzene) DVB Highly Cross Linked Polymer Networks

During the synthesis of styrenic-based polymers, the presence of even very little amount of 0.5% of divinylbenzene [214], indulge in rigid and short crosslinking frameworks. During the process of DVB-styrene copolymerization, permanent porosity was created after the formation of voids, when the diluents are removed. For example, the electron-deficient maleic anhydride is easily copolymerized with electron-rich styrene monomers using different polymerization processes such as suspension polymerization [215], precipitation polymerization [216], and dispersion polymerization [217]. The formed DVB-maleic anhydride networks exist in a highly cross linked manner and have great prospective applications such as magnetite supporting matrix, heavy metal adsorption [214, 218] and gas sorption [219].

4.2.3.3.7 The Effect of External Cross-Linkers on Crosslinking of Polymer Chains for the Preparation of Hyper Cross Linked Polymers

HCP networks with higher hyper cross linked polystyrenes with highly solvated polymer chains having rigid bridges were synthesized through the post-cross linking of bromomethylated polysulfones as external cross linkers via Friedel-Crafts reaction (Figure 4.26) [220]. Similarly, polyaniline with hyper cross linked fashion was synthesized by crosslinking commercially available polyaniline as external cross-linkers with diiodomethane to form methylene links between the N atoms (Figure 4.27) [221]. The BET surface area up to 632 m^2/g was achieved for the obtained hyper cross-linked poly anilines. The additional external crosslinker used in the "knitting" strategy with FeCl3 as the Friedel-Craft catalyst can be used are glyoxylic acid [222], and dibromo-p-xylene [223].

4.2.3.4 Characterization and Applications of HCP

4.2.3.4.1 General Characterization Techniques Adopted for HCP Framework

The samples (HCPs) whose characterization is to be done under vacuum (10^{-5} bar), before the samples must be outgassed for 12 h at 100°C. The specific surface area and total pore volume (Vt) of the HCP samples were analyzed by measuring their BET model.

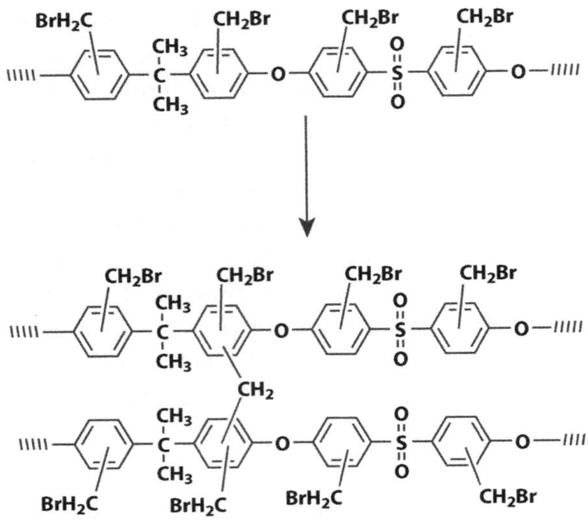

Figure 4.26 Hypercrosslinked polysulfone preparation. (Reproduced from Ref. [203] with permission).

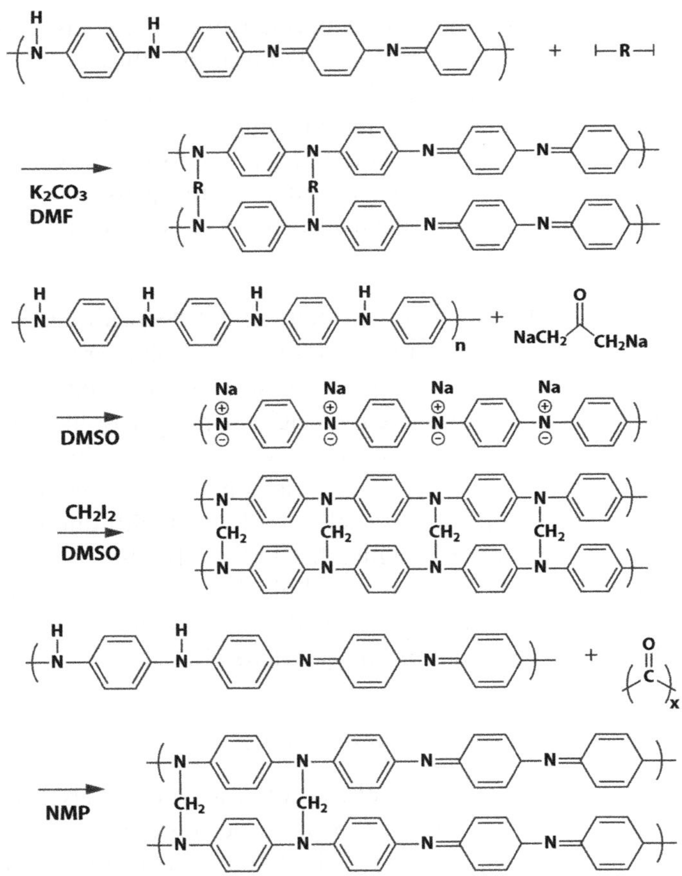

Figure 4.27 Preparation of hypercrosslinked polyanilines by diiodoalkanes. (Reproduced from Ref. [208] with permission).

The Heats of adsorption and pore size distributions of the prepared HCPs were analyzed through the density functional theory (DFT) model and CO_2 adsorption isotherms (pressure of 1.0 bar at both 273K and 298K temperatures) respectively.

Fourier transforms infrared spectroscopies (FTIR) analysis helps to find the functional groups present on the periphery of the HCP surface, thereby finding the efficiencies of HCPs, from that one can find the utility of HCPs in various applications fields. Field emission scanning electron microscope (FE-SEM) micrographs of HCPs analysis help to find the polymer network of HCPs.

From Thermo gravimetric analyses (TGA) under Ar gas (flow rate 20 mL/min) the thermal stability (from 25°C to 800°C) with a flow rate of 2°C/min of the HCPs polymer can be measured. The total elements present in HCPs were measured using Energy Dispersive X-Ray (EDX) elemental analysis.

4.2.3.5 Applications of HCPs

4.2.3.5.1 Adsorption (Gas) Properties of HCPs

HCPs are served as an excellent CO_2 captured material due to the inherent properties of micro porous structures with high BET specific surface area. These HCPs also help to separate CO_2 from flue gases [173, 224]. The high BET-specific surface area of HCPs helps to store both H_2 and methane gases.

4.2.3.5.2 As a Column Packing Material for HPLC

The size-exclusion chromatography (SEC) [225] and high-pressure liquid chromatography (HPLC) adopt HCPs with a highly cross-linked network of macro porous polystyrene as an effective packing material. Since, HCPs hold high sorption capacity and permeability make them as a perfect and efficient column packing material. The eccentric nature of HCPs, such as the ability to withstand the influence of the surrounding pH, eluent concentrations, and polarity helps to maintain mobile phase pressure [205] to hold the column bed with constant weight.

Several companies such as Dow Chemical (Optipore), Jiangsu N&G Environmental Technology (NG-99 and NG-100) and others utilized Hyper cross linked micro porous polystyrene frameworks for most of the industrial applications varying from adsorbents to water treatment technogy [4].

4.2.4 Conjugated Micro Porous Polymers (CMP)

4.2.4.1 Historical Development and Selected Advances of Conjugated Micro Porous Polymers

CMPs are one among the porous materials having a 3D network with extended π-conjugation. The micro porous CTFs [15–17], whose structures are similar to CMPs, possess extended π-conjugation with different formation chemistry. Micro porous poly(arylene ethynylene) frameworks were the initially reported CMPs having BET surface areas in the range of 834 $m^2\ g^{-1}$ during 2007 [12]. Various forms of CMPs were prepared by adopting different synthetic strategies. For example, by varying the monomer alkyne struts range in the synthetic pathway of CMPs, results in a higher surface area up to $1000m^2g^{-1}$. By adopting homo coupling routes for the synthesis of CMPs, it produces amorphous butadiyne containing network [226] with both micro/meso porous hybrid characteristic nature. Weber and Thomas *et al.* [227] prepared CMPs with higher photoluminescence properties. Organo lithium alkylation routes were adapted to synthesize higher hydrophilic natured CMPs [228]. CMPs as light-harvesting materials also had been prepared by the Jiang group in 2010 with the help of poly phenylene. CMPs as molecular sensing devices were also been prepared using carbazole-based precursors. CMPs with hyper-branched structures were also prepared using the tert-butyl functionalized group. "Metal-organic CMPs" were also prepared by post-heating bi pyridine–functionalized CMPs [229] with metal complexes (Rhenium, rhodium, and iridium) or with metal–organic monomers through Sonogashira-Hagihara cross-coupling reactions.

In 2008, Liarg and co-workers produced large-area and defect-free CMP membranes on a bromobenzene-functionalized surface and used them for organic-solvent nanofiltration [230].

4.2.4.2 *Design and Synthetic Strategy Adopted for Synthesizing CMPs*

The options available to customize CMPS are limitless. By simply changing the precursor monomer geometry or by assimilating various metals and heteroatoms, one can tune pore structure, morphology, and optoelectronics. There are a number of coupling reactions had been utilized for synthesizing CMPs. The strategies involved in coupling reactions are discussed in the following sections.

4.2.4.2.1 Sonogashira-Hagihara Coupling

Sonogashira-Hagihara cross-couplings method was first adapted to prepare CMPs. In this reaction, the alkyne-containing monomer is usually coupled with aryl halides in the presence of palladium as the main catalyst and copper as co-catalyst to give tetrakis (triphenylphosphine) palladium. The Sonogashira-Hagihara reaction also adapted homo coupling [226] by using alkyne-containing monomers in the presence of oxygen [231].

4.2.4.2.2 Suzuki-Miyaura Coupling

The commercially available boronic acids with mild base (K_2CO_3) in DMF/water [232] in the presence of Pd(0) (catalyst) are considered favorable conditions to effect Suzuki-Miyaura cross-coupling reaction. The reaction is oxygen-sensitive, hence the reaction should be carried out under oxygen-free conditions [233].

4.2.4.2.3 Yamamoto Coupling

Yamamoto coupling is a very simple reaction that proceeds with halogen–functionalized aryl monomers as a precursor with a minimum of 3 reactive sites and is coupled together with bis(cyclooctadiene)nickel(0) (Ni-(cod)2) [234]. The reaction is very sensitive to water, hence the entire reaction must be carried under the glove box.

4.2.4.2.4 Heck Coupling

This type of coupling reaction involves the utilization of primary alkene with unsaturated halide with K_2CO_3 as base and DMF a solvent in the presence of Pd, ($Pd(Ph_3)_4$) as catalyst under anaerobic conditions [235]. This reaction is also known as the Mizoroki-Heck reaction. The incorporation of C=C bonds take place between a primary alkene and unsaturated halide in the presence of a Pd acts as a catalyst as well as the base.

4.2.4.2.5 Schiff–Base Condensations

Schiff-base reactions involve metal-free synthesis, with the formation of imine based bond between amine and aldehyde–functionalized monomers [236]. Hence, this type of reaction avoids metal residues in the CMP network. These Schiff–based reactions incorporate nitrogen in the CMPs network, which helps to adsorb CO_2 under diverse conditions [237, 238].

4.2.4.2.6 Heterocycle Linkages

Rabbani and El-Kaderi impose benzimidazole linkages in CMPs preparation by condensing ortho diamines in the presence of an aldehyde [239]. In DMF under inert conditions and further react in the presence of atmospheric air to produce the porous polymer. Thiazolothiazole-linked CMPs can be prepared by this method using dithiooxamide and an aldehyde monomer. The reaction can proceed either through the solvothermal method using DMF at 160°C [241] or through ordinary reflux methodology using the same DMF or nitrobenzene at 150°C [240].

4.2.4.3 Characterization of Conjugated Microporous Polymers (CMP)

The characterization of conjugated microporous polymers (CMP), which is in nanoporous regime is not an easy task as in the case of other porous polymers. The most important methods for analyzing CMPs are finding their gas adsorption property [22, 242, 243] to confirm their porosity nature, are positron annihilation lifetime spectroscopy (PALS), scattering methods, and modeling approaches. The total porosity φ and the specific surface area of any system can be accessed through Gas adsorption and the Brunauer–Emmett–Teller (BET) method [242] respectively.

Further, to confirm the micro-porous nature as well as pore shapes and sizes of CMPs, their Density-functional-theory (DFT) measurements like non-linear DFT (NLDFT) and quenched-solid DFT (QSDFT) [244, 245] methods were generally adopted.

Similarly, CO_2 adsorption [246] at 273 K is considered another tool to confirm the micro porous nature of CMPs. Zukal and co-workers confirmed the micro porous natured CMPs, by analyzing their N_2 and CO_2 adsorption [247].

4.2.4.3.1 Spectroscopic Methods of Analysis of CMPs

Spectroscopic methods, especially nuclear magnetic resonance (NMR) and infrared (IR) techniques give valuable information on the porosity such as pore sizes. The use of pulsed-field gradient NMR for the analysis of diffusion in microporous materials also helps to analyze micro porosity present in CMPs [248, 249].

The contribution of modern [250] Xe-NMR technology to analyses the properties of microporous polymers is enormous. The chemisorption of CO_2 over CMPs was confirmed by their FTIR spectral analysis, which also justifies the micro porous nature of CMPs [251–253].

Positron Annihilation Lifetime Spectroscopy (PALS) analysis was helpful to analyses the presence of free volumes present in the polymer space of micro porous CMPs [254, 255].

4.2.4.4 *Applications of CMPs*

The existence of eccentric nature of CMPs, such as their extended conjugation, high porosity and tuneable chemistry extend their applications in many fields. The prime applications of CMPs include gas storage and separations [143, 256], encapsulation of chemicals, [257, 258], heterogeneous catalysis [259, 260], photo redox catalyses [261, 262], light emittance [263], chemosensing, [235, 264], energy storage [265, 266], biological [267], and photo catalytic H_2 evolution [268, 269]. The vast variety of applications of CMPs depends on the presence and nature of extended π-conjugation in these materials.

4.2.4.4.1 Gas Storage and Separation

The synthetic force over structure and composition in CMPs offers them an effective material for the adsorption and storage of gases [270, 271]. By customizing the shape and size of the pores of CMPs, it can also be used for separation applications [272]. By tuning the CMPs structure, its efficiency to gas storage capacity also increases considerably. For example, poly(aryleneethynylene) based CMP network's storage capacity of H_2 is high due to their higher BET surface area of 834 $m^2 g^{-1}$. The higher BET surface area of CMPS extends their ability to CO_2 sorption in the range of 0.97 mmol g^{-1} at 298 K.

4.2.4.4.2 Photoredox Catalysis

In 2013, Vilela *et al.* prepared porosity-dependent [273] benzothiadiazole-based CMPs, which oxidize α-terpinene into caridole in 96% conversion. In general photocatalytic activity is directly related to the surface area. To convert primary amines into mines Shin group [262] suggested that benzodifuran-containing CMPs act as an effective photoredox catalyst.

4.2.4.4.3 Light Emitting Property of CMPs

The presence of extended π-conjugation in CMPs act as luminescent materials. The existence of rigid building blocks with restricted rotation of phenyl rings present in CMPs avoids fluorescent quenching, thereby luminescent properties of CMPs are enhanced considerably. Since, the band gap of the CMPs is tuneable [263], the corresponding luminescent properties are also controlled accordingly.

The above facts were confirmed by the reactions of pyrene-based CMPs with 1,4-dibromobenzene (DB), 1,3,6,8-tetrabromopyrene (TBrPy) and/or 4,4′-dibromobiphenyl (DP) to form YDBPy, YPy, YDPPy and SDBPy. The first letter in all the above abbreviations are, the first letter represents whether a Suzuki-Miyaura or Yamamoto reaction was used (Figures 4.28 a-f). The fluorescence with different emission colors ranging from orange (YDPPy) to red (YPy) and yellow (YDBPy) to green (SDBPy).

4.2.4.4.4 Chemo Sensors

The presence of extended π-conjugation in CMPs considered them an effective luminescent materials. The above properties made CMPs as chemo sensors, since they can sense different types of chemicals by quenching their fluorescence. The prime nature of CMPs as holding chemo sensors ability over nonporous conjugated polymers are mainly because they possess the maximum number of open pores on CMPs' surface. These pores provide space for chemicals to interact with them and this leads to enhanced signal sensitivities.

Figure 4.28 Synthesis of luminescent pyrene-based CMPs. (a) YPy, (b) YDPPy, (c) YDBPy, and (d) SDBPy. (e) images of suspensions of the CMPs in THF (10 mg/10 mL). (f) Photoluminescent spectra of the monomer TBrPy and the formed polymers were recorded in solid-state powder (λexcit = 360 nm). (Adapted from Ref. [263] with permission).

The CMPs' chemo-sensing ability was first invented by Jiang *et al.* [264] with the advent of its, π-conjugated, porous, and photoluminescence properties.

For example, TCB-CMP when excited at 368 nm, it exhibits blue luminescence with a quantum yield of 10%. TCB-CMP's fluorescence quenching is achieved by exposing 2-nitrotoluene, 1,4-benzoquinone, 2,4-dinitrotoluene or nitrobenzene vapors (Figure 4.16) over TCB-CMP. As a result, a higher pore volume of 0.923 $cm^3 g^{-1}$ with a larger BET surface area of 1280 $m^2 g^{-1}$ was achieved for TCB-CMP compared to its linear counterpart.

4.2.5 Porous Aromatic Frameworks (PAFs)

4.2.5.1 Historical Development of PAF

Teng Ben *et al.* [18] in 2009 prepared material with ever high BET surface area of 5600 $m^2 g^{-1}$ among the existing porous materials in highly ordered frameworks is porous aromatic frameworks (PAFs).

"Aromatic", the term arises due to the presence of a large number of aromatic rings in this prototype, called PAF-1. The first PAF was prepared with the help of tetrakis(4-bromophenyl)methane using Yamamoto type Ullmann coupling reactions as shown in Figure 4.29.

Figure 4.29 Yamamoto and Ullmann type coupling reaction for the formation of PAF-1 from tetrahedral building units. (Adapted from Ref. [18] with permission).

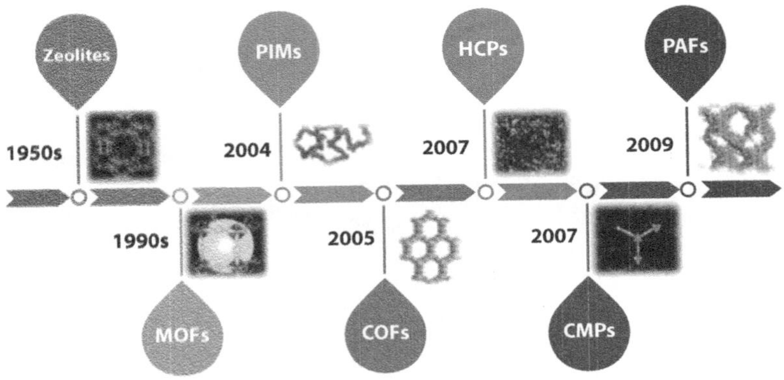

Figure 4.30 A transformation sketch of the time-bound growth of porous solids along with new inventions. (Adapted from Ref. [284] with permission).

In the recent past the development of PAFs and their related studies increased considerably. The other porous organic frameworks related to PAFs are porous organic networks (PONs) [274, 275], porous organic frameworks (POFs) [276, 277], covalent triazine frameworks (CTFs) [15, 278], porous polymer networks (PPNs) [270, 279] porous organic polymers (POPs) [280, 281] and covalent organic polymers (COPs) [282, 283] are among them.

Polymers of porous nature are generally synthesized by random combinations between monomers, whereas to synthesize PAFs, a special synthetic strategy was adapted. PAFs were built with predesigned geometries from firm constructing units. According to the terminological point of view, these PAFs are considered frameworks rather than polymers.

The porosity of PAFs is extremely high, among the existing amorphous porous materials, PAF-1 lead to a major turning point in the history of engineering modern days' porous organic frameworks and are depicted in Figure 4.30. The presence of an extraordinary high surface area amid the amorphous porous materials, PAFs have some unique and significant advantages.

Researchers in all parts of the world focused their research on the porous organic framework and their search increases day by day. According to the Web of Science [284], search results conclude the above fact, such as till September 2019, there are more than 1600 publications that matches either of the following headings as "porous aromatic frameworks", "porous organic frameworks", "porous organic polymers", "porous polymer networks", "covalent triazine frameworks", "porous organic materials", "porous organic networks", "conjugated micro porous polymers", and "microporous polymer networks".

But "metal-organic frameworks" and "covalent organic frameworks" were not included in the search. Similarly, the fast expansion of this field has been evidenced by Web of Science search and the increase in the number of annual citations is from 19 (2008) to 12,000 (2018). The following sections provide the details of the essential development of PAFs in over 10 years.

4.2.5.2 Chemistry of PAF

Porous aromatic frameworks (PAFs) unique nature such as exceptionally high surface areas having rigid frameworks with the existence of networks of carbon–carbon–bond–linked aromatic-based building units, are considered as salient porous solids. By modifying the aromatic motifs in PAF framework, desired functionalities can be established in it. The presence of strong and rigid carbon–carbon bonding frameworks in PAFs makes them stable in severe chemical treatments such as acidic, alkaline, and humid environments. Hence it can be used under any crucial conditions and extend its applications in most of industrial uses.

PAFs can be prepared from diverse building units adapting different types of coupling reactions. Simply performing the combination of distinct shapes and sizes with different building units and exhibits numerous porous skeletons varying in their pore properties such as volume and size etc.

4.2.5.3 Design Principles and Synthetic Strategy Adopted to Synthesize PAFs

PAFs were designed and synthesized based on multiple factors, such as their structure, porosity, a diameter of the pores, reaction mechanisms, the degree of framework of the atoms and the geometries of the building unit. Further, the molecular shapes and steric configurations around the reaction site influence the synthetic pathway.

Design principles of PAFs were evolved by considering the formation of PAF-1 as reference material. The relationships between the surface area of the resultant structure and the reaction mechanism involved in the formation of respective PAFs decide their synthetic strategy. By proper designing of the framework through the processes of topology, design and "reticular chemistry" of the required PAFs one can control the porosity and adsorption properties of the PAFs, which are pre-request qualities and must be adapted to synthesize PAFs with desired properties. The various structure model of synthesized and simulated PAFs are summarized in Figure 4.31.

4.2.5.4 Synthesize of PAFs

4.2.5.4.1 Yamamoto Type Ullmann Coupling Reaction

PAFs were generally constructed with the help of rigid organic building units containing multiple reaction sites. The most accepted synthetic strategy to synthesize PAFs is the Yamamoto type Ullmann reaction. The above type reaction utilizes the de halogenation poly condensation of aryl halides using Ni(0) complex [Ni(0)Lm] as a catalyst. The aryl bromide or aryl chloride undergoes an oxidative addition reaction with the catalyst Ni(0) Lm and forms complexes I and II as shown in Figure 4.32. After some time the disproportionation of I and II takes place and ends in the formation of new complex III and nickle(II) halo genide. The aryl–aryl addition compounds [285] were regenerated from the original Ni(0)Lm complex by the reductive elimination of III. Synthesizing PAFs from this Yamamoto-type Ullmann reactions eliminates the bulky halogen atoms from the building unit of PAFs [18], this ultimately results in unexpectedly high surface area [286] of the prepared PAF. The distinctive nature of Yamamoto-type Ullmann reactions is exhibiting higher yields of the products with high porosity natured PAFs, which makes this reaction very popular among the other methods. Because in this reaction only the nickel-based catalysts were completely removed by simple acid treatments, whereas, this is not possible in another type of coupling reactions. PAFs prepared by this method hold BET surface areas as higher as 4000 $m^2\ g^{-1}$ [18, 287–290] and possess ultrahigh porosity.

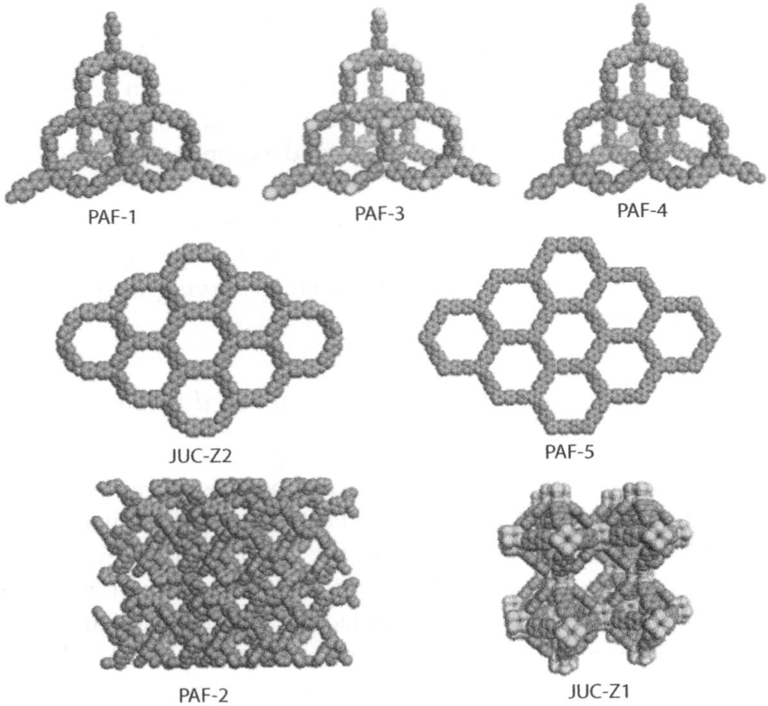

Figure 4.31 Schematic representation of synthesize of PAF and their simulated form (C, purple; N, blue; Si, yellow, O, green, Ge, brown). (Adapted from Ref. [284] with permission).

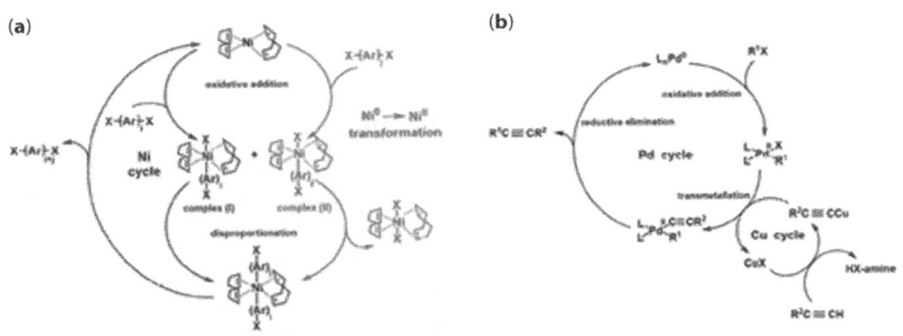

Figure 4.32 Reaction mechanisms involved in the coupling reactions of (a) Ullmann type, Yamamoto coupling and (b) Sonogashira cross-coupling using Pd-catalyst. (Reproduced from Ref. [18] with permission).

4.2.5.4.2 Other Coupling Reactions

Suzuki–Miyaura cross coupling reaction is considered another coupling reaction and it assists the aryl-aryl bond formation between aryl halides and aryl boronic acid in the presence of Pd-based catalysts [291] during the PAF-1 synthesis. Another commonly used coupling reactions involved in PAF synthesize is Sonogashira cross-coupling. The above reaction involves the copolymerization of aryl halides having terminal alkynes and it was activated by a copper species [292] with palladium as catalyst.

The cross-coupling reactions are advantageous compared to homo-coupling Ullmann reactions because it connects two different types of building units with distinct geometries, which helps to syntheses PAFs with diverse topologies.

The number of coupling reactions available to modify the porosity of PAFs is Heck reaction [235, 293, 294], 1,3-diyne formation [269] Stille coupling, [295, 296], and others are also used to form carbon–carbon connections [297–301].

The hydroxyl functionalized PAF-18-OH [302] were synthesized through Sonogashira reaction by copolymerization of hydroxyl functionalized tribromobenzene and the basic building units of ethynyl functionalized tetrahedral tetrakis(4-ethynylphenyl)methane.

The carboxyl-infixed PAF-26-COOH [303] was synthesized by replacing the hydroxyl functional building unit in PAF-18-OH with dibromobenzoic acid. The presence of electron-donating sites in both cases, easily make bonds with light metals, such as Mg, Li and Na and hence enhance their hydrogen storage capacities.

4.2.5.4.3 Cyclo Trimerization Using Cyano Group

In porous organic frameworks, the existence of aromatic cyclic motifs act as a backbone of the most of the precursors, and induces everlasting porosity. PAFs with tetrahedral building skeletons such as PAF-2 and PAF-16 can be obtained from tetrakis(4-cyanophenyl)methane [304] and cyano cyclo trimerization of tetrahedral tetrakis-(4-cyanophenyl)silane [144] respectively. Another attractive route to develop porous frameworks [305–307] of PAFs with Cobalt-catalyzed acetylene trimerization with benzene ring is shown in Figure 4.33.

4.2.5.5 PAF Characterization

4.2.5.5.1 Difficulties and Challenges

The PAF structures are designed according to topology and their simulation properties and these two properties formed a framework of consistent and constant units. The existence of the amorphous nature of PAFs is due to the presence of irreversible nature of the connecting unit present in its frameworks. The XRD pattern of PAF confirmed their amorphous nature. Trewin *et al.* [286] confirmed their amorphous nature by accounting for their disordered three-dimensional networks lead to an energetically most stable structure.

By nature, PAFs possess properties such as, the ability to expand their frameworks extensively and are quite insoluble in most of the common solvents, making them stand different from most of the existing linear polymeric chains or discrete molecules. Hence it is difficult for analyzing PAFs properties. But PAFs are synthesized from the simple composition of reacting molecules and formed aromatic frameworks, many other characterization methods like FTIR, NMR and TGA analysis were performed to confirm the PAF formations.

4.2.5.5.2 Fourier Transform Infrared Spectroscopy

To provide structural information on the synthesized PAFs, their FTIR were recorded and their formation confirmed.

The presence of functional groups on the surface of PAFs was confirmed by recording their FTIR spectral analysis. For example, the existence of the C=O and O–H stretching vibration at 1728 cm^{-1} and 3431 cm^{-1} respectively during the synthesis of the carboxyl group anchored on PAF-26-COOH [303] through the Sonogashira cross-coupling reaction

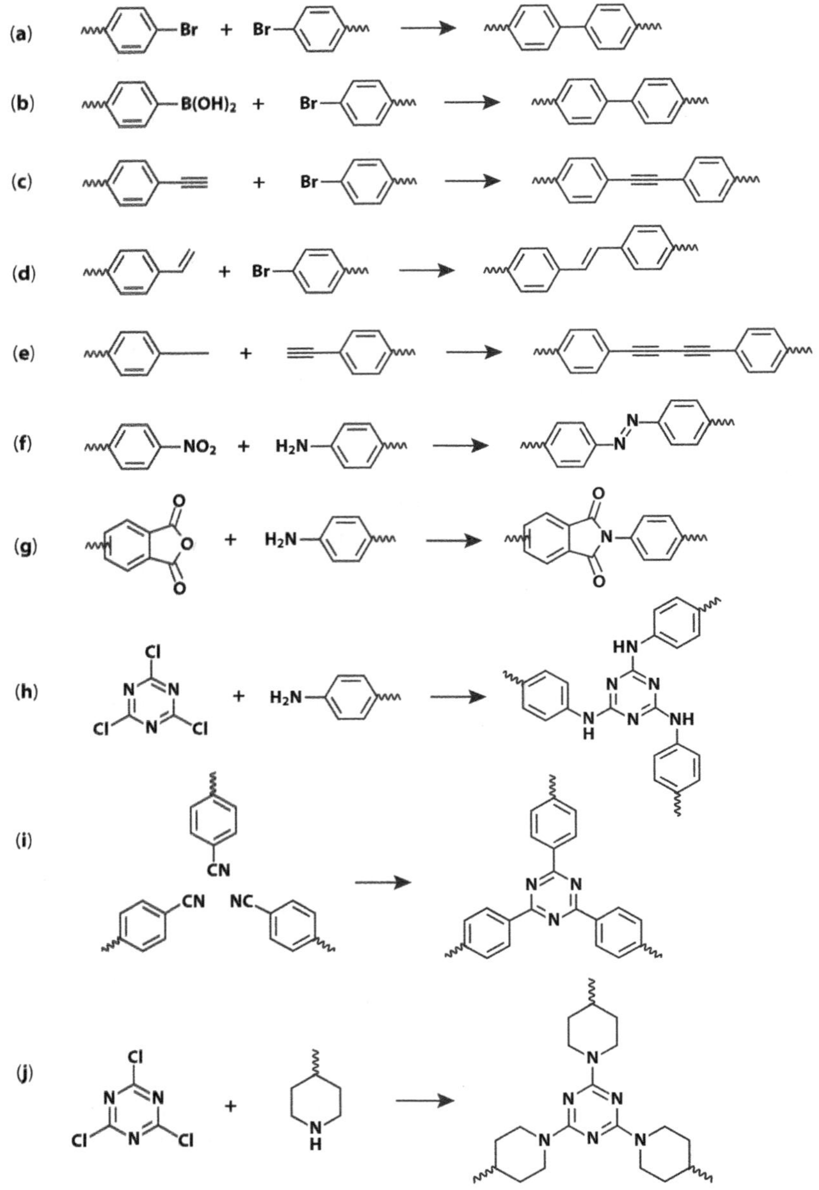

Figure 4.33 Generally adopted coupling reactions involved in porous framework formation (a) Yamamoto-type Ullmann coupling, (b) Suzuki–Miyaura cross coupling, (c) Sonogashira–Hagihara cross coupling, (d) Mizoroki-Heck cross coupling, (e) oxidative Eglinton coupling, (f) base-mediated azo formation, (g) imidization reaction, (h) nucleophilic substitution reaction, (i) cyano cyclotrimerization, and (j) nucleophilic substitution reaction on piperidine. (Reproduced from Ref. [304] with permission).

confirms the presence of carboxylic acid groups during the cross-coupling process and remain unaltered in the entire reaction pathways.

4.2.5.5.3 Nuclear Magnetic Resonance

The magic angle spinning nuclear magnetic resonance (MAS NMR) technology provides information about the existence of short-range order arrangements in the solid state [308] framework of PAFs. Further, the presence of aromatic motifs in the surrounding of PAFs was confirmed by recording their ^{1}H and ^{13}C solid-state NMR studies.

4.2.5.5.4 Thermogravimetric Analysis and Elemental Analysis

The stability of the PAFs framework was analyzed by recording their decomposition temperature through TGA measurements, and it was found that PAFs decompose at 520 °C. TGA analysis also provides the presence of the metal residues in the highly porous framework.

The presence of organic frameworks in PAFs is confirmed by the existence of a sample is almost 0 when the weight residue of PAF-1 [309] reaches a temperature above 600°C. The CHNS analysis of PAFs provide compositional information. The complete quantitative and qualitative information regarding PAFs were obtained from inductively coupled plasma spectroscopy (ICP) and Energy-dispersive X-ray spectroscopy (EDX) analysis.

4.2.5.5.5 Porosity Measurement

The microporous natured PAFs (less than 2nm) emanate from the establishments of structure units at the molecular level. The presence of volatile impurities and solvents in the network of PAFs were eliminated by heating under a vacuum to analyze their pore size, volume and surface area. For PAFs with organic frameworks, heating has to be done carefully to avoid burning the organic framework. For inorganic-based framework, require certain pre-treatments to the complete removal of unreacted impurities. The porosity nature of PAFs framework was analyzed by degassing the sample at a constant temperature through the evaluation of an adsorption–desorption isotherm. The equilibrium adsorption amount of an adsorptive from isotherm analyzed as a function of the relative pressure P/P0 N2 at standard temperature and pressure (STP) [71, 310].

4.2.5.6 Applications

4.2.5.6.1 Gas Adsorption

The inherent nature of PAFs, such as surface area, density of active sites and pore size influence gas adsorption properties in a remarkable manner compared to most of the porous polymer networks. Thus PAFs hold maximum adsorptive property of capturing multiple gases in a concurrent manner [270, 287, 311, 312]. The high porous natured framework along with active absorbing sites of PAFs, makes them an effective gas adsorbent. The presence of covalent linkage between the adjacent frameworks of PAFs justifies their higher thermal stability.

4.2.5.6.2 Methane Adsorption

The high surface area and stability of PAFs, explore them as an efficient methane absorber. PAFs have ability to exhibit hydrogen and CH_4 storage capacity through the heats of adsorption between 15 and 20 kJ mol^{-1} at ambient pressure and temperature of 273 K. PAFs also helps to sequester both CO_2 and CH_4 from the atmosphere, thereby reducing the greenhouse effect [313].

4.2.5.6.3 Adsorptive Separation of Hydrocarbon Mixtures

PAFs are used to separate hydrocarbon mixture by establishing active binding sites into PAFs via their adaptable networks. Since the separation of hydrocarbon mixture mainly depends on the presence of active binding sites on the framework. PAFs are designed and synthesized according to the adsorptive hydrocarbon selectivity. For example porphyrin-based PAFs have a higher affinity towards the hydrocarbon propane than methane due to their micro porosity nature. As a result, these porphyrin-based PAFs offer higher binding affinity to larger molecules, and thus ensure the separation of hydrocarbon mixtures [314].

4.2.5.6.4 Capture of Ammonia

PAFs possesses high physicochemical stability to strong bases and acids, along with effective BET surface area and porosity makes PAF as potential ammonia adsorbents. The presence of ammonia binding sites on the functionalized PAFs increase ΔH_{ads} through the chemical bond formation and it leads to capture ammonia selectively at low concentrations. This binding of ammonia with PAFs provide pathway for multiple chemical interactions. It is achieved by synthesizing functionalized PAF-1 containing functional groups such as sulfonate and carboxyl [315] on the periphery of PAFs framework to capture ammonia. Since ammonia is a Lewis base, it can be easily captured by Lewis acid carboxyl functionalized PAFs.

4.2.6 Porous Organic Cages

Porous materials with discrete molecular structures are considered as real-time alternatives to most of the existing porous materials. Porous organic cage (POC) crystals are considered one of the high specific surface materials in the desolvated state. Research in the corresponding field is progressing to understand their porous nature, structure and wide applications. Supra molecular approach has been adopted to assemble porous organic cage frameworks. The presence of bulky directing groups on the POC conquers the crystal packing. This ultimately creates porosity in an 'extrinsic' manner between the intrinsically porous cage units. Typically 'intrinsic' porosity means porosity existing within the molecule, while 'extrinsic' porosity means porosity existing between the molecules. BET surface area in the order of 854 m^2 g^{-1} exists for typical POCs, which is quite higher than that of most of the un-functionalized cages of the same dimensions.

The existence of pure organic cages is rare compared to coordination cages, which may be due to the formation of covalent bonds formed in the irreversible reactions that do not accept a structural "self-correction" often go along with limited yields in the cyclization step.

In recent times literature survey reveals that there are the large number of covalent bonds involved in the formation of organic cage compounds and are formed in a reversible manner from small and easily available precursors with reasonable to very high yields [316]. Some research findings depict that POC those which exists either in an amorphous or crystalline state will be desolvated in a vacuum [21], and exhibit higher BET surface areas even up to 2071 m^2 g^{-1} [317]. This soluble nature of POC offers some advantages over insoluble network materials [318]. For example, by adjusting the properties of the materials [319] the solubility of porous molecules exhibits some unique properties.

In POCs, the presence of functional groups on the surface enhances the gas adsorption properties by modifying functional groups with suitable chemical reactions. The inner part of the cage molecules due to the presence of functional groups on its periphery enhances the gas sorption properties in a remarkable manner [320]. These POCs also act as a sensor for volatile analytes [321] and they can also be embedded in most of the membranes [322] for various applications.

4.2.6.1 *Characterization of Organic Cages*

POCs being a mesoporous natured material (>2 nm), collapse the formed molecule upon desolvation, hence these POCs could not be in a permanent porous nature [323].

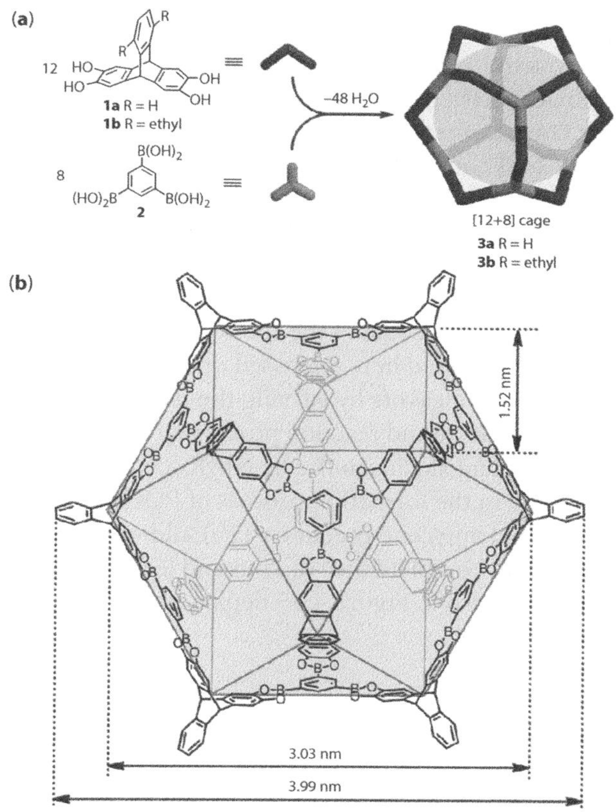

Figure 4.34 (a) Triptycene tetraol (twelve molecules) of 1a or 1b with eight molecules of the triboronic acid 2 to the formation of cuboctahedral [12+8] cages 3a and 3b respectively. (b) The cuboctahedral cage with outer (3.99 nm) and inner (3.03 nm) diameter. Edges of the cuboctahedral geometry are highlighted in red. (Reproduced from Ref. [320] Copyright 2014 Wiley-VCH Verlag GmbH & Co. KGaA, Weinheim).

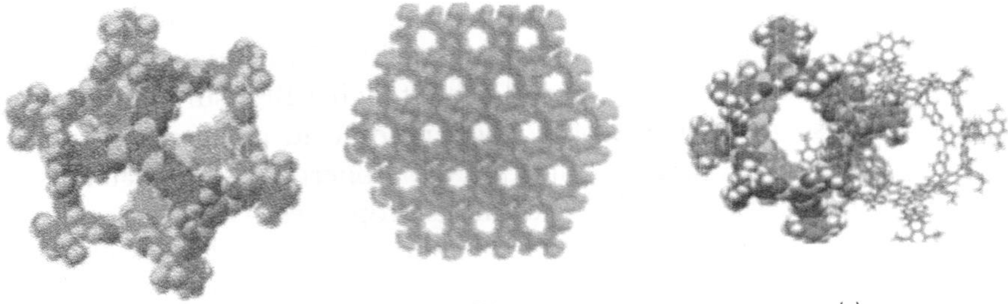

(a) (b) (c)

Figure 4.35 Single-crystal X-ray structure of solvated cage (a) Space-filling model. C gray, H white, O red, B yellow. (b) Molecular packing of cage 3b in solid state. One cage is represented as a stick model. (c) Packing model in the solid state. The yellow surface represents a solvent-accessible surface with a probe radius of 1.8A°. (Reproduced from Ref. [323] with permission).

Diffusion coefficients measurement of DOSY-NMR spectroscopy [323] measurements suggest that organic cage compounds possess solvodynamic diameters, that fall in the range of 4 nm, which was confirmed by Warmuth and co-workers. Single-crystal X-ray analysis of the formed POCs does not provide information regarding porous nature. But Cooper group [324] prepared [4+6] cage compound [325] with an inner and outer diameters as 1.5 nm and 2.9 nm respectively. Gawronski *et al.* prepared POCs with inner and outer diameter in the order of 1.7 nm and 3 nm respectively. These results were confirmed by the combined involvement of dynamic imine bond formation in POCs along with supramolecular chemistry.

The formation of a 48-fold condensation reaction in one-step between eight molecules of the triboronic acid [323, 326] and twelve molecules of triptycene tetraol 1a or 1b to the cuboctahedral [12+8] cages 3a and 3b (Figure 4.34a), respectively. b) The formation of the cuboctahedral cage with selected dimensions as a scaffold, its inner and outer diameter fall in the range of 3.03 nm and 1.52 nm respectively with the existence of open quadric windows of 1.52 nm. In Figure 4.34, the edges of the cuboctahedral geometry are highlighted in red color. The dimensions mentioned here are based on an AM1-optimized model.

However, in all consecutive endeavors by varying the reaction parameters such as nature of reactants strength, solvent nature and reaction processing temperature with the triboronic acid 2 did not give any significant signals in the ^{1}H NMR spectrum of the original product.

The intermediates formed in the formation process of POCs are insoluble, hence further reactions are stopped. In the triptycene precursor (1b) addition of two ethyl chains in the 13th and 16th positions helps to complete the reaction and leads to the formation of cages 3b from the corresponding precursors. Figure 4.35 depicts the single-crystal X-ray structure of solvated cage.

4.3 Conclusions and Perspectives

The resilience nature of Porous Organic Polymers (POP), indulges scientific communities to spread their usage in various fields. Since, porous organic polymers possess the advantages of both porous materials and conventional polymers, expand their widespread applications in the fields of material, chemistry and biology. The distinctive nature of porous polymers, such as the ability to hold high thermo-stability, low density, high surface area and adjustable pore size, are considered new kinds of promising porous materials. These porous organic polymers generally exist in two different morphological arrangements as amorphous and crystalline nature. The present review discussed elaborately the classifications of these POPs, based on their pore size, volume and morphological arrangements of the constituent atoms. The porous organic polymers considered for the present discussions are covalent organic frameworks (COFs), covalent triazine frameworks (CTFs), hypercross-linked polymers (HCPs), porous aromatic frameworks (PAFs), extrinsic and intrinsic porous molecules and porous organic cages. Among the above polymers HCPs and COFs exists in crystalline frameworks, whereas CMPs, CTFs and PAFs exist in the amorphous framework. These POPs' history, chemistry, their inner classifications, the synthetic strategy adopted in each type of POPs and their applications are discussed elaborately. Each kind of porous organic polymers discussed in the present review exhibit one way or another express their distinctive nature and is utilized for various applications, including gas storage, separation of gases mixtures, as heterogeneous catalysis and so on. These POPs with tailored structure and tuneable natured porous size, shape and morphology bring them one of the most important and unavoidable materials in most of industrial applications. These POPs will rule the world in near future and makes the scientific community feel proud.

References

1. Rouquerol, J., Avnir, D., Fairbridge, C.W., Everett, D.H., Haynes, J.H., Pernicone, N., Ramsay, J.D.F., Sing, K.S.W., Unger, K.K., Recommendations for the characterization of porous solids. *Pure Appl. Chem.*, 66, 8, 1739–1758, 1994.
2. Giri, N., Del Pópolo, M.G., Melaugh, G., Greenaway, R.L., Rätzke, K., Koschine, T., Pison, L., Gomes, M.F.C., Cooper, A.I., James, S.L., Liquids with permanent porosity. *Nature*, 527, 216–220, 2015.
3. S.V. Rogozhin, V.A. Davankov, M.P. Tsyurupa, Patent USSR 299165, 1969.
4. Tsyurupa, M.P. and Davankov, V.A., Porous structure of hypercrosslinked polystyrene: State-of-the-art mini-review. *React. Funct. Polym.*, 66, 768–779, 2006.
5. Lee, J.Y., Wood, C.D., Bradshaw, D., Rosseinsky, M.J., Cooper, A.I., Hydrogen adsorption in microporous hypercrosslinked polymers. *Chem. Commun.*, 2670–2672, 2006.
6. Wood, C.D., Tan, B., Trewin, A., Niu, H., Bradshaw, D., Rosseinsky, M.J., Khimyak, Y.Z., Campbell, N.L., Kirk, R., Stöckel, E. *et al.*, Hydrogen storage in microporous hypercrosslinked organic polymer networks. *Chem. Mater.*, 19-8, 2034–2048, 2007.
7. Tan, L. and Tan, B., Hypercrosslinked porous polymer materials: Design, synthesis, and applications. *Chem. Soc. Rev.*, 46, 3322–3356, 2017.
8. (a) McKeown, N.B., Makhseed, S., Budd, P.M., Phthalocyanine-based nanoporous network polymers. *Chem. Commun.*, 2780–2781, 2002. (b) McKeown, N.B., Hanif, S., Msayib, K., Tattershall, C.E., Budd, P.M., Porphyrin-based nanoporous network polymers. *Chem. Commun.*, 2782–2783, 2002.

9. Côté, A.P., Benin, A.I., Ockwig, N.W., O'Keeffe, M., Matzger, A.J., Yaghi, O.M., Porous, crystalline, covalent organic frameworks. *Science*, 310, 1166–1170, 2005.
10. Ding, S.Y. and Wang, W., Covalent organic frameworks (COFs): From design to applications. *Chem. Soc. Rev.*, 42, 548–568, 2013.
11. Huang, N., Wang, P., Jiang, D., Covalent organic frameworks: A materials platform for structural and functional designs. *Nat. Rev. Mater.*, 1, 16068, 2016.
12. Jiang, J.X., Su, F., Trewin, A., Wood, C.D., Campbell, N.L., Niu, H., Dickinson, C., Ganin, A.Y., Rosseinsky, M.J., Khimyak, Y. *et al.*, Conjugated microporous poly(aryleneethynylene) networks. *Angew. Chem. Int. Ed.*, 46, 8574–8578, 2007.
13. Cooper, A.I., Conjugated microporous polymers. *Adv. Mater.*, 21, 1291–1295, 2009.
14. Xu, Y., Jin, S., Xu, H., Nagai, A., Jiang, D., Conjugated microporous polymers: Design, synthesis and application. *Chem. Soc. Rev.*, 42, 8012–8031, 2013.
15. Kuhn, P., Antonietti, M., Thomas, A., Porous, covalent triazine-based frameworks prepared by ionothermal synthesis. *Angew. Chem. Int. Ed.*, 47, 3450–3453, 2008.
16. Ren, S., Bojdys, M.J., Dawson, R., Laybourn, A., Khimyak, Y.Z., Adams, D.J., Cooper, A.I., Porous, fluorescent, covalent triazine-based frameworks via room-temperature and microwave-assisted synthesis. *Adv. Mater.*, 24, 2357–2361, 2012.
17. Katekomol, P., Roeser, J., Bojdys, M., Weber, J., Thomas, A., Covalent triazine frameworks prepared from 1,3,5-tricyanobenzene. *Chem. Mater.*, 25, 1542–1548, 2013.
18. Ben, T., Ren, H., Ma, S., Cao, D., Lan, J., Jing, X., Wang, W., Xu, J., Deng, F., Simmons, J.M. *et al.*, Targeted synthesis of a porous aromatic framework with high stability and exceptionally high surface area. *Angew. Chem. Int. Ed.*, 48, 9457–9460, 2009.
19. Konstas, K., Taylor, J.W., Thornton, A.W., Doherty, C.M., Lim, W.X., Bastow, T.J., Kennedy, D.F., Wood, C.D., Cox, B.J., Hill, J.M. *et al.*, Lithiated porous aromatic frameworks with exceptional gas storage capacity. *Angew. Chem. Int. Ed.*, 51, 6639–6642, 2012.
20. Sozzani, P., Comotti, A., Simonutti, R., Meersmann, T., Logan, J.W., Pines, A., A porous crystalline molecular solid explored by hyperpolarized xenon. *Angew. Chem. Int. Ed.*, 39, 2695–2699, 2000.
21. Tozawa, T., Jones, J.T.A., Swamy, S.I., Jiang, S., Adams, D.J., Shakespeare, S., Clowes, R., Bradshaw, D., Hasell, T., Chong, S.Y. *et al.*, Porous organic cages. *Nat. Mater.*, 8, 973–978, 2009.
22. Sing, K.S.W., Reporting physisorption data for gas/solid systems with special reference to the determination of surface area and porosity (Recommendations 1984). *Pure Appl. Chem.*, 57, 603, 1985.
23. Atwood, J.L., Barbour, L.J., Jerga, A., Storage of methane and freon by interstitial van der Waals confinement. *Science*, 296, 2367–2369, 2002.
24. Tozawa, T. *et al.*, Porous organic cages. *Nat. Mater.*, 8, 973–978, 2009.
25. Thommes, M., Kaneko, K., Neimark, A.V., Olivier, J.P., Rodriguez-Reinoso, F., Rouquerol, J., Sing, K.S.W., Physisorption of gases, with special reference to the evaluation of surface area and pore size distribution (IUPAC Technical Report). *Pure Appl. Chem.*, 87, 9-10, 1051–1069, 2015.
26. Davis, M.E., Ordered porous materials for emerging applications. *Nature*, 417, 813–821, 2002.
27. Lu, G.Q. and Zhao, X.S., Nanoporous materials: An overview. Series on chemical engineering, in: *Science and Engineering*, pp. 1–13, 2004.
28. Wei, Y., Parmentier, T.E., de Jong, K.P., Zečevic, J., Tailoring and visualizing the pore architecture of hierarchical zeolites. *Chem. Soc. Rev.*, 2015.
29. Zhou, H.C. and Kitagawa, Metal–organic frameworks (MOFs). *Chem. Soc. Rev.*, 43, 5415–5418, 2014.

30. Holst, J.R., Trewin, A., Cooper, A.I., Porous organic molecules. *Nat. Chem.*, 2, 915–920, 2010.
31. Jones, J.T.A., Hasell, T., Wu, X., Bacsa, J., Jelfs, K.E., Schmidtmann, M., Chong, S.Y., Adams, D.J., Trewin, A., Schiffman, F. *et al.*, Modular and predictable assembly of porous organic molecular crystals. *Nature*, 474, 367–371, 2011.
32. Hasell, T. and Cooper, A.I., Porous organic cages: Soluble, modular and molecular pores. *Nat. Rev. Mater.*, 1, 16053, 2016.
33. Barrer, R.M., Shanson, V., Dianins, H., Compound as a zeolitic sorbent. *J. Chem. Soc. Chem. Commun.*, 9, 333–334, 1976.
34. Lee, F., Gabe, E., Tse, J.S., Ripmeester, J.A., Crystal-structure, CP/MAS xe-129, and c-13 NMR of local ordering in Dianins compound clathrates. *J. Am. Chem. Soc.*, 110, 6014–6019, 1988.
35. Lloyd, G.O., Bredenkamp, M.W., Barbour, L.J., Enclathration of morpholinium cations by Dianin's compound: Salt formation by partial host-to-guest proton transfer. *Chem. Commun.*, 4053–4055, 2005.
36. Barbour, L.J., Crystal porosity and the burden of proof. *Chem. Commun.*, 1163–1168, 2006.
37. Couderc, G., Hertzsch, T., Behrnd, N.R., Kramer, K., Hulliger, J., Reversible sorption of nitrogen and xenon gas by the guest-free zeolite tris(o-phenylenedioxy)cyclotriphosphazene (TPP). *Microporous Mesoporous Mater.*, 88, 170–175, 2006.
38. Allcock, H.R. and Siegel, L.A., Phosphonitrilic compounds. III.[1]Molecular inclusion compounds of tris(o-phenylenedioxy)phosphonitrile trimer. *J. Am. Chem. Soc.*, 86-23, 5140–5144, 1964.
39. Frampton, C.S., MacNicol, D.D., Mallinson, P.R., White, J.D., Crystal and molecular structure of 4-p-hydroxyphenyl-2,2,4-trimethylthiachroman-1,1-dioxide. *J. Crystallogr. Spectrosc. Res.*, 22, 551–555, 1992.
40. Andreetti, G.D., Ungaro, R., Pochini, A., Crystal and molecular-structure of cyclo(quater (5-tert-butyl-2-hydroxy-1,3-phenylene)methylene) toluene (1-1) clathrate. *J. Chem. Soc. Chem. Commun.*, 1005–1007, 1979.
41. Atwood, J.L., Barbour, L.J., Jerga, A., A new type of material for the recovery of hydrogen from gas mixtures. *Angew. Chem.-Int. Ed.*, 43, 2948–2950, 2004.
42. Dalgarno, S.J., Thallapally, P.K., Barbour, L.J., Atwood, J.L., Engineering void space in organic van der Waals crystals: Calixarenes lead the way. *Chem. Soc. Rev.*, 36, 236–245, 2007.
43. Thallapally, P.K., Dalgarno, S.J., Atwood, J.L., Frustrated organic solids display unexpected gas sorption. *J. Am. Chem. Soc.*, 128, 15060–15061, 2006.
44. Thallapally, P.K. *et al.*, Organic crystals absorb hydrogen gas under mild conditions. *Chem. Commun.*, 5272–5274, 2005.
45. Thallapally, P.K. *et al.*, Gas-induced transformation and expansion of a non-porous organic solid. *Nat. Mater.*, 7, 146–150, 2008.
46. Thallapally, P.K., Wirsig, T.B., Barbour, L.J., Atwood, J.L., Crystal engineering of nonporous organic solids for methane sorption. *Chem. Commun.*, 4420–4422, 2005.
47. Atwood, J.L., Barbour, L.J., Thallapally, P.K., Wirsig, T.B., A crystalline organic substrate absorbs methane under STP conditions. *Chem. Commun.*, 51–53, 2005.
48. Masson, E., Ling, X., Joseph, R., Kyeremeh-Mensah, L., Lu, X., Cucurbituril chemistry: A tale of supramolecular success. *RSC Adv.*, 2, 1213–1247, 2012.
49. Freeman, W.A., Mock, W.L., Shih, N.Y., Cucurbituril. *J. Am. Chem. Soc.*, 103, 7367–7368, 1981.
50. Kudo, H. *et al.*, Molecular waterwheel (Noria) from a simple condensation of resorcinol and an alkanedial. *Angew. Chem.-Int. Ed.*, 45, 7948–7952, 2006.
51. Tian, J., Thallapally, P.K., Dalgarno, S.J., McGrail, P.B., Atwood, J.L., Amorphous molecular organic solids for gas adsorption. *Angew. Chem.-Int. Ed.*, 48, 5492–5495, 2009.

52. Bezzu, C.G., Helliwell, M., Warren, J.E., Allan, D.R., McKeown, N.B., Heme-like coordination chemistry within nanoporous molecular crystals. *Science*, 327, 1627–1630, 2010.
53. Mastalerz, M. and Oppel, I.M., Rational construction of an extrinsic porous molecular crystal with an extraordinary high specific surface area. *Angew. Chem.-Int. Ed.*, 51, 5252–5255, 2012.
54. Kohl, B., Rominger, F., Mastalerz, M., Crystal structures of a molecule designed not to pack tightly. *Chem. Eur. J.*, 21, 17308–17313, 2015.
55. Li, J.-R., Kuppler, R.J., Zhou, H.-C., Selective gas adsorption and separation in metal–organic frameworks. *Chem. Soc. Rev.*, 38, 1477–1504, 2009.
56. Florey, P.J., *Principles of polymer chemistry*, Cornell Univ. Press, Ithaca, NY, 1953.
57. Vodak, D.T., Braun, M., Iordanidis, L., Plévert, J., Stevens, M., Beck, L., Spence, J.C.H., O'Keeffe, M., Yaghi, O.M., One-step synthesis and structure of an oligo (spiro-orthocarbonate). *J. Am. Chem. Soc.*, 124, 18, 4942–4943, 2002.
58. Sonmez, H.B. and Wudl, F., Cross-linked poly(orthocarbonate) as organic solvent sorbents. *Macromolecules*, 38, 1623–1626, 2005.
59. Curtis, S.M., Le, N., Nguyen, T., Ouyang, X., Tran, T., Fowler, F.W., Lauher, J.W., What have we learned about topochemical diacetylene polymerizations? *Supramol. Chem.*, 17, 31, 2005.
60. Hoang, T., Lauher, J.W., Fowler, F.W., The topochemical 1,6-polymerization of a triene. *J. Am. Chem. Soc.*, 124, 10656, 2002.
61. Niu, W., O'Sullivan, C., Rambo, B.M., Smitha, M.D., Lavigne, J.J., Self-repairing polymers: Poly(dioxaborolane)s containing trigonal planar boron. *Chem. Commun.*, 4342–4344, 2005.
62. DeBlase, C.R. and Dichte, W.R., Moving beyond boron: The emergence of new linkage chemistries in covalent organic frameworks. *Macromolecules*, 49, 5297–5305, 2016.
63. Segura, J.L., Mancheñoa, M.J., Zamora, F., Covalent organic frameworks based on Schiff-base chemistry: Synthesis, properties and potential applications. *Chem. Soc. Rev.*, 45, 5635–5671, 2016.
64. Suresh, V.M., Bandyopadhyay, A., Roy, S., Pati, S.K., Maji, T.K., Highly luminescent microporous organic polymer with lewis acidic boron sites on the pore surface: Ratiometric sensing and capture of F^- ions. *Chem. Eur. J.*, 21, 10799–10804, 2015.
65. Zhao, W., Zhuang, X., Wu, D., Zhang, F., Gehrig, D., Laquai, F., Feng, X., Boron–nitrogen-based conjugated porous polymers with multi-functions. *J. Mater. Chem. A*, 1, 13878–13884, 2013.
66. Liu, X., Zhang, Y., Li, H., A, S., Xia, H., Mu, Y., Triarylboron-based fluorescent conjugated microporous polymers. *RSC Adv.*, 3, 44, 21267, 2013.
67. Li, Z., Li, H., Xia, H., Ding, X., Luo, X., Liu, X., Mu, Y., Triarylboron-linked conjugated microporous polymers: Sensing and removal of fluoride ions. *Chem. A Eur. J.*, 17355–17362, 21, 2015.
68. Zhuang, X., Gehrig, D., Forler, N., Liang, H., Wagner, M., Hansen, M.R., Laquai, F., Zhang, F., Feng, X., Conjugated microporous polymers with dimensionality-controlled heterostructures for green energy devices. *Adv. Mater.*, 27, 3789–3796, 2015.
69. Liras, M., Iglesias, M., Sánchez, F., Conjugated microporous polymers incorporating BODIPY moieties as light-emitting materials and recyclable visible-light photocatalysts. *Macromolecules*, 49, 1666–1673, 2016.
70. Fischer, S., Schmidt, J., Strauch, P., Thomas, A., An anionic microporous polymer network prepared by the polymerization of weakly coordinating anions. *Angew. Chem. Int. Ed.*, 52, 12174–12178, 2013.

71. Yan, Z., Yuan, Y., Tian, Y., Zhang, D., Zhu, G., Highly efficient enrichment of volatile iodine by charged porous aromatic frameworks with three sorption sites. *Angew. Chem. Int. Ed.*, 54, 12733–12737, 2015.
72. Van Humbeck, J.F., Aubrey, M.L., Alsbaiee, A., Ameloot, R., Coates, G.W., Dichtel, W.R., Long, J.R., Tetraarylborate polymer networks as single-ion conducting solid electrolytes. *Chem. Sci.*, 6, 5499–5505, 2015.
73. Malek, N., Maris, T., Simard, M., Wuest, J.D., Molecular tectonics. Selective exchange of cations in porous anionic hydrogen-bonded networks built from derivatives of tetraphenylborate. *J. Am. Chem. Soc.*, 127, 5910–5916, 2005.
74. Zhao, W., Han, S., Zhuang, X., Zhang, F., Mai, Y., Feng, X., Cross-linked polymer-derived B/N co-doped carbon materials with selective capture of CO_2. *J. Mater. Chem. A*, 3, 23352–23359, 2015.
75. Hiroyasu, F. and Yaghi, O.M., Storage of hydrogen, methane, and carbon dioxide in highly porous covalent organic frameworks for clean energy applications. *J. Am. Chem. Soc.*, 131, 8875–8883, 2009.
76. Lanni, L.M., William Tilford, R., Bharathy, M., Lavigne, J.J., Enhanced hydrolytic stability of self-assembling alkylated two-dimensional covalent organic frameworks. *J. Am. Chem. Soc.*, 133, 13975–13983, 2011.
77. Li, Y. and Yang, R.T., Hydrogen storage in metal-organic and covalent-organic frameworks by spillover. *AIChE J.*, 54, 269–279, 2008.
78. Fang, Q., Zhuang, Z., Gu, S., Kaspar, R.B., Zheng, J., Wang, J., Qiu, S., Yan, Y., Designed synthesis of large-pore crystalline polyimide covalent organic frameworks. *Nat. Commun.*, 5, 4503, 2014.
79. Chan-Thaw, C.E., Villa, A., Katekomol, P., Su, D., Thomas, A., Prati, L., Covalent triazine framework as catalytic support for liquid phase reaction. *Nano Lett.*, 10-2, 537–541, 2010.
80. Chan-Thaw, C.E., Villa, A., Prati, L., Thomas, A., Triazine-based polymers as nanostructured supports for the liquid-phase oxidation of alcohols. *Eur. J. Chem.*, 1052–1057, 17, 2011.
81. Wang, K., Yang, L.M., Wang, X., Guo, L., Cheng, G., Zhang, C., Jin, S., Tan, B., Cooper, A., Covalent triazine frameworks via a low-temperature polycondensation approach. *Angew. Chem. Int. Ed.*, 56, 14149–14153, 2017.
82. Spitler, E.L. and Dichtel, W.R., Lewis acid-catalysed formation of two-dimensional phthalocyanine covalent organic frameworks. *Nat. Chem.*, 2, 672–677, 2010.
83. Fang, Q., Wang, J., Gu, S., Kaspar, R.B., Zhuang, Z., Zheng, J., Guo, H., Qiu, S., Yan, Y., 3D Porous crystalline polyimide covalent organic frameworks for drug delivery. *J. Am. Chem. Soc.*, 137, 8352–8355, 2015.
84. Zhang, C., Zhang, S., Yan, Y., Xia, F., Huang, A., Xian, Y., Highly fluorescent polyimide covalent organic nanosheets as sensing probes for the detection of 2,4,6-trinitrophenol. *ACS Appl. Mater. Interfaces*, 9, 13415–13421, 2017.
85. Luo, W., Zhu, Y., Zhang, J., He, J., Chi, Z., A dynamic covalent imine gel as a luminescent sensor. *Chem. Commun.*, 50, 11942–11945, 2014.
86. Meyer, C.D., Joiner, C.S., Stoddart, J.F., Template-directed synthesis employing reversible imine bond formation. *Chem. Soc. Rev.*, 36, 1705–1723, 2007.
87. Uribe-Romo, F.J., Hunt, J.R., Furukawa, H., Klöck, C., Keeffe, M.O., Yaghi, O.M., A crystalline imine-linked 3-d porous covalent organic framework. *J. Am. Chem. Soc.*, 131, 4570–4571, 2009.
88. Uribe-Romo, F.J., Doonan, C.J., Furukawa, H., Oisaki, K., Yaghi, O.M., Crystalline covalent organic frameworks with hydrazone linkages. *J. Am. Chem. Soc.*, 133, 11478–11481, 2011.

89. Fang, Q., Gu, S., Zheng, J., Zhuang, Z., Qiu, S., Yan, Y., 3D microporous base-functionalized covalent organic frameworks for size-selective catalysis. *Angew. Chem. Int. Ed.*, 53, 2878–2882, 2014.
90. Banerjee, R., Das, G., Biswal, B.P., Sharah, G.K., Venkatesh, V., Chemical sensing in two dimensional porous covalent organic nanosheets. *Chem. Sci.*, 6, 3931–3939, 2015.
91. Mullangi, D., Nandi, S., Shalini, S., Sreedhala, S., Vinod, C.P., Vaidhyanathan, R., Pd loaded amphiphilic COF as catalyst for multi-fold Heck reactions, C-C couplings and CO oxidation. *Sci. Rep.*, 5, 10876, 2015.
92. Feng, X., Ding, X., Jiang, D., Covalent organic frameworks. *Chem. Soc. Rev.*, 41, 6010–6022, 2012.
93. Beuerle, F. and Gole, B., Covalent organic frameworks and cage compounds: Design and applications of polymeric and discrete organic scaffolds. *Angew. Chem. Int. Ed.*, 57, 4850–4878, 2018.
94. Xiang, Z. and Cao, D., Porous covalent–organic materials: Synthesis, clean energy application and design. *J. Mater. Chem. A*, 1, 2691–2718, 2013.
95. Zhao, F., Liu, H., Mathe, S., Dong, A., Zhang, J., Covalent organic frameworks: From materials design to biomedical application. *Nanomaterials*, 8, 15, 2017.
96. Díaz, U. and Corma, A., Ordered covalent organic frameworks, COFs and PAFs. From preparation to application. *Coord. Chem. Rev.*, 311, 85–124, 2016.
97. Lohse, M.S. and Bein, T., Covalent organic frameworks: Structures, synthesis, and applications. *Adv. Funct. Mater.*, 28, 1705553, 2018.
98. Campbell, N.L., Clowes, R., Ritchie, L.K., Cooper, A.I., Rapid microwave synthesis and purification of porous covalent organic frameworks. *Chem. Mater.*, 21, 204–206, 2009.
99. Dogru, M., Sonnauer, A., Zimdars, S., Döblinger, M., Knochel, P., Bein, T., Facile synthesis of a mesoporous benzothiadiazole-COF based on a transesterification process. *CrystEngComm*, 15, 1500–1502, 2013.
100. Wei, H., Chai, S., Hu, N., Yang, Z., Wei, L., Wang, L., The microwave-assisted solvothermal synthesis of a crystalline two-dimensional covalent organic framework with high CO_2 capacity. *Chem. Commun.*, 51, 12178–12181, 2015.
101. Nüchter, M., Ondruschka, B., Bonrath, W., Gum, A., Microwave assisted synthesis–a critical technology overview. *Green Chem.*, 6, 128–141, 2004.
102. Biswal, B.P., Chandra, S., Kandambeth, S., Lukose, B., Heine, T., Banerjee, R., Mechanochemical synthesis of chemically stable isoreticular covalent organic frameworks. *J. Am. Chem. Soc.*, 135, 5328–5331, 2013.
103. Thote, J., Barike Aiyappa, H., Rahul Kumar, R., Kandambeth, S., Biswal, B.P., Balaji Shinde, D., Chaki Roy, N., Banerjee, R., Constructing covalent organic frameworks in water via dynamic covalent bonding. *IUCrJ*, 3, 402–407, 2016.
104. Bang, J.H. and Suslick, K.S., Applications of ultrasound to the synthesis of nanostructured materials. *Adv. Mater.*, 22, 1039–1059, 2010.
105. Cintas, P. and Luche, J.L., Green chemistry, The sonochemical approach. *Green Chem.*, 1, 115–125, 1999.
106. Yang, S.T., Kim, J., Cho, H.Y., Kim, S., Ahn, W.S., Facile synthesis of covalent organic frameworks COF-1 and COF-5 by sonochemical method. *RSC Adv.*, 2, 10179–10181, 2012.
107. Wang, X., Ma, R., Hao, L., Wu, Q., Wang, C., Wang, Z., Mechanochemical synthesis of covalent organic framework for the efficient extraction of benzoylurea insecticides. *J. Chromatogr. A*, 1–9, 1551, 2018.

108. Zhang, Y., Shen, X., Feng, X., Xia, H., Mu, Y., Liu, X., Covalent organic frameworks as pH responsive signaling scaffolds. *Chem. Commun.*, 52, 11088–11091, 2016b.
109. Doonan, C.J., Tranchemontagne, D.J., Glover, T.G., Hunt, J.R., Yaghi, O.M., Exceptional ammonia uptake by a covalent organic framework. *Nat. Chem.*, 2, 235–238, 2010.
110. El-Kaderi, H.M., Hunt, J.R., Mendoza-Cortés, J.L., Côté, A.P., Taylor, R.E., O'Keeffe, M., Yaghi, O.M., Designed synthesis of 3D covalent organic frameworks. *Science*, 316, 268–272, 2007.
111. Colson, J.W., Mann, J.A., DeBlase, C.R., Dichtel, W.R., Patterned growth of oriented 2D covalent organic framework thin films on single-layer graphene. *J. Polym. Sci. Polym. Chem.*, 53, 378–384, 2015.
112. Wang, H., Ding, H., Meng, X. *et al.*, Two-dimensional porphyrin- and phthalocyanine-based covalent organic frameworks. *Chin. Chem. Lett.*, 27, 1376–1382, 2016.
113. Xia, L. and Liu, Q., Lithium doping on covalent organic framework-320 for enhancing hydrogen storage at ambient temperature. *J. Solid State Chem.*, 244, 1–5, 2016.
114. Yang, Z. and Cao, D., Effect of Li doping on diffusion and separation of hydrogen and methane in covalent organic frameworks. *J. Phys. Chem. C.*, 12591–12598, 116, 2012.
115. Guo, J.H., Zhang, H., Liu, Z.P. *et al.*, Multiscale study of hydrogen adsorption, diffusion, and desorption on Li-doped phthalocyanine covalent organic frameworks. *J. Phys. Chem. C*, 116, 15908–15917, 2012.
116. Shan, M., Seoane, B., Rozhko, E. *et al.*, Azine-linked covalent organic framework (COF)-based mixed-matrix membranes for CO_2/CH_4 separation. *Chem. Eur. J.*, 1–5, 22, 2016.
117. Yin, Z.J., Xu, S.Q., Zhan, T.G. *et al.*, Ultrahigh volatile iodine uptake by hollow microspheres formed from a heteropore covalent organic framework. *Chem. Commun.*, 53, 7266–7269, 2017.
118. Ding, S., Gao, J., Wang, Q., Zhang, Y., Song, W., Su, C., Wang, W.J., Construction of covalent organic framework for catalysis: Pd/COF-LZU1 in Suzuki–Miyaura coupling reaction. *Am. Chem. Soc.*, 133, 19816–1982, 2011.
119. Dalapati, S., Jin, S., Gao, J. *et al.*, An azine-linked covalent organic framework. *J. Am. Chem. Soc.*, 135, 17310–17313, 2013.
120. Ding, S.Y., Dong, M., Wang, Y.W. *et al.*, Thioether-based fluorescent covalent organic framework for selective detection and facile removal of Mercury(II). *J. Am. Chem. Soc.*, 138, 3031–3037, 2016.
121. Li, Z., Zhang, Y., Xia, H. *et al.*, A robust and luminescent covalent organic framework as a highly sensitive and selective sensor for the detection of Cu^{2+} ions. *Chem. Commun.*, 52, 6613–6616, 2016.
122. Zhu, G. and Ren, H., *Gas sorption using porous organic frameworks*, pp. 57–85, 2015.
123. Sharma, R.K., Yadav, P., Yadav, M., Gupta, R., Rana, P., Srivastava, A., Zboril, R., Varma, R.S., Antonietti, M., Gawande, M.B., Recent development of covalent organic frameworks (COFs): Synthesis and catalytic (organic-electro-photo) applications. *Mater. Horiz.*, 1–99, 2019.
124. Bildirir, H., Gregoriou, V.G., Avgeropoulos, A., Scherf, U., Chochos, C.L., Porous organic polymers as emerging new materials for organic photovoltaic applications: Current status and future challenges. *Mater. Horiz.*, 4, 4, 546–556, 2017.
125. Bond, G.C., Louis, C., Thompson, D., *Catalysis by gold*. 384 pp., Imperial College Press, UK, 2006.
126. Puthiaraj, P. and Ahn, W.S., CO_2 Capture by porous hyper-cross-linked aromatic polymers synthesized using tetrahedral precursors. *Ind. Eng. Chem. Res.*, 55, 7917–7923, 2015.
127. Puthiaraj, P., Cho, S.M., Lee, Y.R., Ahn, W.S., Triptycene based 1,2,3-triazole linked network polymers (TNPs): Small gas storage and selective CO_2 capture. *J. Mater. Chem. A*, 3, 6792–6797, 2015.

128. Bi, J., Fang, W., Li, L., Wang, J., Liang, S., He, Y., Liu, M., Wu, L., Covalent triazine-based frameworks as visible light photocatalysts for the splitting of water. *Macromol. Rapid Commun.*, 36, 1799–1805, 2015.
129. Roeser, J., Kailasam, K., Thomas, A., Covalent triazine frameworks as heterogeneous catalysts for the synthesis of cyclic and linear carbonates from carbon dioxide and epoxides. *ChemSusChem*, 5, 1793–1799, 2012.
130. Sun, Q., Dai, Z., Meng, X., Xiao, F.S., Porous polymer catalysts with hierarchical structures. *Chem. Soc. Rev.*, 44, 6018–6034, 2015.
131. Hug, S., Mesch, M.B., Oh, H., Popp, N., Hirscher, M., Senker, J., Lotsch, B.V., A fluorene based covalent triazine framework with high CO_2 and H_2 capture and storage capacities. *J. Mater. Chem. A*, 2, 5928–5936, 2014.
132. Palkovits, R., Antonietti, M., Kuhn, P., Thomas, A., Schüth, F., Solid catalysts for the selective low-temperature oxidation of methane to methanol. *Angew. Chem. Int. Ed.*, 48, 6909–6912, 2009.
133. Kuhn, P., Forget, A., Su, D., Thomas, A., Antonietti, M., From microporous regular frameworks to mesoporous materials with ultrahigh surface area: Dynamic reorganization of porous polymer networks. *J. Am. Chem. Soc.*, 130, 13333–13337, 2008.
134. Gu, C., Liu, D., Huang, W., Liu, J., Yang, R., Synthesis of covalent triazine-based frameworks with high CO_2 adsorption and selectivity. *Polym. Chem.*, 6, 7410–7417, 2015.
135. Kuecken, S., Schmidt, J., Zhi, L., Thomas, A., Conversion of amorphous polymer networks to covalent organic frameworks under ionothermal conditions: A facile synthesis route for covalent triazine frameworks. *J. Mater. Chem. A.*, 3, 24422–24427, 2015.
136. Tao, L.M., Niu, F., Zhang, D., Wang, T.M., Wang, Q.H., Amorphous covalent triazine frameworks for high performance room temperature ammonia gas sensing. *New J. Chem.*, 38, 2774–2777, 2014.
137. Kuhn, P., Thomas, A., Antonietti, M., Toward tailorable porous organic polymer networks: A high-temperature dynamic polymerization scheme based on aromatic nitriles. *Macromolecules*, 42, 319–326, 2009.
138. Bhunia, A., Boldog, I., Moller, A., Janiak, C., Highly stable nanoporous covalent triazine-based frameworks with an adamantane core for carbon dioxide sorption and separation. *J. Mater. Chem. A*, 1, 14990–14999, 2013.
139. Hu, X.M., Chen, Q., Zhao, Y.C., Laursen, B.W., Han, B.H., Straightforward synthesis of a triazine-based porous carbon with high gas-uptake capacities. *J. Mater. Chem. A*, 2, 14201–14208, 2014.
140. Bhunia, A., Vasylyeva, V., Janiak, C., From a supramolecular tetranitrile to a porous covalent triazine-based framework with high gas uptake capacities. *Chem. Commun.*, 49, 3961–3963, 2013.
141. Wu, S., Liu, Y., Yu, G., Guan, J., Pan, C., Du, Y., Xiong, X., Wang, Z., Facile preparation of dibenzoheterocycle-functional nanoporous polymeric networks with high gas uptake capacities. *Macromolecules*, 47, 2875–2882, 2014.
142. Liu, Y., Wu, S., Wang, G., Yu, G., Guan, J., Pan, C., Wang, Z., Control of porosity of novel carbazole-modified polytriazine frameworks for highly selective separation of CO_2–N_2. *J. Mater. Chem. A*, 2, 7795–7801, 2014.
143. Zhao, Y., Yao, K.X., Teng, B., Zhang, T., Han, Y., A perfluorinated covalent triazine-based framework for highly selective and water-tolerant CO_2 capture. *Energy Environ. Sci.*, 6, 3684–3692, 2013.

144. Wang, W., Ren, H., Sun, F., Cai, K., Ma, H., Du, J., Zhao, H., Zhu, G., Synthesis of porous aromatic framework with tuning porosity via ionothermal reaction. *Dalton Trans.*, 41, 3933–3936, 2012.
145. Liu, X., Li, H., Zhang, Y., Xu, B., Xia, S.A.H., Mu, Y., Enhanced carbon dioxide uptake by metalloporphyrin-based microporous covalent triazine framework. *Polym. Chem.*, 4, 2445–2448, 2013.
146. Schwab, M.G., Fassbender, B., Spiess, H.W., Thomas, A., Feng, X., Müllen, K., Catalyst-free preparation of melamine-based microporous polymer networks through schiff base chemistry. *J. Am. Chem. Soc.*, 131, 7216–7217, 2009.
147. Tan, M.X., Zhang, Y., Ying, J.Y., Mesoporous poly(melamine–formaldehyde) solid sorbent for carbon dioxide capture. *ChemSusChem*, 6, 1186–1190, 2013.
148. Kohlmayr, M., Zuckerstätter, G., Kandelbauer, A., Modification of melamine-formaldehyde resins by substances from renewable resources. *J. Appl. Polym. Sci.*, 124, 4416–4423, 2012.
149. Schwab, M.G., Crespy, D., Feng, X., Landfester, K., Müllen, K., Preparation of microporous melamine-based polymer networks in an anhydrous high-temperature mini emulsion. *Macromol. Rapid Commun.*, 32, 1798–1803, 2011.
150. Yang, G., Han, H., Du, C., Luo, Z., Wang, Y., Facile synthesis of melamine-based porous polymer networks and their application for removal of aqueous mercury ions. *Polymer*, 51, 6193–6202, 2010.
151. Tan, M.X., Sum, Y.N., Ying, J.Y., Zhang, Y., A mesoporous poly-melamine-formaldehyde polymer as a solid sorbent for toxic metal removal. *Energy Environ. Sci.*, 6, 3254–3259, 2013.
152. Puthiaraj, P. and Pitchumani, K., Triazine-based mesoporous covalent imine polymers as solid supports for copper-mediated Chan–Lam cross-coupling N-arylation reactions. *Chem.-Eur. J.*, 20, 8761–8770, 2014.
153. Guo, L., Niu, Y., Xu, H., Li, Q., Razzaque, S., Huang, Q., Jin, S., Tan, B., Engineering heteroatoms with atomic precision in donor–acceptor covalent triazine frameworks to boost photocatalytic hydrogen production. *J. Mater. Chem. A*, 6, 19775–19781, 2018.
154. Zhao, H., Jin, Z., Su, H., Jing, X., Sun, F., Zhu, G., Targeted synthesis of a 2D ordered porous organic framework for drug release. *Chem. Commun.*, 47, 6389–6391, 2011.
155. Patel, H.A., Karadas, F., Canlier, A., Park, J., Deniz, E., Jung, Y., Atilhan, M., Yavuz, C.T., High capacity carbon dioxide adsorption by inexpensive covalent organic polymers. *J. Mater. Chem.*, 22, 8431–8437, 2012.
156. Patel, H.A., Karadas, F., Byun, J., Park, J., Deniz, E., Canlier, A., Jung, Y., Atilhan, M., Yavuz, C.T., Highly stable nanoporous sulfur-bridged covalent organic polymers for carbon dioxide removal. *Adv. Funct. Mater.*, 23, 2270–2276, 2013.
157. Puthiaraj, P. and Pitchumani, K., Palladium nanoparticles supported on triazine functionalised mesoporous covalent organic polymers as efficient catalysts for Mizoroki–Heck cross coupling reaction. *Green Chem.*, 16, 4223–4233, 2014.
158. Kundu, S.K. and Bhaumik, A., A triazine-based porous organic polymer: A novel heterogeneous basic organocatalyst for facile one-pot synthesis of 2-amino-4H-chromenes. *RSC Adv.*, 5, 32730–32739, 2015.
159. Zulfiqar, S., Sarwar, M.I., Yavuz, C.T., Melamine based porous organic amide polymers for CO_2 capture. *RSC Adv.*, 4, 52263–52269, 2014.
160. Yu, S.Y., Mahmood, J., Noh, H.J., Seo, J.M., Jung, S.M., Shin, S.H., Im, Y.K., Jeon, I.Y., Baek, J.B., Direct synthesis of a covalent triazine-based framework from aromatic amides. *Angew. Chem. Int. Ed.*, 57, 8438–8442, 2018.

161. Liu, M., Huang, Q., Wang, S., Li, Z., Li, B., Jin, S., Tan, B., Crystalline covalent triazine frameworks by *in situ* oxidation of alcohols to aldehyde monomers. *Angew. Chem. Int. Ed.*, 57, 11968–11972, 2018.
162. Bojdys, M.J., Jeromenok, J., Thomas, A., Antonietti, M., Rational extension of the family of layered, covalent, triazine-based frameworks with regular porosity. *Adv. Mater.*, 22, 2202–2205, 2010.
163. Wang, K., Huang, H., Liu, D., Wang, C., Li, J., Zhong, C., Covalent triazine-based frameworks with ultramicropores and high nitrogen contents for highly selective CO_2 capture. *Environ. Sci. Technol.*, 50, 4869–4876, 2016.
164. Yuan, K., Liu, C., Han, J., Yu, G., Wang, J., Duan, H., Wang, Z., Jian, X., Phthalazinone structure-based covalent triazine frameworks and their gas adsorption and separation properties. *RSC Adv.*, 6, 12009–12020, 2016.
165. Kamiya, K., Kamai, R., Hashimoto, K., Nakanishi, S., Platinum-modified covalent triazine frameworks hybridized with carbon nanoparticles as methanol-tolerant oxygen reduction electrocatalysts. *Nat. Commun.*, 5, 5040–5047, 2014.
166. Iwase, K., Yoshioka, T., Nakanishi, S., Hashimoto, K., Kamiya, K., Copper-modified covalent triazine frameworks as non-noble-metal electrocatalysts for oxygen reduction. *Angew. Chem. Int. Ed.*, 54, 11068–11072, 2015.
167. Chang, F., Guo, J., Wu, G., Liu, L., Zhang, M., He, T., Wang, P., Yu, P., Chen, P., Covalent triazine-based framework as an efficient catalyst support for ammonia decomposition. *RSC Adv.*, 5, 3605–3610, 2015.
168. Puthiaraj, P., Lee, Y.R., Zhang, S.Q., Ahn, W.S., Triazine-based covalent organic polymers: Design, synthesis and applications in heterogeneous catalysis. *J. Mater. Chem. A*, 4, 16288–16311, 2016.
169. He, J., Xu, F.J., Hu, J., Wang, S.L., Hou, X.D., Long, Z., Covalent triazine framework-1: A novel oxidase and peroxidase mimic. *Microchem. J.*, 135, 91–99, 2017.
170. Bavykina, A.V., Olivos-Suarez, A.I., Osadchii, D., Valecha, R., Franz, R., Makkee, M., Kapteijn, F., Gascon, J., Facile method for the preparation of covalent triazine framework coated monoliths as catalyst support: Applications in c1 catalysis. *ACS Appl. Mater. Interfaces*, 9, 26060–26065, 2017.
171. Hao, L., Ning, J., Luo, B., Wang, B., Zhang, Y., Tang, Z., Yang, J., Thomas, A., Zhi, L., Structural evolution of 2D microporous covalent triazine-based framework toward the study of high-performance supercapacitors. *J. Am. Chem. Soc.*, 137, 219–225, 2015.
172. Hao, L., Luo, B., Li, X.L., Jin, M.H., Fang, Y., Tang, Z.H., Jia, Y.Y., Liang, M.H., Thomas, A., Yang, J.H. *et al.*, Terephthalonitrile-derived nitrogen-rich networks for high performance supercapacitors. *Energy Environ. Sci.*, 5, 9747–9751, 2012.
173. Davankov, V. and Tsyurupa, M.P., *Hypercrosslinked polymeric networks and adsorbing materials: Synthesis, properties, structure, and applications*, vol. 56, pp. 3–354, Elsevier, Amsterdam, 2010.
174. Zhang, X., Shen, S., Fan, L., Studies progress of preparation, properties and applications of hyper-cross-linked polystyrene networks. *J. Mater. Sci.*, 42, 18, 7621–7629, 2007.
175. Davankov, V., Rogozhin, S., Tsyurupa, M., New approach to preparation of uniformly crosslinked macroreticular polystyrene structures. *Mezhdunarodnaya Kniga*, 15, 463–465, 1973.
176. Li, B.Y., Gong, R.N., Wang, W., Huang, X., Zhang, W., Li, H.M., Hu, C.X., Tan, B.E., A new strategy to microporous polymers: Knitting rigid aromatic building blocks by external cross-linker. *Macromolecules*, 44, 2410–2414, 2011.

177. Zhang, W., Tang, J., Yu, W., Visible light-driven C-3 Functionalization of indoles over conjugated microporous polymers. *ACS Catal.*, 8, 8084–8091, 2018.
178. Chen, Q., Liu, D.P., Luo, M., Small, nitrogen-containing microporous conjugated polymers via carbazole-based oxidative coupling polymerization: Preparation, porosity, and gas uptake. *Nano. Micro Small*, 10, 308–315, 2014.
179. Zhang, X., Lu, J., Zhang, J., Porosity enhancement of carbazolic porous organic frameworks using dendritic building blocks for gas storage and separation. *Chem. Mater.*, 26, 4023–4029, 2014.
180. Jiang, F., Jin, T., Zhu, X., Substitution effect guided synthesis of task-specific nanoporous polycarbazoles with enhanced carbon capture. *Macromolecules*, 49, 5325–5330, 2016.
181. Jin, T., Xiong, Y., Zhu, X., Rational design and synthesis of a porous, task-specific polycarbazole for efficient CO_2 capture. *Chem. Commun.*, 52, 4454–4457, 2016.
182. Yuan, Y., Huang, H., Chen, L., N,N′-bicarbazole: A versatile building block toward the construction of conjugated porous polymers for CO_2 capture and dyes adsorption. *Macromolecules*, 50, 4993–5003, 2017.
183. Zhang, R.R., Yin, Q., Liang, H.P., Hypercrosslinked porous polycarbazoles from carbazolyl-bearing aldehydes or ketones. *Polymer*, 143, 87–95, 2018.
184. Qiao, S., Li, Z., Zhang, B., Flexible chain & rigid skeleton complementation polycarbazole microporous system for gas storage. *Microporous Mesoporous Mater.*, 284, 205–211, 2019.
185. Staudinger, H. and Heuer, W., Uber hochpolymere Verbindungen, 93. Mitteil.: Uber das Zerreißender FadenMolekdinger, H., styrols. *Ber. Dtsch. Chem. Ges. (A BSeries)*, 67, 7, 1159–1164, 1934.
186. G.F. D'Alelio, Production of synthetic polymeric compositions comprising sulphonated polymerizates of poly-vinyl aryl compounds and treatment of liquid media therewith. U.S. Patent 2,366,007, 1944.
187. Davankov, V.A. and Tsyurupa, M.P., Structure and properties of hypercrosslinked polystyrene-the first representative of a new class of polymer networks. *React. Polym.*, 13, 27–42, 1990.
188. Ahn, J.H., Jang, J.E., Oh, C.G., Ihm, S.K., Cortez, J., Sherrington, D.C., Rapid generation and control of microporosity, bimodal pore size distribution, and surface area in Davankov-type hyper-cross-linked resins. *Macromolecules*, 39, 627–632, 2006.
189. Li, B., Guan, Z., Wang, W., Yang, X., Hu, J., Tan, B., Highly dispersed pd catalyst locked in knitting aryl network polymers for Suzuki–Miyaura coupling reactions of aryl chlorides in aqueous media. *Adv. Mater.*, 24, 3390–3395, 2012.
190. Sekerová, L., Lhotka, M., Vyskočilová, E., Hyper-cross-linked polyacetylene-type microporous networks decorated with terminal ethynyl groups as heterogeneous acid catalysts for acetalization and esterification reactions. *Chem. - Eur. J.*, 24, 14742–14749, 2018.
191. Chaoui, N., Trunk, M., Dawson, R., Trends and challenges for microporous polymers. *Chem. Soc. Rev.*, 46, 3302–3321, 2017.
192. Xu, C., Wang, H., Wang, Q., Ruthenium coordinated with triphenylphosphine-hyper-cross-linked polymer: An efficient catalyst for hydrogen evolution reaction and hydrolysis of ammonia borane. *Appl. Surf. Sci.*, 466, 193–201, 2019.
193. Liang, H.P., Chen, Q., Han, B.H., Cationic polycarbazole networks as visible-light heterogeneous photocatalysts for oxidative organic transformations. *ACS Catal.*, 8, 5313–5322, 2018.
194. Dawson, R., Stevens, L.A., Drage, T.C., Impact of water co adsorption for carbon dioxide capture in microporous polymer sorbents. *J. Am. Chem. Soc.*, 134, 10741–10744, 2012.
195. Zhu, X., Tian, C., Jin, T., Solid-state synthesis of conjugated nanoporous polycarbazoles. *ACS Macro Lett.*, 6, 1056–1059, 2017.

196. Hou, S. and Tan, B., Naphthyl substitution-induced fine tuning of porosity and gas uptake capacity in microporous hyper-cross-linked amine polymers. *ACS Macromolecules*, 51, 2923–2931, 2018.
197. Zhu, T., Xie, F., Huang, T., Preparation of microporous organic polymers via UV-initiated radical copolymerization. *ACS Macro Lett.*, 7, 1283–1288, 2018.
198. Hou, S., Razzaque, S., Tan, B., Effects of synthesis methodology on microporous organic hyper-cross-linked polymers with respect to structural porosity, gas uptake performance and fluorescence properties. *Polym. Chem.*, 10, 1299–1311, 2019.
199. Zhang, C., Zhu, P.C., Tan, L., Triptycene-based hyper-cross-linked polymer sponge for gas storage and water treatment. *Macromolecules*, 48, 8509–8514, 2015.
200. Song, K., Zou, Z., Wang, D., Microporous organic polymers derived microporous carbon supported pd catalysts for oxygen reduction reaction: Impact of framework and heteroatom. *J. Phys. Chem. C*, 120, 2187–2197, 2016.
201. Saleh, M., Lee, H.M., Kemp, K.C., Highly stable CO2/N2 and CO2/CH4 selectivity in hyper-cross-linked heterocyclic porous polymers. *ACS Appl. Mater. Interfaces*, 6, 7325–7333, 2014.
202. Davankov, V.A. and Tsyurupa, M.P., *Comprehensive analytical chemistry*, vol. 56, pp. 1–648, 2011.
203. Veverka, P. and Jerábek, K., Mechanism of hypercrosslinking of chloromethylated styrene–divinylbenzene copolymers. *React. Funct. Polym.*, 41, 21–25, 1999.
204. Davankov, V., Tsyurupa, M., Ilyin, M., Pavlova, L., Hypercross-linked polystyrene and its potentials for liquid chromatography: A mini-review. *J. Chromatogr. A*, 965, 1, 65–73, 2002.
205. Abrams, I.M. and Millar, J.R., A history of the origin and development of macroporous ion-exchange resins. *React. Funct. Polym.*, 35, 1, 7–22, 1997.
206. Fontanals, N., Marce, R.M., Borrull, F., Cormack, P.A.G., Hypercrosslinked materials: Preparation, characterisation and applications. *Polym. Chem.*, 6, 41, 7231–7244, 2015.
207. Davankov, V.A. and Tsyurupa, M.P., Structure and properties of hypercrosslinked polystyrene: The first representative of a new class of polymer networks. *React. Polym.*, 13, 1, 27–42, 1990.
208. V. Davankov, V. Rogozhin, M. Tsjurupa, Macronet polystyrene structures for ionites and method of producing same. US Patent, 1973.
209. C.H. McBurney, Resinous insoluble reaction products of tertiary amines with haloalkylated vinyl aromatic hydrocarbon copolymers. U.S. Patent 2,591,573, 1952.
210. http://www.dow.com/en-us/markets-and solutions/products/AMBERLITEIonExchangeResinsandPolymericAdsorbents/AMBERLITEFPA98Cl.
211. Cjurupa, M., Lalaev, V., Davankov, V., Synthesis and some physicochemical properties of macronet isoporous styrene polymers having crosslinking bridges of the diphenylmethane type. *Acta Polym.*, 35, 6, 451–455, 1984.
212. Pascault, J.P., Sautereau, H., Verdu, J., Williams, R.J., *Thermosetting polymers*, vol. 64, CRC Press, FL, 2002.
213. Turi, E., *Thermal characterization of polymeric materials*, Elsevier, Amsterdam, 2012.
214. Borsig, E., Fiedlerova, A., Haüsler, K., Sambatra, R., Michler, G., Structure and properties of an interpenetrating polymer network-like system consisting of polystyrene-polyethylene:1. Synthesis, elastomeric and thermoanalytical characterization. *Polymer*, 34, 22, 4787–4792, 1993.
215 Maciejewska, M., Szajnecki, Gawdzik, B., Investigation of the surface area and polarity of porous copolymers of maleic anhydride and divinylbezene. *J. Appl. Polym. Sci.*, 125, 1, 300–307, 2012.
216. Frank, R.S., Downey, J.S., Stöver, H., Synthesis of divinylbenzene–maleic anhydride microspheres using precipitation polymerization. *J. Polym. Sci. A Polym. Chem.*, 36, 13, 2223–2227, 1998.

217. Donescu, D., Raditoiu, V., Spataru, C.I., Somoghi, R., Ghiurea, M., Radovici, C., Fierascu, R.C., Schinteie, G., Leca, A., Kuncser, V., Superparamagnetic magnetite–divinylbenzene–maleic anhydride copolymer nanocomposites obtained by dispersion polymerization. *Eur. Polym. J.*, 48, 10, 1709–1716, 2012.
218. Ogawa, N., Honmyo, K., Harada, K., Sugii, A., Preparation of spherical polymer beads of maleic anhydride–styrene–divinylbenzene and metal sorption of its derivatives. *J. Appl. Polym. Sci.*, 29, 9, 2851–2856, 1984.
219. Gonte, R., Balasubramanian, K., Deb, P.C., Singh, P., Synthesis and characterization of mesoporous hypercrosslinked poly(styrene co-maleic anhydride) microspheres. *Int. J. Polym. Mater.*, 61, 12, 919–930, 2012.
220. Tsyurupa, M.P. and Davankov, V.A., Hypercrosslinked polymers: Basic principle of preparing the new class of polymeric materials. *React. Funct. Polym.*, 53, 2–3, 193–203, 2002.
221. Germain, J., Frechet, J.M.J., Svec, F., Hypercrosslinked polyanilines with nanoporous structure and high surface area: Potential adsorbents for hydrogen storage. *J. Mater. Chem.*, 17, 47, 4989–4997, 2007.
222. Wang, H., Pan, L., Deng, W., Yang, G., Liu, X., One-pot synthesis of triptycene-based porous organic frameworks with tailored micropore environments for highly efficient and selective amine adsorption. *Polym. J.*, 48, 7, 787–792, 2016.
223. Bhunia, S., Banerjee, B., Bhaumik, A., A new hypercrosslinked supermicroporous polymer, with scope for sulfonation, and its catalytic potential for the efficient synthesis of biodiesel at room temperature. *Chem. Commun.*, 51, 24, 5020–5023, 2015.
224. Tan, L. and Tan, B., Recent developments of hypercrosslinked microporous organic polymers, in: *Porous Polymers: Design, Synthesis and Applications*, vol. 4, pp. 66–94, The Royal Society of Chemistry, 2016.
225. Galia, M., Svec, F., Frechet, J.M.J., Monodisperse polymer beads as packing material for high-performance liquid chromotography: Effect of divinylbenzene content on the porous and chromatographic properties of poly(styrene-co-divinylbenzene) beads prepared in presence of linear polystyrene as a porogen. *J. Polym. Sci. A Polym. Chem.*, 32, 11, 2169–2175, 1994.
226. Jiang, J.X., Su, F., Niu, H., Wood, C.D., Campbell, N.L., Khimyak, Y.Z., Cooper, A.I., Conjugated microporous poly-(phenylene butadiynylene)s. *Chem. Commun.*, 486–488, 2008.
227. Weber, J. and Thomas, A., Toward stable interfaces in conjugated polymers: Microporous poly(p-phenylene) and poly-(phenyleneethynylene) based on a spirobifluorene building block. *J. Am. Chem. Soc.*, 130, 6334–6335, 2008.
228. Rose, M., Bohlmann, W., Sabo, M., Kaskel, S., Element-organic frameworks with high permanent porosity. *Chem. Commun.*, 2462–2464, 2008.
229. Jiang, J.X., Wang, C., Laybourn, A., Hasell, T., Clowes, R., Khimyak, Y.Z., Xiao, J., Higgins, S.J., Adams, D.J., Cooper, A.I., Metal–organic conjugated microporous polymers. *Angew. Chem. Int. Ed.*, 50, 1072–1075, 2011.
230. Liang, B., Wang, H., Shi, X., Shen, B., He, X., Ghazi, Z.A., Khan, N.A., Sin, H., Khattak, A.M., Li, L. *et al.*, Microporous membranes comprising conjugated polymers with rigid backbones enable ultrafast organic-solvent nanofiltration. *Nat. Chem.*, 10, 961–967, 2018.
231. Thorand, S. and Krause, N., Improved procedures for the palladium-catalyzed coupling of terminal alkynes with aryl bromides (Sonogashira coupling). *J. Org. Chem.*, 63, 8551–8553, 1998.
232. Chen, L., Honsho, Y., Seki, S., Jiang, D., Light-harvesting conjugated microporous polymers: Rapid and highly efficient flow of light energy with a porous polyphenylene framework as antenna. *J. Am. Chem. Soc.*, 132, 6742–6748, 2010.

233. Liu, Q., Tang, Z., Wu, M., Zhou, Z., Design, preparation and application of conjugated microporous polymers. *Polym. Int.*, 63, 381–392, 2014.
234. Schmidt, J., Werner, M., Thomas, A., Conjugated microporous polymer networks via yamamoto polymerization. *Macromolecules*, 42, 4426–4429, 2009.
235. Sun, L., Liang, Z., Yu, J., Xu, R., Luminescent microporous organic polymers containing the 1,3,5-Tri(4-ethenylphenyl)benzene unit constructed by heck coupling reaction. *Polym. Chem.*, 4, 1932–1938, 2013.
236. Pandey, P., Katsoulidis, A.P., Eryazici, I., Wu, Y., Kanatzidis, M.G., Nguyen, S.T., Imine-linked microporous polymer organic frameworks. *Chem. Mater.*, 22, 4974–4979, 2010.
237. Rabbani, M.G., Sekizkardes, A.K., El-Kadri, O.M., Kaafarani, B.R., El-Kaderi, H.M., Pyrene-directed growth of nanoporous benzimidazole-linked nanofibers and their application to selective CO_2 capture and separation. *J. Mater. Chem.*, 22, 25409–25417, 2012.
238. Xu, C. and Hedin, N., Synthesis of microporous organic polymers with high CO_2-over-N_2 selectivity and CO_2 adsorption. *J. Mater. Chem. A*, 1, 3406–3414, 2013.
239. Rabbani, M.G. and El-Kaderi, H.M., Template-free synthesis of a highly porous benzimidazole-linked polymer for CO_2 capture and H_2 storage. *Chem. Mater.*, 23, 1650–1653, 2011.
240. Biswal, B.P., Becker, D., Chandrasekhar, N., Seenath, J.S., Paasch, S., Machill, S., Hennersdorf, F., Brunner, E., Weigand, J.J., Berger, R. *et al.*, Exploration of thiazolo[5,4-d]thiazole linkages in conjugated porous organic polymers for chemoselective molecular sieving. *Chem. - Eur. J.*, 24, 10868–10875, 2018.
241. Zhu, X., Tian, C., Jin, T., Wang, J., Mahurin, S.M., Mei, W., Xiong, Y., Hu, J., Feng, X., Liu, H. *et al.*, Thiazolothiazole-linked porous organic polymers. *Chem. Commun.*, 50, 15055–15058, 2014.
242. Rouquerol, J., Rouquerol, F., Sing, K., *Adsorption by powders and porous solids*, Academic Press, New York, 1998.
243. Rouquerol, F., Rouquerol, J., Sing, K.S.W., *Handbook of porous solids*, vol. 1, F. Schuth, K.S.W. Sing, J. Weitkarnp (Eds.), pp. 236–275, Copyright 0 Wiley-VCH Verlag GmbH, 69469 Weinheirn (Federal Republic of Germany), 2002.
244. Landers, J., Gor, G.Y., Neimark, A.V., Density functional theory methods for characterization of porous materials. *Colloids Surf. A*, 437, 3–32, 2013.
245. Neimark, A.V., Ravikovitch, P.I., Vishnyakov, A., Bridging scales from molecular simulations to classical thermodynamics: Density functional theory of capillary condensation in nanopores. *J. Phys. Condens. Matter.*, 15, 347–365, 2003.
246. Garrido, J., Linares-Solano, A., Martin-Martinez, J.M., Molina-Sabio, M., Rodriguez-Reinoso, F., Torregrosa, R., Use of nitrogen vs. carbon dioxide in the characterization of activated carbons. *Langmuir*, 3, 76–81, 1987.
247. Zukal, A., Slováková, E., Balcar, H., Sedláček, J., Polycyclotrimers of 1,4-diethynylbenzene, 2,6-diethynylnaphthalene, and 2,6-diethynylanthracene: Preparation and gas adsorption properties. *Macromol. Chem. Phys.*, 214, 2016–2026, 2013.
248. Freude, D. and Kärger, J., *Handbook of porous solids*, F. Schüth, K.S.W. Sing, J. Weitkamp (Eds.), pp. 465–505, Wiley- VCH VerlagGmbH, Weinheim,Germany, 2008.
249. Kärger, J., Pfeifer, H., Stallmach, F., Feoktistova, N.N., Zhdanov, S.P., ^{129}Xe and ^{13}C PFG n.m.r, study of the intracrystalline self-diffusion of Xe, CO2, and CO. *Zeolites*, 13, 50–55, 1993.
250. Dlubek, G., *Encyclopedia of polymer science, and technology*, John Wiley & Sons, Inc., Hoboken, N.J., 2008.
251. Soll, S., Zhao, Q., Weber, J., Yuan, J., Activated CO2 sorption in mesoporous imidazolium-type poly(ionic liquid)-based polyampholytes. *Chem. Mater.*, 25, 3003–3010, 2013.
252. Bacsik, Z., Atluri, R., Garcia-Bennett, A.E., Hedin, N., Temperature-induced uptake of CO2 and formation of carbamates in mesocaged silica modified with n-propylamines. *Langmuir*, 26, 10013–10024, 2010, doi.org/10.1021/la1001495.

253. Bacsik, Z., Ahlsten, N., Ziadi, A., Zhao, G., Garcia-Bennett, A.E., Martín-Matute, B., Hedin, N., Mechanisms and kinetics for sorption of CO2 on bicontinuous mesoporous silica modified with n-propylamine. *Langmuir*, 27, 11118–11128, 2011.
254. Yampolskii, Y.P., Methods for investigation of the free volume in polymers. *Russ. Chem. Rev.*, 76, 59–78, 2007.
255. Jansen, J.C., Macchione, M., Tocci, E., Lorenzo, L.D., Yampolskii, Y.P., Sanfirova, O., Shantarovich, V.P., Comparative study of different probing techniques for the analysis of the free volume distribution in amorphous glassy perfluoropolymers. *Macromolecules*, 42, 7589–7604, 2009.
256. Ghanem, B.S., McKeown, N.B., Budd, P.M., High-performance membranes from polyimides with intrinsic microporosity. *Adv. Mater.*, 20, 2766–2771, 2008.
257. Dawson, R., Laybourn, A., Clowes, R., Khimyak, Y.Z., Adams, D.J., Cooper, A.I., Functionalized conjugated microporous polymers. *Macromolecules*, 42, 8809–8816, 2009.
258. Rao, K.V., Mohapatra, S., Maji, T.K., George, S.J., Guest- responsive reversible swelling and enhanced fluorescence in a super-absorbent, dynamic microporous polymer. *Chem. - Eur. J.*, 18, 4505–4509, 2012.
259. Chen, L., Yang, Y., Jiang, D., CMPs as scaffolds for constructing porous catalytic frameworks: A built-in heterogeneous catalyst with high activity and selectivity based on nanoporous metalloporphyrin polymers. *J. Am. Chem. Soc.*, 132, 9138–9143, 2010.
260. Xie, Y., Wang, T.T., Liu, X.H., Zou, K., Deng, W.Q., Capture and conversion of CO_2 at ambient conditions by a conjugated microporous polymer. *Nat. Commun.*, 4, 1960, 2013.
261. Jiang, J.X., Li, Y., Wu, X., Xiao, J., Adams, D.J., Cooper, A.I., Conjugated microporous polymers with rose bengal dye for highly efficient heterogeneous organo-photocatalysis. *Macromolecules*, 46, 8779–8783, 2013.
262. Kang, N., Park, J.H., Ko, K.C., Chun, J., Kim, E., Shin, H.W., Lee, S.M., Kim, H.J., Ahn, T.K., Lee, J.Y. *et al.*, Tandem synthesis of photoactive benzodifuran moieties in the formation of microporous organic networks. *Angew. Chem. Int. Ed.*, 52, 6228–6232, 2013. https://doi.org/10.1002/anie.201300655.
263. Jiang, J.X., Trewin, A., Adams, D.J., Cooper, A.I., Band gap engineering in fluorescent conjugated microporous polymers. *Chem. Sci.*, 2, 1777–1781, 2011.
264. Liu, X., Xu, Y., Jiang, D., Conjugated microporous polymers as molecular sensing devices: Microporous architecture enables rapid response and enhances sensitivity in fluorescence-on and fluorescence-off sensing. *J. Am. Chem. Soc.*, 134, 8738–8741, 2012.
265. Kou, Y., Xu, Y., Guo, Z., Jiang, D., Supercapacitive energystorage and electric power supply using an aza-fused Π-conjugated microporous framework. *Angew. Chem. Int. Ed.*, 50, 8753–8757, 2011.
266. Liao, Y., Wang, H., Zhu, M., Thomas, A., Efficient supercapacitor energy storage using conjugated microporous polymer networks synthesized from Buchwald–Hartwig coupling. *Adv. Mater.*, 30, 1705710, 2018.
267. Gu, C., Huang, N., Gao, J., Xu, F., Xu, Y., Jiang, D., Controlled synthesis of conjugated microporous polymer films: Versatile platforms for highly sensitive and label-free chemo- and biosensing. *Angew. Chem. Int. Ed.*, 53, 4850–4855, 2014.
268. Sprick, R.S., Jiang, J.X., Bonillo, B., Ren, S., Ratvijitvech, T., Guiglion, P., Zwijnenburg, M.A., Adams, D.J., Cooper, A.I., Tuneable organic photocatalysts for visible-light-driven hydrogen evolution. *J. Am. Chem. Soc.*, 137, 3265–3270, 2015.
269. Wang, L., Wan, Y., Ding, Y., Wu, S., Zhang, Y., Zhang, X., Zhang, G., Xiong, Y., Wu, X., Yang, J. *et al.*, Conjugated microporous polymer nanosheets for overall water splitting using visible light. *Adv. Mater.*, 29, 1702428, 2017.

270. Lu, W., Yuan, D., Zhao, D., Schilling, C.I., Plietzsch, O., Muller, T., Bräse, S., Guenther, J., Blümel, J., Krishna, R. *et al.*, Porous polymer networks: Synthesis, porosity, and applications in gas storage/separation. *Chem. Mater.*, 22, 5964–5972, 2010.
271. Dawson, R., Cooper, A.I., Adams, D.J., Nanoporous organic polymer networks. *Prog. Polym. Sci.*, 37, 530–563, 2012.
272. Talapaneni, S.N., Kim, D., Barin, G., Buyukcakir, O., Je, S.H., Coskun, A., Pillar[5]arene based conjugated microporous polymers for propane/methane separation through host–guest complexation. *Chem. Mater.*, 28, 4460–4466, 2016.
273. Zhang, K., Kopetzki, D., Seeberger, P.H., Antonietti, M., Vilela, F., Surface area control and photocatalytic activity of conjugated microporous poly(benzothiadiazole) networks. *Angew. Chem. Int. Ed.*, 52, 1432–1436, 2013.
274. Choi, H.S., Jeon, H.J., Choi, J.H., Lee, G.H., Kang, J.K., Tailoring open metal sites for selective capture of CO2 in isostructural metalloporphyrin porous organic networks. *Nanoscale*, 7, 18923–18927, 2015.
275 Jeon, H.J., Choi, J.H., Lee, Y., Choi, K.M., Park, J.H., Kang, J.K., Highly selective CO2-capturing polymeric organic network structures. *Adv. Energy Mater.*, 2, 225–228, 2012.
276. Perego, J., Piga, D., Bracco, S., Sozzani, P., Comotti, A., Expandable porous organic frameworks with built-in amino and hydroxyl functions for CO2 and CH4 capture. *Chem. Commun.*, 54, 9321–9324, 2018.
277. Ben, T., Shi, K., Cui, Y., Pei, C., Zuo, Y., Guo, H., Zhang, D., Xu, J., Deng, F., Tian, Z. *et al.*, Targeted synthesis of an electroactive organic framework. *J. Mater. Chem.*, 21, 18208–18214, 2011.
278. Liu, M., Guo, L., Jin, S., Tan, B., Covalent triazine frameworks: Synthesis and applications. *J. Mater. Chem. A*, 7, 5153–5172, 2019.
279. Lu, W., Yuan, D., Sculley, J., Zhao, D., Krishna, R., Zhou, H.C., Sulfonate-grafted porous polymer networks for preferential CO2 adsorption at low pressure. *J. Am. Chem. Soc.*, 133, 18126–18129, 2011.
280. Chen, Q., Wang, J.X., Wang, Q., Bian, N., Li, Z.H., Yan, C.G., Han, B.H., Spiro(fluorene-9,9′-xanthene)-based porous organic polymers: Preparation, porosity, and exceptional hydrogen uptake atlow pressure. *Macromolecules*, 44, 7987–7993, 2011.
281. Zhang, Y. and Riduan, S.N., Functional porous organic polymers for heterogeneous catalysis. *Chem. Soc. Rev.*, 41, 2083–2094, 2012.
282. Xiang, Z., Zhou, X., Zhou, C., Zhong, S., He, X., Qin, C., Cao, D., Covalent-organic polymers for carbon dioxide capture. *J. Mater. Chem.*, 22, 22663–22669, 2012.
283. Patel, H.A., Je, S.H., Park, J., Chen, D.P., Jung, Y., Yavuz, C.T., Coskun, A., Unprecedented high-temperature CO2 selectivity in N2-phobic nanoporous covalent organic polymers. *Nat. Commun.*, 4, 1357, 2013.
284 Web of Science, Web of Science Group, 2020, https://clarivate.com/webofsciencegroup/solutions/web-of-science/.
285. Yamamoto, T., Maruyama, T., Zhou, Z.H., Ito, T., Fukuda, T., Yoneda, Y., Begum, F., Ikeda, T., Sasaki, S., pi.-conjugated poly(pyridine-2,5-diyl), poly(2,2'-bipyridine-5,5′-diyl), and their alkyl derivatives. Preparation, linear structure, function as a ligand to form their transition metal complexes, catalytic reactions, n- type electrically conducting properties, optical properties, and alignment on substrates. *J. Am. Chem. Soc.*, 116, 4832–4845, 1994.
286. Trewin, A. and Cooper, A.I., Porous organic polymers: Distinction from disorder? *Angew. Chem. Int. Ed.*, 49, 1533–1535, 2010.
287. Ben, T., Pei, C., Zhang, D., Xu, J., Deng, F., Jing, X., Qiu, S., Gas storage in porous aromatic frameworks (PAFs). *Energy Environ. Sci.*, 4, 3991–3999, 2011.

288. Li, M., Ren, H., Sun, F., Tian, Y., Zhu, Y., Li, J., Mu, X., Xu, J., Deng, F., Zhu, G., Construction of porous aromatic frameworks with exceptional porosity via building unit engineering. *Adv. Mater.*, 30, 1804169, 2018.
289. Jia, J., Chen, Z., Jiang, H., Belmabkhout, Y., Mouchaham, G., Aggarwal, H., Adil, K., Abou-Hamad, E., Czaban-Jóźwiak, J., Tchalala, M.R. *et al.*, Extremely hydrophobic POPs to access highly porous storage media and capturing agent for organic vapors. *Chem.*, 5, 180–191, 2019.
290. Yuan, D., Lu, W., Zhao, D., Zhou, H.C., Highly stable porous polymer networks with exceptionally high gas-uptake capacities. *Adv. Mater.*, 23, 3723–3725, 2011.
291. Miyaura, N., Yanagi, T., Suzuki, A., The palladium-catalyzed cross-coupling reaction of phenylboronic acid with haloarenes in the presence of bases. *Synth. Commun.*, 11, 513–519, 1981.
292. Sonogashira, K., Tohda, Y., Hagihara, N., A convenient synthesis of acetylenes: Catalytic substitutions of acetylenic hydrogen with bromoalkenes, iodoarenes and bromopyridines. *Tetrahedron Lett.*, 16, 4467–4470, 1975.
293. Yuan, Y., Yang, Y., Faheem, M., Zou, X., Ma, X., Wang, Z., Meng, Q., Wang, L., Zhao, S., Zhu, G., Molecularly imprinted porous aromatic frameworks serving as porous artificial enzymes. *Adv. Mater.*, 30, 1800069, 2018.
294. Heck, R.F. and Nolley, J.P., Palladium-catalyzed vinylic hydrogen substitution reactions with aryl, benzyl, and styryl halides. *J. Org. Chem.*, 37, 2320–2322, 1972.
295. Li, P.F., Schon, T.B., Seferos, D.S., Thiophene, selenophene, and tellurophene-based three-dimensional organic frameworks. *Angew. Chem. Int. Ed.*, 54, 9361–9366, 2015.
296. Milstein, D. and Stille, J.K., A general, selective, and facile method for ketone synthesis from acid chlorides and organotin compounds catalyzed by palladium. *J. Am. Chem. Soc.*, 100, 3636–3638, 1978.
297. Park, J.H., Ko, K.C., Park, N., Shin, H.W., Kim, E., Kang, N., Hong Ko, J., Lee, S.M., Kim, H.J., Ahn, T.K. *et al.*, Microporous organic nanorods with electronic push–pull skeletons for visible lightinduced hydrogen evolution from water. *J. Mater. Chem. A*, 2, 7656–7661, 2014.
298. Meng, L., Zou, X., Guo, S., Ma, H., Zhao, Y., Zhu, G., Self- Supported fibrous porous aromatic membranes for efficient CO2/N2 separations. *ACS Appl. Mater. Interfaces*, 7, 15561–15569, 2015.
299. Liu, D.P., Chen, Q., Zhao, Y.C., Zhang, L.M., Qi, A.D., Han, B.H., Fluorinated porous organic polymers via direct C–H arylation polycondensation. *ACS Macro Lett.*, 2, 522–526, 2013.
300. Yuan, R., Gu, Y., Ren, H., Liu, J., Zhu, G., Porous aromatic framework as an efficient metal-free electro-catalyst for nonenzymatic H_2O_2 sensing. *Chem. - Eur. J.*, 23, 9467–9471, 2017.
301. Thirion, D., Lee, J.S., Özdemir, E., Yavuz, C.T., Robust C-C bonded porous networks with chemically designed functionalities for improved CO2 capture from flue gas. *Beilstein J. Org. Chem.*, 12, 2274–2279, 2016.
302. Ma, H., Ren, H., Zou, X., Sun, F., Yan, Z., Cai, K., Wang, D., Zhu, G., Novel lithium-loaded porous aromatic framework for efficient CO_2 and H_2 uptake. *J. Mater. Chem. A*, 1, 752–758, 2013.
303. Ma, H., Ren, H., Zou, X., Meng, S., Sun, F., Zhu, G., Postmetalation of porous aromatic frameworks for highly efficient carbon capture from $CO_2 + N_2$ and $CH_4 + N_2$ mixtures. *Polym. Chem.*, 5, 144–152, 2014.
304. Ren, H., Ben, T., Wang, E., Jing, X., Xue, M., Liu, B., Cui, Y., Qiu, S., Zhu, G., Targeted synthesis of a 3D porous aromatic framework for selective sorption of benzene. *Chem. Commun.*, 46, 291–293, 2010.
305. Xie, Z., Wang, C., deKrafft, K.E., Lin, W., Highly stable and porous cross-linked polymers for efficient photocatalysis. *J. Am. Chem. Soc.*, 133, 2056–2059, 2011.

306. Yuan, S., White, D., Mason, A., Reprogle, B., Ferrandon, M.S., Yu, L., Liu, D.J., Improving hydrogen adsorption enthalpy through coordinatively unsaturated cobalt in porous polymers. *Macromol. Rapid Commun.*, 33, 407–413, 2012.
307. Weston, M.H., Farha, O.K., Hauser, B.G., Hupp, J.T., Nguyen, S.T., Synthesis and metalation of catechol-functionalized porous organic polymers. *Chem. Mater.*, 24, 1292–1296, 2012.
308. Marchetti, A., Chen, J., Pang, Z., Li, S., Ling, D., Deng, F., Kong, X., Understanding surface and interfacial chemistry in functional nanomaterials via solid-state NMR. *Adv. Mater.*, 29, 1605895, 2017.
309. Tian, Y. and Zhu, G., Porous aromatic frameworks (PAFs). *Chem. Rev.*, 120, 16, 8934–8986, 2020.
310. Wang, L., Jia, J., Faheem, M., Tian, Y., Zhu, G., Fabrication of triazine-based porous aromatic framework (PAF) membrane with structural flexibility for gas mixtures separation. *J. Ind. Eng. Chem.*, 67, 373–379, 2018.
311. Zhang, D., Jing, X., Sholl, D.S., Sinnott, S.B., Molecular simulation of capture of sulfur-containing gases by porous aromatic frameworks. *J. Phys. Chem. C*, 122, 18456–18467, 2018.
312. Li, J., Huang, L., Zou, X., Zheng, A., Li, H., Rong, H., Zhu, G., Porous organic materials with ultra-small pores and sulfonic functionality for xenon capture with exceptional selectivity. *J. Mater. Chem. A*, 6, 11163–11168, 2018.
313. Howarth, R.W., Santoro, R., Ingraffea, A., Methane and the greenhouse-gas footprint of natural gas from shale formations. *Clim. Change*, 106, 679, 2011.
314. Meng, S., Ma, H., Jiang, L., Ren, H., Zhu, G., A facile approach to prepare porphyrinic porous aromatic frameworks for small hydrocarbon separation. *J. Mater. Chem. A*, 2, 14536–14541, 2014.
315. Van Humbeck, J.F., McDonald, T.M., Jing, X., Wiers, B.M., Zhu, G., Long, J.R., Ammonia capture in porous organic polymers densely functionalized with brønsted acid groups. *J. Am. Chem. Soc.*, 136, 2432–2440, 2014.
316. Acharyya, K., Mukherjee, S., Mukherjee, P.S., Molecular marriage through partner preferences in covalent cage formation and cage-to-cage transformation. *J. Am. Chem. Soc.*, 135, 2, 554–557, 2013.
317. Schneider, M.W., Oppel, I.M., Ott, H., Lechner, L.G., Hauswald, H.J.S., Stoll, R., Mastalerz, M., Periphery-substituted [4+6] salicylbisimine cage compounds with exceptionally high surface areas: Influence of the molecular structure on nitrogen sorption properties. *Chem. Eur. J.*, 18, 836–847, 2012.
318. Holst, J.R., Trewin, A., Cooper, A.I., Porous organic molecules. *Nat. Chem.*, 2, 915–920, 2010.
319. Hasell, T., Chong, S.Y., Jelfs, K.E., Adams, D.J., Cooper, A.I., Porous organic cage nanocrystals by solution mixing. *J. Am. Chem. Soc.*, 134, 588–598, 2012.
320. Zhang, G., Presly, O., White, F., Oppel, I.M., Mastalerz, M., A permanent mesoporous organic cage with an exceptionally high surface area. *Angew. Chem. Int. Ed.*, 53, 1516–1520, 2014.
321. Brutschy, M., Schneider, M.W., Mastalerz, M., Waldvogel, S.R., Porous organic cage compounds as highly potent affinity materials for sensing by quartz crystal microbalances. *Adv. Mater.*, 24, 6049–6052, 2012.
322. Bushell, A.F., Budd, P.M., Attfield, M.P., Jones, J.T.A., Hasell, T., Cooper, A.I., Bernardo, P., Bazzarelli, F., Clarizia, G., Jansen, J.C., Nanoporous organic polymer/cage compositemem-branes. *Angew. Chem.*, 125, 1291–1294, 2013.
323. Liu, Y., Liu, X., Warmuth, R., Multicomponent dynamic covalent assembly of a rhombicuboc-tahedral nanocapsule. *Chem. Eur. J.*, 13, 8953–8959, 2007.
324. Jelfs, K.E., Wu, X., Schmidtmann, M., Jones, J.T.A., Warren, J.E., Adams, D.J., Cooper, A.I., Large self-assembled chiral organic cages: Synthesis, structure, and shape persistence. *Angew. Chem.*, 123, 10841–10844, 2011.

325. Mastalerz, M., Schneider, M.W., Oppel, I.M., Presly, O., A salicylbisimine cage compound with high surface area and selective CO2/CH4 adsorption. *Chem. Int. Ed.*, 50, 1046–1051, 2011.
326. Pasquale, S., Sattin, S., Escudero-And, E.C., Martinez-Belmonte, M., de Mendoza, J., Giant regular polyhedra from calixarene carboxylates and uranyl. *Nat. Commun.*, 3, 785, 2012.

5

Gel-Type Natural Polymers as Electroconductive Materials

Arshpreet Kaur, Madhvi and Dhiraj Sud*

Department of Chemistry, Sant Longowal Institute of Engineering and Technology, Deemed to be University, Longowal, Sangrur, India

Abstract

Natural polymer derived hydrogels have gained the considerable attention over the past years owing to their 3D cross-linked networks with interstitial spaces, consequently, can hold water molecules and biological fluids within their networking structures. These hydrogels exhibit a broad application spectrum owing to unique properties such as biocompatibility, stimuli responsive nature, hydrophilic properties, biodegradability etc. The treatment of hydrogels with electroactive materials led to endowment of electroconductive, electromechanical, and redox properties, subsequently recognized as smart hydrogel. This chapter delivers a brief overview on the classification of natural polymers and the naturally occurring polymeric materials employed for the hydrogel fabrication. Further, classification of hydrogels along with their various synthesis methods and properties has also been deliberated. In addition, the natural polymer based hydrogels possesses plethora of applications. Nevertheless, current development of natural polymer-based hydrogels as electroconductive materials gained a spotlight which has been discussed in this chapter along with their advantages and shortcomings.

***Keywords*:** Natural polymer, hydrogel, synthesis, properties, electrochemical supercapacitor

5.1 Introduction

In the 21st century, the sole aim of scientists worldwide is towards the exploitation of renewable sources and natural polymers with tailored properties and to target the broader application spectrum for development of mankind [1]. The usage of various natural polymers like starch, cellulose, xanthan gum, carrageenan, alginate etc. in the research and commercial sector are extensively explored for the replacement of widely used petroleum-based polymers [2]. Thus, the natural polymeric materials gained a significant attention over synthetic polymeric materials because of their viability in pharmaceuticals, drug delivery, food industry, cosmetology etc. owing to their easy availability, non-toxicity and pocket friendly nature [3]. Over the past years, the research was fascinated towards the synthesis of natural

**Corresponding author*: suddhiraj@yahoo.com

Inamuddin, Tariq Altalhi and Sayed Mohammed Adnan (eds.) Sustainable Materials for Electrochemical Capacitors, (133–166) © 2023 Scrivener Publishing LLC

polymers derived hydrogels. The hydrophilic nature of natural polymers including polysaccharides, polypeptides, and polynucleotides possesses an ability to form 3D cross-linked networks that have significant water retaining capacity in interstitial sites, thus resulted in naturally derived hydrogel [4]. The formation of 3D networks or gelation process occurred via physical/chemical cross-linking of polymeric chains. These hydrogels showed a remarkable ability for large water uptake because of the hydrophobic properties. In addition, hydrogels gained the commendable interest over the past years because of their biocompatibility, stimuli responsive nature and biodegradability [5]. However, there are some shortcomings associated with the hydrogels which includes

- Unsatisfactory mechanical properties
- Low thermal stability
- Difficult to the control the structural properties

These shortcomings of hydrogels limit their applications in various fields. The research efforts were made for tailored fabrication and modification of hydrogels to eradicate these drawbacks in the future [6].

The natural polymeric materials derived hydrogels acquiring fine 3D cross-linked networks, high water uptake capacity, virtuous biocompatibility, and versatile tailored synthesis methodologies, have emerged as auspicious matrices for the synthesis of electroconductive hydrogels [7]. The chemical modification or blending or doping of hydrogels with electroactive materials made them enriched with electrical properties [8]. The conductivity of the electrochemically active hydrogels relies upon the

- Nature of dopant
- Efficiency of material towards the e^- abstraction and e^- inducing capability with in the matrix
- Mobility of charge carriers

Furthermore, the fabrication of hydrogels with materials having high surface area and significant porosity enhanced their electronic and ionic transporting capabilities [9]. The optimization of mechanical properties of hydrogels broadens their application spectrum. The significant biocompatibility of natural polymeric materials reduces the toxicity of electroactive hydrogels make them accessible to be exposed to the ecosystem [10]. This chapter focuses on the natural polymeric materials that are promising candidates for hydrogels synthesis with desired properties. The classification of hydrogels including synthetic methodologies for fabrication of natural polymer based hydrogels and their exploitation as electrochemical supercapacitors have also been deliberated.

5.2 Natural Polymers

Nature offers abundant examples of polymeric materials from the aspects of both simplicity and complexity that have aroused tremendous interest for their usage in different fields [11]. The classification of natural polymeric materials depending upon their source of origin has been shown in Figure 5.1. Although, these polymers have number of advantages such as

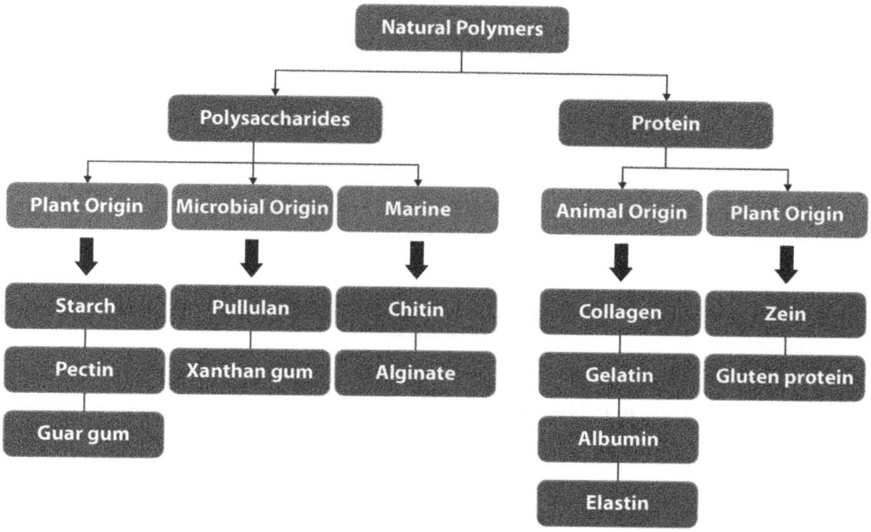

Figure 5.1 Classification of natural polymers based on their origin.

easy availability, economic, non-toxicity etc., also there are some disadvantages associated with them such as [12, 13]

- Microbial contamination
- Batch to batch variation
- Uncontrolled hydration rate
- Slow production rate
- Heavy metal contamination

Natural polymers possess a significant biodegradability and biocompatibility and these can be manipulated which led to tremendous boom in scientific communities related to various research fields during recent times [14]. These materials came out as attractive candidates for drug formulation, pharmacology, cosmetic formulation, drug carrier owing to their mechanical, physico-chemical and biological properties [15]. Natural polymers are one of the promising components used for the cosmetic product formation as these materials are employed as film-formers in nail enamels, thickener and rheology modifier in hair colorants, emulsifiers in lotions and sunscreens [16]. During the past decades, the research is focused on designing the new materials by blending of polymers for refining the properties of modified materials. Hydrogels are one of the designed materials formed by blending of polymers using various techniques. Meanwhile, these materials are widely explored for the electrochemical conductive devices due to their electroactive nature [17].

5.2.1 Hydrogels

Fundamentally, intelligent or smart hydrogels are a kind of peculiar 3D cross-linked networks system of polymeric materials with water as a dispersing medium [18]. Hydrogels

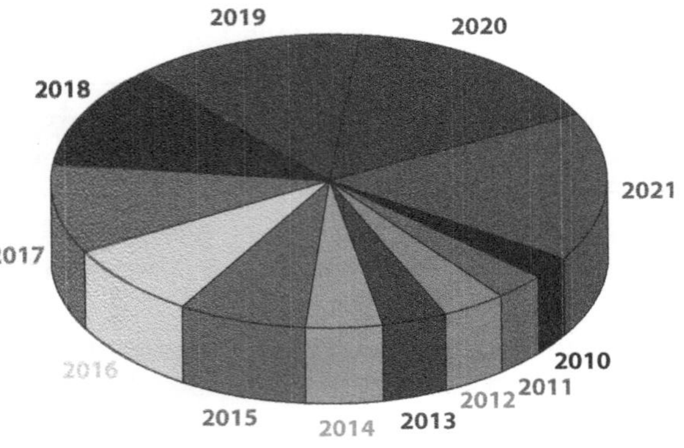

Figure 5.2 Illustrates the increase in published reports related to hydrogels from 2010 to 2021.

are soft in nature and have a considerable absorption ability for water or biological fluids within their interstices without any distortion in network structure. This phenomenon is attributed to presence of polar and hydrophilic functional moieties like hydroxyl group, amine group, carboxylic group, amide group etc. attached with the polymeric chains [19]. Thus, the hydrogels swelling behavior can be controlled by varying the ratios of hydrophilic-to-hydrophobic functional groups. In addition, the cross-linking of polymeric chains helps to maintain the hydrogel networks in a swollen state [20]. Further, the water uptake capability of hydrogels showed the dependency upon the inherit elastic restoring capacity of polymer and osmotic pressure induced by stimuli [21].

The combination of various building blocks viz. cross-linking agent, monomer backbone, dispersing agent and electrolytic monomer generally formed the complete hydrogel structure [22]. Both natural and synthetic polymeric chains contribute to the fabrication of hydrogels. In 1960, Wichterle and Lim reported the first modern hydrogel having imbibing moisture properties synthesized from poly (2-hydroxethyl methacrylate) and explored its utilization in contact lens industry [23]. Afterwards, the published reports on hydrogels showed the extensive growing trends, especially a rapid upsurge from the year 1984 onward. The results have been shown in Figure 5.2 which elaborates the growing trends in published reports related to hydrogels during the year 2010 to 2021 [24].

5.2.2 Classification of Hydrogels

The hydrogels classification depending upon the various factors can be carried out such as source of polymers, ionic charge, synthesis methodology, response to stimuli, cross-linking strategy and type of materials involved [25]. These factors have been summed up briefly in Figure 5.3 and represent the following points [26–28]:

- The polymeric materials such as natural, synthetic or combination of both polymeric materials have a tendency to form well designed hydrogels which may be homopolymer, co-polymers, block co-polymer or terpolymer hydrogels and so on.

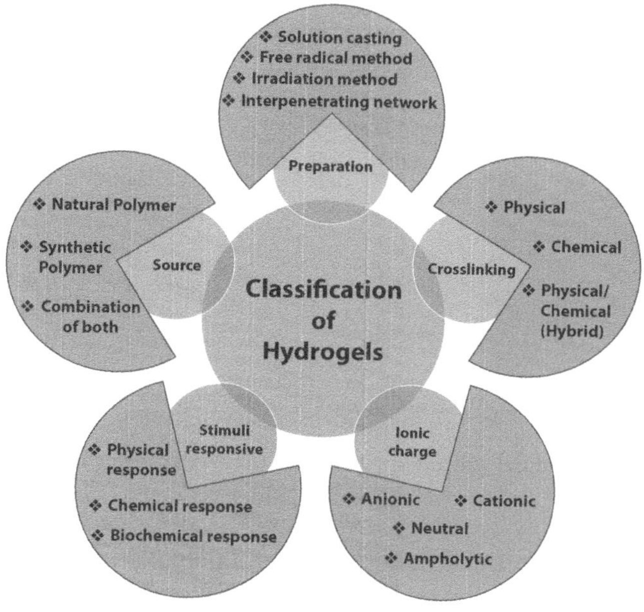

Figure 5.3 Illustrates the classification of hydrogels depending upon various factors.

- The homopolymer, copolymer or multipolymer hydrogels were fabricated using monomer units of single type of hydrophilic material, two types of materials and three or more types of materials respectively.
- The cross-linked hydrogels are formed using a cross-linking agent which can form physically or chemically cross-linked hydrogels or both simultaneously.
- The synthesis of hydrogels can be carried out by adopting several synthetic methodologies such as free radical polymerization, condensation polymerization, by exposure of UV, gamma or microwave radiations, interpenetrating network formation method etc.
- On the basis of physical structural networks, hydrogels are classified into amorphous, semi-crystalline, hydrogen-bonded hydrogels, supermolecules, and hydrocolloid aggregates.
- Depending upon the ionic charges present on the hydrogel network, these can be classified as cationic, anionic, neutral and ampholytic hydrogels. The existence of overall charge on hydrogel network relies on the functional moieties attached to the polymeric chain.

5.2.3 Composition of Hydrogels

Generally, the cross-linked polymers are referred as hydrogels irrespective of type of polymeric material i.e. synthetic, natural or combination of natural and synthetic grafted polymer derivatives [29]. Hydrogels fabricated from naturally occurring polymeric materials are known as natural hydrogels, and possesses a remarkable characteristic of economically safe and pocket-friendly materials [30]. The covalently bonded monomeric units comprised a

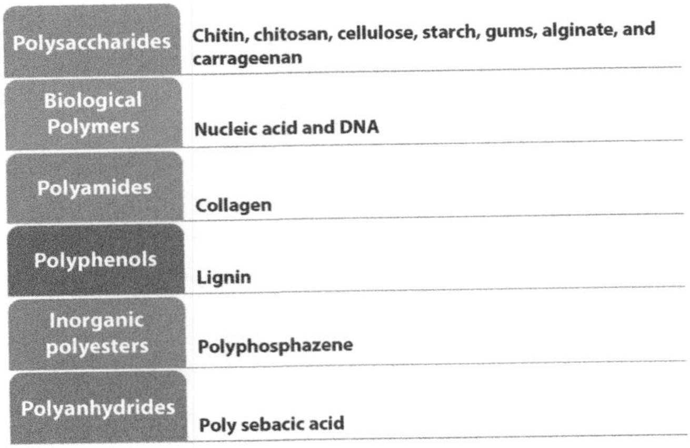

Figure 5.4 Various classes of natural polymeric materials on the basis of their chemical structure.

well-defined and larger chemical structure of natural polymers and based on the chemical structure, natural polymers belong to various classes which have been shown in Figure 5.4.

Polysaccharide and protein derived hydrogels are appeared to be biocompatible with extracellular matrix owing to their natural origin [31]. Nevertheless, the natural polymer derived hydrogels specifically, polysaccharide-based hydrogels, are delicate. In order to ameliorate their mechanical strength, the cross-linking or grafting of natural polymeric chains with monomers or blending with synthetic polymers was carried out [32]. The usage of poly (vinyl alcohol), a synthetic polymer, as a blending agent for enhancement of mechanical properties of naturally derived hydrogels exemplifies the same [33].

Synthetic hydrogels are derived from synthetic polymers which provide the extended flexibility to hydrogels. The widely explored synthetic polymers for hydrogel fabrication are polycaprolactone [34], poly (acrylamide) [35], poly (acrylic acid) [36], poly (ethylene glycol) [37], and poly (vinyl alcohol) [38]. Poly (ethylene glycol) is hydrophilic synthetic polymer which showed the biocompatibility as well as biodegradability and most interestingly, it can go under conjugation with polysaccharides and proteins. Bhattarai *et al.* provided the injectable thermo-responsive grafted hydrogel using poly (ethylene glycol) and chitosan. The synthesized grafted hydrogel solidifies at ambient temperature while behaves as a liquid at low temperature [39].

5.2.4 Natural Polymers Derived Hydrogels

Natural polymers derived hydrogels procure a considerable attention owing to their eco-friendly nature and hydrophilicity. To date, number of reports were published on synthesis of several kind of natural polymeric material-based hydrogels such as polysaccharides [40], protein [41], DNA based hydrogel networks [42] etc. which have been briefly described in the following section. The chemical structures of natural polymers which are widely explored for hydrogel formulations have been compiled in Table 5.1.

Table 5.1 Chemical structures of natural polymeric materials.

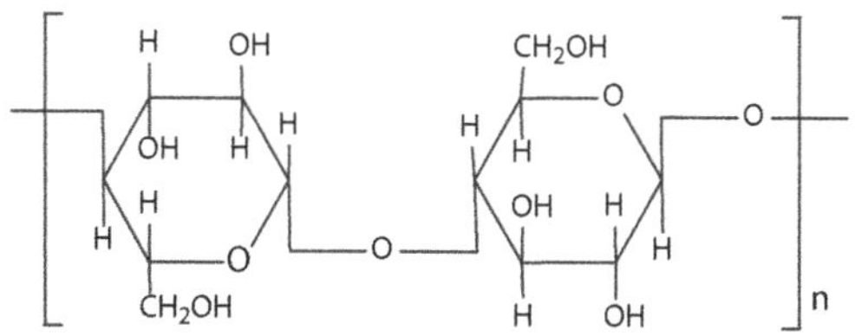 **Cellulose**	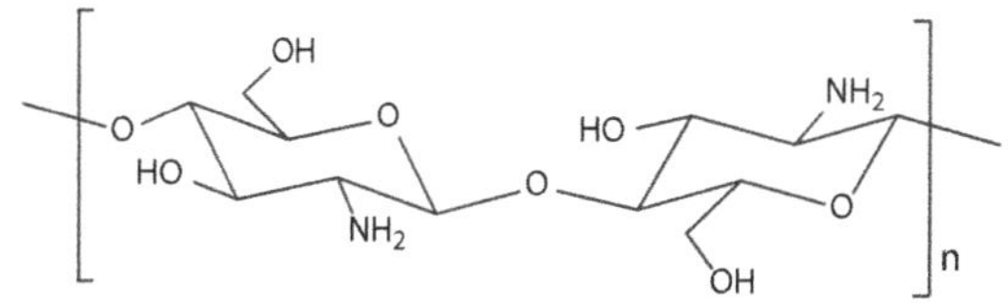 **Chitosan**
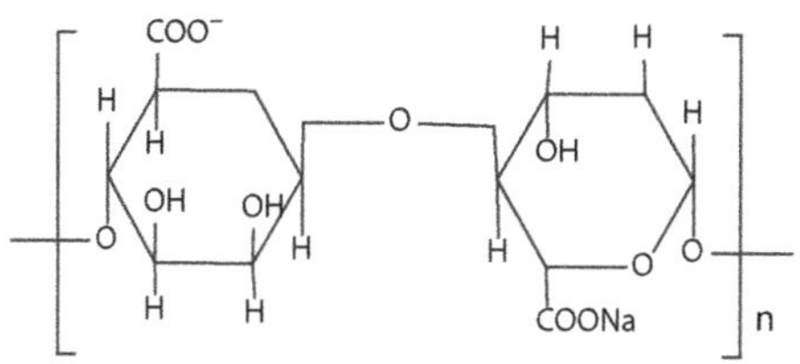 **Sodium Alginate**	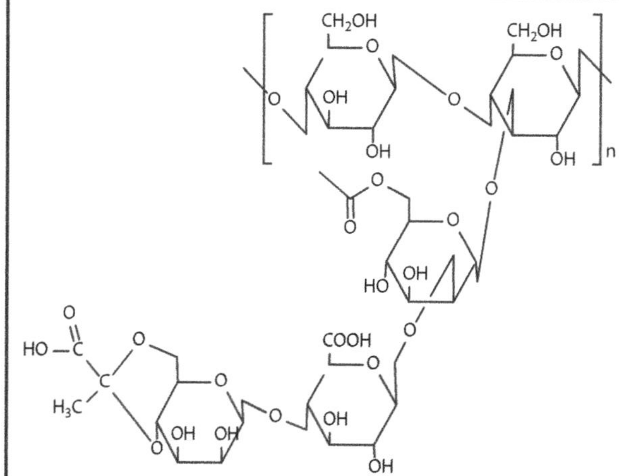 **Xanthan gum**

(Continued)

Table 5.1 Chemical structures of natural polymeric materials. (*Continued*)

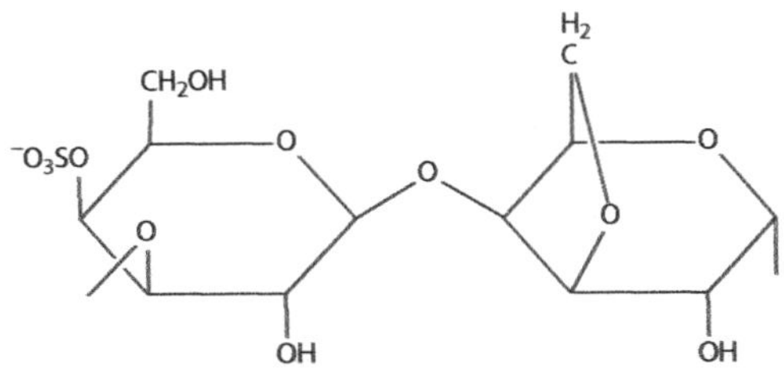

Kappa Carrageenan

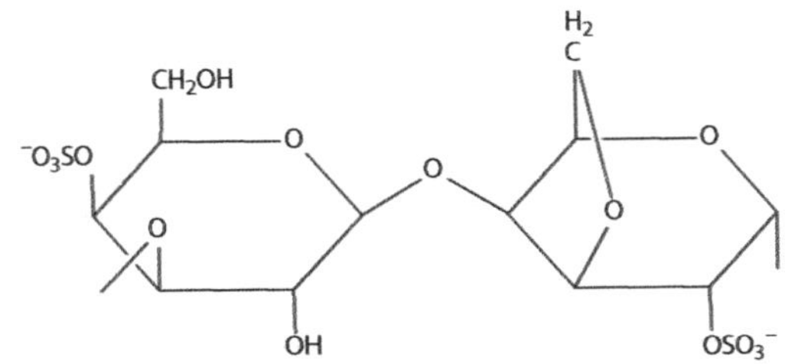

Iota Carrageenan

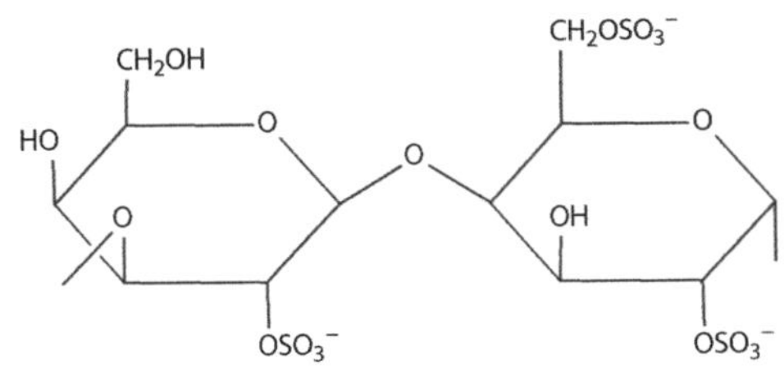

Lambda Carrageenan

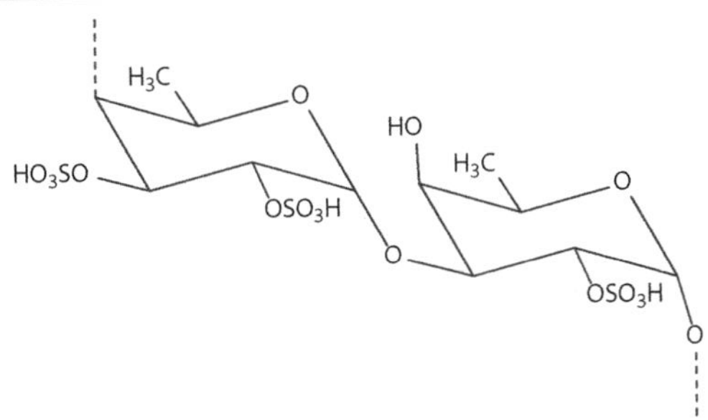

Fucoidan

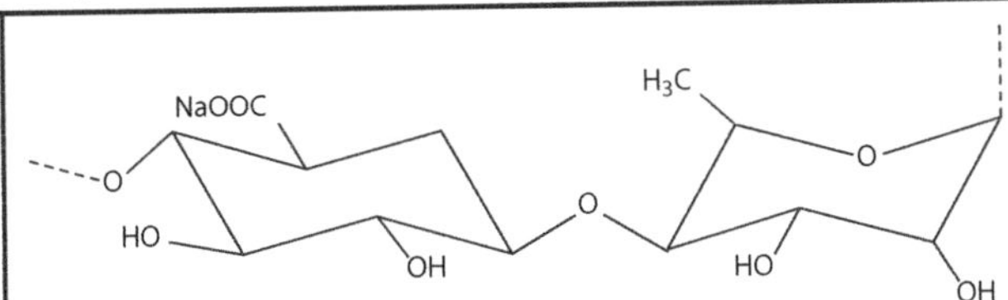

Ulvan

(*Continued*)

Table 5.1 Chemical structures of natural polymeric materials. (*Continued*)

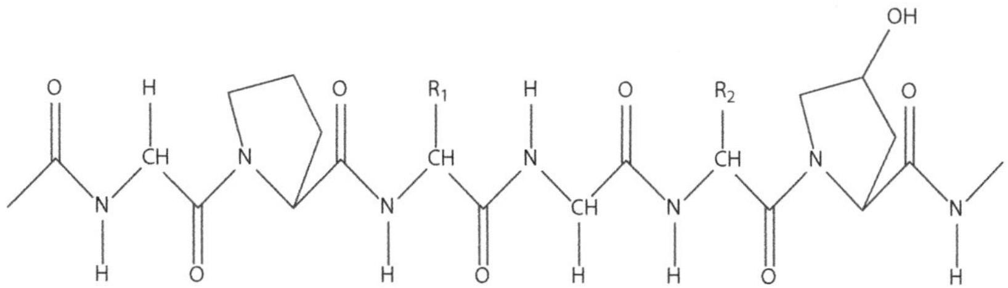

Collagen

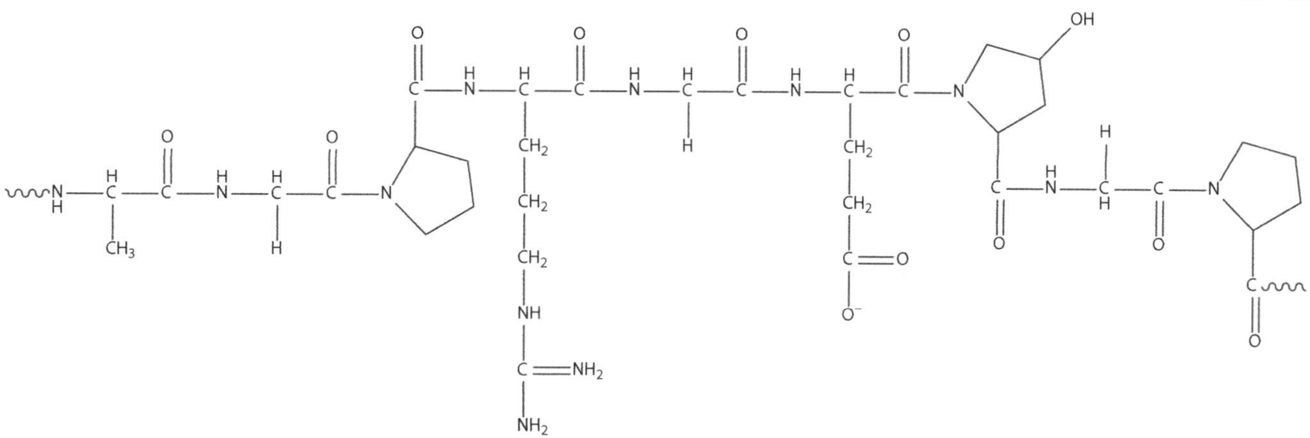

Gelatin

5.2.5 Cellulose-Based Hydrogels

Cellulose, the abundantly distributed biomass material and unbranched polysaccharide constituting β(1-4)-glycosidic-glucose linkages has been extensively explored for environmental to biomedical applications [43]. However, there is an urgency of improvement in mechanical properties of cellulose derived hydrogels in most of instances. The presence of extensive H-bonds in the cellulose structure make it rarely solubilized polymer in most of the common solvents that causes adversity to form hydrogels [44]. Hence, for the fabrication of cellulose-based hydrogels, the electing appropriate solvent is an ardent task. The opportune solvent system furnished the better physical cross-linking which in turn provides the excellent properties to the synthesized hydrogel. Primarily, the chemical cross-linking technique was adopted for the fabrication of cellulose hydrogel having superior mechanical properties [45]. Firstly, the cross-linking between cellulose and NaOH/urea solutions was engendered then subjected to in situ polymerization of acrylamide and accomplished the cellulose-polyacrylamide hydrogel. Zhang and research team reported a strong network structure of cellulose/poly(*N*-isopropylacrylamide) hydrogel possessing high thermal stability and remarkable compressive strength [46]. Similarly, Lu and co-workers in 2018 tailored a cellulose and polyacrylamide based pH responsive hydrogel having dual network structure [47]. Owing to marvelous features of cellulosic hydrogels such as responsiveness to acidic to basic medium, temperature and time, they are considered as smart materials for their applications in multidisciplinary areas such as biomedical field, agriculture and textile industries.

5.2.6 Chitosan-Based Hydrogels

Chitosan, a linear and 2nd most abundant cationic biopolymer is a product of chitin deacetylation. The presence of free -NH_2 and -OH group on the acetylated and deacetylated units expedite the structural modification and derivatization of functional moieties to achieve targeted applications [48]. The appreciable solubility of chitosan in most of the organic acids is attributed to protonation of -NH_2 group [49]. The synthesis of chitosan-based hydrogel was carried out via hydrogen bonded integration of chitosan having short chain length and covalently cross-linked polyacrylamide network exhibiting better tensile strength, improved self-healing and load bearing efficiency, collision and fatigue resistance and significant mechanical properties [50]. Another chitosan-polyacrylamide based cross-linked hydrogel was fabricated employing *N,N*-methylene bisacrylamide as a cross-linker via photopolymerization process. Such hydrogel possessed the 30 times higher tensile and mechanical strength with remarkable compressive strength, elastic modulus and fracture energy [51]. Further, the film forming properties and pH responsiveness of chitosan based hydrogels make it accessible for developing porous network structure via electrodeposition of hydrogels on electrodes which led to entrapment of various biocomponents such as drugs (ibuprofen), radioactive molecules (catechol) etc. in the porous hydrogel network.

5.2.7 Xanthan Gum-Based Hydrogels

Xanthan gum, an anionic exopolysaccharide composed of pentasaccharide units. Xanthan gum possessed the non-Newtonian behavior along with other interesting properties such as thermal stability, high water solubility, high viscosity etc. [52]. Xanthan gum showed the unique rheological properties and results pinpoints that the xanthan gum viscosity has no temperature dependency. However, the grafting of monomer units onto xanthan gum backbone resulted in decrease in viscosity owing to the attainment of flexibility by grafted chains [53]. Due to the presence of diverse functional moieties onto biopolymer xanthan gum, it goes under chemical modifications and facilitates the formation of xanthan gum based grafted or cross-linked hydrogels. Various xanthan gum based hydrogels were synthesized employing acrylic acid [54], acrylamide [55], 4-vinyl pyridine [56] etc. as a monomeric grafting units and the fabricated hydrogels were employed for different applications.

5.2.8 Sea Weed-Derived Polysaccharide-Based Hydrogels

Over the previous decades, the sea weeds-derived polysaccharides came up as ideal candidates for fabrication of hydrogels. Generally, alginate, fucoidan, carrageenan and ulvan are widely explored for their usage in biological and biomedical fields. Meanwhile, the gelling properties of alginate have broadened its application spectrum form tissue engineering to drug delivery [57].

Alginate, an anionic polysaccharide, composed of mannuronic acid and guluronic acid which forms β(1-4) linkages and α(1-4) linkages respectively and thus, commenced a linear, flexible conformations and steric hindrance in the carboxyl groups vicinity respectively. The presence of carboxylate group onto alginate backbone goes under ionic interactions with the cationic cross-linking moiety, consequently led to alginate derived hydrogels [58]. The synthesis of sodium alginate and polyacrylamide based cross-linked hydrogel via ion exchange method employing multivalent cations as a cross-linking agent was carried out. The trivalent cations showed the high binding degree and more compact network with the alginate in contrast to divalent cations owing to the three carboxylic group interactions with the trivalent cations at a time. Further, carrageenan, fucoidan and ulvan based hydrogel were fabricated by ionic cross-linking, chemical cross-linking and photo cross-linking respectively [59].

5.2.9 Protein-Based Hydrogels

The presence of various functional moieties including $-NH_2$, -COOH, -OH, -SH provides the inherent advantages for fabrication of hydrogel as these moieties served the reactive sites for chemical modifications via physical, chemical or enzymatic treatments. Oftentimes, proteins are embedded into hydrogels to furnish the biomimetic materials [60]. However, thermally induced protein-based hydrogels exhibit the poor reswelling capacity and not so good mechanical properties due to high occurrence of protein-protein interactions, electrostatic and hydrophobic interactions because of dehydration. These shortcomings limit their usage in the field which required highly strengthened hydrogels [61].

The proteins that have been mostly used to synthesize hydrogels are Collagen, Gelatin, Fibrin/fibronectin, Elastin, Fibroin, Whey proteins, Sericin. Collagen, a right-handed triple helix constituting a repeated amino acid unit viz. Glycine, proline and hydroxyproline respectively is mainly found in extracellular matrices in a fibrous form. Collagen-based hydrogels network was created by providing the acidic medium that originates aggregating the collagen molecules through covalent bond [62]. Gelatin is derived from the triple-helical structure degradation of collagen into single-strand molecules. Gelatin based hydrogels was synthesized at low temperatures owing to its thermo-responsive nature. The fabricated hydrogels exhibit the poor thermal stability and are nonimmunogenic [63]. Further soya-based hydrogels were synthesized with the aid of chemical cross-linking agent mainly glutaraldehyde which strengthens the structure of hydrogels while, silk based hydrogels were generally prepared by physical cross-linking methods. Keratin, a fibrous protein is one of the promising candidates for gel formation and the keratin derived hydrogels were developed as natural alternative for long-lasting biomaterials [64].

5.2.10 DNA-Based Hydrogels

DNA is a negatively charged double helical structured hydrophilic polyelectrolyte constituting a polar sugar–phosphate backbone. The gel formation capability of DNA attributed to the hydrophilicity which enhanced the bond formation tendency with water molecules. DNA hydrogels possessed a dual property i.e. chemical properties like DNA and physical properties like hydrogels [65]. From past two decades, DNA is emerged as a promising material for fabrication of ordered nanostructures. However, unordered DNA structures led to the formation of hydrogels via physical/chemical entanglement. Tanja *et al.*, presents the overview on first extraction of nucleic acid from complex helical structure to the fundamental developments in DNA hydrogels. The literature data based on DNA hydrogels revealed the formation of hydrogels with positively charged polyelectrolytes was carried out via electrostatic interactions or with the aid of ionic cross linkers which are highly pH, temperature and UV light responsive [66]. Liu and co-workers fabricated a thermal and enzyme responsive DNA based hydrogel series which were employed for multilayer 3D bioprinting [67].

5.3 Synthesis Methods for Fabrication of Natural Polymer-Based Hydrogels

The hydrogels synthesis was basically tailored by following methods:

- ➢ Chemical cross-linking which involves the covalent bonding between the polymeric chains
- ➢ Physical cross-linking exhibiting physical interactions between polymeric chains resulted in prevention from dissolution of hydrogels

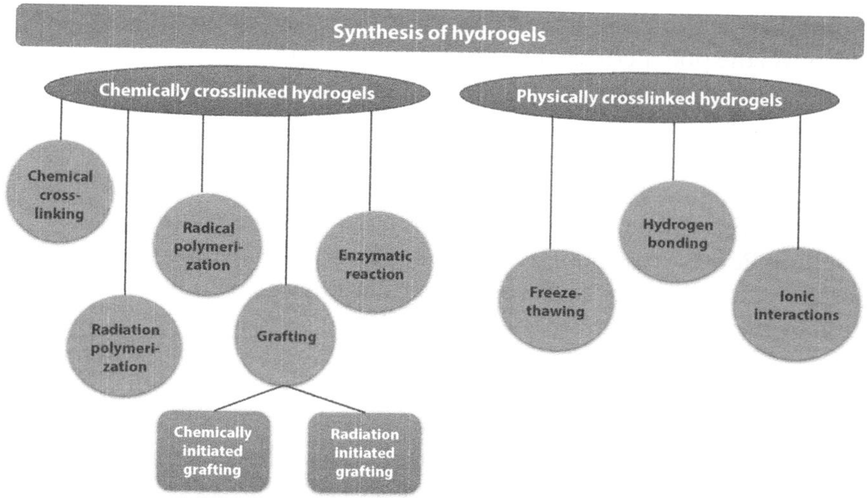

Figure 5.5 Synthesis methods of hydrogels.

The various types of methods fall under chemical and physical cross-linked fabrication techniques have been shown in Figure 5.5 and discussed in this section [68].

5.3.1 Natural Polymer-Based Chemically Cross-Linked Hydrogels

Chemical Crosslinking Method

Crosslinking of polymeric chains via ionic or covalent bonding results in a transformation of liquid polymers into well designed gel networking structures. In chemical cross-linking process, several cross-linking agents such as glutaraldehyde, terephthaldehyde, epichlorhydrin etc. have been used which forms the covalent bonds with the functional moieties present on the polymeric backbone [69]. Bruno M.S. *et al.*, synthesized a xanthan gum and starch based cross-linked gel using citric acid (cross-linker) with significant swelling properties [70]. Cellulose and gelatin cross-linked gel was reported by Buhus G. and co-workers employing glutaraldehyde as a cross-linker for controlled drug release [71]. Further, Yu S. *et al.* introduced a novel carboxymethyl modified chitosan and poloxamer based pH-induced thermosensitive cross-linked hydrogel synthesized using glutaraldehyde cross-linker which exemplify the chemically cross-linked hydrogels [72]. The gel formation with the aid of cross-linking treatment influences the physical properties of the hydrogel depending upon the amorphous/crystalline nature and degree of cross linking which have been discussed below:

- **Elasticity**: The high cross-linking degree reduced the elasticity of hydrogels thereby increasing the strength and durability.
- **Viscosity**: Crosslinking restricts the movement of polymeric chains during the hydrogel formation thus, increases the viscosity which in turn improves the creep behavior.

- **Solubility**: The uncross-linked polymeric chains get dissolved in various solvents. As the polymeric chains get tied with the aid of cross-linker, the resultant cross-linked material is insoluble in solvents. Thus, cross-linking process decreases the solubility.
- **Glass transition temperature**: Crosslinking process decreased the free volume by changing the local molecular packing and formed a strong networking hydrogel which results in high glass transition temperature.
- **Melting Point**: Crosslinking of polymers lowers the melting point of synthesized hydrogels owing to the occurrence of hindrance to the chain orientation by introducing the cross-linking agent.

5.3.2 Grafting Method

The unmodified natural polymeric materials have certain shortcomings such as short shell life, low strength and stability etc. The disadvantages could be overcome by modifying the natural polymers by different methods. Grafting is one of the promising methods which improve the natural polymer properties by addition of other synthetic monomeric units [73]. Generally, the grafting of monomer onto natural polymeric backbone was initiated by generation of free radical by external agent. Then, addition of monomer was carried out by means of grafting. Basis on type of activating agent to initiate the reaction, grafting polymerization reaction is of two types:

(a) Chemical grafting
In chemical grafting, the chemical reagent generates the free radical onto polymeric backbone to initiate the polymerization reaction. Currently, two chemical reagents viz. ammonium persulfate and potassium persulfate are widely explored as initiators for polymerization reaction by means of chemical grafting process. Singh B. *et al.*, reported a chemical grafting of vinyl monomer onto acacia gum backbone by generating a free radical with the aid ammonium persulfate initiator [74]. Further, potassium persulfate initiated polymerization reaction for the synthesis of pullulan based porous hydrogel was carried out [75].

(b) Radiation grafting
Hydrogel synthesis via radiation grafting can be carried out by exposure of different radiations such as microwave radiations, UV radiations, γ-radiations onto aqueous polymeric solution with the usage of various radiation sources. The exposure of radiations generates free radical sites onto polymeric backbone and generation of hydroxyl radicals attributed to the dissociation of water molecules [76]. These free radicals combined and formed the covalent bonding with the extended radicals thus, formed a cross-linked structured hydrogel.

5.3.3 Radical Polymerization Method

Radical polymerization is another approach for natural polymer based hydrogels synthesis involving initiator decomposition by temperature, photon and upon exposure of light for formation of free radicals [77]. Radical polymerization for hydrogel synthesis involves the three steps:

- **Chain initiation**: To general radical with the aid of initiator
- **Chain propagation**: Reaction of radical with the monomeric unit and occurrence of active sites by creation of extended radicals
- **Chain termination**: This is the final step for hydrogel formation comprising of combination of polymeric molecules with existing free radicals.

5.3.4 Irradiation Method

The hydrogel formation by irradiating the highly energetic radiations on long polymeric chains generally involved two phenomena i.e. chemical binding and dissociation of molecules [78]. The polymeric chains subjected to highly energetic gamma/electron beam radiation led to radical formation owing to the hemolytic dissociation of C-H bonds of polymeric chains. In addition, the irradiation process results in the hydroxyl radical formation by decomposition of H_2O molecules. The hydroxyl radical reacts with the polymeric chain and production of macro radicals takes place which further goes under combination reaction and formed the covalent bonded structures of hydrogel [79]. The irradiation methods exhibit the certain advantages of low cost and fast reaction process over conventional method as this method does not entails any additive for initiating the reaction. Tara gum and chitosan/alginate based hydrogels synthesized hydrogels by irradiating gamma radiations with reduction in reaction time exemplify the advantageousness of irradiation method [80].

5.3.5 Enzymatic Reaction Method

Enzymatic treatment approach for hydrogel fabrication is a novel greener concept and better alternate for cross-linking method. This method gives the high efficiency at mild working condition with no formation of toxic by-products. Horseradish peroxide enzyme catalyzed the synthesis reactions of Tragacanth gum and hyaluronic acid-tyramine based hydrogels [81].

5.4 Natural Polymer-Based Physically Cross-Linked Hydrogels

Recently, the fabrication of hydrogels by physically cross-linking method has gained a significant consideration as physical cross-linking technique avoids the usage of toxic chemical additives which acts as a cross-linking agent. The physical cross-linking methods explored in literature for hydrogel synthesis have been discussed here.

5.4.1 By Freezing and Thawing Cycles

Freezing and thawing process emerged as one of the finest synthesis protocols among physical cross-linking techniques. The fabrication of hydrogel via freeze and thawing method was carried out without usage of cross-linking agents and considering the polymer ratio and freezing conditions such as temperature, number of freeze and thaw cycles etc. These conditions may directly affect the properties and targeted applications of synthesized hydrogels. During the recent past years, the synthesis of hyaluronan, starch and salecan based hydrogel was attempted by freeze thawing method and formed a cross-linking of natural polymeric chains with the poly vinyl alcohol by undergoing repetitive freeze thaw cycles [82, 83].

5.4.2 By Hydrogen Bonding

Establishment of hydrogen bonded polymeric hydrogel network exhibited the high gel concentration and low cross-linking density. The H-bond derived non-covalent intermolecular interactions could also participate in gel formation process. The synthesis of H-bond derived physically cross-linked and anti-HIV drug loaded porous hydrogel of hydroxyethyl cellulose and poly acrylic acid was reported by Mabrouk M. and co-workers with improved drug solubility. The high solubility of loaded drug was attributed to the masking of positive charges via hydrogen bonding [84].

5.4.3 By Ionic Interactions

Ionotropic gelation is another promising synthesis protocol for the fabrication of hydrogels. This synthetic approach is dependent upon the polyelectrolyte ability to form the cross-linking networks in the presence of counterions. The natural polyelectrolytes generally possess the negative charged backbones. Thus, have an ability to combine with positively charged or cationic moieties for the formation of hydrogel networks. Das M.K and Senapati P.C. reported a furosemide-loaded alginate microspheres hydrogel with the aid of Calcium, barium and aluminum cations via ionotropic external gelation synthesis protocol [85].

5.5 Properties of Natural Polymer-Based Hydrogels

An ideal hydrogel possessed a numerous characteristic such as ample photostability, high water uptake capacity, appropriate porosity, appreciable sustainability, economically safe, low concentration of monomer, significant biodegradability and biocompatibility, colorless and odorless, pH neutrality etc. However, it is an ardent task to synthesize an ideal hydrogel exhibiting all the characteristics [86]. The natural polymer based hydrogels possessed many of the properties and maintained a balance of properties depending upon the desired applications. In this section, some of the properties of natural polymer derived hydrogels have been discussed.

5.5.1 Mechanical Properties

The evaluation of mechanical strength of natural polymers derived hydrogels is essential for their functioning and sustainability of physical texture while employed for various perspectives. The synthesis of hydrogels with desired mechanical properties was achieved by incorporation of suitable polymeric chains, monomeric units, initiators and cross-linking agents [87]. The high cross-linking degree led to the formation of strong gel network while too high degree of cross-linking results in low elasticity and high brittleness which in turn decreased the flexibility of cross-linked network. Thus, an optimization of cross-linking degree is highly essential to attain the required mechanical and elastic properties of hydrogels. The mechanical properties of hydrogels have been evaluated by numerous techniques such as frequency based sinusoidal testing, compression and tension analysis [88].

5.5.2 Biodegradability

The biodegradability of natural polymer based hydrogels is the key factor for their exploitation in biomedical field. The term 'biodegradability of hydrogels' means the disintegration of hydrogel by the living organism into non-toxic by-products involving hydrolysis and solubilization of bio-entities. The hydrophilicity of natural polymers led to high water uptake capability of naturally derived hydrogels [89]. Thus, the biodegradability of hydrogels relies upon the various factors such as synthesis protocol, presence of hydrophilic moieties in the system, polymer -water interactions etc. The degraded end products may be eradicated from the body by bio-absorption/bio-erosion mechanism. Besides, there are certain polymeric materials which cannot undergo degradation by hydrolysis but required a chemical hydrolysis treatment [90]. Furthermore, enzymatic hydrolysis also results in the degradation of polysaccharide/protein based hydrogels by hydrolase enzyme which enhanced the C-C, C-O, and C-N bond hydrolysis. There are several enzymes such as glycosidase, Lysozyme etc., which catalyze the polysaccharide hydrolysis and led to degradation into harmless by-products [91].

5.5.3 Swelling Characteristics

The dehydrated form of hydrogel is referred as glassy state. When the free interstitial spaces of hydrogel matrix occupied with solvent molecules, the glassy state or dehydrated hydrogel turns into rubbery state which is known as swelling. The swelling behavior of hydrogels can be measured by means of water uptake capacity in the interstitial sites of hydrogels in terms of swelling ratio according to the formula [92]

Swelling ratio = weight of swollen gel/ weight of dry gel

The swelling mechanism of hydrogels generally involved three steps:

(1) Diffusion of water into the interstitial hydrogel network
(2) Loosening up of polymer chains
(3) Expansion of hydrogel network

The water uptake efficiency of hydrogels depends upon the numerous factors such as (a) degree of cross-linking, (b) chemical structure, (c) thermal conditions, (d) pH, etc. The cross-linking degree has significant impact on the swelling capacity of the hydrogels as cross-linked hydrogels with high degree of cross-linking possesses the lower swelling capacity and vice-versa [93]. The hydrogels containing more hydrophilic moieties in contrast to hydrophobic groups in their chemical structures showed the tremendous swelling properties. Further, pH responsive hydrogels exhibit the high swelling capacity owing to the ionization of hydrophilic moieties by varying the pH conditions.

5.6 Stimuli Sensitivity of Hydrogels

Stimuli responsive polymeric hydrogel networks have gained tremendous attention in various fields owing to their responsiveness towards environmental changes, such as light sources, thermal conditions, enzymes, pH, ionic strength, shear stress etc. Some of these stimuli-sensitive properties of hydrogels have been discussed over here.

pH-responsive hydrogel networks restraining pendant acidic groups (carboxylic acid, sulfonic acid etc.) and basic groups (amines) on the polymeric backbone named as anionic and cationic hydrogels respectively [94]. Depending upon the nature of pendant group, H^+ leave or attached with the polymeric chain with varying pH conditions. Polymeric chains exhibiting abundance of cationic and anionic functional moieties are referred as polyelectrolytes. The pendants of polyelectrolytes undergo ionization with pH variation and effects the swelling properties as well as dissociation constants [95]. pH-responsive hydrogels were widely explored for biomedical applications especially for targeted and controlled release of drug owing to the pH difference in body parts. The hydrogels comprised of cationic polymeric chains released the drug at low pH levels and these kinds of hydrogels mainly exploited as stomach ($pH<3$) drug delivery carrier [96].

Temperature is a widely explored stimulus in environmentally sensitive natural polymer based hydrogels. Thermal responsive polymeric hydrogels experienced a volume phase transformation with change in temperature conditions [97]. The solubility of polymeric materials also showed dependency on the thermal conditions which can be recognized from lower critical solution temperatures (low solubility at high temperature) and upper critical solution temperatures (high solubility at high temperature). The hydrogels having lower critical solution temperatures exhibiting the low water uptake capacity at low temperatures and shrinks at extended temperatures are referred as negative temperature-responsive systems [98].

5.7 Application of Hydrogels as Electrochemical Supercapacitors

In today's world, the human's dependence on electronic gadgets and electric vehicles is increasing rapidly, due to which need of energy sources is rising. Subsequently, many research communities are focusing on energy storage devices and conversion technologies [99]. Among various energy storage and electrochemical technologies including fuel cells, batteries and electrochemical supercapacitors have become most popular because they are renewable energy sources [100]. Due to the tunable properties, electrochemical supercapacitors come out as attractive strategy and its applications are extended for hybrid or electric vehicles, electronics, smart grids, and aircrafts [101]. Apart from no greenhouse gases release, there are several advantages of using electrochemical supercapacitors like high power density (3000–40,000 W kg^{-1}), fast charging (charge time 1-10 s), high efficiency, wide operating temperature range (-40 to 85 °C), lifetime (> 100,000 cycles), explosion risk free, inexpensive maintenance and eco-friendly [102]. The electrochemical supercapacitors are also popularly called as ultracapacitors and supercapacitors [103].

Supercapacitors are achieving attention on commercial basis due to benefits which are complementary to the other energy storage devices. Supercapacitors are having power densities more than fuel cells and batteries as well as energy densities higher than traditional capacitors. The supercapacitor technology is a bridge between the fuel cells/batteries and conventional capacitors regarding power and energy density [104].

5.7.1 Types of Supercapacitors

The Supercapacitors can be categorized into three classes (shown in Figure 5.6) that are electrochemical double layer capacitors, pseudo capacitors and asymmetric or hybrid capacitors based on the charge/energy storage mechanism involved [99, 105].

5.7.2 Electrochemical Double-Layer Capacitor (EDLC)

EDLCs have alike mechanism to that of conventional capacitors. The only difference is that EDLCs store charges in the interface of electrode and electrolyte instead of dielectric layer. The potential difference in EDLCs induces different polarity at the electrodes which leads to the movement of electrolyte ions to the porous network of electrodes. In traditional capacitors, the capacitance depends upon the separator thickness while in case of EDLCs, capacitance depends upon the width of double layer at the interface of electrode and electrolyte. Figure 5.7 clearly shows that the thickness of the separator is much larger than the thickness of double layer (few angstroms). From which, it can be easily concluded that EDLCs will have higher capacitance than the conventional ones. Besides this, EDLCs have larger interface area between the electrode and electrolyte comparative to the conventional capacitors

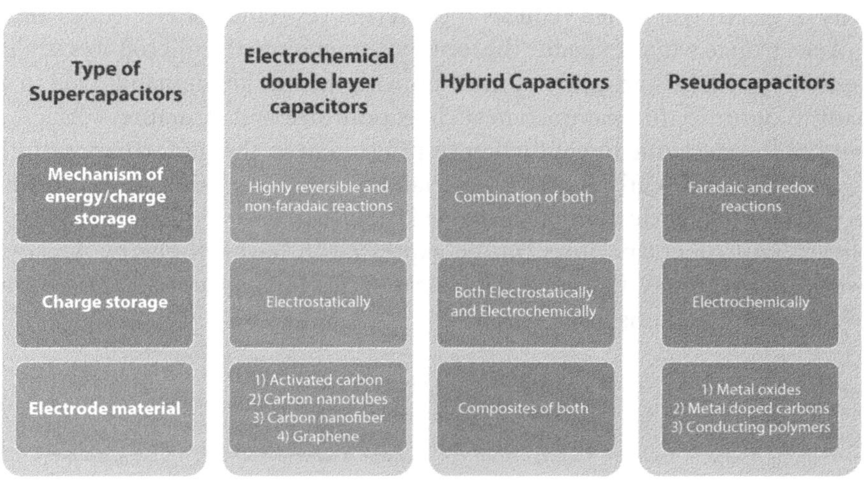

Figure 5.6 Classification of supercapacitors.

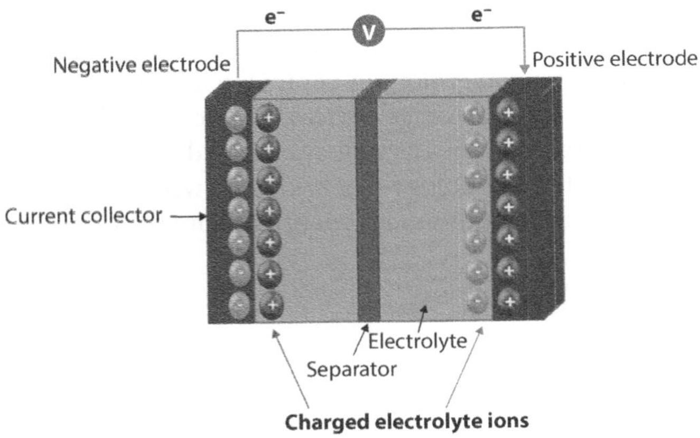

Figure 5.7 Pictorial depiction of electrochemical double layer capacitor.

which further enhances the capacitance of EDLCs. The capacitance of EDLCs can be computed from the following equation:

$$C = \frac{\varepsilon_r \varepsilon_0 A}{d}$$

Where C represents the capacitance, ε_r represents dielectric constant of electrolyte, ε_0 is vacuum permittivity, A is the electrode's surface area when dipped in the electrolyte, and d is the width of the double layer at the electrode-electrolyte interface. When the charging process occurs, anions of the electrolyte gathered at the anode and cations of the electrolyte get attracted towards the cathode. The charge or energy storage mechanism in EDLCs involves reversible and non-faradaic reactions at the interface. Due to physical charge transfer, EDLCs have comparatively longer lifetime.

The other factors which amplify the performance of EDLCs are magnificent electrical conductivity of electrolyte which reduces the internal resistance of electrodes and wetting tendency of electrolyte which expedite the ions movement into the micropores of electrodes.

Despite this, the EDLCs are still not meet up with the market demand. Due to which, it is important to optimize the parameters such as pore size and structure, surface area and conductivity of the electrode to get further improvement in the performance of EDLCs in terms of capacitance. The carbon based materials have high surface area due to which they have been used for the fabrication of electrode [103].

The efficiency of supercapacitors can be expressed in terms of energy density, E (Wh kg^{-1}), power density, P (kW kg^{-1}) and equivalent series resistance, ESR (Ω) which can be calculated from the equations given below:

$$E = \frac{C \times \Delta V^2}{2} \times \frac{1000}{3600}$$

$$P = \frac{I \times \Delta V}{2 \times m_{ac}} \times 1000$$

$$ESR = \frac{iR_{drop}}{2 \times I}$$

where 'I' is the discharge current (A), ΔV is the change in potential after charging or discharging process (V), m_{ac} is the weight of electrode (g), and during charging-discharging, iR_{drop} (V) is the difference in electrical potential between the periphery of conducting phase [105].

5.7.3 Pseudo Capacitor

In pseudo capacitors, charge transfer mechanism is based on faradaic process which involves redox reactions on the surface of electrode, contrasting EDLCs, in which physical charge accumulation occurs. During charging-discharging process, the material for the electrode should have redox capability for sustaining charge balance via chemisorption of electrolyte ions. Through the faradaic reaction, electrons are tunneled through the interface of current collector and electrode and this charge transfer mechanism is of three types –

i) Underpotential deposition (UPD)
ii) Redox pseudo capacitance
iii) Intercalation pseudo capacitance

In UPD charge transfer mechanism, metal ions or cations adsorb on the metal surface to form a monolayer beyond the redox potential, therefore UPD mechanism is also known as 2D adsorption pseudo capacitance. Whereas in case of redox pseudo capacitance, ions adsorbed electrochemically at the electrode material surface followed by charge transfer process. The electroactive materials like metal doped carbons, transition metal oxides and conducting polymers are explored to demonstrate redox pseudo capacitance. The valence states of these electroactive materials change momentarily accompanied by charge transfer during the charging-discharging process. The third mechanism is intercalation pseudo capacitance, in which, metal ions or cations intercalate in the layers of electrode material followed by charge transfer. The intercalation pseudo capacitance involves no phase (crystallographic) change, for which, rigid materials are required and there should be minimal distortion in structure during ions transportation in 2D pathway. The pseudo capacitors have 10 times higher specific capacitance than that of EDLCs. But pseudo capacitors generally have low power density and less stability than EDLCs due to low electrical conductivity during life cycles [103].

5.7.4 Asymmetric or Hybrid Supercapacitors

These supercapacitors are called as asymmetric supercapacitor because of distinct electrodes, one is battery-alike Faradaic electrode, and another is capacitive carbonaceous electrode. This type of supercapacitor is a fusion of both EDLC and pseudo capacitor thence also known as hybrid supercapacitors. This supercapacitor is designed in such fashion that it has advantages of both EDLCs (high power density) and pseudo capacitors (high energy density and high capacitance). Consequently, optimization of individual electrode leads to the better performance of the supercapacitor. Still, there is a disadvantage of hybrid supercapacitor that it may have lessened cycle stability [99].

The supercapacitor's efficiency can be boosted with the evolution of its design. The evolution of the technology behind the supercapacitors is greatly dependent upon the deliberate selection of supercapacitor components such as electrodes, electrolyte, conductive current collectors, and sealants and modification in material of components are the foremost priority. Thus, the main components of electrochemical supercapacitor are briefly discussed below.

5.8 Conducting Polymer Hydrogels as Electrode Materials

Electrodes are the fundamental components of supercapacitor and play vital role in controlling the specific energy and power values of the supercapacitor. Therefore, electrodes used in the fabrication of supercapacitor should be of electroactive materials. The specific energy of supercapacitors is less comparative to batteries and fuel cells due to which intensive innovation is going on to fabricate such electrode material which fills this energy density gap while maintaining energy storing capacity [104].

Mainly three type of materials have been used for electrode fabrication in supercapacitors that are carbon materials conducting polymer hydrogels and transition metal oxides. In EDLCs, generally carbon based materials (Activated carbon, templated carbons, carbon nanotubes, carbon nanofiber, graphene) have been investigated for electrodes fabrication because of high abundance, cost effectiveness, electroactivity, tunable morphology and high surface area [105]. The charge produced in carbon based materials deposited in electrical double layer produced at the electrolyte/electrode interface. Consequently, the specific conductance reckons on the surface area which is approachable to ions of the electrolyte. Despite the advantages, these materials have limitations of low specific capacitance value. However, on the other side, conducting polymer hydrogels can overpower the above limitations [101]. With tunable properties, conducting polymer hydrogels can be engineered to satisfy the required demands of electrodes and upgrade the performance of supercapacitors (Figure 5.8).

The cross-linked structure of polymer hydrogel furnished with water insolubility and endows mechanical and physical strength. The electrodes should have ionic and electronic conductivity to expedite the movement of ions or electrons of the electrochemical reaction process. The hydrogels are synthesized by introducing conjugated polymers for conducting electrons along the conductive chains of the polymer and ions through the electrolyte in between the hydrogel networks [106]. The specific type of cross-linking agent in the

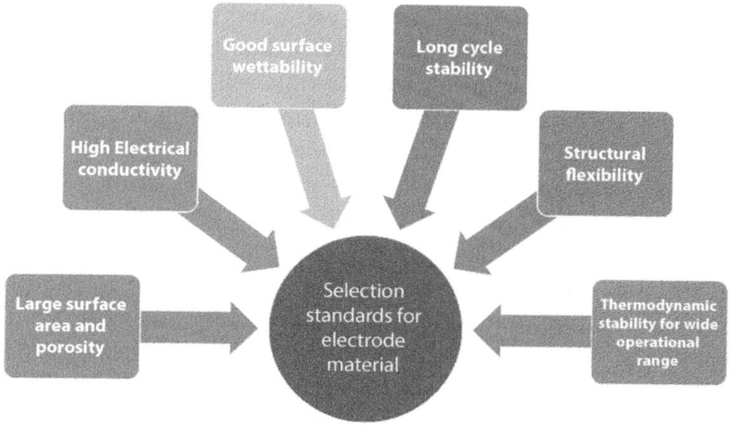

Figure 5.8 Selection standards for electrode material.

synthesis of conducting polymers is selected to get the specific morphology and electrochemical property of the electrode for supercapacitor devices [100]. Besides high electrical conductivity, conducting polymers have other superior features also like high pseudo capacitance, structural flexibility, fast redox reactivity, eco-friendly and cost effective [107]. The reversible redox reactions occurred in polymer matrix generate pseudo capacitance. The most frequently employed conducting polymers for electrode fabrication are polyaniline [108], polythiophene [109], polypyrrole [110] and poly (3,4-ethylenedioxythiophene) [111] etc.

The conducting polymer synthesized in the film form deposited on the electrode surface by requisite potential. The various factors like monomer concentration, electrolyte concentration, solvent nature, and applied potential during deposition of polymer film affect the morphological flexibility and conductivity of the electrode. The major limitation of using conducting polymers is low cycle and mechanical stability due to shrinking-swelling phenomena occur during charging-discharging of electrode. Therefore, substantial research has been done to overcome this limitation of using conducting polymer in supercapacitors [112]. In the last decade, noteworthy improvement has been done in cyclic as well as mechanical stability of polymer based supercapacitors and it has been emerged as appealing candidate for supercapacitor devices [113].

The conducting polymer hydrogels exhibit advantages of both organic conductors and hydrogels that are electrical and electrochemical properties [114]. Therefore, many research efforts have been done for the fabrication of conducting polymer hydrogels which are broadly divided into three categories – polymerization via chemical or electrochemical means, copolymerization of conducting with non-conducting one and introduction of cross-linking agents into the polymer [115]. The third approach of synthesizing conducting polymer based hydrogels in the presence of cross-linkers is less beneficial because cross-linkers are generally of non-conductive nature due to which electrical conductivity as well as electrochemical activity reduces. Consequently, self-cross-linking method is better to employ in which there is no need of cross-linking agent [116].

Guo *et al.* employed this self-cross-linking approach to synthesize polyaniline hydrogel from aniline hydrochloric salt precursor and using oxidizing agent, ammonium persulphate.

The synthesized polyaniline hydrogel electrode exhibited maximal 750 F g^{-1} specific conductance [117]. This ensures that self-cross-linked polyaniline can be promising candidate for electrode in energy storage gadgets. Despite this, self-cross-linking technique is not appropriate for the synthesis of all conducting polymer hydrogels and other treatments have also been given in combination such as supercritical CO_2 drying and treatment with acid, which limits its usage. The Inganas group directly applied conducting polymer hydrogels in supercapacitors by simple preparation method where poly (3, 4-ethylene-dioxythiophene)-poly (styrenesulphonate) is ionically cross-linked through the aqueous dispersion [118]. The synthesized PEDOT-PSS hydrogel electrode exhibited low mechanical firmness and less specific energy due to high monomeric unit mass of PEDOT-PSS. Then, the PEDOT-PSS hydrogel electrode was coated electrochemically with another conducting polymer, polypyrrole with large water content (75%) and show improved mechanical strength. Before cross-linking, water content of the hydrogel electrode can be controlled by partial drying dispersion. Furthermore, with both electrodes, very high specific power values can be attained without reducing much extractable energy of supercapacitors. The polyaniline hydrogel among the conducting polymer hydrogels, received much attention due to its high conductivity, tunable morphology, high pseudocapacitance, thermodynamic stability, good optical properties, and cheap synthesis methods [119]. These polyaniline hydrogels can be synthesized via different means such as chemical or electrochemical polymerization. In [120], synthesis of polyaniline based hydrogel using phytic acid as a dopant was reported. The polyaniline hydrogel electrode shows high capacitance, cyclic stability and high electrical conductivity. Pan *et al.* synthesized 3D polyaniline hydrogel with the similar method which involves gelation mechanism. The prepared polyaniline hydrogel exhibited 0.11 S cm^{-1} ionic conductivity, 480 F g^{-1} specific capacitance at 0.2 A g^{-1} current density, 83% capacitance up to 10,000 charge-discharge cycles at 5 A g^{-1} current density [121]. Xiang *et al.* report the synthesis of 3D polyaniline/phytic acid through electrochemical polymerization means which is free from frameworks as well as initiating agents. Due to high conductivity, better contact of electrode connecting electronic and ionic transporting phase, the supramolecular hydrogel shows large specific capacitance (311.3 F g^{-1} at 1 A g^{-1}) and areal capacitance (561.6 mF cm^{-2} at 1mA cm^{-1}). Moreover, solid state micro supercapacitor provides remarkable energy density (49.9 μWh cm^{-2} at 0.4 mW cm^{-2} power density). The micro-supercapacitor retains 76% capacitance after 10,000 charge-discharge cycles [122].

Shi *et al.* report the synthesis of nanostructured polypyrrole hydrogel by adopting interfacial polymerization method. The phytic acid is used in the preparation of 3D nanostructured hydrogels and it plays vital role in controlling morphological, mechanical, and electrochemical properties by varying the phytic acid to pyrrole ratio. The solid state supercapacitor prepared using polypyrrole electrode exhibit 380F g^{-1} specific capacitance, 6.4 F cm^{-2} areal capacitance at 20mg cm^{-2} mass loading and high rate capability [123]. The magnificent electrochemical performance and flexibility of prepared electrode material boost up the development of conducting polymer hydrogels for flexible supercapacitors [124].

5.9 Conducting Polymer Hydrogels as Electrolyte Materials

Besides electrodes, the electrolytes are also the basic component of supercapacitor and thence, major factor in deciding the performance of electrochemical devices. The standard features for electrolyte material shown in Figure 5.9. Although, the electrochemical performance of supercapacitors based on aqueous electrolyte prone to deterioration due to external conditions like bending, stretching, twisting, and shearing. Supercapacitors having liquid electrolytes, still endured with irreversible functional disorders, electrolyte leakage, explosion, and environmental issues. To overcome these issues, there is dire need of developing better electrolytes. The polymer hydrogel based electrolytes are emerging as advanced class of electrolyte materials. In this liquid electrolyte ions serves as conductive fillers, dispersed in polymer network which act as scaffolds. Generally, polymers used for this purpose are polyethylene oxide, polyvinyl alcohol, polyacrylic acid, polyacrylamide. The polymer hydrogel electrolytes exhibit numerous benefits such as high ionic conductivity, excellent mechanical flexibility, non-corrosiveness, low volatility, low leakage, and high safety. The higher ionic conductivity with polymer hydrogel electrolytes is because of easy diffusion of ions in the micropores of the polymer hydrogel. Thus, polymer hydrogel electrode is investigated as an appealing candidate for electrolyte material in flexible electrochemical supercapacitors [125].

Lewandowski *et al.* report the fabrication of supercapacitor using polyethylene oxide hydrogel as an electrolyte and activated carbon powder as an electrode. It exhibits good mechanical strength and 93 F g^{-1} specific capacitance almost similar with KOH electrolyte capacitor which ensures the feasibility of hydrogel electrolyte in supercapacitors. But its low melting point, high crystallinity and low electrode/electrolyte contact area reduces its electrochemical performance of supercapacitors [126].

The polyvinyl alcohol (PVA) based hydrogel electrolytes are comparatively possess high ionic conductivity, non-corrosiveness, non-toxicity, superior safety, and biocompatibility. The supercapacitor developed from PVA-LiCl hydrogel electrolyte exhibit high performance with 2.4 V operating voltage, superior rate capability, coulombic efficiency, and long life cycles with 95.5% capacity retention [127]. In hydrogel based electrolytes, water freezes and limit ion diffusion at subzero temperatures and their structures degrade at high temperature and therefore, have narrow operating range. Chao *et al.* [128] used montmorillonite (MMT) material as a dopant in PVA hydrogel based electrolyte to boost up the thermal

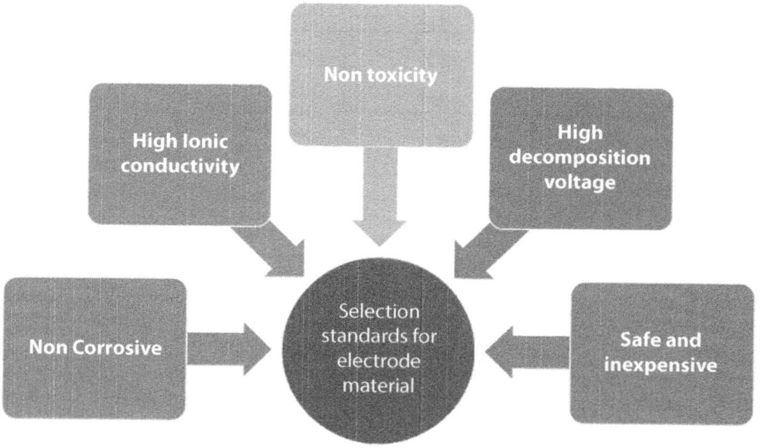

Figure 5.9 Selection standards for electrolyte material.

property of the polymer hydrogel electrolyte. Because to its lamellar structure formation, active ions transport through the conductive path and thus, ionic conductivity increases. The DMSO mixed with water, hydrogen bonding occurs between the DMSO and water molecules. Due to which, hydrogen bonding in between the water molecules restrained and consequently, reduces the freezing point. Thus, the prepared MMT/PVA hydrogel based electrolyte was anti-freezing as well as thermally more stable. The collective impact of using MMT and DMSO on the electrolyte properties is extension in its ionic conductivity up to 0.17×10^{-4} and 0.76×10^{-4} S cm^{-1} at −50 and 90 °C temperature, respectively. Generally, the hydrogel electrolytes are synthesized by physical blending method. But, due to this approach, mechanical strength and structural integrity reduces. Therefore, cross-linking strategy is more appealing approach for hydrogel electrolyte synthesis to improve the mechanical strength. Wang *et al.* [129] fabricate chemically cross-linked PVA hydrogel electrolyte using glutaraldehyde as cross-linking agent. The covalently cross-linked hydrogel electrolyte exhibits 0.082 S cm^{-1} ionic conductivity, 488 mF cm^{-2} areal specific capacitance,

Table 5.2 Comparison of different polymer electrodes or electrolytes based supercapacitor performances.

Polymer electrodes or electrolytes based supercapacitors	**Specific capacitance**	**Charge-discharge cycles**	**Capacitance retention**	**References**
Graphene/Polyaniline Nanofiber	210 F g^{-1}	800	73.8%	[108]
Polythiophene-tartaric acid nanoparticles	156 F g^{-1}	1000	97%	[109]
Polypyrrole electrode	170 F g^{-1}	10,000	98%	[110]
Polypyrrole nanowire arrays	566 F g^{-1}	300	70%	[112]
Polyvinyl alcohol/H_3PO_4 gel electrolyte	81 F g^{-1}	5000	94.5%	[113]
Polyaniline hydrogel electrode	430 F g^{-1}	1000	86%	[114]
Polyaniline nanofibers/Graphene hydrogel	492 F g^{-1}	1000	97.71%	[115]
Polyaniline hydrogel	862 F g^{-1}	5000	64%	[119]
3D Polyaniline hydrogel electrode	480 F g^{-1}	10,000	83%	[121]
3D Polyaniline/Phytic acid hydrogel	311.3 F g^{-1}	10,000	76%	[122]
Polypyrrole hydrogel	380 F g^{-1}	3000	90%	[123]
Polyvinyl alcohol polymer electrolytes	110-112 F g^{-1}	1000	95.5%	[127]

300% mechanical stretchability, superior elasticity, high mechanical strength which endures 290% tensile strain.

Although there is remarkable improvement in PVA electrolytes, they have limitations. Thence, it is important to design hydrogel electrolytes in such fashion that it exhibit better properties. The polyacrylic acid shows better ionic conductivity, mechanical flexibility and superior water loving property which improves its conductivity. Furthermore, carboxyl groups of PAA form hydrogen and coordination bonds with metal ions which makes it mechanically more flexible. The superior mechanical performance of the hydrogel electrolyte ensured by high cross-linking density of non-covalently bonded hydrogel. Yingxi *et al.* [130] further improved the performance of PAA hydrogel electrolyte by copolymerizing with stearyl methacrylate monomers in ferric nitrate solution accompanied by immersing in sulphuric acid. The prepared hydrogel exhibits >30 mS cm^{-1} ionic conductivity and >2400% stretchability. The flexible supercapacitors developed by above mentioned electrolyte and carbon with polyaniline based electrode shows 217.4 F g^{-1} specific capacitance at 1.67 A g^{-1} current density, superior cyclic stability up to 2000 life cycles with 80% capacitance retention. The performance of different supercapacitors based on polymer electrodes or electrolytes compared in Table 5.2.

The polyacrylamide (PAM) is also a better alternative to polyvinyl alcohol and because of amide groups, generally have weak hydrogen bonding comparative to polyvinyl alcohol. Thus, there is need to ameliorate the mechanical properties of PAM hydrogel based electrolytes. To achieve better mechanical stretchability, cross-linking agent is introduced into the polymer network system. Huang *et al.* [131] prepared the PAM based hydrogel electrolyte by introducing vinyl hybrid silica nanoparticles which act as cross-linking sites and release energy during stretching and compressing of hydrogel electrolytes. The supercapacitor based on this electrolyte exhibit better performance.

5.10 Conclusion

Natural polymer based hydrogels are soft in nature and have a considerable absorption ability for water or biological fluids within their interstices without any distortion in network structure owing to the presence of polar and hydrophilic functional moieties such as hydroxyl group, amine group, carboxylic group, amide group etc. attached with the polymeric chains. This chapter overviewed the diverse natural polymeric materials involved for the fabrication of hydrogels including various synthesis techniques. Further, classification of hydrogels depending upon type of polymeric material, way of cross-linking, stimuli responsiveness, synthesis method and ionic charge present on hydrogel. The natural polymer derived hydrogels possess abundant properties such as biodegradability, biocompatibility, mechanical properties, swelling characteristics, pH and thermal sensitivity etc. along with significant electrochemical supercapacitor activity. However, in this chapter, we have further closely focused on the electrochemical supercapacitors, their mechanism of energy storage and research work recently conducted on the synthesis of hydrogels derived from polymers that are polyaniline, poly (3, 4-ethylenedioxythiophene), polyvinyl alcohol and polypyrrole and their modified form for the synthesis of electrode and electrolyte material to improve the electrochemical activity or efficiency of supercapacitors.

References

1. Ahmed, E.M., Hydrogel: Preparation, characterization, and applications: A review. *J. Adv. Res.*, 6, 105, 2015.
2. Alonso-Sande, M., Teijeiro-Osorio, D., Remuñán-López, C., Alonso, M.J., Glucomannan, a promising polysaccharide for biopharmaceutical purposes. *Eur. J. Pharm. Biopharm.*, 72, 453, 2009.
3. Pandey, R. and Khuller, G., Polymer based drug delivery systems for mycobacterial infections. *Curr. Drug Delivery*, 1, 195, 2004.
4. Chatterjee, A.N., Yu, Q., Moore, J.S., Aluru, N.R., Mathematical modeling and simulation of dissolvable hydrogels. *J. Aerosp. Eng.*, 16, 55, 2003.
5. Alinejad, Y., Adoungotchodo, A., Hui, E., Zehtabi, F., Lerouge, S., An injectable chitosan/chondroitin sulfate hydrogel with tunable mechanical properties for cell therapy/tissue engineering. *Int. J. Biol. Macromol.*, 113, 132, 2018.
6. Barrett-Catton, E., Ross, M.L., Asuri, P., Multifunctional hydrogel nanocomposites for biomedical applications. *Polymers*, 13, 856, 2021.
7. Annabi, N., Tamayol, A., Uquillas, J.A., Akbari, M., Bertassoni, L.E., Cha, C., Khademhosseini, A., 25th Anniversary article: Rational design and applications of hydrogels in regenerative medicine. *Adv. Mater.*, 26, 85, 2013.
8. Dai, L., Conjugated and fullerene-containing polymers for electronic and photonic applications: Advanced syntheses and microlithographic fabrications. *J. Macromol. Sci., Part C: Polym. Rev.*, 39, 273, 1999.
9. Li, Y., Rodrigues, J., Tomás, H., Injectable and biodegradable hydrogels: Gelation, biodegradation and biomedical applications. *Chem. Soc. Rev.*, 41, 2193, 2012.
10. Malliaras, G. and Friend, R., An organic electronics primer. *Phys. Today*, 58, 53, 2005.
11. Satturwar, P.M., Fulzele, S.V., Dorle, A.K., Biodegradation and in vivo biocompatibility of rosin: A natural film-forming polymer. *AAPS PharmSciTech*, 4, 1, 2003.
12. Jani, G.K., Shah, D.P., Prajapati, V.D., Jain, V.C., Gums and mucilages: Versatile excipients for pharmaceutical formulations. *Asian J. Pharm. Sci.*, 5, 309, 2009.
13. Shirwaikar, A., Prabu, S.L., Kumar, G.A., Herbal excipients in novel drug delivery systems. *Indian J. Pharm. Sci.*, 70, 415, 2008.
14. Li, Y., Yang, H.Y., Lee, D.S., Advances in biodegradable and injectable hydrogels for biomedical applications. *J. Control. Release*, 330, 151, 2021.
15. Anita Lett, J., Sagadevan, S., Fatimah, I., Hoque, M.E., Lokanathan, Y., Léonard, E., Oh, W.C., Recent advances in natural polymer-based hydroxyapatite scaffolds: Properties and applications. *Eur. Polym. J.*, 148, 110360, 2021.
16. Zhao, X., Chen, X., Yuk, H., Lin, S., Liu, X., Parada, G., Soft materials by design: Unconventional polymer networks give extreme properties. *Chem. Rev.*, 121, 4309, 2021.
17. Shi, Z., Gao, X., Ullah, M.W., Li, S., Wang, Q., Yang, G., Electroconductive natural polymer-based hydrogels. *Biomaterials*, 111, 40, 2016.
18. Samadian, H., Maleki, H., Allahyari, Z., Jaymand, M., Natural polymers-based light-induced hydrogels: Promising biomaterials for biomedical applications. *Coord. Chem. Rev.*, 420, 213432, 2020.
19. Jose, G., Shalumon, K.T., Chen, J.-P., Yu, C., Natural polymers based hydrogels for cell culture applications. *Curr. Med. Chem.*, 27, 2734, 2020.
20. Cheng, F., Chen, H., Li, H., Recent progress on hydrogel actuators. *J. Mater. Chem. B*, 9, 1762, 2021.
21. Hajikhani, A., Wriggers, P., Marino, M., Chemo-mechanical modelling of swelling and cross-linking reaction kinetics in alginate hydrogels: A novel theory and its numerical implementation. *J. Mech. Phys. Solids*, 153, 104476, 2021.

22. Summonte, S., Racaniello, G.F., Lopedota, A., Denora, N., Bernkop-Schnürch, A., Thiolated polymeric hydrogels for biomedical application: Cross-linking mechanisms. *J. Control. Release*, 330, 470, 2021.
23. Ghorpade, V.S., Dias, R.J., Mali, K.K., Mulla, S.I., Citric acid crosslinked carboxymethylcellulose-polyvinyl alcohol hydrogel films for extended release of water-soluble basic drugs. *J. Drug Deliv. Sci. Technol.*, 52, 421, 2019.
24. Bashir, S., Hina, M., Iqbal, J., Rajpar, A.H., Mujtaba, M.A., Alghamdi, N.A., Ramesh, S., Fundamental concepts of hydrogels: Synthesis, properties, and their applications. *Polymers*, 12, 2702, 2020.
25. Ullah, F., Othman, M.B.H., Javed, F., Ahmad, Z., Akil, H.M., Classification, processing and application of hydrogels: A review. *Mater. Sci. Eng. C*, 57, 414, 2015.
26. Qiu, Y. and Park, K., Environment-sensitive hydrogels for drug delivery. *Adv. Drug Deliv. Rev.*, 53, 321, 2001.
27. Cong, H.P., Wang, P., Yu, S.H., Stretchable and self-healing graphene oxide–polymer composite hydrogels: A dual-network design. *Chem. Mater.*, 25, 3357, 2013.
28. Jumel, K. and Kennedy, J.F., Polysaccharides: Syntheses, modifications and structure/property relations. *Carbohydr. Polym.*, 11, 253, 1989.
29. Tsirigotis-Maniecka, M., Szyk-Warszyńska, L., Maniecki, Ł., Szczęsna, W., Warszyński, P., Wilk, K.A., Tailoring the composition of hydrogel particles for the controlled delivery of phytopharmaceuticals. *Eur. Polym. J.*, 151, 110429, 2021.
30. Ali, A. and Ahmed, S., Recent advances in edible polymer based hydrogels as a sustainable alternative to conventional polymers. *J. Agric. Food Chem.*, 66, 6940, 2018.
31. Mahinroosta, M., Jomeh Farsangi, Z., Allahverdi, A., Shakoori, Z., Hydrogels as intelligent materials: A brief review of synthesis, properties and applications. *Mater. Today Chem.*, 8, 42, 2018.
32. Islam, A., Yasin, T., Bano, I., Riaz, M., Controlled release of aspirin from pH-sensitive chitosan/poly(vinyl alcohol) hydrogel. *J. Appl. Polym. Sci.*, 124, 4184, 2011.
33. Kirschner, C.M. and Anseth, K.S., Hydrogels in healthcare: From static to dynamic material microenvironments. *Acta Mater.*, 61, 931, 2013.
34. Vijayavenkataraman, S., Vialli, N., Fuh, J.Y., Lu, W.F., Conductive collagen/polypyrrole-b-polycaprolactone hydrogel for bioprinting of neural tissue constructs. *Int. J. Bioprinting*, 5, 229, 2019.
35. Venkatesan, J., Lowe, B., Anil, S., Manivasagan, P., Kheraif, A.A.A., Kang, K.H., Kim, S.K., Seaweed polysaccharides and their potential biomedical applications. *Starch/Staerke*, 66, 1, 2015.
36. Rocha, P.M., Santo, V.E., Gomes, M.E., Reis, R.L., Mano, J.F., Encapsulation of adipose-derived stem cells and transforming growth factor-β1 in carrageenan-based hydrogels for cartilage tissue engineering. *J. Bioact. Compat. Polym.*, 26, 493, 2011.
37. Santo, V.E., Frias, A.M., Carida, M., Cancedda, R., Gomes, M.E., Mano, J.F., Reis, R.L., Carrageenan-based hydrogels for the controlled delivery of PDGF-BB in bone tissue engineering applications. *Biomacromolecules*, 10, 1392, 2009.
38. Thébaud, N.B., Pierron, D., Bareille, R., Le Visage, C., Letourneur, D., Bordenave, L., Human endothelial progenitor cell attachment to polysaccharide-based hydrogels: A pre-requisite for vascular tissue engineering. *J. Mater. Sci.: Mater. Med.*, 18, 339, 2007.
39. Dash, M., Samal, S.K., Bartoli, C., Morelli, A., Smet, P.F., Dubruel, P., Chiellini, F., Biofunctionalization of ulvan scaffolds for bone tissue engineering. *ACS Appl. Mater. Interfaces*, 6, 3211, 2014.
40. Zhu, T., Mao, J., Cheng, Y., Liu, H., Lv, L., Ge, M., Lai, Y., Recent progress of polysaccharide-based hydrogel interfaces for wound healing and tissue engineering. *Adv. Mater. Interfaces*, 1900761, 1, 2019.

41. Katyal, P., Mahmoudinobar, F., Montclare, J.K., Recent trends in peptide and protein-based hydrogels. *Curr. Opin. Struct. Biol.*, 63, 97, 2020.
42. Morya, V., Walia, S., Mandal, B.B., Ghoroi, C., Bhatia, D., Functional DNA based hydrogels: Development, properties and biological applications. *ACS Biomater. Sci. Eng.*, 6, 6021, 2020.
43. Zainal, S.H., Mohd, N.H., Suhaili, N., Anuar, F.H., Lazim, A.M., Othaman, R., Preparation of cellulose-based hydrogel: A review. *J. Mater. Res. Technol.*, 10, 935, 2021.
44. Dutta, S.D., Patel, D.K., Lim, K.T., Functional cellulose-based hydrogels as extracellular matrices for tissue engineering. *J. Biol. Eng.*, 13, 1, 2019.
45. Astrini, N., Anah, L., Haryono, A., Crosslinking parameter on the preparation of cellulose based hydrogel with divynilsulfone. *Proc. Chem.*, 4, 275, 2012.
46. Chang, C. and Zhang, L., Cellulose-based hydrogels: Present status and application prospects. *Carbohydr. Polym.*, 84, 40, 2011.
47. Lin, F., Lu, X., Wang, Z., Lu, Q., Lin, G., Huang, B., Lu, B., In situ polymerization approach to cellulose–polyacrylamide interpenetrating network hydrogel with high strength and pH-responsive properties. *Cellulose*, 26, 1825, 2018.
48. Pellá, M.C.G., Lima-Tenório, M.K., Tenório-Neto, E.T., Guilherme, M.R., Muniz, E.C., Rubira, A.F., Chitosan-based hydrogels: From preparation to biomedical applications. *Carbohydr. Polym.*, 196, 233, 2018.
49. Shariatinia, Z. and Jalali, A.M., Chitosan-based hydrogels: Preparation, properties and applications. *Int. J. Biol. Macromol.*, 115, 194, 2018.
50. Li, S.N., Li, B., Gong, L.X., Yu, Z.R., Feng, Y., Jia, D., Tang, L.C., Enhanced mechanical properties of polyacrylamide/chitosan hydrogels by tuning the molecular structure of hyperbranched polysiloxane. *Mater. Des.*, 162, 162, 2018.
51. Risbud, M.V. and Bhonde, R.R., Polyacrylamide-chitosan hydrogels: *In vitro* biocompatibility and sustained antibiotic release studies. *Drug Delivery*, 7, 69, 2000.
52. Kaur, A., Singh, D., Sud, D., A review on grafted, crosslinked and composites of biopolymer Xanthan gum for phasing out synthetic dyes and toxic metal ions from aqueous solutions. *J. Polym. Res.*, 27, 1, 2020.
53. Zhong, L., Oostrom, M., Truex, M.J., Vermeul, V.R., Szecsody, J.E., Rheological behavior of xanthan gum solution related to shear thinning fluid delivery for subsurface remediation. *J. Hazard. Mater.*, 244-245, 160, 2013.
54. Zheng, M., Lian, F., Zhu, Y., Zhang, Y., Liu, B., Zhang, L., Zheng, B., pH-responsive poly (xanthan gum-g-acrylamide-g-acrylic acid) hydrogel: Preparation, characterization, and application. *Carbohydr. Polym.*, 210, 38, 2019.
55. Jindal, R., Kaith, B.S., Mittal, H., Rapid synthesis of acrylamide onto xanthan gum based hydrogels under microwave radiations for enhanced thermal and chemical modifications. *Polym. Renew. Resour.*, 2, 105, 2011.
56. Kumar, A., Deepak, Sharma, S., Srivastava, A., Kumar, R., Synthesis of xanthan gum graft copolymer and its application for controlled release of highly water soluble Levofloxacin drug in aqueous medium. *Carbohydr. Polym.*, 171, 211, 2017.
57. Beaumont, M., Tran, R., Vera, G., Niedrist, D., Rousset, A., Pierre, R., Forget, A., Hydrogel-forming algae polysaccharides: From seaweed to biomedical applications. *Biomacromolecules*, 22, 1027, 2021.
58. Abasalizadeh, F., Moghaddam, S.V., Alizadeh, E., Akbari, E., Kashani, E., Fazljou, S.M.B., Akbarzadeh, A., Alginate-based hydrogels as drug delivery vehicles in cancer treatment and their applications in wound dressing and 3D bioprinting. *J. Biol. Eng.*, 14, 1, 2020.
59. İsmail, O. and Gökçe Kocabay, Ö., Absorption and adsorption studies of polyacrylamide/sodium alginate hydrogels. *Colloid Polym. Sci.*, 299, 783, 2021.

60. Jonker, A.M., Löwik, D.W.P.M., van Hest, J.C.M., Peptide- and protein-based hydrogels. *Chem. Mater.*, 24, 759, 2011.
61. Sui, Z., King, W.J., Murphy, W.L., Protein-based hydrogels with tunable dynamic responses. *Adv. Funct. Mater.*, 18, 1824, 20082012.
62. Valero, C., Amaveda, H., Mora, M., García-Aznar, J.M., Combined experimental and computational characterization of crosslinked collagen-based hydrogels. *PLoS One*, 13, e0195820, 2018.
63. Jaipan, P., Nguyen, A., Narayan, R.J., Gelatin-based hydrogels for biomedical applications. *MRS Commun.*, 7, 416, 2017.
64. Kim, S.Y., Park, B.J., Lee, Y., Park, N.J., Park, K.M., Hwang, Y.S., Park, K.D., Human hair keratin-based hydrogels as dynamic matrices for facilitating wound healing. *Ind. Eng. Chem. Res.*, 73, 142, 2019.
65. Lee, J.B., Peng, S., Yang, D., Roh, Y.H., Funabashi, H., Park, N., Luo, D., A mechanical metamaterial made from a DNA hydrogel. *Nat. Nanotechnol.*, 7, 816, 2012.
66. Tang, H., Duan, X., Feng, X., Liu, L., Wang, S., Li, Y., Zhu, D., Fluorescent DNA–poly(phenylenevinylene) hybrid hydrogels for monitoring drug release. *Chem. Commun.*, 6, 641, 2009.
67. Li, C., Chen, P., Shao, Y., Zhou, X., Wu, Y., Yang, Z., Liu, D., A writable polypeptide-DNA hydrogel with rationally designed multi-modification sites. *Small*, 11, 1138, 2014.
68. Hennink, W. and van Nostrum, C., Novel crosslinking methods to design hydrogels. *Adv. Drug Delivery Rev.*, 54, 13, 2002.
69. Barbucci, R., Pasqui, D., Favaloro, R., Panariello, G., A thixotropic hydrogel from chemically cross-linked guar gum: Synthesis, characterization and rheological behaviour. *Carbohydr. Res.*, 343, 3058, 2008.
70. Simões, B.M., Cagnin, C., Yamashita, F., Olivato, J.B., Garcia, P.S., de Oliveira, S.M., Eiras Grossmann, M.V., Citric acid as crosslinking agent in starch/xanthan gum hydrogels produced by extrusion and thermopressing. *LWT - Food Sci. Technol.*, 125, 108950, 2019.
71. Buhus, G., Popa, M., Desbrieres, J., Hydrogels based on carboxymethylcellulose and gelatin for inclusion and release of chloramphenicol. *J. Bioact. Compat. Polym.*, 24, 525, 2009.
72. Yu, S., Zhang, X., Tan, G., Tian, L., Liu, D., Liu, Y., Pan, W., A novel pH-induced thermosensitive hydrogel composed of carboxymethyl chitosan and poloxamer cross-linked by glutaraldehyde for ophthalmic drug delivery. *Carbohydr. Polym.*, 155, 208, 2017.
73. Kaith, B.S., Jindal, R., Kapur, G.S., Kumar, V., Synthesis, characterization and evaluation ofGum tragacanthand acrylic acid based hydrogel for sustained calcium chloride release – enhancement of water-holding capacity of soil. *J. Chin. Adv. Mater. Soc.*, 2, 40, 2014.
74. Singh, B., Sharma, S., Dhiman, A., Acacia gum polysaccharide based hydrogel wound dressings: Synthesis, characterization, drug delivery and biomedical properties. *Carbohydr. Polym.*, 165, 294, 2017.
75. Saber-Samandari, S. and Gazi, M., Pullulan based porous semi-IPN hydrogel: Synthesis, characterization and its application in the removal of mercury from aqueous solution. *J. Taiwan Inst. Chem. Eng.*, 51, 143, 2015.
76. Abdel Ghaffar, A.M., El-Arnaouty, M.B., Abdel Baky, A.A., Shama, S.A., Radiation-induced grafting of acrylamide and methacrylic acid individually onto carboxymethyl cellulose for removal of hazardous water pollutants. *Des. Monomers Polym.*, 19, 706, 2016.
77. Karadağ, E., Ödemiş, H., Kundakçi, S., Üzüm, Ö. B., Swelling characterization of acrylamide/zinc acrylate/xanthan gum/sepiolite hybrid hydrogels and its application in sorption of Janus Green B from aqueous solutions. *Adv. Polym. Technol.*, 35, 248, 2015.
78. El-Naggar, A.A., Radiation synthesis of superabsorbent hydrogels based on carboxymethyl cellulose/sodium alginate for absorbent of heavy metal ions from waste water. *J. Thermoplast. Compos. Mater.*, 29, 16, 2014.

79. Abd Alla, S.G., Sen, M., El-Naggar, A.W.M., Swelling and mechanical properties of superabsorbent hydrogels based on Tara gum/acrylic acid synthesized by gamma radiation. *Carbohydr. Polym.*, 89, 478, 2012.
80. Fan, L., Yang, H., Yang, J., Peng, M., Hu, J., Preparation and characterization of chitosan/gelatin/PVA hydrogel for wound dressings. *Carbohydr. Polym.*, 146, 427, 2016.
81. Tavakol, M., Vasheghani-Farahani, E., Soleimani, M., Mohammadifar, M.A., Hashemi-Najafabadi, S., Hafizi, M., Synthesis and characterization of an enzyme mediated in situ forming hydrogel based on gum tragacanth for biomedical applications. *Iran. J. Biotechnol.*, 12, e15811, 2014.
82. Kenawy, E.-R., Kamoun, E.A., Mohy Eldin, M.S., El-Meligy, M.A., Physically crosslinked poly(vinyl alcohol)-hydroxyethyl starch blend hydrogel membranes: Synthesis and characterization for biomedical applications. *Arab. J. Chem.*, 7, 372, 2014.
83. Qi, X., Hu, X., Wei, W., Yu, H., Li, J., Zhang, J., Dong, W., Investigation of Salecan/poly(vinyl alcohol) hydrogels prepared by freeze/thaw method. *Carbohydr. Polym.*, 118, 60, 2015.
84. Mabrouk, M., Chejara, D.R., Mulla, J.A.S., Badhe, R.V., Choonara, Y.E., Kumar, P., Pillay, V., Design of a novel crosslinked HEC-PAA porous hydrogel composite for dissolution rate and solubility enhancement of efavirenz. *Int. J. Pharm.*, 490, 429, 2015.
85. Das, M.K. and Senapati, P.C., Evaluation of furosemide-loaded alginate microspheres prepared by ionotropic external gelation technique. *Acta Pol. Pharm.*, 64, 253, 2007.
86. Guo, W., Lu, C.H., Orbach, R., Wang, F., Qi, X.J., Cecconello, A., Seliktar, D., Willner, I., pH-Stimulated DNA hydrogels exhibiting shape-memory properties. *Adv. Mater.*, 27, 73, 2015.
87. Bashir, S., Teo, Y.Y., Ramesh, S., Ramesh, K., Mushtaq, M.W., Rheological behavior of biodegradable N-succinyl chitosan-g-poly (acrylic acid) hydrogels and their applications as drug carrier and *in vitro* theophylline release. *Int. J. Biol. Macromol.*, 117, 454, 2018.
88. Oyen, M., Mechanical characterisation of hydrogel materials. *Int. Mater. Rev.*, 59, 44, 2014.
89. Tan, H. and Marra, K.G., Injectable, biodegradable hydrogels for tissue engineering applications. *Materials*, 3, 1746, 2010.
90. Gao, X., He, C., Xiao, C., Zhuang, X., Chen, X., Biodegradable pH-responsive polyacrylic acid derivative hydrogels with tunable swelling behavior for oral delivery of insulin. *Polymer*, 54, 1786, 2013.
91. Das, R., Panda, A., Pal, S., Synthesis and characterization of a novel polymeric hydrogel based on hydroxypropyl methyl cellulose grafted with polyacrylamide. *Cellulose*, 19, 933, 2012.
92. Bashir, S., Teo, Y.Y., Ramesh, S., Ramesh, K., Synthesis and characterization of karaya gum-g-poly (acrylic acid) hydrogels and *in vitro* release of hydrophobic quercetin. *Polymer*, 147, 108, 2018.
93. Tanan, W., Panichpakdee, J., Saengsuwan, S., Novel biodegradable hydrogel based on natural polymers: Synthesis, characterization, swelling/reswelling and biodegradability. *Eur. Polym. J.*, 112, 678, 2019.
94. Vaghani, S.S., Patel, M.M., Satish, C., Synthesis and characterization of pH-sensitive hydrogel composed of carboxymethyl chitosan for colon targeted delivery of ornidazole. *Carbohydr. Res.*, 347, 76, 2012.
95. Mura, C., Manconi, M., Valenti, D., Manca, M.L., Díez-Sales, O., Loy, G., Fadda, A.M., *In vitro* study of N-succinyl chitosan for targeted delivery of 5-aminosalicylic acid to colon. *Carbohydr. Polym.*, 85, 578, 2011.
96. Bilia, A., Carelli, V., Di Colo, G., Nannipieri, E., *In vitro* evaluation of a pH-sensitive hydrogel for control of GI drug delivery from silicone-based matrices. *Int. J. Pharm.*, 130, 83, 1996.
97. Chen, H., Gu, Y., Hu, Y., Qian, Z., Characterization of pH-and temperature-sensitive hydrogel nanoparticles for controlled drug release. *PDA J. Pharm. Sci. Technol.*, 61, 303, 2007.

98. Katono, H., Maruyama, A., Sanui, K., Ogata, N., Okano, T., Sakurai, Y., Thermo-responsive swelling and drug release switching of interpenetrating polymer networks composed of poly (acrylamide-co-butyl methacrylate) and poly (acrylic acid). *J. Control. Release*, 16, 215, 1991.
99. Kim, B.K., Sy, S., Yu, A., Zhang, J., Electrochemical supercapacitors for energy storage and conversion, in: *Handbook of Clean Energy Systems*, pp. 1–25, 2015.
100. Mirzaeian, M., Abbas, Q., Ogwu, A., Hall, P., Goldin, M., Mirzaeian, M., Jirandehi, H.F., Electrode and electrolyte materials for electrochemical capacitors. *Int. J. Hydrog. Energy, 42*, 40, 25565–25587, 2017.
101. Fic, K., Platek, A., Piwek, J., Frackowiak, E., Sustainable materials for electrochemical capacitors. *Mater. Today, 21*, 4, 437–454, 2018.
102. Dai, H., Zhang, G., Rawach, D., Fu, C., Wang, C., Liu, X., Dubois, M., Lai, C., Sun, S., Polymer gel electrolytes for flexible supercapacitors: Recent progress, challenges, and perspectives. *Energy Storage Mater., 34*, 320–355, 2021.
103. Wu, Z., Li, L., Yan, J.M., Zhang, X.B., Materials design and system construction for conventional and new-concept supercapacitors. *Adv. Sci., 4*, 6, 1600382, 2017.
104. Yan, J., Wang, Q., Wei, T., Fan, Z., Recent advances in design and fabrication of electrochemical supercapacitors with high energy densities. *Adv. Energy Mater., 4*, 4, 1300816, 2014.
105. Suriyakumar, S., Bhardwaj, P., Grace, A.N., Stephan, A.M., Role of polymers in enhancing the performance of electrochemical supercapacitors: A review. *Batter. Supercaps, 4*, 4, 571–584, 2021.
106. Guo, Y., Bae, J., Zhao, F., Yu, G., Functional hydrogels for next-generation batteries and supercapacitors. *Trends Chem., 1*, 3, 335–348, 2019.
107. Lu, Y., He, W., Cao, T., Guo, H., Zhang, Y., Li, Q., Shao, Z., Cui, Y., Zhang, X., Elastic, conductive, polymeric hydrogels and sponges. *Sci. Rep., 4*, 1, 1–8, 2014.
108. Wu, Q., Xu, Y., Yao, Z., Liu, A., Shi, G., Supercapacitors based on flexible graphene/polyaniline nanofiber composite films. *ACS Nano, 4*, 4, 1963–1970, 2010.
109. Gnanakan, S.R.P., Murugananthem, N., Subramania, A., Organic acid doped polythiophene nanoparticles as electrode material for redox supercapacitors. *Polym. Adv. Technol., 22*, 6, 788–793, 2011.
110. Huang, Y., Tao, J., Meng, W., Zhu, M., Huang, Y., Fu, Y., Gao, Y., Zhi, C., Super-high rate stretchable polypyrrole-based supercapacitors with excellent cycling stability. *Nano Energy, 11*, 518–525, 2015.
111. Vázquez, M., Danielsson, P., Bobacka, J., Lewenstam, A., Ivaska, A., Solution-cast films of poly (3, 4-ethylenedioxythiophene) as ion-to-electron transducers in all-solid-state ion-selective electrodes. *Sens. Actuators B: Chem., 97*, 2-3, 182–189, 2004.
112. Huang, J., Wang, K., Wei, Z., Conducting polymer nanowire arrays with enhanced electrochemical performance. *J. Mater. Chem., 20*, 6, 1117–1121, 2010.
113. Purkait, T., Singh, G., Kumar, D., Singh, M., Dey, R.S., High-performance flexible supercapacitors based on electrochemically tailored three-dimensional reduced graphene oxide networks. *Sci. Rep., 8*, 1, 1–13, 2018.
114. Wang, K., Zhang, X., Li, C., Zhang, H., Sun, X., Xu, N., Ma, Y., Flexible solid-state supercapacitors based on a conducting polymer hydrogel with enhanced electrochemical performance. *J. Mater. Chem. A, 2*, 46, 19726–19732, 2014.
115. Moussa, M., El-Kady, M.F., Dubal, D., Tung, T.T., Nine, M.J., Mohamed, N., Kaner, R.B., Losic, D., Self-assembly and cross-linking of conducting polymers into 3D hydrogel electrodes for supercapacitor applications. *ACS Appl. Energy Mater., 3*, 1, 923–932, 2019.
116. Xu, Z., Chu, X., Wang, Y., Zhang, H., Yang, W., Three-dimensional polymer networks for solid-state electrochemical energy storage. *Chem. Eng. J., 391*, 123548, 2020.

117. Guo, H., He, W., Lu, Y., Zhang, X., Self-crosslinked polyaniline hydrogel electrodes for electrochemical energy storage. *Carbon*, *92*, 133–141, 2015.
118. Ghosh, S.O.U.M.Y.A.D.E.B. and Inganäs, O., Conducting polymer hydrogels as 3D electrodes: Applications for supercapacitors. *Adv. Mater.*, *11*, 14, 1214–1218, 1999.
119. Liu, Y., Weng, B., Razal, J.M., Xu, Q., Zhao, C., Hou, Y., Seyedin, S., Jalili, R., Wallace, G.G., Chen, J., High-performance flexible all-solid-state supercapacitor from large free-standing graphene-PEDOT/PSS films. *Sci. Rep.*, *5*, 1, 1–11, 2015.
120. Eftekhari, A. and Jafarkhani, P., Polymerization of aniline through simultaneous chemical and electrochemical routes. *Polym. J.*, *38*, 7, 651–658, 2006.
121. Pan, L., Yu, G., Zhai, D., Lee, H.R., Zhao, W., Liu, N., Wang, H., Tee, B.C.K., Shi, Y., Cui, Y., Bao, Z., Hierarchical nanostructured conducting polymer hydrogel with high electrochemical activity. *Proc. Natl. Acad. Sci.*, *109*, 24, 9287–9292, 2012.
122. Chu, X., Huang, H., Zhang, H., Zhang, H., Gu, B., Su, H., Liu, F., Han, Y., Wang, Z., Chen, N., Yan, C., Electrochemically building three-dimensional supramolecular polymer hydrogel for flexible solid-state micro-supercapacitors. *Electrochim. Acta*, *301*, 136–144, 2019.
123. Shi, Y., Pan, L., Liu, B., Wang, Y., Cui, Y., Bao, Z., Yu, G., Nanostructured conductive polypyrrole hydrogels as high-performance, flexible supercapacitor electrodes. *J. Mater. Chem. A*, *2*, 17, 6086–6091, 2014.
124. Sardana, S., Gupta, A., Singh, K., Maan, A.S., Ohlan, A., Conducting polymer hydrogel based electrode materials for supercapacitor applications. *J. Energy Storage*, 32, 103510, 2021.
125. Cheng, T., Zhang, Y.Z., Wang, S., Chen, Y.L., Gao, S.Y., Wang, F., Lai, W.Y., Huang, W., Conductive hydrogel-based electrodes and electrolytes for stretchable and self-healable supercapacitors. *Adv. Funct. Mater.*, *31*, 24, 2101303, 2021.
126. Lewandowski, A., Zajder, M., Frąckowiak, E., Beguin, F., Supercapacitor based on activated carbon and polyethylene oxide–KOH–H2O polymer electrolyte. *Electrochim. Acta*, *46*, 18, 2777–2780, 2001.
127. Yang, C.C., Hsu, S.T., Chien, W.C., All solid-state electric double-layer capacitors based on alkaline polyvinyl alcohol polymer electrolytes. *J. Power Sources*, *152*, 303–310, 2005.
128. Lu, C. and Chen, X., All-temperature flexible supercapacitors enabled by antifreezing and thermally stable hydrogel electrolyte. *Nano Lett.*, *20*, 3, 1907–1914, 2020.
129. Wang, K., Zhang, X., Li, C., Sun, X., Meng, Q., Ma, Y., Wei, Z., Chemically crosslinked hydrogel film leads to integrated flexible supercapacitors with superior performance. *Adv. Mater.*, *27*, 45, 7451–7457, 2015.
130. Lin, Y., Zhang, H., Liao, H., Zhao, Y., Li, K., A physically crosslinked, self-healing hydrogel electrolyte for nano-wire PANI flexible supercapacitors. *Chem. Eng. J.*, *367*, 139–148, 2019.
131. Huang, Y., Zhong, M., Shi, F., Liu, X., Tang, Z., Wang, Y., Huang, Y., Hou, H., Xie, X., Zhi, C., An intrinsically stretchable and compressible supercapacitor containing a polyacrylamide hydrogel electrolyte. *Angew. Chem. Int. Ed.*, *56*, 31, 9141–9145, 2017.

6

Ionic Liquids for Supercapacitors

Guocai Tian

State Key Laboratory of Complex Non-Ferrous Metal Resource Clean Utilization, Faculty of Metallurgical and Energy Engineering, Kunming University of Science and Technology, Kunming, China

Abstract
Supercapacitor is a new generation of energy storage device between rechargeable battery and common capacitor. They have been widely concerned and applied because of their fast charging and discharging speed, high-energy storage density, small volume, long cycle life and little environmental pollution. However, the low-energy density of supercapacitors limits their further applications. Improving energy density has become a challenge for its further development and application. Electrolyte is one of the most important factors in super capacitors. The research and development of new electrolytes with high efficiency, stability, green and low price to improve the energy density of supercapacitors is the main research hotspot in this field. In the past decades, many new high-performance electrolytes have been developed and used to improve the performance of supercapacitors. In this chapter, we mainly focus on the research progress and future research direction of ionic liquids as new green electrolytes for supercapacitors. We briefly introduced the related knowledge and properties of supercapacitors and ionic liquids. The research progress of pure ionic liquids, functional ionic liquids, mixed systems of ionic liquids and ionic liquids, mixtures of ionic liquids and organic solvents or salts as supercapacitor electrolytes are summarized and discussed, and future research directions are discussed.

Keywords: Supercapacitor, ionic liquid, electric double-layer capacitors (EDLC), proton ionic liquids, functionalized ionic liquids, aprotic ionic liquids, ionic liquid mixture electrolyte

6.1 Introduction

With the progress of society, the deepening of industrialization and the rapid expansion of population, mineral energy is increasingly exhausted, and mankind is facing a serious energy crisis. Meanwhile, the combustion of fossil energy and materials will produce a lot of smoke and harmful gas, which will make the human living environment worse and worse. Therefore, the increasing depletion of mineral energy and environmental degradation is two major problems that need to be solved urgently. The development and utilization of

Email: tiangc@kust.edu.cn; tiangc@iccas.ac.cn; tiangc01@gmail.com

Inamuddin, Tariq Altalhi and Sayed Mohammed Adnan (eds.) Sustainable Materials for Electrochemical Capacitors, (167–204) © 2023 Scrivener Publishing LLC

new renewable and clean energy are an effective way to solve these problem. Solar energy, tidal energy, geothermal energy, wind energy and other renewable green energy have gradually attracted great attention. However, many green renewable-energy sources are highly dependent on the environment and climate, and they have great discontinuity. In order to make full use of the renewable energy, innovative energy storage equipment and technologies play an important role in building a new power system. A large number of energy storage devices, such as traditional lead-acid batteries, capacitors, metal ion batteries (such as sodium ion batteries and lithium-ion batteries) and other devices, are emerging and widely used. Supercapacitors have been widely concerned and applied as additional energy storage devices because of their excellent cycle performance and ultra-high power performance [1, 2].

Supercapacitor is an electrochemical capacitor (EC). As a new energy storage device among traditional capacitors and batteries, it contains the following advantages [3, 4]. (1) Faster charging speed and high-power density. The energy storage mechanism of supercapacitors belongs to a physical process rather than a chemical process, and will not cause changes in new and old substances due to chemical reactions. Therefore, it has characteristics similar to the conventional flat-panel capacitor. It can charge and discharge large current and provide hundreds of amperes of the instantaneous current. (2) High charge/discharge efficiency and long cycle life. The energy storage mechanism of the supercapacitor is a physical process, which mainly occurs on the electrode surface, has a little effect on the interior of the electrode, and has limited effect on the surface morphology of the electrode. Therefore, supercapacitors have good charge/discharge efficiency and high cycle life. (3) Green environmental protection and safety. Since the supercapacitor does not involve a chemical reaction, it will not produce harmful substances during use. Its electrode materials are generally carbon or inert materials, which has a wide variety of raw materials and low prices. In general, their production process will not have a negative effect on consumers and the environment.

Because its electrolyte is liquid, and the energy storage mechanism is a physical process, which can eliminate local hot spots without violent explosion. Compared with traditional batteries, it has higher security. Therefore, in many cases, it has unique application advantages. Supercapacitors can be used as backup power for computers, cameras, timers, etc. It can also be used in high-tech weapons (such as laser weapons, electric guns, etc.) that require a continuous high-current pulse electric energy. However, the most important application of super capacitors lies in the current booming electric vehicles. The composite power supply composed of supercapacitor and secondary battery or fuel cell in parallel can respond to the peak power demand of electric vehicles during starting and climbing. It can be used as an accumulator to recover energy when the vehicle goes downhill and brakes. In recent years, super capacitors have attracted extensive attention [5–16].

Supercapacitors usually consist of two electrodes, electrolyte and diaphragm material. Among them, electrode materials and electrolyte are the main components of supercapacitors. Therefore, the research on capacitors is mostly focused on these two aspects. Significant research progress has been achieved in the preparation of electrode materials, which has been summarized and reviewed in various literatures [9–34]. In this chapter, we will mainly discuss the research progress and development direction of new electrolytes in supercapacitors, which is also one of the research hotspots of capacitor.

The commonly used electrolytes for supercapacitors include aqueous and non-aqueous electrolytes. Non-aqueous is including the organic electrolytes and polymer electrolytes, and so on [17]. Among them, water electrolyte has high conductivity, but is constrained by water decomposition voltage. Its working voltage is usually not larger than 1 V, so its energy density is very low. Organic electrolyte has a stable wide electrochemical window and can obtain a greater specific energy. However, they are easy to volatilize or degrade at a higher temperature, flammable and explosive when overcharged, and there are potential safety hazards, which have become the main factor limiting their application in electric vehicles and other relative devices and fields [18]. Therefore, the research and development of stable, efficient and green electrolytes is very important in current research.

In recent years, the emergence of ionic liquids as green solvents and electrolytes has aroused great interest of researchers. Depending on the requirements of physical and electrochemical properties of capacitors, researchers have explored the application of ionic liquid as the electrolyte in capacitors. As the electrolyte of a supercapacitor, ionic liquid has a higher-energy density and a wider electrochemical stable window than the aqueous solution. Compared with traditional organic electrolytes, ionic liquids are not volatile, have no ignition point and are not flammable even under high pressure. It is considered to be a new generation of high safety electrolyte for supercapacitors. At present, numerous research works have been performed, and significant progress has been achieved. This chapter mainly focuses on the research progress of ionic liquids and related electrolytes formed with ionic liquid for the supercapacitors. The outline of this chapter is as follows. Firstly, a brief introduction of the supercapacitors, the advantage and disadvantage of electrolysis currently used in supercapacitors is given. Secondly, the basic knowledge of ionic liquids and their unique properties are provided. Thirdly, a comprehensive review of the research progress, advantages and disadvantages of ionic liquids as supercapacitor electrolytes and supporting materials are given. Finally, the future directions of research in this field is presented.

6.2 Brief Introduction of Supercapacitor

6.2.1 Supercapacitor and Its Classification

Supercapacitor is a new generation of the energy storage devices among the battery and common flat-panel capacitor. Supercapacitors usually store charge by physical means or energy by the reversible rapid oxidation-reduction reaction on the surface. The chemical storage method depends on the chemical reaction rate, while the physical storage method does not have these limitations, so the life of the capacitor is almost unlimited (millions of charge and discharge cycles). Table 6.1 shows the performance comparison of different energy storage methods [19–30]. It can be seen from Table 6.1 that supercapacitors have both the high-power density characteristics of capacitors and the high-energy density characteristics of batteries, and their power efficiency and cycle life are significantly better than batteries. There are many different ways to classify supercapacitors [19–30]. Supercapacitors can be divided into Faraday quasi-capacitors or pseudo-capacitors (PC), hybrid capacitors (HCS) and electric double-layer capacitors (EDLC), due to the mechanisms of energy storage. According to the difference in the composition of positive and negative electrodes

Table 6.1 Comparison of different electrochemical energy storage methods.

Characteristic parameter	Traditional capacitor	Supercapacitor	Battery
Charging time	10^{-6}–10^{-3} s	<1 min	1–5 h
Power density (W/kg)	>>10000	500-10000	<1000
Discharge time	10^{-6}–10^{-3} s	<1 min	0.3–3 h
Cycle life	∞	>500000	≈1000
Coulomb efficiency	≈100	85–98	70–85
Energy density (Wh/kg)	<0.1	1–10	10–100

and the different reactions between electrodes, capacitors can be divided into asymmetric capacitors and symmetric capacitors (also known as hybrid supercapacitors). According to the difference of electrode materials, it can be classified as carbon electrode capacitor, transition metal oxide capacitor and conductive polymer capacitor. According to different electrolytes, it can also be divided into water-based electrolyte capacitors, non-aqueous or organic electrolyte capacitors, and solid electrolyte capacitors.

The most commonly used classification is based on the energy storage mechanism [19–30], as showed in Figure 6.1. Electric double layer capacitor (EDLC) is a kind of super capacitor composed of two high surface area electrodes and electrolyte. During charging, the charge is separated on both sides of the electrode/electrolyte interface through electrostatic action, thus forming a kind of electrochemical energy storage system. The charge of the electric double-layer capacitor is stored in the electric double-layer, and the bulk phase of the electrode material does not undergo chemical reactions and physical changes. Therefore, the charging and discharging process is highly reversible and can realize rapid

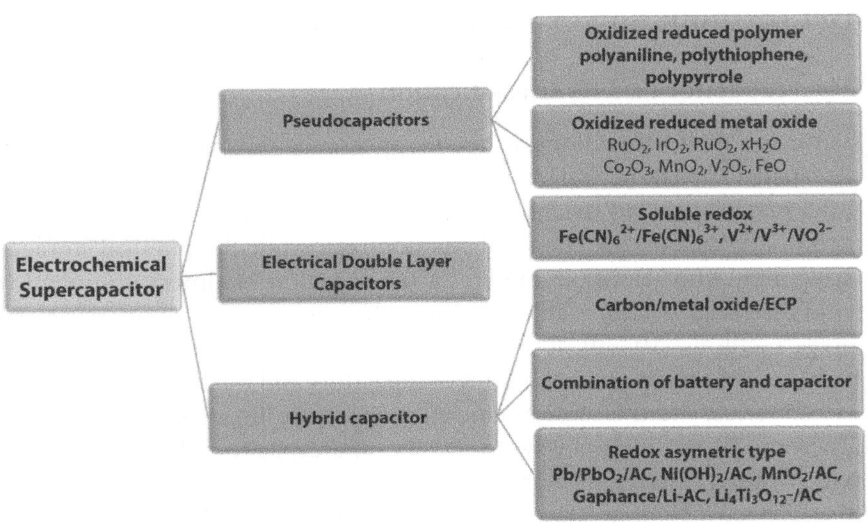

Figure 6.1 Classification of supercapacitors. (The figures are self-drawn.)

charging and discharging. Electrode materials of EDLC usually use porous carbon materials with large specific surfaces, such as carbon nano-tubes, graphite and activated carbon. Faraday quasi-capacitor, also known as pseudo-capacitors, mainly deposits electroactive substances on the surface of oxide electrode or two-dimensional or quasi two-dimensional space of bulk phase through under potential to produce highly reversible chemical adsorption, desorption or redox reaction, so as to produce Faraday quasi-capacitor related to electrode charging potential [31]. Therefore, it can obtain higher-energy density and capacitance (10 - 100 times) than an electric double-layer capacitor. The main difference between PC and EDLC capacitor is that the former is based on Faraday's reaction, which involves the rapid reversible redox reaction between electroactive substances and electrolyte on the surface of an electrode. The most active substances are ruthenium oxide [32], manganese oxide [33], vanadium dioxide [33], conductive polymers [34] and functional groups containing oxygen or nitrogen atoms on the surface [35].

Hybrid supercapacitor (HSC) combines the characteristics of high-power density and low energy density of electric double-layer capacitor and complements the high-energy density and low power density behavior of battery, so as to overcome the performance difference between battery and electric double-layer capacitor to achieve higher performance. The cathode of HSC adopts the material of a quasi-capacitor which can provide high-energy density through Faraday's reaction. Its anode used a material of an electric double-layer capacitor, which can provide high-power density. At present, the cathodes used in HSC include carbon-based electrodes (including graphite, carbon nanotubes, activated carbon (AC), graphite), polymer based carbon composites, different metal oxides and three-dimensional mesoporous carbon. HSC combines the advantages of PCs and EDLC to further improve its working voltage and enhance its higher-energy density.

6.2.2 Electrolyte of Supercapacitor

The performance of electrolyte is a part of the key factors affecting the electrochemical performance of capacitors. As an important part of a supercapacitor, the performance of electrolyte greatly determines the performance of the supercapacitor. The main function of the electrolyte is to provide ionic conductivity, so as to promote compensation of charges on the electrode in the supercapacitor. The behaviors of electrolyte mainly include the size and the type of ions, the interaction between solvents and ions, the concentration of ions and solvents, the interaction between electrolytes and electrodes, as well as the potential window [36]. These characteristics affect the capacitance, energy density, cycle life, and power density of capacitors. Figure 6.2 shows the effect of electrolyte on the performance of a capacitor [10]. The working voltage of a supercapacitor is directly determined by the potential window of electrolyte, and then affects its power density and energy density. The equivalent series resistance (ESR) of a supercpacitor has a great effect on the power density. The power density is very closely determined by the conductivity of ions in the electrolyte. In addition, interaction between solvents and ions, electrode materials and electrolytes will influence the self-discharge and cycle life of supercapacitors. Therefore, the development of novel high-performance electrolytes with high thermal stability, high ionic conductivity, wide operating potential, low viscosity, and is the research focus and hotspot in this field.

The good electrolyte used in supercapacitor will meet the following requirements: high purity, wide electrochemical windows, low resistivity, high working voltage, small radius

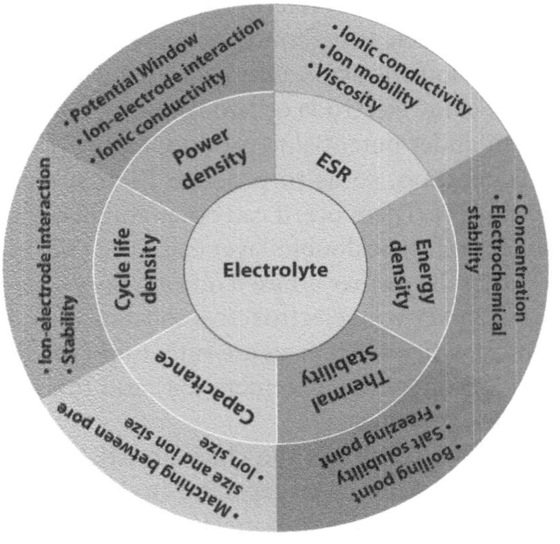

Figure 6.2 Effect of electrolytes for supercapacitor. (The figures are self-drawn.)

and gap of solvated ions, low toxicity, high ion concentration and low cost. However, so far, there is no perfect electrolyte that can fully meet the actual need. Each electrolyte has its advantages and disadvantages. In general, the electrolyte used in supercapacitors can be classified as three categories, aqueous electrolyte, organic or polymer electrolyte, and ionic liquid electrolyte. Figure 6.3 indicates the type of electrolyte and its typical representative.

Table 6.2 shows the performance of three typical electrolytes commonly used in supercapacitors. Aqueous electrolyte usually includes sulfuric acid solution or potassium hydroxide solution. They have high conductivity and low cost, but their decomposition voltage

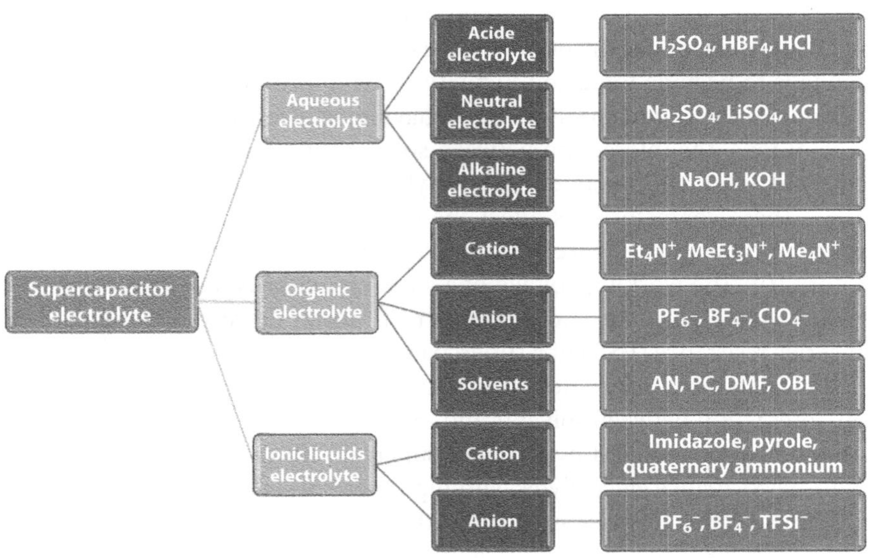

Figure 6.3 The type of electrolytes and the representations. (The figures are self-drawn.)

Table 6.2 Property comparison of typical electrolytes.

Electrolyte	EW (V)	σ(mS/cm)	η (cP.s)	Assembly gas	Cost	Toxicity	Advantages	Disadvantage
Aqueous electrolyte	≤1.2	High	Low	Air	Low	Low	High conductivity large capacitance	Low EW
Organic electrolyte	2.5–2.8	Low	Medium/high	Inert gas	Medium/high	Medium/high	High EW, almost non corrosive	Large electrolyte ions, low conductivity, high demand for external environment
Ionic liquid	2–6	Very low	Very high	Inert gas	Very high	No and low vapor pressure	High thermal stability, wide EW and chemical stability etc.	High viscosity, high demand for external environment

is low (generally no more than 1.0 V), which leads to minimal electrochemical window and limited the application of capacitors containing these electrolytes. In addition, they are highly corrosive and easy to solidify and gasify, which also limits their application. Organic electrolyte has become the main commercial electrolyte for supercapacitors because of its high electrochemical potential window of 3–5 V, high conductivity and a wide range of working temperature. Compared with aqueous electrolytes, organic electrolytes have lower conductivity, smaller specific capacitance, and higher cost. There are safety problems such as flammability, volatilization and toxicity that need further attention. Due to the extraordinary environmental requirements, their preparation and device assembly process usually need to be carried out in a strict environment to avoid introducing impurities and affecting its performance.

Therefore, it is of great significance to developing new electrolyte systems with high conductivity, wide electrochemical window, highly conductive ion concentration and good stability. As a kind of green electrolyte and solvent, ionic liquids (ILs) have aroused considerable interest to researchers and become a current research hotspot. ILs are commonly used in various fields due to their high conductivity, good chemical and thermal stability, low melting point and easy solubility in most organic solvents [37–48]. Ionic liquid as the electrolyte has the characteristics of high conductivity, a non-volatile, wide electrochemical window, flame retardant and good thermal stability [38, 39], which is very suitable for supercapacitor electrolyte [40, 41].

6.3 Ionic Liquids and Its Unique Properties

Ionic liquid is also known as a type of room temperature or a kind of low-temperature molten salts. It is an ionic system formed by small inorganic or organic anions and large organic cations, which is liquid at room temperature or near room temperature [42–48]. At room temperature or below 100°C, most ionic liquids are liquid, which is mainly determined by its chemical composition and structure. Ionic liquids with different structures and functions can be synthesized by using different combinations of the anions and cations. Therefore, ionic liquids are also known as "designable solvents" [44–48]. Table 6.3 displays the typical anions and cations of ionic liquids. Currently known anions and cations can be formed about 10^{18} kinds of ionic liquids However, there are no universal and specific rules for the design of ionic liquids. But it can be generally achieved by balancing ion-ion interaction and ion symmetry in a relatively large range of ion structures.

ILs are called green electrolytes and solvents because of their unique properties [44–46]. In recent years, it has attracted extensive attention and has been successfully applied in various fields [44–48]. Thanks to its thermal stability, wide electrochemical window, high conductivity and low vapor pressure make ionic liquids more widely used in electrochemical fields such as lithium batteries, capacitors, sensors and electrodeposition of active metal. The electrochemical window (EW) is the maximum voltage range that ionic liquids can withstand without electrochemical decomposition [44–48]. The high electrochemical window (EW) of the ionic liquid can improve the stability of the battery using a high voltage. Tables 6.4–6.6 shows the electrochemical windows of imidazole, pyridine, pyrrole, ammonium, sulfur and other systems. From Tables 6.4 to 6.6, we can find that the different cations and anions' types have a profound influence on the electrochemical window of the

Table 6.3 The common used anions and cations for ionic liquids.

Types	Chemical formula or structure
Cations	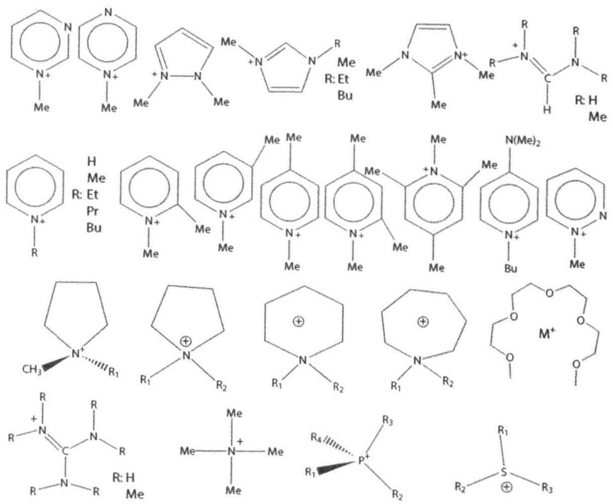
Anions	F^-, I^-, $AlCl_4^-$, Br^-, $A1_2C1_7^-$, Cl^-, $A1_3Cl_{10}^-$, $Cu_2Cl_3^-$, $Fe_2Cl_7^-$, ClO_4^-, NO_3^-, NO_2^-, SO_4^{2-}, CH_3COO^-, $N(CF_3SO_2)_2^-$, $CH_3SO_3^-$, HSO_4^-, BF_4^-, PO_4^{3-}, $CF_3SO_3^-$, $C_3F_7COO^-$, $(C_2H_5SO_3)_2N^-$, $C_4F_9SO_3^-$, $C_2H_5BF_3^-$, $C_3H_7BF_3^-$, $C_4H_9BF_3^-$, $C_5H_{11}BF_3^-$, $CH_2CHBF_3^-$, FSI^-, $CH_3BF_3^-$, $N(CN)_2^-$, $C_3F_7BF_3^-$, $C_4F_7BF_3^-$, SCN^-

Table 6.4 Electrochemical window of non-haloaluminate imidazolium ionic liquids.

ILs	WE	EW (V)	Reference
[Emim]BF_4	Pt	4.40	[49]
[Emim]CH_3SO_3	Pt	4.10	[49]
[Emim]NTf_2	GC	4.50	[49]
[Emim]TSAC	GC	4.10	[49]
[Emim]$F(HF)_{2.3}$	GC	3.10	[50]
[Pmim]$F(HF)_{2.3}$	GC	3.20	[50]
[Emim]NbF_6	GC	3.50	[51]
[Emim]CF_3BF_3	GC	4.63	[52]

(Continued)

Table 6.4 Electrochemical window of non-haloaluminate imidazolium ionic liquids. (*Continued*)

ILs	WE	EW (V)	Reference
[Emim]$C_2F_5BF_3$	GC	4.65	[52]
[Emim]$nC_3F_7BF_3$	GC	4.71	[52]
[Emim]$nC_4F_9BF_3$	GC	4.67	[52]
[Bmim]BF_4	GC	4.68	[52]
[Bmim]NTf_2	GC	4.62	[52]
[Emim]SbF_6	GC	4.00	[53]
[Emim]TaF_6	GC	4.50	[53]
[Pmim]BF_4	GC	4.40	[54]
[Emim]OTf	Pt	4.30	[55]
[Emim]CF_3COO	Pt	3.40	[56]
[Emim]$C(CN)_3$	Pt	2.90	[56]
[Emim]DCA	Pt	3.30	[56]
[Emim]NTf_2	Pt	4.60	[56]
[Emim]NTf_2	Pt	4.90	[56]
[Emim]$F(HF)_{2.3}$	Pt	3.50	[57]
[Emim]NTf_2	Au-micro	4.78	[57]
[Bmim]NTf_2	Au-micro	4.76	[57]
[C_8mim]NTf_2	Au-micro	4.83	[57]
[C_{10}mim]NTf_2	Au-micro	4.89	[57]
[Bmim]BF_4	Au-RED	4.20	[58]
[Bmim]BF_4	Pt-RED	4.60	[58]
[Bmim]BF_4	Pt	4.00	[58]
[Bmim]BF_4	W-RED	6.10	[58]
[Bmim]PF_6	Au-RED	5.95	[58]
[Bmim]PF_6	GC-RED	6.23	[58]
[Bmim]PF_6	Pt-RED	5.70	[58]
[Bmim]BF_4	Pt	5.10	[59]

(*Continued*)

Table 6.4 Electrochemical window of non-haloaluminate imidazolium ionic liquids. (*Continued*)

ILs	WE	EW (V)	Reference
[Bmim]NTf_2	Pt	5.0	[59]
[Bmim]PF_6	Pt	5.00	[59]
[Bmim]NTf_2	Pt	5.20	[59]
[Hmim]Cl	Pt	3.30	[59]
[Hmim]FAP	Pt	4.80	[59]
[Pmim]NTf_2	GC	4.30	[60]
[Pmmim]AsF_6	GC	4.42	[60]
[Pmmim]CTf_3	GC	5.25	[60]
[Pmmim]PF_6	GC	4.31	[60]
[Pmmim]NTf_2	GC	5.20	[60]
[Bmim]BF_4	Ag	4.20	[61]
[Bmim]BF_4	Au	4.60	[62]
[Bmim]BF_4	Ta	5.20	[62]
[Bmmim]BF_4	Au	4.60	[63]
[Bmmim]BF_4	GC	4.20	[63]
[Bmmim]BF_4	Pt	4.50	[63]
[Bmmim]BF_4	Ta	5.40	[62]
[Bmim]BF_4	Pt-micro	4.10	[63]
[Mdim]BF_4	Pt-micro	3.80	[63]
[Bmim]PF_6	Au	4.50	[64]
[Emim]BF_4	GC	4.55	[65]

ionic liquids. Quaternary ammonium salt cations are more stable than those of other ionic liquids. Stability of cation is ammonium cation ≈ phosphonium cation > pyrrolidinium cation > piperidinium cation ≈ morpholinium cation> azepanium cation ≈ imidazolium cation > sulfonium cation > pyridinium cation. For similar ionic liquids, the stability of cations increases with the growth of side chains. When side chains of cation isomers are different, the electrochemical windows (EW) of ionic liquids will also be very different. The stability of anions and cations of the ionic liquids is restrained and influenced by each other. The improvement of the stability of anions (or cations) of ionic liquids will promote the stability of the corresponding cations (or anions) coexisting with them, and the EW

Table 6.5 Electrochemical window of piperidinium, morpholinium and pyridinium ionic liquids.

ILs	WE	EW (V)	Reference
[BPy]NTf_2	GC	3.80	[49]
[EmPy]BF_4	Pt	4.30	[49]
[Dmfp]BF_4	GC	4.00	[49]
[BPy]BF_4	Pt	3.40	[55]
[PMPip]NTf_2	GC	5.90	[66]
[PMPip]TSAC	GC	5.00	[66]
[BMPy]DCA	GC	2.96	[67]
[HPy]NTf_2	GC	3.96	[67]
[HPPy]NTf_2	GC	3.81	[67]
[MOEMMor]TPTP	GC	3.98	[67]
[MOEMMor]NTf_2	GC	4.68	[67]
[BMPy]NTf_2	GC	5.56	[68]
[Pip]HCOO	GC	3.23	[69]
[MPip]HCOO	GC	3.25	[69]
[MPip]CF_3COO	GC	3.25	[69]
[MPip]HSO_4	GC	3.99	[69]
[Azep]HCOO	GC	3.26	[69]
[Azep]CF_3COO	GC	3.29	[69]
[AEPip]NTf_2	GC	5.32	[70]
[AMPip]NTf_2	GC	5.32	[70]
[AMMor]NTf_2	GC	5.31	[70]
[AEMor]NTf_2	GC	5.28	[70]
[EMPip]$F(HF)_{2.3}$	GC	5.30	[71]
[PMPip]$F(HF)_{2.3}$	GC	5.22	[71]
[BMPip]$F(HF)_{2.3}$	GC	5.33	[71]
[BMPip]NTf_2	GC	5.23	[72]

Table 6.6 Electrochemical window of phosphonium, ammonium, sulfonium and pyrrolidinium ionic liquids.

ILs	WE	EW (V)	Reference
$[N_{4444}]$TSAC	GC	4.60	[49]
$[S_{222}]NTf_2$	GC	4.87	[49]
$[S_{444}]NTf_2$	GC	4.90	[49]
[BMPyr]NTf_2	Graphite	6.00	[54]
$[P_{1116}]NTf_2$	GC	5.60	[55]
$[P_{1116}]$TSAC	GC	4.90	[55]
$[N_{1113}]NTf_2$	GC	5.60	[55]
$[N_{1113}]$TSAC	GC	4.60	[55]
$[N_{2226}]NTf_2$	Au micro	5.99	[57]
$[N_{5555}]NTf_2$	GC	5.30	[57]
$[P_{2225}]NTf_2$	GC	6.40	[73]
[BMPyr]DCA	GC	4.62	[67]
[BMPyr]OTf	GC	4.83	[67]
[BMPyr]TFA	GC	4.20	[67]
[HMPyr]NTf_2	GC	5.68	[67]
$[N_{2226}]NTf_2$	Pt	5.40	[59]
$[P_{2228}]NTf_2$	GC	6.60	[73]
$[P_{2224}]$FSA	GC	5.10	[74]
$[P_{2225}]$FSA	GC	5.10	[75]
$[N_{1114}]NTf_2$	GC	5.85	[76]
$[N_{1224}]CF_3BF_3$	GC	5.75	[76]
$[N_{1224}]C_2F_5BF_3$	GC	5.64	[76]
$[N_{1224}]NTf_2$	GC	5.69	[76]
$[N_{2226}]NTf_2$	GC	5.95	[77]
[MPyr]HCOO	GC	3.22	[77]
[MPyr]CF_3COO	GC	3.28	[77]
[MPyr]HSO_4	GC	3.98	[77]

(Continued)

Table 6.6 Electrochemical window of phosphonium, ammonium, sulfonium and pyrrolidinium ionic liquids. (*Continued*)

ILs	WE	EW (V)	Reference
[MPyr]HCOO	GC	3.22	[77]
[AMPyr]NTf_2	GC	5.55	[78]
[PMPyr]NTf_2	GC	5.64	[78]
[PMPyr]$F(HF)_{2.3}$	GC	5.34	[79]
[PeMPyr]NTf_2	Pt	5.91	[80]
[HMPyr]NTf_2	Pt	5.95	[80]
[C_7MPyr]NTf_2	Pt	5.96	[80]
[PMPyr]NTf_2	Pt	5.84	[81]
[BMPyr]NTf_2	Pt	5.86	[81]
[BMPyr]NTf_2	GC	5.85	[82]

of ionic liquids will become larger. The electrochemical behavior of ionic liquids is firmly related to the electrode materials used. The measured results vary greatly with the reference electrode, and the working electrode used in electrochemical measurement. The most frequently used working electrodes are tungsten (W), glass carbon (GC), Pt and Au. The order of the measured electrochemical window is W > GC > Pt Au. The electrochemical window measurement with tungsten (W) as the working electrode is much wider than that of other working electrodes, so tungsten is a useful working electrode materials.

According to different compositions, ionic liquids can basically be classified as proton type, aprotic type and amphoteric type [44]. In supercapacitor, ionic liquids can replace some traditional organic or aqueous electrolyte, improving the electrochemical performance, which is one of the excellent solvents or substitutes of organic electrolyte of a supercapacitor [19–30]. It is not difficult to see from the published literature that the ionic liquids used in supercapacitors are only a small part of ILs family. The common ionic liquids used in supercapacitors mainly include imidazole, pyrrolidine and tetraethylammonium based ionic liquids since they have large electrochemical stability window and higher ionic conductivity [83, 84]. The most-used cations are 1-butyl-3-methylimidazolium cation ($[Bmim]^+$), 1-ethyl-3-methylimidazolium cation ($[Emim]^+$), tetraethylammonium cation (Et_4N^+), n-propyl-n-methylpyrrolidine cation ($[Pyr_{13}]^+$), 1-butyl-1-methylpyrrolidine cation ($[Pyr_{14}]^+$), Common anions mainly used in supercapacitors are tetrafluoroborate (BF_4^-), chloride ion (Cl^-), hexafluorophosphate (PF_6^-), bromine ion (Br^-), bis(trifluoromethylsulfonyl) imine anion ($TFSI^-$), bis(fluorosulfonyl) imine anion (FSI^-) etc. In addition, ionic liquids with almost no vapor pressure have unique advantages in solving the safety problems of volatilization and flammability related to organic solvents. This makes the ionic liquids based supercapacitor more suitable for high-temperature applications.

6.4 Application of Ionic Liquids in Supercapacitors

In terms of safety, the main problem of the energy storage device is that after the voltage exceeds the stable electrochemical window, the electrolyte may decompose and form harmful products, such as explosive gas. As mentioned above, ionic liquids provide a wider electrochemical window than traditional organic electrolytes. The significant difference of the electrochemical stability window will result in the significant difference of energy density. For example, the energy density of ionic liquid [Emim]BF_4 electrolyte is higher than that of KOH solution with one order of magnitude [85]. Romann *et al.* [86] recently discovered that the ionic liquid electrolyte [Emim]TCB still has good safety characteristics even under the extreme conditions by using a wider potential than its stability window of 2.4 V with the Ag/AgCl pair. The ionic liquid is polymerized to form a barrier layer and doped on the surface of the graphite electrode, which makes the graphite electrode behave as an insulator, which can prevent any electrochemical reaction that may be harmful to the electrochemical battery.

6.4.1 Pure Ionic Liquid as Electrolyte

Ionic liquids have unlimited possibilities due to their unique properties. In fact, only a few kinds of ionic liquids are employed in numerous applications, especially in electrochemical systems. Although anions and cations have many choices, the synthesis process is difficult and cheap. In addition, regardless of the fact that most kinds of ionic liquids are liquids, they are not suitable for most applications because of their high viscosity. The charge, size and shape of anions and cations of ionic liquids directly affect the movement of ions and the physical properties of the electrolytes in ionic liquid based supercapacitors [87], and also lead to different structures of electrical double layer. The capacitance of the supercapacitor can be adjusted or enhanced by selecting specific anions [88] or cations [89]. Compared with the solvated ions in conventional electrolytes, the ions in ionic liquids are more complex. Therefore, the capacitance performance improvement with ionic liquid electrolyte is tightly related to the selection of cations and anions [90]. In recent years, various ionic liquids were used as the electrolyte for supercapacitors, and good results have been obtained, and significant progress has been achieved.

6.4.1.1 Aprotic Ionic Liquids

Sun *et al.* [91] studied the behavior of EDLC with 1-ethyl-3-methylimidazolium thiocyanate ([Emim]SCN) ionic liquid electrolyte and activated carbon-based electrode. It helps reduce viscosity and higher ionic conductivity than that of 1-ethyl-3-methylimidazolium boron-tetrafluoride ([Emim]BF_4) ionic liquid. Pandey [92] proved that 1-ethyl-3-methylimidazolium tetracyano borate ([Emim]TCB) can be used as the electrolyte of activated carbon-based supercapacitors and has a low viscosity and high ionic conductivity. Matsumoto *et al.* [93] found that when the charging voltage of 2.5 V is utilized, a higher capacitance was obtained for 1-ethyl-3-methylimidazole difluorophosphate ([Emim]PO_2F_2) than that of [Emim]BF_4. However, the voltage of the electric double-layer capacitor with [Emim]PO_2F_2 is less than 3 V, which is lower than that of [Emim]BF_4. Shi *et al.* [94]

studied the performance of a series of 1-ethyl-3-methylimidazolium ionic liquids with BF_4^-, DCA^-, NTf_2^-, OAC^- and $EtSO_4^-$ as the anion in graphite based supercapacitors. It was pointed out that the viscosity of ILs has relation to the acceptance ability of hydrogen bonds in anions and cations. The supercapacitor with [Emim]DCA ionic liquid shows high specific capacitance and charge discharge capacity under high current. At the same time, it also has the minimum resistance, which is mainly determined by low viscosity and small ion size. However, the electrochemical stability window of a 2.3 V for [Emim]DCA is much smaller than that of ionic liquids [Emim] BF_4 of a 4 V.

The performance of a super-carbon electrode was investigated by Senda *et al.* [95]. Five fluorohydrides based ionic liquids with the cation of 1, 3-dimethylimidazole cation $[Emim]^+$, $[Dmim]^+$, $[Bmim]^+$, 1-methoxy-3-methylpyrrolidine ($[Mompyr]^+$) and 1-ethyl-3-methylpyrrolidine cation ($[Empyr]^+$) was studied. The obtained specific capacitances are higher than that of pure $[Emim]BF_4$ or the organic electrolyte of $TEABF_4$/PC with an operating voltage of 1–3.2 V. The largest specific capacitances of three fluorinated hydride ionic liquid based on imidazole cations is 178 $F.g^{-1}$ for $[Dmim](HF)_{2.3}F$, followed by 162 $F.g^{-1}$ for $[Emim](HF)_{2.3}F$, and the smaller is 135 $F.g^{-1}$ for $[Bmim](HF)_{2.3}F$. It showed that the larger the cation volume, the smaller the specific capacitance. Activated carbon based supercapacitors with some low melting-point fluorinated hydride (e.g. $[Emim](HF)_{2.3}F$) ionic liquids still have high specific capacitance of 64 $F.g^{-1}$ even at -40°C, which is much higher than that of $[TEA]BF_4$/PC of 20 $F.g^{-1}$.

The electrochemical behavior of a graphite based supercapacitors with three ionic liquids of $[Emim]BF_4$, [Emim]TFSI, and 1-methyl-1-propylpiperidine bis(trifluoromethylsulfonyl) imide ([Mppi]TFSI) ILs as the electrolyte, as well as the interaction between ILs electrolyte and graphite was studied by Li *et al.* [96]. When applied voltage on two electrodes is in the range of 3.7–4.6 V, there is an electrochemical reaction to [Emim]TFSI or $[Emim]BF_4$ on the graphite electrode. When the applied voltage on two electrodes is in the range of 4.0–5.8 V, electrochemical reaction to [Mppi]TFSI will occur on graphite electrodes. The maximum energy density of a supercapacitor with [Mppi]TFSI can be achieved when applied voltage reaches 4.4 V. The specific capacitance of [Mppi]TFSI based capacitor is lower than that of [Emim]TFSI or $[Emim]BF_4$ based capacitor. However, the electrochemical stability window and working voltage of [Mppi]TFSI electrolytes are large, resulting in the highest energy density of 169 $W.h.kg^{-1}$

It was found that the graphite nano-sheets based supercapacitor with ionic liquid $[C_1ClPyrr]N(CN)_2$ has a higher power density and energy density than that of $[C_1ClPyrr]$ Ntf_2. This is mainly because the smaller $N(CN)_2$ ionic liquid has a lower viscosity, higher conductivity and capacitance, which compensate for the reduction of electrochemical potential [97].

Rennie [98] studied the electrochemical behavior of ionic liquids containing sulfur cations as a supercapacitor electrolyte and the effect of cation structure on capacitor performance. It was found that the working voltage of $[S_{223}]Ntf_2$ is 2.6 V. The operating voltage shows that $[S_{222}]Ntf_2$ of 2.8 V is more stable than that of $[S_{221}]Ntf_2$ of 2.7 V, which may be related to the charge density of cations. Although the working voltage of $[S_{221}]Ntf_2$ is small, it has a higher power density and energy density.

Lewandowski *et al.* [99] studied the activated carbon based supercapacitor based on [Pmpi]TFSI. It was found that an energy density of 240 $kJ.kg^{-1}$ can be reached, which is the obtained energy density level of lead-acid battery. After that, they detailed studied $[Emim]BF_4$,

[Emim]TFSI, [Pmpi]TFSI, and [Pmpy]TFSI ionic liquids as well as their mixture with PC as the electrolyte for supercapacitor [100]. It was found that high-energy density can be achieved for pure ILs electrolyte. When ILs is properly mixed with organic solvent, the power density can be significantly improved when the loss of energy density is relatively small.

Simon *et al.* [101] studied the microporous activated carbon-based supercapacitor assembled with [Bmpy]TFSI electrolyte. The working voltage of 5 V can be obtained at 60 °C. The performance of supercapacitors combined with [Bmpy]TFSI and TiC-CDC was also studied. The capacitance of 130 $F.g^{-1}$ can be realized at a high temperature of 100°C. The highest working temperature range of the supercapacitor was further expanded [102]. They further studied the performance of supercapacitors assembled by the mixture of [Pmpi] FSI and [Bmpy]TFSI with OLC and CNT based electrodes [103]. A wide operating temperature range of 50–100 °C was achieved, and the lowest operating temperature range was expanded for supercapacitors. By using chemically activated stripped graphite oxide electrode, Zhu *et al.* [104] found that the capacitance is from 30 to 160 $F.g^{-1}$, and even in the temperature range of -40–50 °C, a specific capacitance of up to 100 $F.g^{-1}$ can be obtained.

6.4.1.2 *Proton Ionic Liquids*

Proton ionic liquids (PILs) include proton pyrrolidine nitrate ([Pyr]NO_3), diisopropyl ethyl ammonium bis (trifluoromethylsulfonyl) imide ([Dipea]TFSI), triethylammonium bis (trifluoromethylsulfonyl) imide ([Et_3NH]TSFI), and pyrrolidinium bis (trifluoromethylsulfonyl) imide ([Pyrr]TFSI). Like aprotic ionic liquids, proton ionic liquids provide a high stability window, and they show higher conductivity than aprotic ionic liquids in all temperature ranges. As the electrolyte of supercapacitors, proton ionic liquids have not attracted extensive attention from researchers since their low operating windows. However, proton ionic liquids also have some advantages. For example, proton type ionic liquids are not difficult to synthesize and cheaper than non-proton type ionic liquids. Proton type ionic liquids have protons that can be utilized in the redox process. Faraday reaction on the electrolyte/electrode interface can be observed when ruthenium oxide is used as the electrode material for a pseudo capacitor [105]. When porous carbon is used as electrode material, the functional groups containing oxygen on the surface can react with the protons in the electrolyte to produce higher capacitance. However, it still stores far less energy than organic electrolyte based capacitors.

Sato *et al.* [106] combined quaternary ammonium cations containing methoxy groups with BF_4^- and $TFSI^-$ to form the new ionic liquids of [Deme]TFSI and [Deme]BF_4, and evaluated the redox potential, conductivity and other physicochemical properties. It was found that [Deme]BF_4 has an electrochemical window with 6.0 V and a conductivity of 4.8×10^{-3} $S.cm^{-1}$ at 25 °C. These ionic liquids were used in an activated carbon-based EDLC, which are compared with the organic electrolyte system of triethylamine tetrafluoroboric acid/PC. It was found that EDLC using [Deme]BF_4 with working voltage at 2.5 V had good discharge/charge stability and higher capacity at 100 °C. The electrochemical property of EDLC assembled by activated carbon, and a mesoporous nickel based mixed rare-earth oxide (NMO) electrode, and ionic liquid electrolyte was studied by Liu *et al.* [107]. A good electrochemical performance was obtained. The charge/discharge voltage and energy density are up to 3 V and 50 $Wh.kg^{-1}$ respectively. There is almost no capacity attenuation by the constant current charge discharge of 500 times.

Kim *et al.* [108] studied electrochemical performance of [Deme]TFSI and [Deme]BF_4 based electrochemical capacitors. Among them, [Deme]TFSI and [Deme]BF_4 had a higher-energy density and wide electrochemical window than that of [TEA]BF_4/PC and [Emim] BF_4. Under the conditions of charging voltage of 3.5 V and 1 $mA.cm^{-2}$, the capacitance of 50 $F.g^{-2}$ can be achieved. The specific capacitance rises with the increase of charging voltage, which is better than that of a capacitor with $[TEA]_4$/ PC.

Brandt *et al.* [109] assembled three proton-ionic liquids [Et_3NH]TFSI, trimethylammonium bis (trifluoromethylsulfonyl) imide ([Me_3NH]TFSI) and pyrrolidine nitrate ([Pyr] NO_3) with activated carbon for supercapacitor. It was found that the working voltage of 2.1 V can be realized for proton ionic liquid as the electrolyte. They have good cycle stability in a wide range of temperatures. Demarconnay [110] studied the electrochemical behavior of the supercapacitor with the mixture electrolyte of [Et_3NH]TFSI and water based. It was found that an operating voltage of 2.4 V can be applied.

6.4.1.3 Functionalized Ionic Liquids

Functionalized ILs are ILs that introduce one or more functional groups into anions and cations, or have some special functions due to the specific structure of anions and cations, or can be designed and synthesized as solvents or catalysts in a reaction. The emergence of functionalized ILs makes ILs more attractive and has become the leading direction of the development of ionic liquids. At present, functional ionic liquids are in a period of rapid progress, and the synthetic route is becoming more and more mature. The application field has gradually expanded from the simple use as solvent to the synthesis of numerous functional materials, the separation and purification of substances, the dissolution of mineral samples, electrochemical applications and other fields. However, there are still some deficiencies in functionalized ionic liquids. The introduction of functional groups of imidazole side chains in ionic liquids will lead to changes in their chemical properties, thermal stability, electrochemical window and viscosity, and is not conducive to purification. Therefore, the main problem to be solved in the field of functional ionic liquids application is to construct the structure-activity relationship (QSAR) between the structure of functional groups and their physicochemical properties. Thus, more effective functional groups can be selected and introduced according to actual need to give full play to their performance advantages.

The behavior and property of activated carbon based supercapacitors with [C_3mim]SCN, [C_4mim]SCN and [C_5mim] SCN as electrolytes were investigated by Jin *et al.* [111]. It is found that the internal resistances of supercapacitors with [C_3mim]SCN, [C_4mim]SCN and [C_5mim]SCN are 0.3 Ω, 0.35 Ω and 0.19 Ω, respectively, which is significantly less than the internal resistance of [Bmim]BF_4 and [Bmim]PF_6. The orders of electrochemical windows of three room temperature ionic liquids are: [C_3mim]SCN > [C_4mim]SCN > [C_5mim]SCN. The carbon chain length of the alkyl group on the cationic substituent has a great effect on the stable electrochemical window of ILs. Generally speaking, the shorter the carbon chain of alkyl, the larger the EW of ionic liquid, which is primarily due to the volume and migration speed of ions in an ionic liquid. The carbon chain of alkyl on the imidazole ring of ionic liquid also affects the charge/discharge characteristics of the capacitor. [C_3mim]SCN shows more stable capacitive behavior, wide electrochemical window and better conductivity, which are more suitable for working electrolytes of supercapacitors. The electrochemical performance of activated carbon supercapacitors with [Pmim]ClO_4 ionic liquids electrolyte was studied [112]. It is found that the electrochemical stability window of 5.0 V can be reached and there is no obvious redox peak on the positive and negative scanning cyclic voltammetry curve of the electrolyte, indicating that the activated carbon electrode and [Pmim]ClO_4 electrolyte has a good stability and double-layer capacitance characteristics. Internal resistance of 0.21 Ω is smaller than [Emim]BF_4 and [Emim]PF_6, indicating that it has better capacitor characteristics and greater potential in the electrochemical application. The specific capacitance of the capacitor decreased by only about 1.8% after 1000 cycles, which show that [Pmim]ClO_4 has good cycle performance and is suitable to be used as the electrolyte of the supercapacitor.

It was shown from the above work that ILs with the unique properties, especially the wide electrochemical window, high thermal stability, low vapor pressure and a wide range of liquid temperature, make them perform well at a high or low temperature. The progress of other related works is provided in Table 6.7. It should be pointed out that ionic liquids have high viscosity and low conductivity. In order to achieve better capacitance characteristics, further exploration and study is needed to decrease the viscosity and to enhance the conductivity.

Table 6.7 Performance of various IL-based supercapacitors.

Systems	Material	Specific capacitance/ $F g^{-1}$)	Specific energy/ $Wh kg^{-1}$	Cell voltage	Ref.
[Emim]BF_4	3D graphite-based porous materials	231	29.4	3.5	[113]
[C_2mim]BF_4	Polypyrrole- derived AC	290	16	2.3	[114]
[Emim]BF_4	3D cross-linked graphite material	213	27.2	3.5	[115]
[C_2mim]BF_4	C/C composite	95	12.1	3.5	[116]
[Emim]BF_4	N-doped mesoporous carbon spheres	170	15.9	3	[117]
[C_2mim]BF_4	SiC-CDC	138	12.9	3	[118]
[Emim]TFSI	AC	110	7.2	2.5	[119]
[Emim]FSI	AC	110	7.2	2.5	[119]
[C_2mim]BF_4	AC	220	28.1	3.5	[120]
[Bmim]BF_4	AC	170	21.7	3.5	[120]
[Emim]BF_4	Derived AC	144	2.7	1.35	[121]
[Emim]DCA	AC	41.7	2.7	2.5	[122]

(Continued)

Table 6.7 Performance of various IL-based supercapacitors. (*Continued*)

Systems	Material	Specific capacitance/ F g^{-1})	Specific energy/ Wh kg^{-1}	Cell voltage	Ref.
[C_3mim]TFSI	AC	10.7	0.8	2.7	[123]
[Emim]BF_4	Mesoporous tubular graphite	78	3.3	4	[124]
[C_2mim]BF_4	Graphite sheets	120	20	4	[125]
[Emim]BF_4	Onion-like carbon	25	3.2	3.5	[126]
[C_2mim]TFSI	Onion-like carbon	19	2.4	3.5	[126]
[Emim]BF_4	Chemically converted graphite	273.1	45.3	4	[127]
[Emim]TFSI	3Dom carbon	160	6.7	2	[128]
[Hmim]TFSI	3Dom carbon	89	3.7	2	[128]
[Omim]TFSI	3Dom carbon	60	2.5	2	[128]
[Emmim]TFSI	3Dom carbon	101	4.2	2	[128]
[N_{1114}]TFSI	3Dom carbon	81	3.4	2	[128]
[N_{2444}]TFSI	3Dom carbon	21	0.9	2	[128]
[$Pyr_{1,4}$]TFSI	3Dom carbon	108	4.5	2	[128]
[Emim]TFSI	Micro-meso porous carbon	140	13.1	3	[129]
[C_2mim]TFSI	Activated mesophase pitch carbon	180	23	3.5	[130]
[Emim]TFSI	GPC (graphite-based hierarchically porous carbon)	212	27.1	3.5	[131]
[C_2mim]TFSI	Carbide-derived carbon	160	15	3	[132]
[Emim]TFSI	Reduced graphite oxide	112.6	12	3.2	[133]
[Emim]$B(CN)_4$	AC	173	16.2	3	[134]
[Emim]Ac	AC	122	2.9	1.5	[135]
[$Pyr_{1,3}$]FSI	AC	181.2	11.8	2.5	[136]
[Bmim]BF_4	AC	140	21	3.8	[137]
[Hmim]BF_4	AC	104	6.8	2.5	[138]
[Bmim]TFSI	Porous carbon	86	8.1	3	[139]
[P_{1114}]TFSI	Mesoporous carbon	32	2.1	2.5	[139]

(*Continued*)

Table 6.7 Performance of various IL-based supercapacitors. (*Continued*)

Systems	Material	Specific capacitance/ $F\ g^{-1}$)	Specific energy/ $Wh\ kg^{-1}$	Cell voltage	Ref.
$[Pip_{14}]TFSI$	Mesoporous carbon	58	3.5	3	[139]
$[N_{1122}]BF_4$	AC	152	14.3	3	[140]
$[Pyr_{11}]BF_4$	AC	155	14.5	3	[140]
$[N_{2224}]DCA$	GNS	42	3.9	3	[141]
$[Pyr_{13}]TFSI$	AC	131.6	9.3	2.6	[142]
$[Pyr_{15}]TFSI$	AC	122.4	8.6	2.6	[142]
$[Pyr_{14}]TFSI$	AC	104	2.7	2.5	[143]
$[Pip_{14}]C(CN)_3$	AC	85	3.5	2	[144]
$[Pip_{14}]B(CN)_4$	AC	30	2.8	3	[145]

6.4.2 Mixture Electrolyte of Ionic Liquids

Ionic liquids have high electrochemical stability and can achieve the maximum voltage of 3.2–3.5 V in electric double-layer capacitors. Compared with organic electrolytes, energy density of ionic liquids is considerably increased. In addition, ionic liquids are usually non-flammable, and their vapor pressure is insignificant. Compared with volatile solvent such as an acetonitrile, ionic liquid electrolyte is an opportunity to prepare safer devices. However, there are some disadvantages for ionic liquids. They are generally more expensive than the organic solvents, low conductivity and high viscosity, which will enlarge the resistance of devices and limit the power output of electric double-layer capacitors. One way to address the problem of low transport performance of pure ILs is to use the mixture of organic solvents and ILs. Commonly used organic solvents in this field are γ-Butyrolactone, propylene carbonate, acetonitrile, Hexamethylene chloride (ADN), methoxypropionitrile. Another method is to use a eutectic mixture of different ionic liquids.

6.4.2.1 Binary of Ionic Liquids

Although ionic liquid electrolyte has obvious advantages, there continues being abundant room for improvement. Eutectic mixture of proton ionic liquid is also one of the possible options for supercapacitor electrolytes. Because of their low viscosity, they can be applied in a wide range of temperatures. Eutectic ionic liquid mixtures are usually composed of the same anions and different cations [146, 147]. A common example is that FSI anions form eutectic ionic liquid mixtures with pyrrolidine and piperidine cations with the same structural formula, which can avoid the formation of an ordered lattice structure of cations. Another method is to mix ionic liquids with traditional organic solvents, which can reduce system viscosity and increase system conductivity with more than one order of magnitude.

However, this method will reduce the electrochemical window of ionic liquid, but the system continues to have good performance. Lian *et al.* [146] studied the capacitance behavior of onion-like carbon based capacitors with two similar ILs and their mixtures as electrolytes. The results showed that the capacitance of the capacitor with mixed ILs electrolyte is significantly higher than that of single pure ionic liquid electrolyte.

6.4.2.2 *Mixed Electrolyte of Organic Solvent and Ionic Liquids*

To reduce the viscosity and enlarge its conductivity of ionic liquid, especially used for low operating temperatures, the mixed solution containing organic solvent and ionic liquid has been used for the electrolyte of the supercapacitor. Because of relatively high ionic conductivity, imidazole ILs have been applied to form the hybrid electrolyte with organic solvents.

Wang *et al.* [148] added [Emim]BF_4 ionic liquid into the organic solvents such as AN, PC and DMF to form three new electrolytes for activated carbon-based EDLC. The electrochemical property of the capacitor was studied. It was found that [Emim]BF_4 has large solubility (>6 mol.L^{-1}) in AN, PC and DMF. When the concentrations of [Emim]BF_4 in [Emim]BF_4/AN, [Emim]BF_4/PC and [Emim]BF_4/DMF at 25°C is attained to 2.4, 2.7 and 2.6 mol.L^{-1}, the conductivity of systems attain its largest value as 66.3, 20.3 and 35.3 mS.cm^{-1}, respectively. Corresponding electrochemical windows of these three systems are 4.0, 3.7 and 3.6 V, respectively. There is no sharp rise of capacitor voltage in the latter charging period of EDLC. Among them, the capacitor using [Emim]BF_4/AN electrolyte with concentration of 2.4 mol.L^{-1} has a relatively high working voltage and the most stable charge/discharge performance. Orita *et al.* [33, 149] studied the performance of supercapacitors of a series of mixed electrolytes of alkylated ionic liquids and organic solvent of propylene carbonate. Dialkyl-Imidazole ionic liquids of [Daim]TFSA and [Emim]FSA were used. It was found that [Daim]TFSA can provide lower resistance and higher capacitance than that of [Emim]FSA in a wider operating temperature range. However, the stability of a supercapacitor with [Daim]TFSA/PC is lower than that of a supercapacitor with [Emim]TFSI/PC. The stability can be improved by adding dimethyl carbonate to propylene carbonate. Lin *et al.* [150] studied the properties of the mixed system of [Emim]TFSI ionic liquid with almost similar size of anion and anion with acetonitrile solvent. It was found that although the sizes of bare molecular of $TFSI^-$ and $[Emim]^+$ were almost same the sizes of solvated ion is different in acetonitrile, and the sizes enlarged in order of $[TFSI]^-$ (in acetonitrile) > $[Emim]^+$ (in acetonitrile) > $[Emim]^+$ > $TFSI^-$. Krause *et al.* [151] studied the high-voltage EDLC using the mixture electrolyte of [Pyr_{11}]TFSI ionic liquid and propylene carbonate (PC). It was found that the voltage window of the system is 3.5 V. Using [Pyr_{11}]TFSI/PC electrolyte can keep excellent cycle stability. The loss of capacitance is only 5% after 100000 cycles with a voltage of 3.5 V.

Janes *et al.* [152] added 1,2-ethylene glycol dimethyl ether with a volume fraction of 0–90% into [Emim]TFSI as a new electrolyte for supercapacitor. The electrochemical performance was studied by constant current charge/discharge, electrochemical reactance and cyclic voltammetry methods. It was found that the conductivity of the mixing electrolyte can reach 24.21 mS.cm^{-1}, which is higher than 5.67 mS.cm^{-1} of pure ionic liquid [Emim]TFSI. When the volume fraction of 1, 2-ethylene glycol dimethyl ether increases to 40%, the stored power in the supercapacitor is enhanced from 13 to 20.5 kW.kg^{-1}. Under constant energy density, the supercapacitor containing 40% ethylene glycol dimethyl ether in

Table 6.8 Mixed ionic liquids with organic solvents for supercapacitors.

Mixed system	Electrode material	Concentration	Gravimetric specific capacitance/Fg^{-1}	Energy density of cells/ $Wh.kg^{-2}$	Cell voltage/V	Ref.
$[N_{2222}]BF_4/AN$	3D graphite-based porous materials	1.0M	202	15.3	2.7	[113]
$[N_{2222}]BF_4/AN$	3Dcross-linked graphite material	1.0M	190	14.4	2.7	[115]
$[N_{2222}]BF_4/PC$	C/C composite	1.0M	92.1	8.6	3	[116]
$[Emim]BF_4/PC$	AC	1.0M	91	5.9	2.5	[117]
[Emim]TFSI/AN	Graphite-derived carbons	1.3M	174	22.2	3.5	[156]
$[N_{2222}]BF_4/AN$	AC	1.0M	96	7.3	2.7	[157]
$[Pyr_{11}]BF_4/PC$	AC	1.9M	85	9	3.19	[157]
$[Pyr_{14}]TFSI/PC$	AC	1.5M	86	10.9	3.49	[157]
$[Bmim]BF_4/AN$	Porous carbon (a-MEGO)	2.0M	166	21.2	3.5	[158]
$[N_{2222}]BF_4/AN$	Porous carbon	1.0M	150	11.4	2.7	[158]
$[EmmIm]BF_4/AN$	AC	0.8M	142.6	10.8	2.7	[159]
$[N_{1113}]BF_4/AN$	AC	1.0M	135	12.7	3	[160]
$N_{111(1O1)}]BF_4/AN$	AC	1.0M	142	13.3	3	[160]
$[N_{2221}]BF_4/AN$	AC	1.0M	136	3.2	1.5	[161]
$[N_{2221}]BF_4/PC$	AC	2.0M	110	4.6	2	[162]

(*Continued*)

Table 6.8 Mixed ionic liquids with organic solvents for supercapacitors. (*Continued*)

Mixed system	Electrode material	Concentration	Gravimetric specific capacitance/Fg^{-1}	Energy density of cells/ $Wh.kg^{-2}$	Cell voltage/V	Ref.
$[N_{2221}]BF_4$/FEC	AC	3.0M	112.4	4.7	2	[162]
$[N_{2222}]BF_4$/AN	Carbon nanosheets	1.0M	134	5.6	2	[163]
$[N_{2222}]BF_4$/AN	AC	1.0M	105.7	8	2.7	[164]
$[Pyr_{11}]BF_4$/AN	AC	1.0M	115.7	8.8	2.7	[164]
$[Pyr_{11}]BF_4$/PC	AC	1.0M	105.7	8	2.7	[164]
$[N_{2222}]BF_4$/AN	AC	1.0M	55	5.2	3	[165]
$[N_{2222}]BF_4$/AN	Carbon nanosheets	1.0M	100	7	2.6	[166]
$[N_{2222}]BF_4$/AN	AC	1.0M	99.2	7.5	2.7	[167]
$[AP]BF_4$/AN	AC	1.0M	127.4	9.7	2.7	[167]
$[N_{2222}]BF_4$/AN	Nanodiamond	1.5M	38	2.1	2.3	[168]
$[N_{2222}]BF_4$/PC	AC	1.0M	130	9.9	2.7	[169]
$[N_{2222}]$TFSI/PC	AC	1.0M	120	10.5	2.9	[169]
$[Pyr_{11}]BF_4$/PC	AC	1.0M	130	13.9	3.2	[169]
$[Pyr_{14}]$TFSI/PC	AC	1.0M	108	13.8	3.5	[169]
$[N_{2222}]BF_4$/PC	AC/G-CNT	1.0M	44	1.3	1.7	[170]
$[Pyr_{11}]BF_4$/PC	AC	2.3M	92	9.8	3.2	[171]

(*Continued*)

Table 6.8 Mixed ionic liquids with organic solvents for supercapacitors. (*Continued*)

Mixed system	Electrode material	Concentration	Gravimetric specific capacitance/Fg^{-1}	Energy density of cells/ $Wh.kg^{-2}$	Cell voltage/V	Ref.
$[N_{2222}]BF_4$/EC/DEC	2D porous carbon	1.0M	180	10.8	2.4	[172]
$[N_{2222}]PF_6$/AN	AC	1.0M	106	13.5	3.5	[173]
$[N_{2222}]PF_6$/PC	AC	1.0M	108.4	13.8	3.5	[173]
$[N_{2222}]$TF/AN	AC	1.7M	125	5.2	2	[174]
$[N_{2222}]$TF/PC	AC	1.0M	115	4.8	2	[174]
$[Pyr_{14}]$TFSI/BTN	AC	1.0M	84	9.7	3.5	[175]
$[Pyr_{14}]$TFSI/ADN	AC	1.2M	84	7	3.5	[175]
[SBP]BF_4/PC/DMC/EC	AC	1.5M	85	8	3	[176]
[SBP]BF_4/PC/DMC	AC	1.5M	90.7	8.5	3	[176]
$[Azp_{14}]$TFSI/PC	AC	1.0M	72	7.8	3.5	[177]
$[P_{666(14)}]$TFSI/AN	AC	25*wt*%	70	8.3	3.4	[178]

their study can supply the highest power density. The results show that this supercapacitor has excellent performance. Anouti *et al.* [153] studied trimethylthiobis (trifluoromethylsulfonyl) imide ([Me_3S]TFSI) electrolyte for activated carbon-based EDLC. The stable temperature of [Me_3S] TFSI can reach up to 280 °C. It was found that [Me_3S]TFSI has an electrochemical stability window of about 5 V on the Pt electrode, and the working voltage is up to 3 V at 80 °C. The capacities of the EDLC using [Me_3S]TFSI ionic liquid as an electrolyte can attain to 150 F.g^{-1} with a good cycling ability, which shows that the capacitor is suitable for high-temperature applications.

Schutter *et al.* [154] reported the performance of a supercapacitor with a binary mixture electrolyte of [$Pyrr_{11}$]TFSI ionic liquid and adiponitrile and butyronitrile. It is found that both mixture electrolytes have good transmission properties and high electrochemical stability and. At 20 °C, the viscosity and conductivity of butyronitrile mixture can reach a 2.46 and 17.14 mS.cm^{-1} respectively, which is better than the most advanced electrolyte.

In addition, mixed electrolytes with organic solvents and imidazole ionic liquids are also used as electrolytes for hybrid supercapacitors and pseudo-capacitors. It was shown that addition of dimethylformamide into [Bmim]PF_6 ionic liquid for asymmetric AC/MnO_2 based capacitor. The capacitance increased and the internal resistance reduced [155], which is attributable to the improved ion mobility and electrolytic permeability compared with pure ionic liquid. Another research progress in this field is listed in Table 6.8.

6.4.2.3 *Mixed Electrolyte of Ionic Liquid and Ionic Salt*

Compared with mixtures of electrolytes of ionic liquid and organic solvent, higher capacitive performance can be obtained by adding salts into ionic liquids. This is because ions of adding salts can improve the compatibility between ILs and the electrode at the electrolyte/electrode interface. The addition of different types of ions, such as transition metal ions, can change the arrangement of ions in ILs and can improve the performance of capacitors.

The electric double-layer supercapacitor is the first device which uses the metal salts in its electrolyte [179]. Using [Emim]BF_4 electrolyte containing Cu (II), the specific capacitance of the capacitor was twice that of pure [Emim]BF_4. Xie *et al.* [180] studied the performance imidazole ionic liquid based capacitor with ferrocene addition. It is found that ferrocene anion is a charged specially attached to the surface of an electrode. When the EDLC is charged, the ferrocene redox system is the same as the direct Faraday reaction, which can produce a uniform plane battery material.

Quaternary ammonium salt is one of the commonly used salts in electrolyte supercapacitors. The electrochemical performance of a supercapacitor composed of [$(C_4H_8)_2N$]BF_4 ammonium salt and [Bmim]PF_6 ionic liquid was studied [23, 181]. It is found that too high or too low concentrations of [$(C_4H_8)_2N$]BF_4 and [Bmim]PF_6 are not conducive to ion migration, so the conductivity of the system is low. The electrolyte with medium concentration of [$(C_4H_8)_2N$]BF_4 and [Bmim]PF_6 (ratio 4:6) has a high conductivity, the electrochemical window of the system can reach 4.8 V, and the specific capacitance is only attenuated by 1.5% with 1000 cycles of charge/discharge. When the charging voltage is 0–2.7 V, the specific capacitance of the supercapacitor with this concentration of electrolyte is 456.786 F.cm^{-3}.

The electrochemical properties of activated carbon based supercapacitors using the mixed electrolyte with quaternary ammonium salt [$(C_4H_8)_2N$]BF_4 and [Bmim]CF_3CO_2 ionic liquids were studied [23]. It was found that the mixed electrolyte has good cycle reversibility and good wettability. With the continuous increase of quaternary ammonium salt concentration, the electrochemical window of mixed electrolyte increases. When ammonium salt concentration applied is 2.06 mol.L^{-1}, the electrochemical window of the electrolyte reaches a maximum of 2.7 V, and then the electrochemical window will become smaller when the concentration of ammonium salt continues to increase. The capacity decreases by only 2% after 1000 cycles of discharge/charge. After 30 cycles of the test of charge/discharge, the charging efficiency is 96.59%, and the specific capacitance is 464.82 F.cm^{-3}.

The performance of activated carbon with mixed electrolyte of lithium perchlorate ($LiClO_4$) and [Emim]ES ionic liquid system was studied [23]. It is found that the mixed system is an electrolyte with good solubility, appropriate viscosity and stable electrochemical properties. When $x_{LiClO4} = 0.05$, the conductivity of the mixed system reaches the maximum of 8.3 mS.cm^{-1}, and the electrochemical window of the mixed system at this concentration is 5.1 V. The specific capacitance of the system reduced by only 1.9% after 1000 cycles, indicating that the mixed electrolyte has excellent charge/discharge performance and cycle life. When the charge/discharge current is 100 mA, the working voltage can reach 3 V and the specific capacitance of a single electrode is 446.04 F.cm^{-3}. These results show that the [Emim]ES/$LiClO_4$ system has the potential for large-scale application in supercapacitors.

6.5 Conclusion and Prospective

With the stimulating demand for sustainable and clean energy, electrochemical supercapacitors have become one of the emerging main power supply and energy storage devices with its advantages of long life, high-power density and high efficiency. However, one of the key challenges faced by supercapacitors is its low-energy density and stability, which hinder its large-scale application in the energy storage field. In order to solve this problem, one of the principal directions of research and development of supercapacitors should be to improve the capacity and electrochemical window. As an important component of the supercapacitor, the performance of electrolyte determines the behavior of the supercapacitor to a great extent. Ionic liquid electrolyte has excellent properties and stable electrochemical performance, which are conducive to solve the problem of low energy density of supercapacitors. We can prepare different electrolyte components by selecting different combinations of cations and anions.

Although various scientific researches on the performance of electrolyte in supercapacitor were carried out, there are still some problems to be solved, such as the mismatch between the newly developed electrode materials and electrolyte, low-energy density, the lack of standard methods for electrolyte performance evaluation, high cost of some electrolytes, and lack of in-depth view of the charge storage mechanism of complex electrodes. To solve these problems, the future development direction of supercapacitor electrolyte should mainly concentrate on the following points [182].

(1) Look for new low resistance materials with high specific surface, appropriate structure, pore size and distribution to promote ion diffusion. Carbon materials, metal oxides, conductive polymers, mixture of electrolytes with ionic liquids with solvent or salts, and other materials are composited and doped to prepare new materials to improve the conductivity

of the capacitor. Constructing a surface chemical structure compatible with the electrolyte and improving the cycle life of supercapacitors will be an important direction of electrode material research.

(2) Developing novel ionic liquid electrolytes to improve the electrochemical working window of the electrolyte, so as to enhance the safety, energy density and stability of the capacitor. Exploring the efficient purification method of ionic liquid to improve the purity of the electrolyte and reduce the influence of impurities on the working voltage window and cycle life of ionic liquid. Reducing the resistance and impedance of the supercapacitor and further enhancing the power density of the supercapacitor. Building a simple and low-cost preparation method for new electrolyte of ionic liquids with a high conductivity and a low viscosity. Decreasing the solvation radius of cation and anion to make full use of the specific surface area of materials. Optimizing the matching of electrode materials and electrolyte and to enhance the overall behavior of supercapacitors. Establishing the standard evaluation method of electrolyte performance will also be an important research direction in the future.

(3) Exploring the appropriate assembly process and mode of different electrode materials and electrolytes for supercapacitors. In order to enhance the behavior of the capacitor, we must consider the local properties of the electrode material and the electrolyte, as well as pay more attention to improving the matching between the electrolyte and the material of the electrode. Using the structure information of electrode material, the size of ions, the interaction between the salt, electrolyte and solvent, the relationship between the chemical reactivity of electrolyte and electrode, as well as the performance of the capacitor to build the best assembly process and method is also a research direction in the future.

(4) Study of the structure and properties of solution/electrode interface, charge transfer, ion transport and micro-mechanism of charge and discharge. It is noted that the adsorption and desorption of free ions in ionic liquids in carbon materials are closely related to the microstructure of materials. Therefore, electrolytes in ionic liquids will show high specific capacitance and good cycling performance in carbon materials with appropriate pore structure and pore size. However, the effects of electrode microstructure, electrode/solution interface structure and characteristics, ion and charge transport on capacitance, and the mechanism of material adsorption and desorption at the electrode interface are not obvious. It is very important to make full use of the first principle and molecular simulation technology combined with experiments to deeply recover the structure and properties of the solution/electrode interface, as well as reveal the mechanism of charge transfer, ion transport and charge/discharge.

Acknowledgments

This work are partially supported by Back-up Personnel Foundation of Academic and Technology Leaders of Yunnan Province (2011HR013) and the National Natural Science Foundation of China (51774158, 51264021).

References

1. Miller, J.R. and Simon, P., Electrochemical capacitors for energy management. *Science*, 321, 5889, 651–652, 2008.
2. Liu, W.W., Yan, X.Q., Lang, J.W., Electrochemical behavior of graphite nano sheets in alkylimidazolium tetrafluoroborate ionic liquid electrolytes: Influences of organic solvents and the alkyl chains. *J. Mater. Chem.*, 21, 13205–13212, 2011.
3. Zhang, Z., Deng, M., Hu, Y., Characteristics and applications of electrochemical capacitors. *Electron. Comp. Mater.*, 22, 11, 01–05, 2003.
4. Shi, Z., *Preparation and capacitance properties of carbon based electrochemical capacitor electrode materials*, Tianjin University, Tianjin, 2007.
5. Arbizzani, C., Mastragostino, M., Soavi, F., New trends in electrochemical supercapacitors. *J. Power Sources*, 100, 164–170, 2001.
6. Huggins, R.H., Supercapacitors and electrochemical pulse sources. *Solid State Ionics*, 134, 179–195, 2000.
7. Chu, A. and Braatz, P., Comparison of commercial supercapacitors and high-power lithium-ion batteries for power-assist applications in hybrid electric vehicles. I. Initial characterization. *J. Power Sources*, 112, 236–246, 2002.
8. Kotz, R. and Carbon, M., Principles and applications of electrochemical capacitors. *Electrochim. Acta*, 45, 2483–2498, 2000.
9. Biniak, S., Dzielerdziak, B., Siedlewski, J., The electrochemical behavior or carbon fiber electrodes in various electrolytes, double layer capacitance. *Carbon*, 33, 1255–1263, 1995.
10. Simon, P. and Gogotis, Y., Materials for electrochemical capacitors. *Nat. Mater.*, 7, 845–854, 2008.
11. Liu, S., Wei, L., Wang, H., Review on reliability of supercapacitors in energy storage applications. *Appl. Energy*, 278, 115436, 2020.
12. Zhang, L., Hu, X., Wang, Z., Sun, F., Dorrell, D.G., A review of supercapacitor modeling, estimation, and applications: A control/management perspective. *Renewable Sustainable Energy Rev.*, 81, 1868–1878, 2018.
13. Raza, W., Ali, F., Raza, N., Luo, Y., Kim, K.H., Yang, J., Kwon, E.E., Recent advancements in supercapacitor technology. *Nano Energy*, 52, 441–473, 2018.
14. Miller, E.E., Hua, Y., Tezel, F.H., Materials for energy storage: Review of electrode materials and methods of increasing capacitance for supercapacitors. *J. Energy Storage*, 20, 30–40, 2018.
15. Salanne, M., Ionic liquids for supercapacitor applications, in: *Ionic Liquids II*, Inamuddin (Ed.), pp. 29–53, Springer, UK, 2017.
16. Yin, L., Li, S., Liu, X., Yan, T., Ionic liquid electrolytes in electric double layer capacitors. *Sci. China Mater.*, 62, 1537–1555, 2019.
17. Xu, K., Ding, M.S., Jow, T.R., A better quantification of electrochemical stability limits for electrolytes in double layer capacitors. *Electrochim. Acta*, 46, 1823–1827, 2001.
18. Mastragstino, M. and Soavi, F., Strategies for high-performance supercapacitors for HEV. *J. Power Sources*, 174, 8993, 2007.
19. González, A., Goikolea, E., Barrena, J.A., Review on supercapacitors: Technologies and materials. *Renewable Sustainable Energy Rev.*, 58, 1189–1206, 2016.
20. Khalid, M., Arshid, N., Grace, N., *Advances in supercapacitor and supercapattery: Innovations in energy storage devices*, pp. 10–98, Elsevier, AE Amsterdam, 2020.
21. Frackowiak, E. and Béguin, F., *Supercapacitors: Materials, systems and applications*, pp. 1–215, Wiley-VCH Verlag GmbH & Co., Poznan, 2013.
22. Ruan, D., *Principle, manufacture and application of dynamic electric double-layer capacitor*, pp. 56–325, Science Press, Beijing, 2018.

23. Wei, Y., Zhang, G., Lang, X., *Preparation and application of key materials for supercapacitors*, pp. 1–46, Chemical Industry Press, Beijing, 2018.
24. Liang, S., *Synthesis of high concentration electrolyte and its application in supercapacitor*, Tianjin University of Technology, Tianjin, 2021.
25. Li, G., *Design and performance research of high specific energy electric double layer supercapacitor*, pp. 1–35, Wuhan Textile University, Wuhan, Hubei, 2021.
26. Ye, M., *Construction and performance study of ionic liquid based supercapacitor*, pp. 1–25, Lanzhou University of Technology, Gansu: Lanzhou, 2020.
27. Zhang, W., *Synthesis and electrochemical properties of high voltage electrolytes for supercapacitors*, pp. 1–28, Tianjin University of Technology, Tianjing, 2020.
28. Zhang, K., *Research on supercapacitors based on imidazole based ionic liquid mixture electrolyte*, pp. 1–38, Harbin Institute of Technology, Heilongjiang, Harbin, 2019.
29. Xu, L., *Research on electrochemical capacitors based on ionic liquids and their influencing factors*, pp. 1–26, Shanghai Normal University, Shanghai, 2009.
30. Wang, Y., *Preparation and electrochemical properties of imidazole ionic liquid electrolytes*, pp. 1–18, Xiangtan University, Hunan, Xiangtan, 2005.
31. Conway, B., Transition from supercapacitor to battery behavior in electrochemical energy storage, in: *Power Sources Symposium. Proceedings of the 34th International Volume*, New York, pp. 319–327, 1990.
32. Hu, C.C., Chang, K.H., Lin, M.C., Design and tailoring of the nano tubular arrayed architecture of hydrous RuO_2 for next generation supercapacitors. *Nano Lett.*, 6, 2690–2695, 2006.
33. Zhang, H., Cao, G., Wang, Z., Growth of manganese oxide nano-flowers on vertically-aligned carbon nanotube arrays for high-rate electrochemical capacitive energy storage. *Nano Lett.*, 8, 9, 2664–2668, 2008.
34. Fan, L.Z., Hu, Y.S., Maier, J., High electroactivity of polyaniline in supercapacitors by using a hierarchically porous carbon monolith as a support. *Adv. Funct. Mater.*, 17, 16, 3083–3087, 2007.
35. Seredych, M., Hulicova-Jurcakova, D., Lu, G.Q., Surface functional groups of carbons and the effects of their chemical character, density and accessibility to ions on electrochemical performance. *Carbon*, 46, 11, 1475–1488, 2008.
36. Zhong, C., Deng, Y., Hu, W., Qiao, J., Zhang, J., A review of electrolyte materials and compositions for electrochemical supercapacitors. *Chem. Soc. Rev.*, 44, 21, 7484–7539, 2015.
37. Michel, A., Endres, F., Douglas, R., Ionic-liquid materials for the electrochemical challenges of the future. *Nat. Mater.*, 8, 621–629, 2009.
38. McFarlane, D.R., Sun, J., Golding, J., High conductivity molten salts based on the imide ion. *Electrochim. Acta*, 45, 1271–1278, 2000.
39. Rikah, Y.I., Room temperature ionic liquids of alkylimidazolium cations and fluoroanions. *J. Fluor. Chem.*, 105, 221–227, 2000.
40. McEwen, A.B., McDevitt, S.F., Koch, V.R., Nonaqueous electrolytes for electrochemical capacitors: Imidazolium canons and inorganic fluorides with organic carbonates. *J. Electrochem. Soc.*, 144, 84–86, 1997.
41. McEwen, A.B., Ngo, H.L., Goldman, J.L., Electrochemical properties of imidazolium salt electrolytes for electrochemical capacitor applications. *J. Electrochem. Soc.*, 146, 1687–1695, 1999.
42. Rogers, R.D. and Voth, G.A., Ionic liquids. *Acc. Chem. Res.*, 40, 1077–1078, 2007.
43. Freemantle, M., Designer solvents-ionic liquids may boost clean technology development. *Chem. Eng. News*, 76, 13, 32–37, 1998.
44. Jiang, D.E. and Wu, J., Microscopic insights into the electrochemical behavior of nonaqueous electrolytes in electric double-layer capacitors. *J. Phys. Chem. Lett.*, 4, 8, 1260–1267, 2013.
45. Deng, Y.Q., *Ionic liquids: Properties, preparation and application*, pp. 20–45, China SINO-PEC Press, Beijing, 2006.

46. Li, R.X., *Green solvents: Synthesis and application of ionic liquids*, pp. 27–39, Chemical Industry Engineering Press, Beijing, 2004.
47. Zhang, S.J. and Lv, X.M., *Ionic liquids from fundamental study to industrial application*, pp. 35–95, Science Press, Beijing, 2006.
48. Zhang, S.J., *Nano-micro structure of ionic liquid in process intensification*, pp. 242–245, Chemical Industry Engineering Press, Beijing, 2020.
49. Ohno, H., *Electrochemical aspects of ionic liquids*, pp. 23–152, Wiley, New Jersey, 2008.
50. Hagiwara, R., Matsumoto, K., Nakamori, Y., Physicochemical properties of 1, 3-dialkylimidazolium fluorohydrogenate room-temperature molten salts. *J. Electrochem. Soc.*, 150, 12, D195–D199, 2003.
51. Sutto, T.E. and Duncan, T.T., Electrochemical and structural characterization of Mg ion intercalation into Co_3O_4 using ionic liquid electrolytes. *Electrochim. Acta*, 80, 1, 413–417, 2012.
52. Zhou, Z.B., Matsumoto, H., Tatsumi, K., Low-melting, low-viscous, hydrophobic ionic liquids: 1-alkyl (alkyl ether) -3-methylimidazolium perfluoroalkyl trifluo-trifluoroborate. *Chem. Eur. J.*, 10, 24, 6581–6591, 2004.
53. Matsumoto, K., Hagiwara, R., Yoshida, R., Syntheses, structures and properties of 1-ethyl-3-methylimidazolium salts of fluorocomplex anions. *Dalton Trans.*, 7, 1, 144–149, 2004.
54. Nishida, T., Tashiro, Y., Yamamoto, M., Physical and electrochemical properties of 1-alkyl-3-methylimidazolium tetrafluoroborate for electrolyte. *J. Fluor. Chem.*, 120, 2, 135–141, 2003.
55. Galiński, M., Lewandowski, A., Stepniak, I., Ionic liquids as electrolytes. *Electrochim. Acta*, 51, 26, 5567–5580, 2006.
56. Bonhote, P., Dias, A.P., Papageorgiou, N., Hydrophobic, highly conductive ambient-temperature molten salts. *Inorg. Chem.*, 35, 5, 1168–1178, 1996.
57. Hagiwara, R., Hirashige, T., Tsuda, T., Acidic1-ethyl-3-methylimidazolium fluoride: A new room temperature ionic liquid. *J. Fluor. Chem.*, 99, 1, 1–3, 1999.
58. Suarez, P.A.Z., Selbach, V.M., Dullius, J.E., Enlarged electrochemical window in dialkyl-imidazolium cation based room-temperature air and water table molten salts. *Electrochim. Acta*, 42, 16, 2533–2535, 1997.
59. Mahony, A.M., Silvester, D.S., Aldous, L., Effect of water on the electrochemical window and potential limits of room-temperature ionic liquids. *Chem. Eng.*, 53, 12, 2884–2891, 2008.
60. Koch, V.R., Dominey, A., Nanlundiah, C., The intrinsic anodic stability of several anions comprising solvent-free ionic liquids. *J. Electrochem. Soc.*, 143, 3, 798–803, 1996.
61. NuLi, Y.N., Yang, J., Wang, P., Electrodeposition of magnesium film from BMIMBF4 ionic liquid. *Appl. Surf. Sci.*, 252, 23, 8086–8090, 2006.
62. Moganty, S.S., Baltus, R.E., Roy, D., Electrochemical windows and impedance characteristics of $[Bmim^+][BF_4^-]$ and $[Bdmim^+][BF_4^-]$ ionic liquids at the surfaces of Au, Pt, Ta and glassy carbon electrodes. *Chem. Phys. Lett.*, 483, 1/3, 90–94, 2009.
63. Schroder, U., Wadhawan, J.D., Compton, R.G., Water-induced accelerated ion diffusion: Voltammetric studies in1-methyl-3-2,6-(s)-dimethylocten-2-ylimidazolium tetrafluoroborate, 1-butyl-3-methylimidazolium tetrafluoroborate and hexafluorophosphate ionic liquids. *New J. Chem.*, 24, 12, 1009–1015, 2000.
64. Endres, F., Ionic liquids: Solvents for the electrodeposition of metals and semiconductors. *ChemPhysChem*, 3, 2, 145–154, 2002.
65. Li, X. and Johnson, K.E., Electrochemistry of 1-butyl-3-methyl-1himidazolium tetrafluoroborate ionic liquid. *J. Electrochem. Soc., 150*, 6, E307–E311, 2003.
66. Matsumoto., H., Sakaebe, H., Tatsumi, K., Preparation of room temperature ionic liquids based on aliphatic onium cations and asymmetric amide anions and their electrochemical properties as a lithium battery electrolyte. *J. Power Sources*, 146, 1/2, 45–50, 2005.

67. Hayyan, M., Mjalli, F.S., Hashima, M.A., Investigating the electrochemical windows of ionic liquids. *Ind. Eng. Chem.*, 19, 1, 106–112, 2013.
68. Rao, C.J., Venkatesan, K.A., Tata, B.V.R., Radiation stability of some room temperature ionic liquids. *Radiat. Phys. Chem.*, 80, 5, 643–649, 2011.
69. Wu, T.Y., Su, S.G., Lin, Y.C., Electrochemical and physicochemical properties of cyclic amine-based bronsted acidic ionic liquids. *Electrochim. Acta*, 56, 2, 853–862, 2010.
70. Wu, T.Y., Su, S.G., Wang, H.P., Electrochemical studies and self-diffusion coefficients in cyclic ammonium based ionic liquids with allyl substituents. *Electrochim. Acta*, 56, 9, 3209–3218, 2011.
71. Matsumoto, K., Hagiwara, R., Ito, Y., Room-temperature ionic liquids with high conductivities and wide electrochemical windows, N-Alkyl-N-methylpyrrolidinium and N-Alkyl-N-methylpiperidinium fluorohydrogenates. *Electrochem. Solid-State Lett.*, 7, 10, E29–E44, 2004.
72. Yuan, L.X., Feng, J.K., Ai, X.P., Improved discharge ability and reversibility of sulfur cathode in a novel ionic liquid electrolyte. *Electrochem. Commun.*, 8, 4, 610–614, 2006.
73. Tsunashima, K., Physical and electrochemical characterization of ionic liquids based on quaternary phosphonium cations containing a carbon-carbon double bond. *Electrochim. Acta*, 56, 11, 4351–4355, 2011.
74. Tsunashima, K. and Sugiya, M., Physical and electrochemical properties of low-viscosity phosphonium ionic liquids as potential electrolytes. *Electrochem. Commun.*, 9, 9, 2353–2358, 2007.
75. Tsunashima, K., Kawabata, A., Matsumiya, M., Low viscous and highly conductive phosphonium ionic liquids based on bis (fluorosulfonyl) amide anion as potential electrolytes. *Electrochem. Commun.*, 13, 2, 178–181, 2011.
76. McFarlane, D.R., Sun, J., Golding, J., High conductivity molten salts based on the imide ion. *Electrochim. Acta*, 45, 8/9, 1271–1278, 2000.
77. Zhou, Z.B., Matsumoto, H., Tatsumi, K., Low-melting, low-viscous, hydrophobic ionic liquids: Aliphatic quaternary ammonium salts with perfluoro-alkyltrifluoroborates. *Chem.-Eur. J.*, 11, 2, 752–766, 2005.
78. Wu, T.Y., Su, S.G., Lin, Y.C., Electrochemical and physicochemical properties of cyclic amine-based bronsted acidic ionic liquids. *Electrochim. Acta*, 56, 2, 853–862, 2010.
79. Wasserscheid, P. and Welton, T., *Ionic liquids in synthesis*, pp. 141–150, Wiley-Vch, Weinheim, 2008.
80. Matsumoto, K., Hagiwara, R., Y., Room-temperature ionic liquids with high conductivities and wide electrochemical windows, N-Alkyl-N-methylpyrrolidinium and N-Alkyl-N-methylpiperidinium fluoro-hydrogenates. *Electrochem. Solid-State Lett.*, 7, 10, E29–E44, 2004.
81. Howlett, P.C. and Izgorodina, E.I., Electrochemistry at negative potentials in bis (trifluoromethanesulfonyl) amide ionic liquids. *Z. Phys. Chem.*, 220, 10, 3083–3098, 2006.
82. Appetecchia, G.B., Montaninoa, M., Zaneb, D., Effect of the alkyl group on the synthesis and the electrochemical properties of N-Alkyl-N-methyl-pyrrolidinium bis (trifluoromethanesulfonyl) imide ionic liquids. *Electrochim. Acta*, 54, 4, 1325–1332, 2009.
83. MacFarlane, D.R., Meakin, P., Sun, J., Pyrrolidinium imides: A new family of molten salts and conductive plastic crystal phases. *J. Phys. Chem. B*, 103, 20, 4164–4170, 1999.
84. Qi, C., Hua, Y., Xu, C., Li, J., Zhang, Q., Li, Y., Research advances in the electrochemical window of ionic liquids. *Chin. J. Process Eng.*, 14, 4, 694–707, 2014.
85. Lei, Z., Liu, Z., Wang, H., Sun, X., Lu, L., Zhao, X.S., A high-energy-density supercapacitor with graphite–CMK-5 as the electrode and ionic liquid as the electrolyte. *J. Mater. Chem. A*, 1, 2313–2321, 2013.
86. Romann, T., Anderson, E., Pikma, P., Tamme, H., Möller, P., Lust, E., Reactions at graphite tetracyanoborate ionic liquid interface. New safety mechanisms for supercapacitors and batteries. *Electrochem. Commun.*, 74, 38–41, 2017.

87. Sillars, F.B., Fletcher, S.I., Mirzaeian, M., Hall, P.J., Variation of electrochemical capacitor performance with room temperature ionic liquid electrolyte viscosity and ion size. *Phys. Chem. Chem. Phys.*, 14, 17, 6094–6100, 2012.
88. Liu, X., Wang, Y., Li, S., Effects of anion on the electric double layer of imidazolium-based ionic liquids on graphite electrode by molecular dynamics simulation. *Electrochim. Acta*, 184, 164–170, 2015.
89. Liu, W.W., Yan, X.B., Lang, J.W., Pu, J.B., Xue, Q.J., Supercapacitors based on graphite nanosheets using different non-aqueous electrolytes. *New J. Chem.*, 37, 7, 2186–2195, 2013.
90. Vadiyar, M.M., Patil, S.K., Bhise, S.C., Ghule, A.V., Han, S.H., Kolekar, S.S., Improved electrochemical performance of a $ZnFe_2O_4$ nanoflake-based supercapacitor electrode by using thiocyanate-functionalized ionic liquid electrolytes. *Eur. J. Inorg. Chem.*, 36, 2015, 5832–5838, 2015.
91. Sun, G.H., Li, K.X., Sun, C.G., Application of 1-ethyl-3-methylimidazolium thiocyanate to the electrolyte of electrochemical double layer capacitors. *J. Power Sources*, 162, 2, 1444–1450, 2006.
92. Pandey, G.P. and Hashmi, S.A., Studies on electrical double layer capacitor with a low-viscosity ionic liquid 1-ethyl-3-methylimidazolium tetracyanoborate as electrolyte. *Bull. Mater. Sci.*, 36, 4, 729–733, 2013.
93. Matsumoto, K. and Hagiwara, R., Electrochemical properties of the ionic liquid 1-ethyl-3-methylimidazolium difluorophosphate as an electrolyte for electric double-layer capacitors. *J. Electrochem. Soc.*, 157, 5, A578, 2010.
94. Shi, M., Kou, S., Yan, X., Engineering the electrochemical capacitive properties of graphite sheets in ionic-liquid electrolytes by correct selection of anions. *ChemSusChem*, 7, 11, 3053–3062, 2014.
95. Li, J., Tang, J., Yuan, J., Zhang, K., Shao, Q., Sun, Y., Qin, L.C., Interactions between graphite and ionic liquid electrolyte in supercapacitors. *Electrochim. Acta*, 197, 84–91, 2016.
96. Huang, P.L., Luo, X.F., Peng, Y.Y., Pu, N.W., Ger, M.D., Yang, C.H., Chang, J.K., Ionic liquid electrolytes with various constituent ions for graphite-based supercapacitors. *Electrochim. Acta*, 161, 371–377, 2015.
97. Rennie, A.J., Martins, V.L., Torresi, R.M., Hall, P.J., Ionic liquids containing sulfonium cations as electrolytes for electrochemical double layer capacitors. *J. Phys. Chem. C*, 119, 42, 23865–23874, 2015.
98. Rochefort, D. and Pont, A.L., Pseudocapacitive behavior of RuO_2 in a proton exchange ionic liquid. *Electrochem. Commun.*, 8, 9, 1539–1543, 2006.
99. Lewandowski, A. and Olejniczak, A., N-methyl-N-propylpiperidinium bis(trifluo-romethanesulphonyl)imide as an electrolyte for carbon-based double layer capacitors. *J. Power Sources*, 172, 1, 487–492, 2007.
100. Lewandowski, A., Olejniczak, A., Galinski, M., Performance of carbon-carbon supercapacitors based on organic, aqueous and ionic liquid electrolytes. *J. Power Sources*, 195, 17, 5814–5819, 2010.
101. Balducci, A., Dugas, R., Taberna, P.L., High temperature carbon-carbon supercapacitor using ionic liquid as electrolyte. *J. Power Sources*, 165, 2, 922–927, 2007.
102. Largeot, C., Taberna, P.L., Gogotsi, Y., Microporous carbon-based electrical double layer capacitor operating at high temperature in ionic liquid electrolyte. *Electrochem. Solid State Lett.*, 14, 12, A174–A176, 2011.
103. Lin, R., Taberna, P.L., Fantini, S., Capacitive energy storage from -50 to 100 °C using an ionic liquid electrolyte. *J. Phys. Chem. Lett.*, 2, 19, 2396–2401, 2011.
104. Zhu, Y., Murali, S., Stoller, M.D., Carbon-based supercapacitors produced by activation of graphite. *Science*, 332, 6037, 1537–1541, 2011.

105. Mysyk, R., Raymundo-Piñero, E., Anouti, M., Lemordant, D., Béguin, F., Pseudo-capacitance of nanoporous carbons in pyrrolidinium-based protic ionic liquids. *Electrochem. Commun.*, 12, 3, 414–417, 2010.
106. Sato, T., Masuda, G., Takagi, K., Electrochemical properties of novel ionic liquids for electric double layer capacitor applications. *Electrochim. Acta*, 49, 21, 36033611, 2004.
107. Liu, H.T., He, P., Li, Z.Y., A novel nickel-based mixed rare-earth oxide/activated carbon supercapacitor using room temperature ionic liquid electrolyte. *Electrochim. Acta*, 51, 10, 1925–1931, 2006.
108. Kim, Y.J., Matsuzawa, Y., Ozaki, S., High energy-density capacitor based on ammonium salt type ionic liquids and their mixing effect by propylene carbonate. *J. Electrochem. Soc.*, 152, 4, A710–A715, 2005.
109. Brandt, A., Pires, J., Anouti, M., Balducci, A., An investigation about the cycling stability of supercapacitors containing protic ionic liquids as electrolyte components. *Electrochim. Acta*, 108, 226–231, 2013.
110. Demarconnay, L., Calvo, E.G., Timperman, L., Anouti, M., Lemordant, D., Raymundo-Piñero, E., Béguin, F., Optimizing the performance of supercapacitors based on carbon electrodes and protic ionic liquids as electrolytes. *Electrochim. Acta*, 108, 361–368, 2013.
111. Jin, Z., Tian, Y., Zhang, Q., Cai, K., He, T., Applied research of 1-butyl--3-methylimidazolium ionic liquids for supercapacitor. *Electron. Comp. Mater.*, 29, 9, 66–69, 2010.
112. Béguin, F., Presser, V., Balducci, A., Frackowiak, E., Carbons and electrolytes for advanced supercapacitors. *Adv. Mater.*, 26, 14, 2219–2251, 2014.
113. Zhang, L., Zhang, F., Yang, X., Porous 3d graphite-based bulk materials with exceptional high surface area and excellent conductivity for supercapacitors. *Sci. Rep.*, 3, 1408, 2013.
114. Wei, L., Sevilla, M., Fuertes, A.B., Polypyrrole-derived activated carbons for high-performance electrical double-layer capacitors with ionic liquid electrolyte. *Adv. Funct. Mater.*, 22, 827–834, 2012.
115. Zhang, F., Lu, Y., Yang, X., A flexible and high-voltage internal tandem supercapacitor based on graphite-based porous materials with ultrahigh energy density. *Small*, 10, 2285–2292, 2014.
116. Chang, P.P., Wang, C.Y., Kinumoto, T., Frame-filling C/C composite for high-performance EDLC with high withstanding voltage. *Carbon*, 131, 184–192, 2018.
117. Kumagai, S., Hatomi, M., Tashima, D., Electrochemical performance of microporous and mesoporous activated carbons in neat and diluted 1-ethyl-3-methylimidazolium tetrafluoroborate. *J. Power Sources*, 343, 303–315, 2017.
118. Tang, J., Wang, J., Shrestha, L.K., Activated porous carbon spheres with customized mesoporous through assembly of diblock copolymers for electrochemical capacitor. *ACS Appl. Mater. Interfaces*, 9, 18986–18993, 2017.
119. Tee, E., Tallo, I., Thomberg, T., Supercapacitors based on activated silicon carbide-derived carbon materials and ionic liquid. *J. Electrochem. Soc.*, 163, A1317–A1325, 2016.
120. Yamagata, M., Hirayama, M., Nishishita, S., Impact of lithium salt addition to ionic liquid electrolytes for high-performance electric double-layer capacitors. *Electrochemistry*, 81, 857–862, 2013.
121. Zhang, L., Yang, X., Zhang, F., Controlling the effective surface area and pore size distribution of sp2 carbon materials and their impact on the capacitance performance of these materials. *J. Am. Chem. Soc.*, 135, 5921–5929, 2013.
122. Tooming, T., Thomberg, T., Kurig, H., High power density supercapacitors based on the carbon dioxide activated d-glucose derived carbon electrodes and 1-ethyl-3-methylimidazolium tetrafluoroborate ionic liquid. *J. Power Sources*, 280, 667–677, 2015.

123. Sillars, F.B., Fletcher, S.I., Mirzaeian, M., Hall, P.J., Variation of electrochemical capacitor performance with room temperature ionic liquid electrolyte viscosity and ion size. *Phys. Chem. Chem. Phys.*, 14, 6094–6100, 2012.
124. Tian, J., Cui, C., Zheng, C., Qian, W., Mesoporous tubular graphite electrode for high performance supercapacitor. *Chin. Chem. Lett.*, 29, 599–602, 2018.
125. Shi, M., Kou, S., Yan, X., Engineering the electrochemical capacitive properties of graphite sheets in ionic liquid electrolytes by correct selection of - anions. *ChemSusChem*, 7, 3053–3062, 2014.
126. Van Aken, K.L., Beidaghi, M., Gogotsi, Y., Formulation of ionic-liquid electrolyte to expand the voltage window of supercapacitors. *Angew. Chem.*, 127, 4888–4891, 2015.
127. Yang, X., Zhu, J., Qiu, L., Li, D., Bioinspired effective prevention of restacking in multilayered graphite films: Towards the next generation of high-performance supercapacitors. *Adv. Mater.*, 23, 2833–2838, 2011.
128. Mousavi, M.P.S., Wilson, B.E., Kashefolgheta, S., Ionic liquids as electrolytes for electrochemical double-layer capacitors: Structures that optimize specific energy. *ACS Appl. Mater. Interfaces*, 8, 3396–3406, 2016.
129. Fuertes, A.B. and Sevilla, M., High-surface area carbons from renewable sources with a bimodal micro-mesoporosity for high-performance ionic liquid-based supercapacitors. *Carbon*, 94, 41–52, 2015.
130. Huang, H.C., Yen, Y.C., Chang, J.C., An ether bridge between cations to extend the applicability of ionic liquids in electric double layer capacitors. *J. Mater. Chem. A*, 4, 19160–19169, 2016.
131. Hao, X.F., Yan, Y., Gao, L.G., Ionothermal synthesis of graphite-based hierarchically porous carbon for high-energy supercapacitors with ionic liquid electrolyte. *Electrochim. Acta*, 241, 124–131, 2017.
132. Largeot, C., Portet, C., Chmiola, J., Relation between the ion size and pore size for an electric double-layer capacitor. *J. Am. Chem. Soc.*, 130, 2730–2731, 2008.
133. Lin, J.H., The anionic surfactant/ionic liquids intercalated reduced graphite oxide for high-performance supercapacitors. *Nanoscale Res. Lett.*, 13, 215, 2018.
134. Pandey, G.P. and Hashmi, S.A., Studies on electrical double layer capacitor with a low-viscosity ionic liquid 1-ethyl-3-methylimidazolium tetracyanoborate as electrolyte. *Bull. Mater. Sci.*, 36, 729–733, 2013.
135. Haque, M., Li, Q., Smith, A.D., Thermal influence on the electrochemical behavior of a supercapacitor containing an ionic liquid electrolyte. *Electrochim. Acta*, 263, 249–260, 2018.
136. Lee, J.H., Ryu, J.B., Lee, A.S., High-voltage ionic liquid electrolytes based on ether functionalized pyrrolidinium for electric double-layer capacitors. *Electrochim. Acta*, 22, 1847–1852, 2016.
137. Aref, A.R., Chou, C.C., Rajagopalan, R., Randall, C., Bimodal porous carbon electrodes derived frompolyfurfuryl alcohol/phloroglucinol for ionic liquid based electrical double layer capacitors. *J. Mater. Res.*, 33, 1189–1198, 2017.
138. Kado, Y., Imoto, K., Soneda, Y., Yoshizawa, N., Highly enhanced capacitance of mgo-templated mesoporous carbons in low temperature ionic liquids. *J. Power Sources*, 271, 377–381, 2014.
139. Rennie, A.J., Sanchez-Ramirez, N., Torresi, R.M., Hall, P.J., Ether-bond-containing ionic liquids as supercapacitor electrolytes. *J. Phys. Chem. Lett.*, 4, 2970–2974, 2013.
140. Hong, J., Hwang, B., Lee, J., Kim, K., Effects of cyclic structure of ammonium ions on capacitance in electrochemical double layer supercapacitors. *J. Electrochem. Sci.*, 8, 1–6, 2017.
141. Zarrougui, R., Hachicha, R., Rjab, R., Physicochemical characterizations of novel dicyanamide-based ionic liquids applied as electrolytes for supercapacitors. *RSC Adv.*, 8, 31213–31223, 2018.

142. Cho, J., Shin, W.K., Kim, D.W., Electrochemical characterization of electric double layer capacitors assembled with pyrrolidinium-based ionic liquid electrolytes. *J. Electrochem. Sci.*, 7, 199–205, 2016.
143. Pohlmann, S., Lobato, B., Centeno, T.A., Balducci, A., The influence of pore size and surface area of activated carbons on the performance of ionic liquid based supercapacitors. *Phys. Chem. Chem. Phys.*, 15, 17287–17294, 2013.
144. Martins, V.L., Rennie, A.J.R., Torresi, R.M., Hall, P.J., Ionic liquids containing tricyanomethanide anions: Physicochemical characterisation and performance as electrochemical double-layer capacitor electrolytes. *Phys. Chem. Chem. Phys.*, 19, 16867–16874, 2017.
145. Martins, V.L., Rennie, A.J.R., Sanchez-Ramirez, N., Improved performance of ionic liquid supercapacitors by using tetracyanoborate anions. *ChemElectroChem*, 5, 598–604, 2017.
146. Lian, C., Liu, K., Van Aken, K.L., Gogotsi, Y., Wesolowski, D.J., Liu, H.L., Wu, J.Z., Enhancing the capacitive performance of electric double-layer capacitors with ionic liquid mixtures. *ACS Energy Lett.*, 1, 1, 21–26, 2016.
147. Timperman, L., Vigeant, A., Anouti, M., Eutectic mixture of protic ionic liquids as an electrolyte for activated carbon-based supercapacitors. *Electrochim. Acta*, 155, 164–173, 2015.
148. Wang, Y., Gao, D., Li, C., Wang, G., Liu, J., Su, G., Application of ionic liquids in electric double layer capacitors. *Power Supply Technol.*, 29, 7, 466–469, 2005.
149. Orita, A., Kamijima, K., Yoshida, M., Allyl-functionalized ionic liquids as electrolytes for electric double-layer capacitors. *J. Power Sources*, 195, 21, 7471–7479, 2010.
150. Lin, R., Huang, P., Segalini, J., Largeot, C., Taberna, P.L., Chmiola, J., Simon, P., Solvent effect on the ion adsorption from ionic liquid electrolyte into sub-nanometer carbon pores. *Electrochim. Acta*, 54, 27, 7025–7032, 2009.
151. Krause, A. and Balducci, A., High voltage electrochemical double layer capacitor containing mixtures of ionic liquids and organic carbonate as electrolytes. *Electrochem. Commun.*, 13, 8, 814–817, 2011.
152. Jänes, A., Eskusson, J., Thomberg, T., Romann, T., Lust, E., Ionic liquid-1, 2-dimethoxyethane mixture as electrolyte for high power density supercapacitors. *J. Energy Chem.*, 25, 4, 609–614, 2016.
153. Anouti, M., Timperman, L., El Hilali, M., Boisset, A., Galiano, H., Sulfonium bis (trifluorosulfonimide) plastic crystal ionic liquid as an electrolyte at elevated temperature for high-energy supercapacitors. *J. Phys. Chem. C*, 116, 17, 9412–9418, 2012.
154. Schütter, C., Neale, A.R., Wilde, P., Goodrich, P., Hardacre, C., Passerini, S., Balducci, A., The use of binary mixtures of 1-butyl-1-methylpyrrolidinium bis {(trifluoromethyl) sulfonyl} imide and aliphatic nitrile solvents as electrolyte for supercapacitors. *Electrochim. Acta*, 220, 146–155, 2016.
155. Zhang, X., Zhao, D., Zhao, Y., Tang, P., Shen, Y., Xu, C., Xiao, Y., High performance asymmetric supercapacitor based on MnO_2 electrode in ionic liquid electrolyte. *J. Mater. Chem. A*, 1, 11, 3706–3712, 2013.
156. Kim, T., Jung, G., Yoo, S., Activated graphite-based carbons as supercapacitor electrodes with macro- and mesopores. *ACS Nano*, 7, 6899–6905, 2013.
157. Brandt, A. and Balducci, A., Theoretical and practical energy limitations of organic and ionic liquid-based electrolytes for high voltage electrochemical double layer capacitors. *J. Power Sources*, 250, 343–351, 2014.
158. Zhu, Y., Murali, S., Stoller, M.D., Carbon-based supercapacitors produced by activation of graphite. *Science*, 332, 1537–1541, 2011.
159. Zhang, Q., Yang, H., Lang, X., 1-ethyl-2,3-dimethylimidazolium tetrafluoroborate ionic liquid mixture as electrolyte for high-voltage supercapacitors. *Ionics*, 25, 231–239, 2019.

160. Han, T., Park, M.-S., Kim, J., The smallest quaternary ammonium salts with ether groups for high-performance electrochemical double layer capacitors. *Chem. Sci.*, 7, 1791–1796, 2016.
161. Härmas, M., Thomberg, T., Kurig, H., Microporous–mesoporous carbons for energy storage synthesized by activation of carbonaceous material by zinc chloride, potassium hydroxide or mixture of them. *J. Power Sources*, 326, 624–634, 2016.
162. Nambu, N., Takahashi, R., Takehara, M., Electrolytic characteristics of fluoroethylene carbonate for electric double-layer capacitors at high concentrations of electrolyte. *Electrochemistry*, 81, 817–819, 2013.
163. Sevilla, M. and Fuertes, A.B., Direct synthesis of highly porous interconnected carbon nanosheets and their application as high-performance supercapacitors. *ACS Nano*, 8, 5069–5078, 2014.
164. Nguyen, H.V.T., Kwak, K., Lee, K.K., 1, 1-dimethylpyrrolidinium tetrafluoroborate as novel salt for high-voltage electric double-layer capacitors. *Electrochim. Acta*, 299, 98–106, 2019.
165. Ruther, R.E., Sun, C.N., Holliday, A., Stable electrolyte for high voltage electrochemical double-layer capacitors. *J. Electrochem. Soc.*, 164, A277–A283, 2017.
166. Fuertes, A.B. and Sevilla, M., Hierarchical microporous/mesoporous carbon nanosheets for high-performance supercapacitors. *ACS Appl. Mater. Interfaces*, 7, 4344–4353, 2015.
167. DeRosa, D., Higashiya, S., Schulz, A., High performance spiro ammonium electrolyte for electric double layer capacitors. *J. Power Sources*, 360, 41–47, 2017.
168. Portet, C., Yushin, G., Gogotsi, Y., Electrochemical performance of carbon onions, nanodiamonds, carbon black and multiwalled nanotubes in electrical double layer capacitors. *Carbon*, 45, 2511–2518, 2007.
169. Pohlmann, S., Kühnel, R.S., Centeno, T.A., Balducci, A., The influence of anion–cation combinations on the physicochemical properties of advanced electrolytes for supercapacitors and the capacitance of activated carbons. *ChemElectroChem*, 1, 1301–1311, 2014.
170. Ye, S., Kim, I.J., Yang, S.H. *et al.*, CV grown graphite/CNT composite as additive material to improve the performance of electric double layer capacitors (EDLCs). *J. Mater. Sci: Mater. Electron.*, 28, 6592–6600, 2017.
171. Pohlmann, S. and Balducci, A., A new conducting salt for high voltage propylene carbonate-based electrochemical double layer capacitors. *Electrochim. Acta*, 110, 221–227, 2013.
172. Xu, J., Gao, Q., Zhang, Y., Preparing two-dimensional microporous carbon from pistachionut-shell with high areal capacitance as supercapacitor materials. *Sci. Rep.*, 4, 5545, 2014.
173. Hu, C., Qu, W., Rajagopalan, R., Randall, C., Factors influencing high voltage performance of coconut char derived carbon based electrical double layer capacitor made using acetonitrile and propylene carbonate based electrolytes. *J. Power Sources*, 272, 90–99, 2014.
174. Gamby, J., Taberna, P.L., Simon, P., Studies and characteristics of various activated carbons used for carbon/carbon supercapacitors. *J. Power Sources*, 101, 109–116, 2001.
175. Schütter, C., Neale, A.R., Wilde, P., The use of binary mixtures of 1-butyl-1-methylpyrrolidinium bis{(trifluoromethyl)sulfonylimide and aliphatic nitrile solvents as electrolyte for supercapacitors. *Electrochim. Acta*, 220, 146–155, 2016.
176. Yu, X., Wang, J., Wang, C., Shi, Z., A novel electrolyte used in high working voltage application for electrical double-layer capacitor using spiro-(1,1′)-bipyrrolidinium tetrafluoroborate in mixtures solvents. *Electrochim. Acta*, 182, 1166–1174, 2015.
177. Pohlmann, S., Olyschläger, T., Goodrich, P., Mixtures of azepanium based ionic liquids and propylene carbonate as high voltage electrolytes for supercapacitors. *Electrochim. Acta*, 153, 426–432, 2015.
178. Frackowiak, E., Lota, G., Pernak, J., Room-temperature phosphonium ionic liquids for supercapacitor application. *Appl. Phys. Lett.*, 86, 164104, 2005.

179. Roldán, S., Blanco, C., Granda, M., Menéndez, R., Santamaría, R., Towards a further generation of high-energy carbon-based capacitors by using redox-active electrolytes. *Angew. Chem. Int. Ed.*, 50, 7, 1699–1701, 2011.
180. Xie, H.J., Gélinas, B., Rochefort, D., Redox-active electrolyte supercapacitors using electroactive ionic liquids. *Electrochem. Commun.*, 66, 42–45, 2016.
181. Zhang, Z., Gao, X., Yang, L., Electrochemical properties of BF_4^- and $TFSI^-$ series room temperature ionic liquid green electrolytes. *Sci. Bull.*, 50, 15, 1584–1588, 2005.
182. Anouti, M. and Timperman, L., A pyrrolidinium nitrate protic ionic liquid-based electrolyte for very low-temperature electrical double-layer capacitors. *Phys. Chem. Chem. Phys.*, 15, 17, 6539–6548, 2013.

7
Functional Binders for Electrochemical Capacitors

Purnima Baruah and Debajyoti Mahanta*

Department of Chemistry, Gauhati University, Guwahati, Assam, India

Abstract
Electrochemical capacitors or supercapacitors are becoming more popular day by day in the global energy scenario due to their unique position between batteries and capacitors in terms of power and energy. Active material, conducting agent, binder, and current collector are the four essential components of a supercapacitor electrode. Among these components, binder is used in the lowest possible ratio, but it acts as a crucial component in maintaining the performance of the electrode by holding all other components together. Generally, synthetic fluorinated polymers are used as binder materials in supercapacitors for their excellent binding ability. But, use of these fluoropolymers is not sustainable from economic as well as environmental point of view. To resolve these issues, different eco-friendly, cheap materials are being explored by the researchers to use as binders. In this book chapter, we have discussed the role of binders, issues related to the conventional binders and different types of sustainable binding materials explored by the researchers for supercapacitor electrode preparation.

Keywords: Binder, sustainable, eco-friendly, energy storage, supercapacitor

7.1 Introduction

Entire world is presently experiencing a critical time due to exponentially increasing utilization of depleting fossil fuel based resources and massive environmental pollution caused as a consequence. In order to meet the global energy demand in such alarming condition, the scientific community across the world is focusing on development of suitable energy storage technology. Among different types of energy storage devices, electrochemical capacitors or supercapacitors have the ability to bridge the broad gap of power and energy between batteries and capacitors [1]. Due to high power density, fast charge-discharge behaviour and excellent cyclic performance supercapacitors are becoming more popular in global energy domain [2]. There are four essential functioning components in a supercapacitor device- active electrode material, conducting agent, binder and the current collector. Among these components, binder is used in the lowest mass fraction. Though present in very less amount, binder is an indispensable component of supercapacitor as it bestows structural integrity to the electrode [3].

The major function of binder is to hold all other important components of the device i.e., the active electrode material, the conductive agent and the current collector strongly

**Corresponding author*: debam@gauhati.ac.in

Inamuddin, Tariq Altalhi and Sayed Mohammed Adnan (eds.) Sustainable Materials for Electrochemical Capacitors, (205–220) © 2023 Scrivener Publishing LLC

together. It regulates the pore volume in the electrode which can be accessed by electrolyte ions and influences the conductivity at the interface [4]. The performance of the supercapacitor is greatly influenced not only by the amount of binder used, but also by its various properties. Electrochemical as well as thermal stability, inertness and solubility are such properties, which determine the usability of a binder material [5]. Synthetic fluoropolymers like Polyvinylidene fluoride (PVDF), Polytetrafluoroethylene (PTFE), poly(acrylic acid) (PAA), poly(vinylalcohol) (PVA), Nafion etc. are the most commonly used binders in supercapacitor electrode fabrication [6–8]. But, a number of environmental and economic issues are associated with the utilization of these synthetic binders. Their high cost, water-insolubility and toxicity prevent us from getting an eco-friendly, sustainable supercapacitor. Environmental remediation has become the foremost demand throughout the globe and in order to achieve this, focus has been given on utilization of sustainable materials. The sustainability of a material can be determined by evaluating the functioning of the material as well as environmental and economic impact of its utilization. The physical, electrical, mechanical, and thermal behaviors of a material are key facets for its sustainability. It should be eco-friendly and non-toxic to different components of the environment. Moreover, the cost of the raw materials as well as cost involved in the manufacturing process should be low for economic sustainability of a material [9].

In the recent times, a number of bio-based materials have been used by the researchers as binders for supercapacitors. In this chapter, we have discussed about those green binding materials effectively used for getting sustainable supercapacitors with excellent performance.

7.2 Characteristics of Binder

A material can be considered an ideal binder if it can hold all the components of the electrode paste together without lowering the porosity and conductivity of the electrode [10–12]. In order to be an effective binder, a material must possess the following requirements-

1) It must provide consistent coating on the electrode surface. Non-uniform coating is not desirable.
2) The binding material must not undergo thermal degradation while heating the prepared electrode at high temperatures. In other words, it must have very good thermal stability.
3) A binder should be electrochemically inert. Otherwise, it will imperil the electrochemical experiments of the supercapacitor electrodes.
4) Slurry should be prepared using eco-friendly solvent or without any solvent. Water is the most desirable solvent in this case.
5) The binding material must be insoluble in the electrolyte employed.
6) It must ensure strong cohesion between the active material and the conducting agent.
7) It should not hamper transportation of ions in the electrolyte.

7.3 Method of Fabricating Supercapacitor Electrode

The first step of electrode fabrication is preparation of slurry. In general, slurry is made by blending definite proportions of active material, binder and conducting additive using proper solvent and this electrode paste is coated on current collector with a coloring brush. Nickel foam, stainless steel sheet, graphite paper etc. are most commonly used current collectors. The dry weight of electrode paste deposited on the electrode is calculated by measuring the weight of electrode before coating and after coating and drying. Steps involved in supercapacitor electrode fabrication are showed in Figure 7.1. The homogeneity of the electrode paste on the supercapacitor electrodes greatly depends on the weight ratio of the binder used. Use of higher percentage of binder in the paste may lead to agglomeration which can adversely affect the performance of the constructed electrode. For this reason, it is vital to determine the proper weight percentage of the binder that can offer optimum electrochemical performance to the electrode.

7.4 Mechanism of Binding Process

A number of theoretical models are available in literature to explain the mechanistic pathway of binding. Theory of mechanical interlocking, theory of diffusion, theory of chemical bonding, thermodynamic theory, electronic theory, acid-base theory are some of them [13–18].

At the initial step of binding process, binder undergoes diffusion into the pores available in the active electrode material. In the second step, mechanical interlocking of the binder takes place. In case of non-reactive binders, mechanical interlocking happens due to drying whereas reactive binders get mechanically interlocked due to formation of chemical bonds induced by polymerization reactions [19, 20]. Schematic representation of the steps involved in binding process is showed in Figure 7.2 [21].

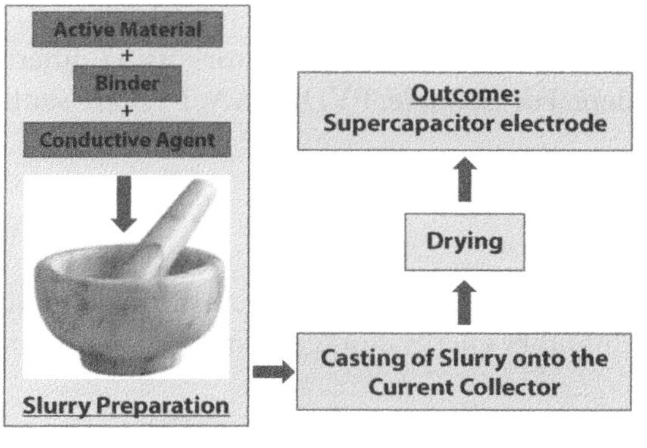

Figure 7.1 Steps involved in supercapacitor electrode fabrication.

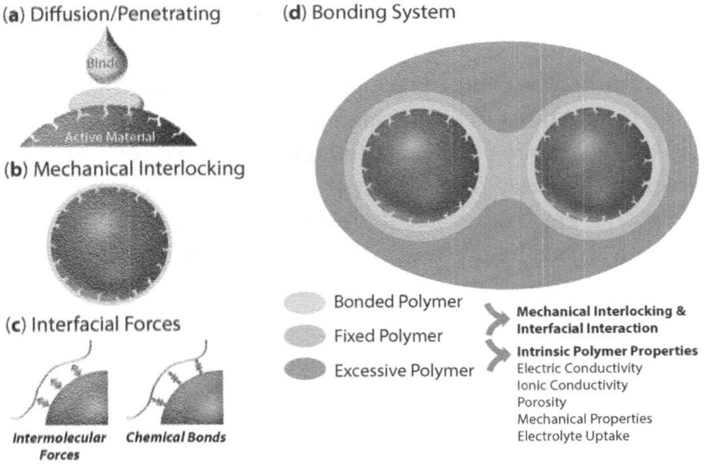

Figure 7.2 Schematic presentation of the steps involved in binding process. [Reprinted with permission from Chen, H., Ling, M., Hencz, L., Ling, H. Y., Li, G., Lin, Z., Liu, G., Zhang, S. Exploring Chemical, Mechanical, and Electrical Functionalities of Binders for Advanced Energy-Storage Devices. *Chem. Rev.*, 118, 8936, 2018] [21].

7.5 Classification of Binders

7.5.1 On the Basis of Origin

a) Natural Binders

Those binders which can be derived from various bio-resources like plants, animals etc. are termed as natural binders. Majority of them are used directly after extracting from the respective bio-resources. A few other natural binders need purification or chemical modification after extraction. For example, gum Arabic can be used directly after extraction whereas another natural binder carboxyl methyl cellulose is synthesized by reacting cellulose with chloroacetic acid [22, 23].

b) Synthetic Binders

Those binders which are synthesized in the laboratory using different chemical reactions are called synthetic binders. For example, PVDF, PAA, PVA are synthetic binders [24, 25].

7.5.2 On the Basis of Reactivity

a) Non-reactive Binder

These binders are mixed directly with the active electrode material to make the paste. Most of binders used in contemporary energy storage devices belong to this category. Ex.: PVDF, PAA

b) Reactive Binder

Reactive binder is formed via chemical reaction after mixing binder precursor with active electrode material. Ex.: Cross linked poly(vinylalcohol)-polyethylenimine (PVA-PEI) binder [26].

7.6 Characterization Techniques

In order to find out the right binder for an active material quite a few bulk properties like toxicity, mechanical strength, adhesive strength, flexibility, hydrophilic nature, electrical conductivity etc. should be taken into account [27–30]. These properties are firmly dependent on chemical composition of the material. Different spectroscopic techniques like Fourier-transform infrared spectroscopy (FTIR), X-ray diffraction analysis (XRD), Raman, Nuclear magnetic resonance (NMR), X-ray photoelectron spectroscopy (XPS) etc. are used for evaluating chemical composition of the binder material [27, 31]. Thermal stability as well as degradation pattern of binder is investigated using Thermogravimetric Analysis (TGA) and Differential Scanning Calorimetry (DSC) techniques [32, 33]. Mechanical behaviors like tensile strength, hardness etc. are examined by stress-strain experiment, scratch test and nano-indentation techniques [34]. Rheometer is used to verify viscosity and other rheological properties of the electrode paste [35]. Scanning electron microscopy (SEM), Transmission electron microscopy (TEM), Atomic force microscopy (AFM) techniques are used to investigate the surface as well as morphology of the materials [36, 37]. Effect of the binder material in the electrochemical performance of the active electrode material is examined by cyclic voltammetry (CV), galvanostatic charge–discharge (GCD) experiments, and electrochemical impedance spectroscopic (EIS) techniques [38].

7.7 Conventional Binders and Related Issues

Different polymeric binders are most commonly used for fabrication of supercapacitor electrodes. These are synthetic in nature and multiple economic and environmental issues are related to their utilization. Some of these are discussed below-

a) PVDF

PVDF is the most popularly used binder in contemporary energy storage devices. It is basically a fluoropolymer containing CH_2-CF_2 units well-known for its adequate electrochemical stability and good binding ability. Optimally performing supercapacitor electrodes can be fabricated using PVDF and therefore it is extensively used as binder [39]. But it is water-insoluble and requires a costly, harmful, non-volatile organic solvent, N-methyl-2-pyrrolidone (NMP) (b. pt: 203°C) to make the slurry [40, 41].

b) PTFE

PTFE is also a fluoropolymer containing -CF_2-CF_2- unit structure which is used as binder due to its high chemical stability. But its hydrophobic nature restricts in proper wetting of the electrode material by the electrolyte which finally results in high internal resistance [42].

c) Nafion

Nafion is nothing but a derivative of PTFE. In order to overcome the limitations of PTFE, it is polymerized with Perfluorosulfonic acid and the resultant copolymer "Nafion" is used as binder. The structure of Nafion consists of two segments: hydrophobic segment with PTFE skeletal unit and hydrophilic branches with ion exchangeable sulfonic acid groups. Due to the presence of ion exchange units, Nafion imparts excellent chemical stability as well as hydrophilic property, but its high price restricts its utility as binder [43].

7.8 Sustainable Binders

For reaching at the goal of sustainable development in the true sense, use of greener materials for designing present day supercapacitors has become a critical aspect. A number of bio-based materials have been explored by the scientific community for using as binders in supercapacitors (Table 7.1). Some of them are discussed below-

a) Carboxymethyl Cellulose (CMC)

Among different bio-based materials, the most frequently used binder is CMC. It is free from fluorine, water soluble and available at very low price. It is a linear macromolecule with glucose units connected to each other by glycosidic linkages [44]. The structure of CMC is shown in Figure 7.3. From different published research works, it was observed that supercapacitor electrodes fabricated with CMC binder showed similar electrochemical performance along with excellent cyclic stability [45, 46].

b) Starch

In the year 2015, Varzi and his co-workers introduced starch as a water-processible efficient binder for supercapacitors [47]. Starch is the second highest available bio-based macromolecule well-known for its biodegradability and low cost. It consists of two types of molecular units- the linear part consisting of amylose units and the branched part consisting of amylopectin units. Amylose units provide active sites for hydrophobic bonding whereas amylopectin units offer resistance to prevail over shrinkage and brittleness of the electrode coating. Due to this particular structural combination, starch is regarded as ideal binder for energy storage devices [48, 49]. Varzi and his co-workers revealed that starch required lesser

Table 7.1 A list of green binders used in electrochemical capacitors along with their sources.

Binders	Sources	Solubility	Reference
Cellulose	Jute, wood, whole grains, cotton, straw, bacteria, algae etc.	Soluble in ionic liquids	[56, 57, 59]
Chitosan	Different shellfish species such as crab, lobster, shrimp, crayfish etc.	Soluble in hot water and dilute acids.	[62–65]
Carboxymethyl cellulose (CMC)	Manufactured from cellulose (cotton, jute etc.)	Water soluble	[44, 45]
Starch	Potato, maize, wheat etc.	Soluble in hot water	[48–50]
Tragacanth gum	Astragalus gummifer shrubs	Water soluble	[66]
Arabic gum	Plants of Acacia Senegal and Acacia seyal species	Water soluble	[67, 68]
Xanthan gum	Manufactured by fermenting carbohydrate in presence of bacteria	Water soluble	[68]
Sodium alginate	Algae, brown seaweeds	Water soluble	[69]

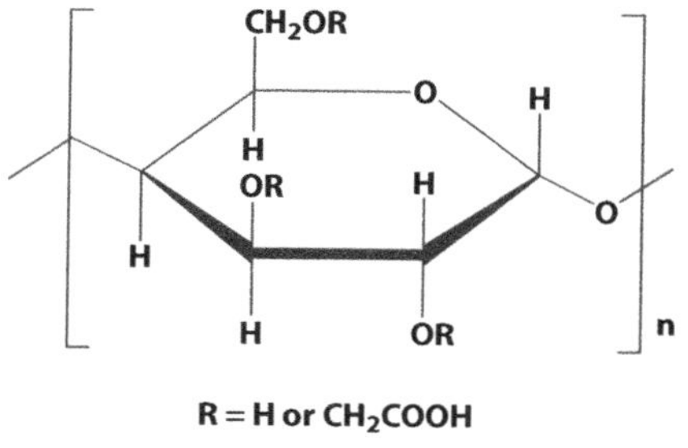

Figure 7.3 Structure of CMC macromolecule. [Reprinted with permission from Lee, B. R., Oh, E. S., Effect of Molecular Weight and Degree of Substitution of a Sodium-Carboxymethyl Cellulose Binder on $Li_4Ti_5O_{12}$ Anodic Performance. *J. Phys. Chem. C*, 117(9), 4404, 2013] [44].

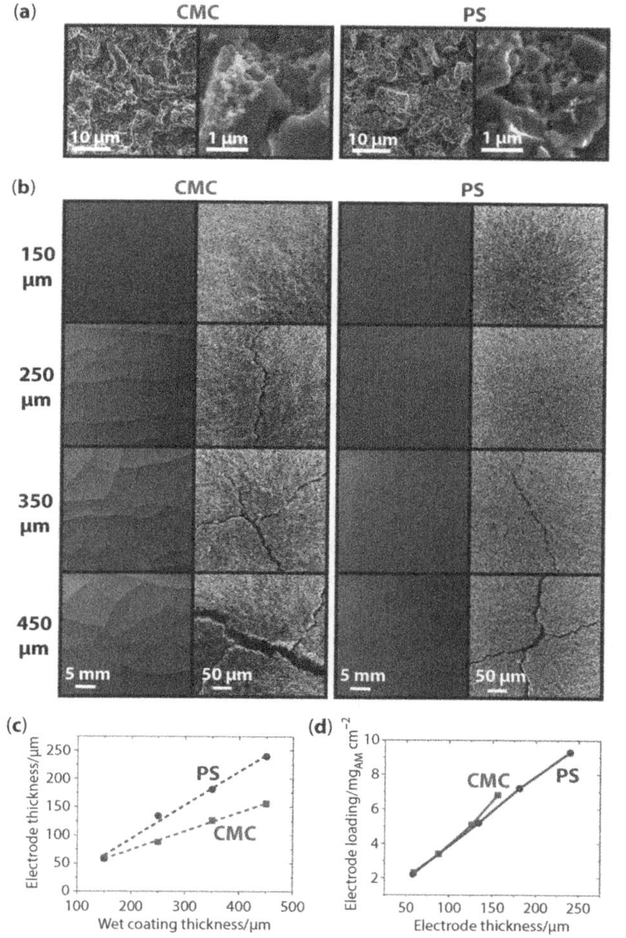

Figure 7.4 (a) SEM micrographs of the electrode fabricated with CMC or Potato Starch (PS) at 1µm and 10 µm resolution (b) Photos and SEM micrographs of the electrode coated with increasing wet thickness from 150 to 450 mm, (c) Plot representing thickness of the dry electrode vs. increasing wet coating, (d) Plot of loaded active material vs. dry electrode thickness. [Reprinted with permission from Varzi, A., Passerini, S., Enabling high areal capacitance in electrochemical double layer capacitors by means of the environmentally friendly starch binder. *J. Power Sources*, 300, 216, 2015] [47].

amount of water for slurry preparation in comparison to CMC binder. When electrodes were dried to remove water content, those prepared using starch shrinks very less creating less cracking than CMC (Figure 7.4). Thus, more uniform electrodes can be prepared with more active material loading per unit area using starch as binder in place of CMC. This resulted in higher areal capacitance in case of electrodes fabricated with starch. The same research group, in the year 2020, carried out a comparative study to analyze effectiveness of three natural polymeric compounds- potato starch, guar gum and wheat starch for using as binder to replace CMC [50]. CMC is not suitable binder for flexible supercapacitors because of its high tendency towards cracking. In their study, it was found that a mixture of potato starch and guar gum in 3:1 weight ratio give most propitious electrochemical performance analogous to that of CMC. It is possible to construct highly flexible electrodes using this mixture binder as it can endure bending tests up to mass loading of 7 mg cm^{-2}. But CMC binder cannot resist bending at such high mass loading.

In 2019, Jezowski and Kowalczewski developed a starch-based glue for using as binder. The supercapacitor devices fabricated with this glue binder exhibited excellent charge propagation along with very good capacitance retention [51].

c) Graphene Oxide-Based Binders

B. Xua and his coworkers reported reduced graphene oxide as an efficient functional binder for fabricating supercapacitor electrodes [52]. Commonly used fluorinated polymeric binders are insulating in nature addition of which lower down the conductivity of the electrode decreasing its capacitance [53]. Reduced graphene oxide is a superior material having intrinsic conductivity and hence it can act as both binder and conducting additive. B. Xua and his coworkers used mixtures of graphene oxide and highly porous carbon materials having different morphologies like powder, fiber, sphere etc. to get conducting 3D electrodes of reduced graphene oxide/porous carbon via thermal reduction method. All of these supercapacitors exhibit higher capacitance along with excellent rate capability in comparison to supercapacitors fabricated using PVDF and PTFE binders. This study established reduced graphene oxide as superior environmentally sustainable binder which also fulfill the functions of conducting agent, active material as well as flexible substrate [52].

Mario Rapisarda and his coworkers developed a hybrid gel by mixing graphene oxide and starch to utilize as eco-friendly binder [54]. The gel was mixed with active electrode material and carbon black in water to make the slurry. It provided a 3-D morphology to the fabricated electrode offering very good charge transfer phenomena. The electrode showed good value of specific capacitance (173.8 F g^{-1}) along with outstanding cyclic stability of 93.1% after 17000 cycles. The thermal stability of the supercapacitor electrode was also very high due to extensive hydrogen bonding among the functional groups of starch and graphene oxide.

d) Cellulose

Most extensively available biopolymer, cellulose has also been emerged as a potent alternative for replacing fluoro-polymeric binders. In comparison to PVDF, cellulose binder can retain capacitance of supercapacitor to significantly higher level [55]. The quality as well as dimension of cellulose fibers varies within a very broad range and these variations can alter the performance of supercapacitors fabricated using cellulose binder. In order to investigate this, B. Andres and his coworkers examined the electrochemical efficiency of different types of cellulose fibers which include raw kraft pulp, purified cellulose fibers, cellulose microfibrils and cellulose nanofibrils [56]. Though they obtained almost similar values of specific capacitance for all types of electrodes, electrical resistance was found to be lowest for electrodes constructed with raw kraft pulp and purified fibers. Natural cellulose is soluble neither in water nor in most of the common organic solvents. This issue creates difficulty in using cellulose as binder for electrode fabrication. But it is soluble in ionic liquids such as 1-ethyl-3-methylimidazolium acetate [57–59]. Therefore, in order to employ cellulose as binder, either ionic liquid should be used or process should be modified to make it soluble in aqueous medium. In the year 2013, Bockenfeld and his co-workers reported natural cellulose as effective binder for supercapacitors containing activated carbon as the active material [60]. They had used two different processes for slurry preparation. In the first method, 2 wt% cellulose solution was allowed to get dissolved in 1-ethyl-3-methylimidazolium acetate at 60 °C by stirring for half an hour. Final slurry was obtained after adding dry mixture of the activated carbon and the conducting additive to this solution and mixing for 12 h. In the other method, 3 wt % microcrystalline cellulose suspension was made in deionized water and a 2 wt % polyethylene oxide solution dissolved in deionized water was added to that suspension. Then, dry mixture of the active material and the conducting agent was added to the cellulose solution. Finally, 4 wt % Triton X-100 was added as surfactant.

Getting inspiration from the unique architecture of spider-web, a 3-dimentional binder was designed by Q. Guan and his co-workers in 2021 [61]. They adopted bio-synthetic pathway to get highly flexible 3-dimentional supercapacitor electrodes using the networking structure of bacterial cellulose as binder. Electrodes were obtained after deposition of aerogel consisting of activated carbon and carbon black which were entangled within the 3-dimentional frame of bacterial cellulose. These electrodes exhibit great flexibility and can be bent into any shape without any structural damage and loss of electrochemical performance. Their study revealed that bacterial cellulose network binder provides excellent, mechanically stable, robust framework to get highly flexible supercapacitor which is not possible using conventional PVDF binder.

e) Chitosan

Chitosan is a polysaccharide usually found in the exoskeleton of many shellfish such as crab, lobster, shrimp etc. and in many fungi as well as insects. Its linear structure primarily comprises of de-acetylated and acetylated groups: de-acetylated group is β-(1–4)-linked D-glucosamine and acetylated group is N- acetyl-D-glucosamine [62]. On account of widescale availability, low price, flexibility, excellent thermo-stability and mechanical strength, chitosan is a promising candidate of sustainable binders to be used in supercapacitors. This biopolymer also exhibits outstanding retention capacity as well as ion adsorption ability [63]. In 2021, Nor Azmira Salleh and the coworkers reported the effect of variation in the amount of chitosan binder on the electrochemical properties of graphene based asymmetric supercapacitor using five different weight percentage of it: 5%, 10%, 15%, 20% and 25% [64]. Their investigation revealed that increasing wt% of chitosan from 15%, 20% to 25% decreases the surface area of the active material- graphene and thus lowers the active sites on it bringing about subsequent lowering in the specific capacitance. When 5 wt% of chitosan was used, they observed agglomeration of graphene sheets due to strong van der Waals force (Figure 7.5a). Excellent electrochemical performance was obtained by using 10 wt% of chitosan. At this wt%, chitosan can diffuse into the graphene matrix very well imparting strong adhesion and high surface area (Figure 7.5b). Moreover, the fabricated electrode is resistant to bending operations without any negative impact on its electrochemical performance. It can retain capacitance up to 87.2% of its initial value after carrying out charge–discharge for 1000 cycles.

In a similar study, M. Raja and his coworkers reported bio-based blend of Chitosan and poly(ethylene glycol)-ran-poly(propylene glycol) as a highly effective green binder for high performing activated carbon based supercapacitor [65]. For preparing the polymer blend, they had used 1:1 ratio (wt/wt) of chitosan and poly(ethylene glycol)-ran-poly(propylene

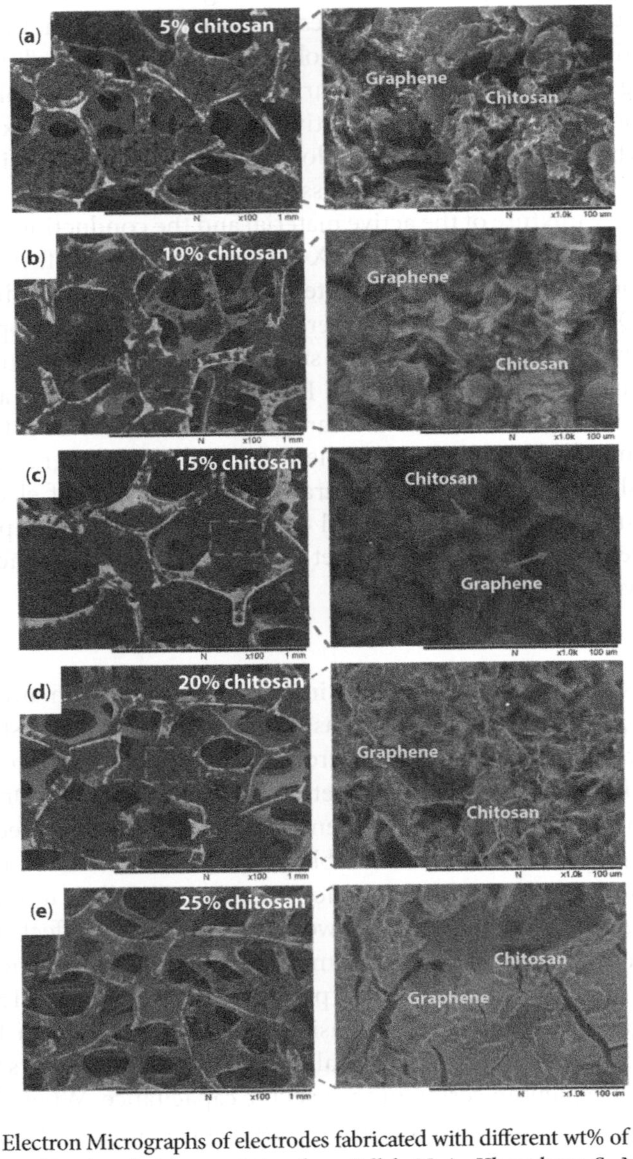

Figure 7.5 Scanning Electron Micrographs of electrodes fabricated with different wt% of Chitosan: (a) 5, (b) 10, (c) 15, (d) 20, and (e) 25. [Reprinted with permission from Salleh, N. A., Kheawhom, S., Mohamad, A. A., Chitosan as biopolymer binder for graphene in supercapacitor electrode. *Results Phys.*, 25, 104244, 2021] [64].

glycol) in 1% acetic acid. Symmetric supercapacitors fabricated using this environmentally sustainable binder showed excellent electrochemical properties along with 99% capacitance retention after 6000 cycles owing to its outstanding wetting nature. Their study revealed that this bio-based blend can be used as a potent substitute of conventional polymeric binders like Nafion, PVDF etc. without any compromise in the electrochemical performance.

f) Natural Gums

In 2021, Scalia and coworkers reported applicability of naturally available tragacanth gum as binder for supercapacitors for the first time [66]. This gum is present in the indigenous gummifer shrubs of west parts of Asia. They had carried out a comparative study on the efficiency of two sustainable aqueous binders -Na-carboxymethyl cellulose (CMC) and tragacanth gum and found that electrodes fabricated using tragacanth binder exhibit superior electrochemical performance as well as greater thermal stability. This variation in the properties of the active material with the change of binder can be explained on the basis of surface area. When the material was prepared with CMC binder, micropores on the material were heavily blocked by CMC lowering its surface area. On the other hand, pore blocking nature of tragacanth gum is lower than CMC which offers higher pore volume as well as surface area to the material prepared with tragacanth.

Naturally available Arabic gum and Xanthan gum have also attracted the researches for using as binders for supercapacitor electrode fabrication. Both of these biomacromolecules contain negatively charged functional groups- hydroxyl and carboxyl anions. As a consequence of these existing functional groups, Arabic gum and Xanthan gum can enhance interaction between electrode and electrolyte providing flexibility as well as good adhesive strength. In simple words, these binders have the ability to boost the transportation of electrolyte ions [67]. In 2020. S. Bayatpour and his coworkers used both of these natural binders for supercapacitor fabrication [68].

In their investigation, they had prepared Arabic gum solution by dissolving 800 mg of it in 15 ml distilled water. Similarly, Xanthan gum solution was prepared by dissolving 400 mg of it in 30 ml water. 40 mg of active electrode material was then added to each of the gum solutions separately. Finally electrodes were coated by dipping Ni foams (surface area: 1 cm^2) in either one of the gum solutions and sonicating at 40 °C for 20 min. Electrochemical experiments revealed that electrode prepared without binder showed specific capacitance 575 F g^{-1} whereas electrode prepared with Xanthan gum showed 1735 F g^{-1} and that with Arabic gum showed 2150 F g^{-1}. Hence, these gums can be regarded as efficient green binder for supercapacitors with excellent performance.

g) Sodium Alginate

Sodium alginate is a biopolymer belonging to polysaccharide type, widely available in different algae. This linear polymer is a major cell wall- component of brown seaweeds and contains a high percentage of alginic acid [69]. The structure of Sodium Alginate is shown in Figure 7.6.

Due to the presence of alginic acid, each monomeric unit of sodium alginate possesses carboxylate functional units and hence can form extensive hydrogen bondings with the active electrode material when it is used as binder in energy storage devices [70, 71]. In 2017, Juan Zeng and his coworkers reported an eco-friendly solid state supercapacitor using sodium alginate binder [72]. They used hydrogel of sodium alginate as electrolyte

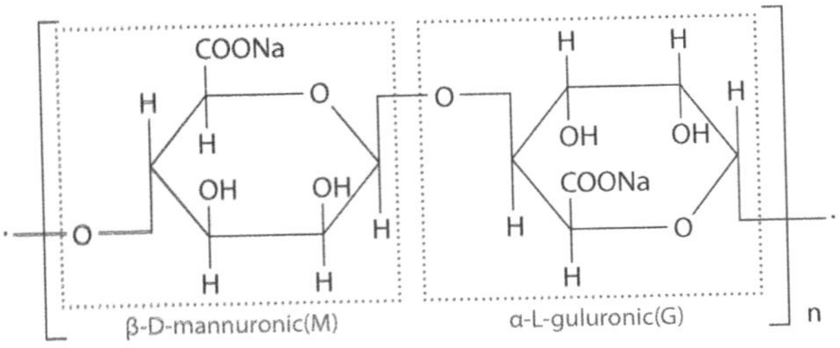

Figure 7.6 Chemical composition of Sodium Alginate binder. [Reprinted with permission from Xia, L., Huang, H., Fan, Z., Hu, D., Zhang, D., Khan, A. S., Pan, L., Hierarchical macro-/meso-/microporous oxygen-doped carbon derived from sodium alginate: A cost-effective biomass material for binder-free supercapacitors. *Mater. Des.*, 182, 108048, 2019] [69].

and the constructed 3-dimensional solid supercapacitor did not require any additional conducting agent because of the intrinsic conducting properties of sodium alginate. Kelp chips were used as precursor for synthesizing sodium alginate via decalcification. For electrode fabrication, they used kelp derived active carbon and mixed it with binder in 94:6 weight ratio using deionized water as solvent. The device showed specific capacitance of 277 F g^{-1} and volumetric capacitance of 88.2 F cm^{-3} along with excellent cyclic stability.

Studies revealed that alginate binder does not block pores on the electrode like many other binders and hence the porosity of the electrode can be accessed to greater extent if it is fabricated with alginate binder. Therefore, it has the capability to accelerate the energy delivery process of the supercapacitor. Moreover, alginate binder provides highly stable rheological properties to the electrode paste [73, 74]. All these observations reflect that alginate can be an effective binder candidate for high performing sustainable supercapacitors.

7.9 Conclusion

The chapter provides a brief report on the binders used in the fabrication of electrodes for constructing supercapacitors. The discussion includes different conventional binders and the issues related to their utilization followed by a number of sustainable binders explored by researchers till now. It was revealed that a variety of green materials have promising properties to replace polymeric binders and generate high performing supercapacitors. In order to design more eco-friendly supercapacitor devices, we must say good bye to all types of toxic components used during fabrication and we must concentrate our focus in utilizing biodegradable, renewable and water processible components. We finish the chapter with an expectation that more and more of such green materials will be explored by the researchers of the world in the near future to substitute the conventional binders so that development can be attained in the true sense without affecting the environment.

References

1. Wang, Y., Chang, Z., Qian, M., Lin, T., Huang, F., A bridge between battery and supercapacitor for power/energy gap by using dual redox-active ions electrolyte. *Chem. Eng. J.*, 375, 122054, 2019.

2. Pandolfo, A. and Hollenkamp, A., Carbon properties and their role in supercapacitors. *J. Power Sources*, 157, 11, 2006.
3. Liu, J., Zhang, Q., Zhang, T., Li, J.T., Huang, L., Sun, S.G., A robust ion-conductive biopolymer as a binder for Si anodes of lithium-ion batteries. *Adv. Funct. Mater.*, 25, 3599, 2015.
4. Zhu, Z., Tang, S., Yuan, J., Qin, X., Deng, Y., Qu, R., Haarberg, G.M., Effects of various binders on supercapacitor performances. *Int. J. Electrochem. Sci.*, 11, 8270, 2016.
5. Marshall, J.E., Zhenova, A., Roberts, S., Petchey, T., Zhu, P., Dancer, C.E.J., McElroy, C.R., Kendrick, E., Goodship, V., On the solubility and stability of polyvinylidene fluoride. *Polymers*, 13, 1354, 2021.
6. Yang, S., Sun, L., An, X., Qian, X., Construction of flexible electrodes based on ternary polypyrrole@cobalt oxyhydroxide/cellulose fiber composite for supercapacitor. *Carbohydr. Polym.*, 229, 115455, 2020.
7. Vetter, J., Novák, P., Wagner, M.R., Veit, C., Moller, K.C., Besenhard, J.O., Hammouche, A., Ageing mechanisms in lithium-ion batteries. *J. Power Sources*, 147, 269, 2005.
8. Subramaniam, C.K., Ramya, C.S., Ramya, K., Performance of EDLCs using Nafion and Nafion composites as electrolyte. *J. Appl. Electrochem.*, 41, 197, 2011.
9. Ahmad, Z., Farooq, R., Rasheed, K., Khan, A.U., Fatima, U., Harnessing green engineering for eco-friendly housing and utilities in South-Asian Countries. *J. Pet. Environ. Biotechnol.*, 7, 295, 2016.
10. Kouchachvili, L., Maffei, N., Entchev, E., Novel binding material for supercapacitor electrodes. *J. Solid State Electrochem.*, 18, 2539, 2014.
11. Chen, H., Ling, M., Hencz, L., Ling, H.Y., Li, G.R., Lin, Z., Liu, G., Zhang, S.Q., Exploring chemical, mechanical, and electrical functionalities of binders for advanced energy-storage devices. *Chem. Rev.*, 118, 8936, 2018.
12. Schutter, C., Pohlmann, S., Balducci, A., Industrial requirements of materials for electrical double layer capacitors: Impact on current and future applications. *Adv. Energy Mater.*, 9, 1900334, 2019.
13. McBain, J. and Hopkins, D., On adhesives and adhesive action. *J. Phys. Chem.*, 29, 188, 1924.
14. Pizzi, A., Mtsweni, B., Parsons, W., Wood-induced catalytic activation of PF adhesives autopolymerization vs. PF/wood covalent bonding. *J. Appl. Polym. Sci.*, 52, 1847, 1994.
15. Derjaguin, B., Aleinikova, I., Toporov, Y.P., On the role of electrostatic forces in the adhesion of polymer particles to solid surfaces. *Powder Technol.*, 2, 154, 1969.
16. Sharpe, L. and Schonhorn, H., Theory gives direction to adhesion work. *Chem. Eng. News*, 41, 67, 1963.
17. Drago, R.S., Vogel, G.C., Needham, T.E., Four-parameter equation for predicting enthalpies of adduct formation. *J. Am. Chem. Soc.*, 93, 6014, 1971.
18. Bikerman, J.J., *The science of adhesive joints*, Elsevier, New York, 2013.
19. Venables, J.D., Adhesion and durability of metal-polymer bonds. *J. Mater. Sci.*, 19, 2431, 1984.
20. Burkstrand, J.M., Metal-polymer interfaces: Adhesion and x-ray photoemission studies. *J. Appl. Phys.*, 52, 4795, 1981.
21. Chen, H., Ling, M., Hencz, L., Ling, H.Y., Li, G., Lin, Z., Liu, G., Zhang, S., Exploring chemical, mechanical, and electrical functionalities of binders for advanced energy-storage devices. *Chem. Rev.*, 118, 8936, 2018.
22. Ling, M., Xu, Y., Zhao, H., Gu, X., Qiu, J., Li, S., Wu, M., Song, X., Yan, C., Liu, G., Dual-functional gum arabic binder for silicon anodes in lithium ion batteries. *Nano Energy*, 12, 178, 2015.
23. Adinugraha, M.P. and Marseno, D.W., Synthesis and characterization of sodium carboxymethylcellulose from cavendish banana pseudo stem (Musa cavendishii LAMBERT). *Carbohydr. Polym.*, 62, 164, 2005.

24. Markevich, E., Salitra, G., Aurbach, D., Influence of the Pvdf binder on the stability of $LiCoO_2$ electrodes. *Electrochem. Commun.*, 7, 1298, 2005.
25. Koo, B., Kim, H., Cho, Y., Lee, K.T., Choi, N.S., Cho, J., A highly cross-linked polymeric binder for high-performance silicon negative electrodes in lithium ion batteries. *Angew. Chem. Int. Ed.*, 51, 8762, 2012.
26. Liu, Z., Han, S., Xu, C., Luo, Y., Peng, N., Qin, C., Zhou, M., Wang, W., Chen, L., Okada, S., *Situ* crosslinked PVA–PEI polymer binder for long-cycle silicon anodes in Li-ion batteries. *RSC Adv.*, 6, 68371, 2016.
27. Drofenik, J., Gaberscek, M., Dominko, R., Poulsen, F.W., Mogensen, M., Pejovnik, S., Jamnik, J., Cellulose as a binding material in graphitic anodes for Li ion batteries: A performance and Degradation study. *Electrochim. Acta*, 48, 883, 2003.
28. Li, G., Ling, M., Ye, Y., Li, Z., Guo, J., Yao, Y., Zhu, J., Lin, Z., Zhang, S., Acacia senegal–inspired bifunctional binder for longevity of lithium–sulfur batteries. *Adv. Energy Mater.*, 5, 1500878, 2015.
29. Ling, M., Qiu, J., Li, S., Yan, C., Kiefel, M.J., Liu, G., Zhang, S., Multifunctional SA-PProDOT binder for lithium ion batteries. *Nano Lett.*, 15, 4440, 2015.
30. Chen, W., Qian, T., Xiong, J., Xu, N., Liu, X., Liu, J., Zhou, J., Shen, X., Yang, T., Chen, Y., A new type of multifunctional polar binder: Toward practical application of high energy lithium sulfur batteries. *Adv. Mater.*, 29, 1605160, 2017.
31. Komaba, S., Yabuuchi, N., Ozeki, T., Han, Z.J., Shimomura, K., Yui, H., Katayama, Y., Miura, T., Comparative study of sodium polyacrylate and poly (vinylidene fluoride) as binders for high capacity Si–graphite composite negative electrodes in Li-ion batteries. *J. Phys. Chem. C*, 116, 1380, 2012.
32. Gunasekaran, S.S., Elumalali, S.K., Kumaresan, T.K., Meganathan, R., Ashok, A., Pawar, V.,
33. Vediappan, K., Ramasamy, G., Karazhanov, S.Z., Raman, K., Bose, R.S., Partially graphitic nanoporous activated carbon prepared from biomass for supercapacitor application. *Mater. Lett.*, 218, 165, 2018.
34. Aslan, M., Weingarth, D., Jackel, N., Atchison, J.S., Grobelsek, I., Presser, V., Polyvinylpyrrolidone as binder for castable supercapacitor electrodes with high electrochemical performance in organic electrolytes. *J. Power Sources*, 266, 374, 2014.
35. Deng, L. and Zhang, L.M., Rheological characteristics of chitin/ionic liquid gels and electrochemical properties of regenerated chitin hydrogels. *Colloids Surf. A Physicochem. Eng. Asp.*, 586, 124220, 2020.
36. Na, R., Lu, N., Li, L., Liu, Y., Luan, J., Wang, G., A robust conductive polymer network as a multi-functional binder and conductive additive for supercapacitors. *ChemElectroChem*, 7, 3056, 2020.
37. Suneetha, R.B., Selvi, P., Vedhi, C., Synthesis, structural and electrochemical characterization of Zn doped iron oxide/grapheneoxide/chitosan nanocomposite for supercapacitor application. *Vacuum*, 164, 396, 2019.
38. Staiti, P. and Lufrano, F., Study and optimisation of manganese oxide-based electrodes for electrochemical supercapacitors. *J. Power Sources*, 187, 1, 284, 2009.
39. Cetinkaya, T. and Dryfe, R.A.W., Electrical double layer supercapacitors based on graphene nanoplatelets electrodes in organic and aqueous electrolytes: Effect of binders and scalable performance. *J. Power Sources*, 408, 91, 2018.
40. Ortega, P.F.R., Trigueiro, J.P.C., Silva, G.G., Lavall, R.L., Improving supercapacitor capacitance by using a novel gel nanocomposite polymer electrolyte based on nanostructured SiO2, PVDF and imidazolium ionic liquid. *Electrochim. Acta*, 188, 809, 2016.
41. Fic, K., Lota, G., Meller, M., Frackowiak, E., Novel insight into neutral medium as electrolyte for high-voltage supercapacitors. *Energy Environ. Sci.*, 5, 5842, 2012.

42. Qu, Q.T., Wang, B., Yang, L.C., Shi, Y., Tian, S., Wu, Y.P., Study on electrochemical performance of activated carbon in aqueous Li_2SO_4, Na_2SO_4 and K_2SO_4 electrolytes. *Electrochem. Commun.*, 10, 1652, 2008.
43. Zhong, S., Zhan, C., Cao, D., Zeolitic imidazolate framework-derived nitrogen-doped porous carbons as high performance supercapacitor electrode materials. *Carbon*, 85, 51, 2015.
44. Lee, B.R. and Oh, E.S., Effect of molecular weight and degree of substitution of a sodium-carboxymethyl cellulose binder on $Li_4Ti_5O_{12}$ anodic performance. *J. Phys. Chem. C*, 117, 4404, 2013.
45. Krause, A., Kossyrev, P., Oljaca, M., Passerini, S., Winter, M., Balducci, A., Electrochemical double layer capacitor and lithium-ion capacitor based on carbon black. *J. Power Sources*, 196, 8836, 2011.
46. Brandt, A., Isken, P., Lex-Balducci, A., Balducci, A., Adiponitrile-based electrochemical double layer capacitor. *J. Power Sources*, 204, 213, 2012.
47. Varzi, A. and Passerini, S., Enabling high areal capacitance in electrochemical double layer capacitors by means of the environmentally friendly starch binder. *J. Power Sources*, 300, 216, 2015.
48. Xu, X., Liu, W., Zhong, J., Luo, L., Liu, C., Luo, S., Chen, L., Binding interaction between rice glutelin and amylose: Hydrophobic interaction and conformational changes. *Int. J. Biol. Macromol.*, 81, 942, 2015.
49. Selvaraj, T., Perumal, V., Fhan Khor, S., Anthony, L.S., Gopinath, S.C.B., Muti Mohamed, N., The recent development of polysaccharides biomaterials and their performance for supercapacitor applications. *Mater. Res. Bull.*, 126, 110839, 2020.
50. Ruschhaupt, P., Varzi, A., Passerini, S., Natural polymers as green binders for high-loading supercapacitor electrodes. *ChemSusChem*, 13, 763, 2020.
51. Jeżowski, P. and Kowalczewski, P.L., Starch as a green binder for the formulation of conducting glue in supercapacitors. *Polymers*, 11, 1648, 2019.
52. Xua, B., Wanga, H., Zhua, Q., Suna, N., Anasoric, B., Hua, L., Wanga, F., Guanb, Y., Gogotsi, Y., Reduced graphene oxide as a multi-functional conductive binder for supercapacitor electrodes. *Energy Storage Mater.*, 12, 128, 2018.
53. Abbas, Q., Pajak, D., Frackowiak, E., Beguin, F., Effect of binder on the performance of carbon/carbon symmetric capacitors in salt aqueous electrolyte. *Electrochim. Acta*, 140, 132, 2014.
54. Rapisarda, M., Marken, F., Meo, M., Graphene oxide and starch gel as a hybrid binder for environmentally friendly high-performance supercapacitors. *Commun. Chem.*, 4, 1, 2021.
55. Varzi, A., Balducci, A., Passerini, S., Natural cellulose: A green alternative binder for high voltage electrochemical double layer capacitors containing ionic liquid-based electrolytes. *J. Electrochem. Soc.*, 161, A368, 2014.
56. Andres, B., Dahlstrom, C., Blomquist, N., Norgren, M., Olin, H., Cellulose binders for electric double-layer capacitor electrodes: The influence of cellulose quality on electrical properties. *Mater. Des.*, 141, 342, 2018.
57. Phillips, D.M., Drummy, L.F., Conrady, D.G., Fox, D.M., Naik, R.R., Stone, M.O., Trulove, P.C., De Long, H.C., Mantz, R.A., Dissolution and regeneration of Bombyx mori silk fibroin using ionic liquids. *J. Am. Chem. Soc.*, 126, 14350, 2004.
58. Swatloski, R.P., Spear, S.K., Holbrey, J.D., Rogers, R.D., Dissolution of cellose with ionic liquids. *J. Am. Chem. Soc.*, 124, 4974, 2002.
59. Heinze, T., Dorn, S., Schobitz, M., Liebert, T., Kohler, S., Meister, F., Interactions of ionic liquids with polysaccharides – 2: Cellulose. *Macromol. Symp.*, 262, 8, 2008.
60. Bockenfeld, N., Jeong, S.S., Winter, M., Passerini, S., Balducci, A., Natural, cheap and environmentally friendly binder for supercapacitors. *J. Power Sources*, 221, 14, 2013.

61. Guan, Q.F., Ling, Z.C., Han, Z.M., Luo, T.T., Yang, H.B., Yang, K.P., Yin, C.H., Yu, S.H., Sustainable 3D structural binder for high-performance supercapacitor by biosynthesis process. *Adv. Funct. Mater.*, 31, 2105070, 2021.
62. Chai, L., Qu, Q., Zhang, L., Shen, M., Zhang, L., Zheng, H., Chitosan, a new and environmental benign electrode binder for use with graphite anode in lithium-ion batteries. *Electrochim. Acta*, 105, 378, 2013.
63. He, X., Song, P., Shen, X., Sun, Y., Ji, Z., Zhou, H., Li, B., Chitosan-assisted synthesis of wearable textile electrodes for high-performance electrochemical energy storage. *Cellulose*, 26, 9349, 2019.
64. Salleh, N.A., Kheawhom, S., Mohamad, A.A., Chitosan as biopolymer binder for graphene in supercapacitor electrode. *Results Phys.*, 25, 104244, 2021.
65. Raja, M., Sadhasivam, B., Dhamodharan, R., Ramanujam, K., A chitosan/poly (ethylene glycol)-ran-poly (propylene glycol) blend as an eco-benign separator and binder for quasi-solid-state supercapacitor applications. *Sustainable Energy Fuels*, 3, 760, 2019.
66. Scalia, A., Zaccagnini, P., Armandi, M., Latini, G., Versaci, D., Lanzio, V., Lamberti, A., Varzi, A., Passerini, S., Lamberti, A., Tragacanth gum as green binder for sustainable water processable electrochemical capacitor. *ChemSusChem*, 14, 356, 2021.
67. Jeong, Y.K., Kwon, T., Lee, I., Kim, T.S., Coskun, A., Choi, J.W., Millipede-inspired structural design principle for high performance polysaccharide binders in silicon anodes. *Energy Environ. Sci.*, 8, 1224, 2015.
68. Bayatpour, S., Afsharpour, M., Dini, Z., Naderi, H.R., High performance supercapacitor electrodes using functionalized CNTs/MoO_3 with natural polysaccharide binders. *J. Mater. Sci.: Mater. Electron.*, 31, 6150, 2020.
69. Xia, L., Huang, H., Fan, Z., Hu, D., Zhang, D., Khan, A.S., Pan, L., Hierarchical macro-/meso-/microporous oxygen-doped carbon derived from sodium alginate: A cost-effective biomass material for binder-free supercapacitors. *Mater. Des.*, 182, 108048, 2019.
70. Kovalenko, B., Zdyrko, A., Magasinski, B., Hertzberg, Z., Milicev, R., Burtovyy, I., Luzinov, G., Yushin, A., Major constituent of brown algae for use in high-capacity Li-Ion batteries. *Science*, 75, 334, 2011.
71. Shi, N., Li, X., Fan, T., Zhou, H., Ding, J., Zhang, D., Zhu, H., Biogenic N-I-codoped TiO_2 photocatalyst derived from kelp for efficient dye degradation. *Energy Environ. Sci.*, 4, 172, 2011.
72. Zeng, J., Wei, L., Guo, X., Bio-inspired high-performance solid-state supercapacitors with the electrolyte, separator, binder and electrodes entirely from kelp. *J. Mater. Chem. A*, 5, 25282, 2017.
73. Tran, H.Y., Wohlfahrt-Mehrens, M., Dsoke, S., Influence of the binder nature on the performance and cycle life of activated carbon electrodes in electrolytes containing Li-salt. *J. Power Sources*, 342, 301, 2017.
74. Chou, S.L., Pan, Y., Wang, J.Z., Liu, H.K., Dou, S.X., Small things make a big difference: Binder effects on the performance of Li and Na batteries. *Phys. Chem. Chem. Phys.*, 16, 20347, 2014.

8

Sustainable Substitutes for Fluorinated Electrolytes in Electrochemical Capacitors

Sina Yaghoubi[1], Seyyed Mojtaba Mousavi[2], Seyyed Alireza Hashemi[3], Aziz Babapoor[1] and Chin Wei Lai[4]*

[1]Department of Chemical Engineering, University of Mohaghegh Ardabili, Ardabil, Iran
[2]Department of Chemical Engineering, National Taiwan University of Science and Technology, Taipei, Taiwan
[3]Nanomaterials and Polymer Nanocomposites Laboratory, School of Engineering, University of British Columbia, Kelowna, BC, Canada
[4]Nanotechnology and Catalysis Research Centre (NANOCAT), Institute for Advanced Studies (IAS), Universiti Malaya (UM), Kuala Lumpur, Malaysia

Abstract
Nowadays, researchers are working hard to design and develop methods to store electricity for the future demands of zero-emission energy production methods and transportation. Many efforts have been made to improve the safety of portable electronic devices. In this regard, many researchers have studied different materials to develop non-flammable electrolytes for electrochemical double-layer capacitors (EDLC) and lithium-ion batteries (LIBs), which power electronic devices. The need to develop efficient and fire-safe energy storage devices has led to the widespread use of fluorinated electrolytes as a substitute for ionic liquid-based electrolytes. Although fluorinated electrolytes have many advantages over other electrolytes, their widespread usage created new challenges, especially environmental problems. Thus, the electrolytes of devices that store electricity are widely studied. Since lithium-ion batteries contain fluorinated organic compounds, many studies have focused on them. Thus, in this chapter, fluorinated electrolytes and their advantages and disadvantages will be overviewed, and at the end, environment-friendly and fluorine-free alternatives to fluorinated electrolytes will be discussed.

Keywords: Electrochemical supercapacitors, fluorinated electrolytes, EDLC

8.1 Introduction

Nowadays, researchers are working hard to design and develop methods to store electricity for the future demands of zero-emission energy production methods and transportation. The reason for the growing interest in these criteria is the scarcity of fossil fuel resources and the environmental problems created by conventional methods of producing electricity. Therefore, the development of batteries, fuel cell capacitors, and supercapacitors is gradually

**Corresponding author*: cwlai@um.edu.my

Inamuddin, Tariq Altalhi and Sayed Mohammed Adnan (eds.) Sustainable Materials for Electrochemical Capacitors, (221–246) © 2023 Scrivener Publishing LLC

becoming more vital [1–3]. Supercapacitors are powerful in terms of power density, speed of charge and discharge processes, and life cycle. As a result, their usage has grown in academic research and industries. Fundamentals, like the existence of cathodes and anodes and separators soaked in electrolytes, make rechargeable batteries and supercapacitors look like each other. However, supercapacitors are generally considered to be between batteries and ordinary dielectric capacitors. Supercapacitors are used in applications such as the fabrication of electronic devices, transport industry, aerospace, and military industries, balancing the electrical grids. They are also used in electrical energy storage units to protect and amplify electric current or as a replacement for batteries [4–6]. Supercapacitors can store more power (watts) inside themselves, but unlike batteries, they cannot store much energy per unit volume. An unlimited charge/discharge cycle and the ability to generate more significant currents than batteries are some of the outstanding features of supercapacitors that make them perfect for usage in regenerative braking systems. The only drawback of supercapacitors is their lower energy density compared to batteries. This fact makes replacing batteries with them unfeasible [7, 8]. In recent years, electrochemical supercapacitors (ESs) have become the center of attention of many researchers. ESs have a good power density, long life cycle, and performance (power per energy). In the case of performance, ESs are somewhat between fuel cells, dielectric capacitors, and batteries [9–11].

Electrochemical manners of storing electricity, such as electrochemical capacitors and lithium-ion batteries (LIBs) [13–17], can store produced electricity through renewable energies, like wind power and solar energy, and provide the stored electricity when needed. Thus, they can be considered a promising solution to the problems faced by the renewable energy industry today. As mentioned, Electrochemical energy storage devices generally include a cathode, an anode, and an electrolyte solution that permits ions to transfer from one electrode to another, a separator to separate the anode and cathode, and collectors of current that permit the electricity to flow out of the cell [13–16, 18–21]. Figure 8.1 shows the specific energy and power in

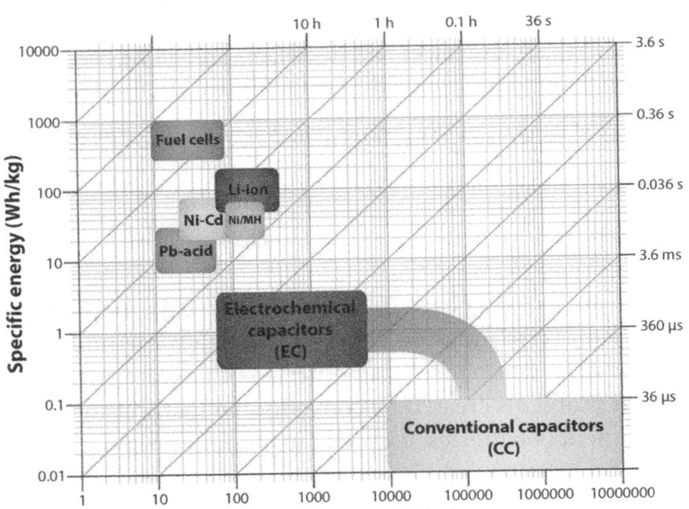

Figure 8.1 Specific energy and power in different methods of storing electricity (reprinted from [12], Copyright © 2018, Elsevier).

different methods of storing electricity. Generally, electrodes of electrochemical supercapacitors are made out of materials with high porosity and surface area and are nanoscale. Conductive solid particles of the electrodes (such as carbon particles and metal oxides) and the electrolyte are separated by an interface, shown in Figure 8.2. This interface makes storage and separation of charges possible. The mentioned interface is technically an EDLC [22].

The efficiency of electrochemical methods of storing electricity is influenced by the properties of the materials of their fabrication; Hence, materials science studies are vital for developing new methods of electrochemically storing electricity. Numerous materials have been synthesized and fabricated for different parts of such devices, such as binders, separators, anodes, cathodes, and various electrolytes, to improve the performance of electrochemical capacitors and LIBs [13–16, 18–21]. Currently, the main barriers to the development and expansion of the usage of electrochemical methods of storing electrical energy are poor performance and safety of such methods, high overall costs of manufacturing and operating energy storage devices, inflexibility of such devices, and problems that have to be faced during fabrication, identification, and development of environment-friendly materials, and production processes that are easily recyclable and up-scalable [24–28].

Many efforts have been made to improve the safety of portable electronic devices. In this regard, many researchers have studied different materials to develop non-flammable electrolytes for electrochemical double-layer capacitors (EDLC) and lithium-ion batteries (LIBs), which power electronic devices. One of these materials is ionic liquids (ILs) which are usually non-flammable. Numerous publications proposed the usage of ILs as safe electrolytes [29–33]. The main disadvantage of using ILs as electrolytes is that they are highly viscous and poor conductors of electricity (usually between 0.1 and 14 mS.cm-1) [32]. Since IL-based electrolytes are not flawless, highly fluorinated solvents are proposed

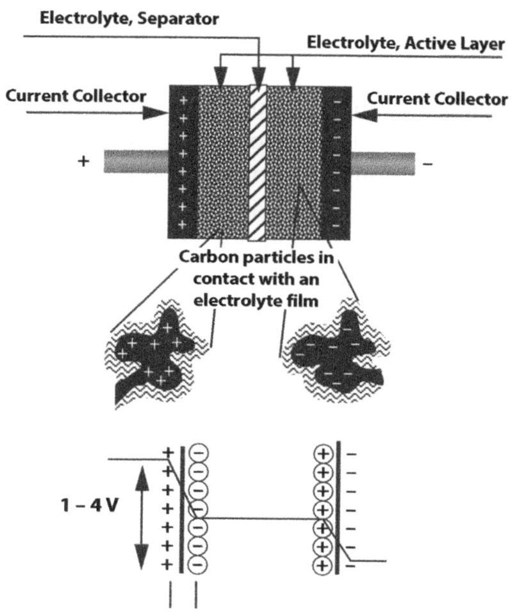

Figure 8.2 Schematic of an EDLC (reprinted from [23], Copyright © 2000, Elsevier).

as an alternative to them. Fluorinated solvents have low flammability or are not flammable [34]. This low or non-flammability of highly fluorinated solvents originates from the fact that replacing Hydrogen in C-H bonds with Fluorine atoms mostly leads to significant stability in chemical and electrochemical activities [35, 36]. Although fluorinated electrolytes have many advantages over other electrolytes, their widespread usage created new challenges, especially environmental problems. Since electronic devices have a limited lifespan, the growth in their production exponentially increased the number of spent batteries and supercapacitors and created a massive amount of electronic waste that must be processed subsequently to avoid catastrophic environmental damage. In addition, although supercapacitors have a longer lifespan than batteries and other methods of storing electrical energy, when devices that rely on them reach their end of life, bulk discarding of them is more likely [37, 38].

Thus, in this chapter, fluorinated electrolytes and their advantages and disadvantages will be overviewed, and at the end, environment-friendly and fluorine-free alternatives to fluorinated electrolytes will be discussed.

8.2 Fluorinated Electrolytes

The need to develop efficient and fire-safe energy storage devices has led to the widespread use of fluorinated electrolytes as a substitute for ionic liquid-based electrolytes. Thus, the electrolytes of devices that store electricity have been widely studied. Since lithium-ion batteries contain fluorinated organic compounds, most studies have focused on them. These studies have proposed materials for use as electrolytes, including salts of ligands that contain fluorine, fluorinated polymer electrolytes, and partially fluorinated solvents [36, 39–41]. Despite the importance of developing fluorinated solvents and, efficient and safe electrolytes for EDLCs, such materials have rarely been studied by researchers, and only a few papers are available compared to the considerable number of ongoing studies about electrode materials [42–50]. EDLCs have a much lower capacity for storing energy (usually smaller than 30 W.h/kg in an aqueous environment) than rechargeable batteries. By utilizing electrostatic interactions amongst electrodes and ions inside the electrolyte, EDLCs can store electricity. EDLCs can quickly discharge electrical current and have a power value greater than 10 kilowatts per kilogram, a good energy efficiency, which is nearly equal to 100%, and a long life span and can withstand up to 500,000 cycles of charging and discharging. Thus, EDLCs seem suitable for uses that demand highly efficient advanced energy storage management methods [51, 52]. A higher energy density and power can be achieved in electrochemical energy storage (EES) methods by increasing the operating voltage [53–56]. An approach like this requires developing new electrode materials and electrolytes that are chemically, physically, and electrochemically stable and can operate at high voltages. Non-flammability and effective interactions with the electrodes are necessary for the new liquid electrolytes that are currently used [51].

Francke *et al.* proposed some novel electrolytes for EDLCs and used fluorinated solvent named 1,1,1,3,3,3-Hexafluoropropan-2-ol [34]. 1,1,1,3,3,3-Hexafluoroisopropan-2-ol (HFIP) is a fluorinated alcohol that is frequently used in the synthesis of organic compounds. It is used as an additive or a reaction medium that promotes the formation of radical or cationic intermediates [34, 57–59]. Many studies reported that using fluorinated alcohols in homogeneous catalyzed reactions positively affects the selectivity and efficiency of the reaction [60, 61]. The dipole moment sharply increases thanks to the electron-withdrawing sections of trifluoromethyl, which makes trifluoromethyl a stronger acid than other organic solutions (pKa = 9 [62] vs. isopropanol: pKa = 17). Thus, using trifluoromethyl in solutions makes the formation of hydrogen bonds easier. Since fluorinated alcohols are hardly oxidizable and can amplify the Oxygen-Oxygen bond gap in peroxides and the Carbon-Oxygen gap in epoxides, they are frequently used as a solvent in ring-opening and oxidation reactions [63]. Alcohols and their corresponding alkoxides, on the other hand, show weaker nucleophilicity and are used as precursors for new superweak anions [64–67]. Usage of HFIP as an important additive in electro-oxidative reactions of derivatives of phenols around diamond electrodes that are doped with boron has been reported by Kriste *et al.* [68, 69]. Based on performed experiments by Franke *et al.*, it was found that HFIP is not anodically oxidizable. However, since HFIP is highly polar and has a low nucleophilicity, it can help radical, cationic intermediates to stabilize. Since EES devices, such as EDLCs, require highly polar, chemically inert, and electrochemically stable solvents, using HFIP as a solvent in electrolytes was found to be very appealing; thus, it is worth a detailed description.

Studies about using HFIP as a solvent in EES methods are relatively limited. So far, HFIP has been utilized in limited applications, such as anion receptors of boronic acid ester in Li/

Table 8.1 Solubility and EC of several conducting salts in HFIP at T = 25 °C [34].

Formula of salt	Solubility (mol/L)	Molecular weight (g/mol)	Electrical conductivity [1 M] (mS/cm)
$LiBF_4$	<0.2	93.8	–
$NaBF_4$	<0.2	109.8	–
TEABF	≥1.5	217.1	15.1
TBABF	≥1.3	329.3	7.4
$LiCl_{O4}$	<0.2	106.4	–
TBAPC	≥1.3	41.9	6.8
NMe_4PF_6	≈ 0.4	219.1	–
TBAPF	≥1	387.4	7.1
LiF	<0.1	25.9	–
NaF	<0.1	42	–
KF	≥1.3	58.1	1.6
RbF	≥1.3	104.4	3
CsF	≥1.3	151.9	4
Li[HFIP][a]	≥1	174	0.0043

[a] Conductivity of a 0.9 M solution was measured.

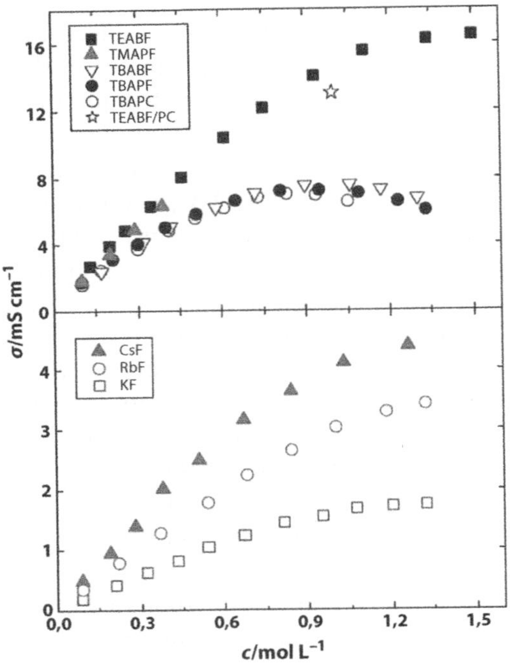

Figure 8.3 EC of different concentrations of conducting salts in HFIP at T = 25 °C; top: tetraalkylammonium salts; bottom: alkali metal fluorides [34, 76] (reprinted from [34], Copyright © 2012, Elsevier).

Table 8.2 Comparing the properties of HFIP and other conventional solvents of EDLCs [75, 76].

Parameter	PC	AN	HFIP
Density [g/mL] at 25 °C	1.2	0.79	1.61
Molecular weight (g/mol)	102.09	41.05	168.05
$T_{Melting\ Point}$ (°C)	-54	-45	-3
$T_{Boiling\ Point}$ (°C)	242	82	59
Flashpoint (°C)	132-135	2-6	-
η (mPa.s)	2.51	0.37	1.62
ε at 25 °C	64.4	36	16.7
Surface tension ($mN.m^{-1}$) at 25 °C	42	28.66	16.14
EC_{TEABF} in a 1 M solution ($mS.cm^{-1}$)	13	56	15

$(CFx)_n$ cells [34, 70–74]. Generally, due to the strong acidity and nucleophilicity of alcohols, they are not considered to be fit for utilization as a solvent in energy storage applications, and this fact means that they do not meet the requirements of the existing standards for electrolytes of LIBs and EDLCs and reduces the potential window of utilizing them in energy storage devices. However, since the nucleophilicity of HFIP is relatively low and it has a high dipole moment, the conducting salts dissolve well in it; Thus, the discussed problems cannot be justified. HFIP's compatibility with LIB components is difficult to predict due to the existence of complex and sensitive processes and components in LIBs.

Nevertheless, its utilization in EDLCs is predictable. Since HFIP is highly polar and acidic, it should make rapid rearrangement of the ions near the electrode surfaces easier, making achieving rapid charge/discharge rates possible. HFIP has low nucleophilicity, making it electrochemically more stable than solvents such as isopropanol. In addition, since carbon electrodes have a high potential to be utilized in hydrogen fuel cells, it is expected that HFIP is compatible with the materials of current electrodes (activated carbon and other various modifications applied to the arrangement of carbon atoms) [34]. Solubility and electrical conductivity of the solution of several salts and their electrical conductivity in electrolytes based on HFIP are given in Table 8.1 and Figure 8.3 for different concentrations of conducting salts.

Table 8.2 represents some of the important features of HFIP and compares them with other prevalent solvents [75].

8.3 Sustainable Substitutes for Fluorinated Electrolytes

8.3.1 Aqueous Electrolytes

Among different types of electrolytes, aqueous electrolytes are frequently used in supercapacitors due to having inherent properties like high ionic conductivity, low toxicity, low viscosity, and low manufacturing and operating costs thanks to being able to operate in different conditions and their assembly. These properties make them an attractive option for commercial applications [77, 78]. In particular, implementing aqueous electrolytes in supercapacitors makes them deliver higher amounts of electricity. Delivering higher amounts of electricity and power is an essential feature of supercapacitors. Since aqueous electrolytes have a higher electrical conductivity than organic electrolytes, they seem fit enough to be utilized in high-power supercapacitors. They are considered desirable electrolytes due to their safety, cost-effectiveness, and minimal environmental impact [77, 79–86]. Some of the aqueous electrolytes are described below.

8.3.1.1 Seawater

Seawater is a good conductor of electricity, meaning it can be utilized as an electrolyte in electrochemical energy storage methods [87–89]. Seawater consists of about 3.5% sodium chloride (35 g/l, 0.6 M NaCl), and its chemical composition is well-known [89]. Not long ago, researchers investigated the performance of electrochemical methods of energy storage that utilize seawater as an electrolyte [85, 90–94]. For instance, Zhao *et al.* electrochemically investigated the function of electrodes in supercapacitors using synthetic seawater as an electrolyte [28, 93]. However, little research is available about utilizing natural seawater and its electrochemical performance as an electrolyte in supercapacitors. Challagulla *et al.* investigated an easy-to-make, cheap, environment-friendly, and stable design of supercapacitors. Their proposed design used high-area activated carbon fibers (2350 m^2/g) as the electrodes of the supercapacitor, which were extracted from the waste of the industrial production of cotton. The electrochemical behavior of electrodes in supercapacitors was studied using solutions of neutral salts such as KCl, NaCl, and LiCl as an electrolyte. The results were compared with an unmodified sample of seawater that was used as an aqueous electrolyte. They proposed using seawater, a 0.6 M solution of NaCl in water, as an economical and stable electrolyte for the first time [85, 95].

8.3.1.2 *Aqueous Solution of Redox-Active Ligands as Electrolytes*

Aqueous electrolytes can electrochemically perform better by adding redox-active ligands to the electrolyte. Electrochemical capacitors have included this charge storage mechanism since the late 2000s, and the goal of this mechanism is to enhance specific energy, which is achieved through rapid and reversible redox processes in the electrolyte solution. Iodides [96] or hydroquinones [97] are the most prevalent compounds studied to be used as additives in water-based electrochemical capacitors. The long time function of such electrolytes is astonishingly not prevented, specifically for electrolytes that contain $\frac{I_2}{I^-}$ species; But, since the calculation methods can lead to an overestimation of the stability of electrolytes that contain redox-active ligands, the parameters of the aging test for such compounds should not be ignored [98]. Since Electrostatic attractions and faradaic processes determine such systems' electrochemical performance, galvanostatic characteristics are no longer determined by a constant voltage gradient within charge and discharge processes. Since iodides have a high redox activity, the positive electrodes of systems that contain iodides perform better since the ongoing reactions occur around them [12]:

$$2I^- \rightleftharpoons I_2 + 2e^- \quad (8.1)$$

$$I^- + I_2 \rightleftharpoons I_3^- \quad (8.2)$$

$$2I_3^- \rightleftharpoons 3I_2 + 2e^- \quad (8.3)$$

$$I_3^- + I_2 \rightleftharpoons I_5^- \quad (8.4)$$

Usage of iodides as an additive for aqueous sulfate-based electrolytes such as $MnSO_4$ [99], Na_2SO_4 [100], or H_2SO_4 [101] or nitrate-based electrolytes, such as $Mg(NO_3)_2$ has been studied for symmetric electrochemical capacitors [102]. In every studied case, it was observed that using iodides as additives increases the current density, which means that iodide additives can improve the energy output of the devices. Figure 8.4 shows an example of cyclic voltammograms in symmetric electrochemical capacitors with and without the

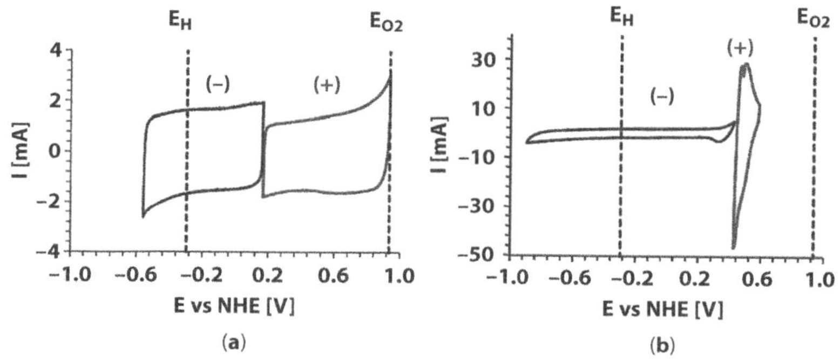

Figure 8.4 CV (*scan rate* = 1 $mV.s^{-1}$) of capacitors' electrodes that were charged up to 1.5 V: (a) in 2 mol/L $MnSO_4$; (b) in 2 mol/L $MnSO_4$ + 0.5 mol/L KI (reprinted from [99], © 2017, Elsevier).

presence of an iodide (such as potassium iodide) as an additive in the electrolyte solution. Based on Figure 8.4, a peak can be observed for the amount of electrical current in the positive electrode, which appears as a reversible redox peak. In addition, it can be deduced that the existence of iodides does not affect the maximum reachable electrochemical voltage, which is reported to be 1.5 V for a solution that includes both potassium iodide (KI) and manganese (II) sulfate ($MnSO_4$) [99]. Studies that used a solution that consists of 0.06 M KI in 1 M H_2SO_4 for cyclability tests [101] showed that a regime of 4 A/g makes reactions with iodide activity very difficult to achieve. Thus, proper evaluation of the performance of the redox-active electrolyte should be seriously considered. The addition of potassium iodide to Nitrate-based electrolytes [102] makes obtaining a voltage of 1.8 V feasible.

Using Quinone-based electrolytes is hypothesized to improve how the capacitor performs electrochemically. Adding redox-active ligands to the electrolyte is a cheaper and more straightforward procedure than sophisticatedly modifying the surface of carbon electrodes but similarly improves the performance of capacitors [103]. Hydroquinones are hypothesized to accommodate inside the carbon structure, leading to improved capacitance. However, the type of the used electrode and its physicochemical properties are important factors that affect the amount of contribution of quinone/hydroquinone (Q/HQ) to the overall performance improvement of the system [104]. In all redox-active electrolytes, it is most likely to only achieve an enhanced performance on one electrode; Nevertheless, these electrolytes also make the symmetric cells electrochemically perform better. The performance of Q/HQ additives can also be improved by using polymers that are conductors of electricity [105–108]; Thus, using different compounds in quinone-based systems has been proposed. The most interesting proposed method that can make quinone-based systems perform better is using electrolyte gels that include reducing compounds as additives [109–111]. The utilization of other redox-active species in such systems requires additional research. A recent study proposed using a combination of ferricyanide/ferrocyanide as reducing agents to diminish the undesirable water decomposition reaction [112].

$$Fe(CN)_6^{4-} \rightleftharpoons Fe(CN)_6^{3-} + e^- \tag{8.5}$$

The ferricyanide/ferrocyanide redox pair systems hugely suffer from the self-discharge phenomenon. The high activity of redox-active ligands may be beneficial, as keeping the reaction rate as high as possible is essential. However, high activity in redox-active species increases the chance of occurrence of the shuttle effect, which can significantly increase the energy loss and self-discharge in ferricyanide/ferrocyanide redox pair systems. To fix this problem, preventing the redox sections from moving at respective electrodes seems sensible. Another way to diminish the self-discharge problem is using ion-selective membranes, which are very expensive, instead of conventional cellulose separators. However, in real-life applications, this solution may be problematic [12, 113, 114]. Based on prior research, iodides are reported to be much more stable than quinones, bromides [115], thiocyanates [116], or ferricyanide [12], and this makes them the best option to be utilized in electrochemical supercapacitors among all redox-active compounds.

8.3.2 Organic Electrolytes

Having a high EC and a good electrochemical window is essential in electrolytes that are used in supercapacitors. These electrolyte features affect supercapacitors' power and energy capacity, respectively. Organic electrolytes, which consist of a solution of salts in organic solvents, compared to aqueous electrolytes, offer wide electrochemical windows (>2.8 V), and this makes achieving higher specific energy possible in devices that utilize them [80, 117, 118]. Conventional organic electrolytes that are soluble in acetonitrile (AN), such as tetraethylammonium tetrafluoroborate ($TEABF_4$) and propylene carbonate (PC) are often utilized in commercially manufactured, and research EDLCs and usually work in a voltage range of up to 2.8 volts. EDLCs that utilize electrolytes that are prepared using AN as a solvent are more powerful in comparison to ones that utilize electrolytes that are prepared using PC as a solvent and perform better in low-temperature conditions [119–121]. NASA reported an organic electrolyte that freezes at -85.7 °C in its Tech Briefs published in 2010. In this electrolyte, $TEABF_4$ was added to a mixture of AN and 1,3-dioxolan (DOL) in a volume ratio of 1:1. Cells that use this type of electrolyte exhibited linear discharge curves in an extensive range of temperatures [122]. AN is toxic and has a low flashpoint, so it has not been used in some countries, like Japan, for many years. Therefore, using PC as a solvent in commercial EDLCs is considered a promising option as an alternative for AN-based electrolytes. Apart from not being as toxic as AN, they have a substantial electrochemical window, a high electrical conductivity, a high boiling point, and a low melting point, which means that they are liquid at extensive range temperatures and are hydrolysis resistant. Thus their usage is preferred [76, 123]. The properties of the dissolved salt that exist in electrolytes may also be necessary for developing high-performance EDLCs; Thus, many researchers are focused on selecting and synthesizing supporting salts. Et_4NBF_4 is a highly soluble salt with a wide electrochemical window, and its solution, consisting of conventional solvents, has high ionic conductivity. Among other known salts, Et_4NBF_4 has become the most common supporting salt used in the organic electrolytes of EDLCs [76, 124, 125]. However, many conventional organic solvents cannot dissolve more than 1.0 mol/L of $TEABF_4$ and this concentration of $TEABF_4$, is insufficient for maintaining an optimal high conductivity. Therefore, the usage of salts of cyclic quaternary ammonium and asymmetric salts of tetraalkylammonium, such as triethylmethylammonium tetrafluoroborate ($TEMABF_4$), 1-ethyl-1-methyl-pyrrolidinium tetrafluoroborate ($MEPYBF_4$) and tetramethylene pyrrolidine (TMP) were investigated. A solution with a higher concentration of these salts can be prepared; therefore, higher conductivity can be achieved [126–130]. $TEMABF_4$ is highly soluble in PC; Thus, it can be utilized as a $TEABF_4$ substitute. Increasing the voltage is the best strategy for increasing power and energy density in EDLCs, and this comes from the fact that the energy density in supercapacitors is related to V^2 and capacitance [131]. Many studies showed that in every commercially known electrolyte for EDLCs, increasing the operating voltage by more than 3 volts is very challenging. The electrochemical window of organic electrolytes can be significantly affected by the chosen solvents, supporting salts, and impurities in the electrolyte [76, 132]. Studies have also shown that the size of ions and the difference between the properties of different salts affects the capacitance and power performance of EDLCs significantly [133–135]. Investigations showed that a high specific capacity could be achieved in EDLC by utilizing quaternary ammonium salts, which have small cations [133]. Salts also affect the electrochemical window in organic electrolytes [127, 136, 137]. EDLCs utilize a solution of N-butyl-N-methylpyrrolidinium bis (trifluoromethanesulfonyl)imide ($PYR_{14}TFSI$) in PC can operate at voltages up to 3.5 volts and have excellent cycle stability. After 100,000 cycles that were performed at 3.5 volts, only a 5% loss of capacitance was observed in EDLCs that were using such electrolytes [138, 139]. In addition, performed tests on spiro-(1,10)-bipyrolidinium tetrafluoroborate ($SBP - BF_4$) confirmed that the novel $SBP - BF_4$/PC electrolytes could withstand an operating voltage of up to 3.2 volts in activated carbon EDLCs, and the capacitor behaved well. Sadly, these newly proposed organic salts are more expensive than $TEABF_4$ making their utilization in current applications impractical [140]. Currently, the typical commercially available organic electrolytes of EDLCs include salts, such as tetraethylammonium tetrafluoroborate ($TEABF_4$), dissolved in PC or AN. For conventional AN or PC-based EDLCs, using the capacitor at an operating voltage greater than 2.7 volts may severely decompose the electrolyte and the impurities, like water, and cause irreversible unwanted reactions in the activated carbon electrodes. These adverse reactions lead to gas evolution reactions and form a passive film on the electrodes [141–144]. In PC-based electrolytes, an H-type cell that can collect the formed gases in positive and negative compartments was used to experimentally investigate the occurrence of the gas evolution phenomenon [145]. Examinations showed that in voltages above 3.0 V, CO_2, and CO are released in positrode, and Hydrogen and gases like propylene, carbon dioxide, ethylene, and carbon monoxide are released around the negatrod [145, 146].

Many researchers investigated the implementation of new solvents that can be used in a wide range of voltages. In 2011, new linear sulfonate-based electrolytes were found. Investigations showed that the electrolytes based on ethyl isopropyl sulfonate (EiPS) are more durable at high voltages up to 3.7 V and have high cycle stability [147]. It is also observed that the alkylated cyclic carbonates can withstand voltages above 3.0 V [148]. In particular, the 2,3-butylene (2,3-BC) carbonate electrolyte can resist voltages as high as 3.5 volts. This durability comes from the prominent resistance of 2,3-BC against oxidation. As briefly mentioned, thanks to fluorine's low polarization and high electronegativity, fluorinated solvents are more stable in the case of chemical and electrochemical activities [34]. One of the investigated fluorinated solvents is fluorostonitrile (FAN), which was tested, and it was found that it can provide a broad electrochemical window; However, it was also shown that a 1.0 molar solution of $TEABF_4$ in *FAN* has a less ionic conductivity than a 1.0 mol/L solution of $TEABF_4$ in *AN* [43]. Similar investigations have been performed to solve other electrolyte problems using AN as a solvent, such as their low flashpoint and relative electrochemical instability. Adiponitrile (ADN) is a compound that can be a possible solvent in EDLCs [149]. EDLCs that utilize a 0.7 mol/L solution of TEABF4 in ADN as an electrolyte are reported to withstand operating voltages as high as 3.75 volts and can withstand beyond 35,000 cycles at cell voltages that are as high as 3.5 volts. However, this new electrolyte has less ionic conductivity than AN-based electrolytes; Thus, it needs further improvement. Although the utilization of newly proposed organic electrolytes can significantly boost the voltage of EDLCs, they have low ionic conductivity and are highly viscose, especially at room temperature, and these facts negatively impact the performance of EDLCs that utilize them. Some researchers studied the so-called lithium-ion capacitors (LICs). By using these types of capacitors, the operating voltage can be increased. A LIC is made by combining an electrode of a LIB with an EDLC electrode. Therefore, LICs are essentially a supercapacitor. LICs usually have an operating voltage above ~4.0 V, which creates a higher energy capacity (> 30 $W.h/kg$). Since Li^+ ions dissolved in organic solvents instead of water have a smaller size, organic electrolytes consisting of $LiClO_4$ or $LiPF_6$ and a mixture of two or more carbonate solvents (such as $EC + DMC$) have been frequently used in LICs. Degradation mechanisms in electrolytes have been previously investigated. These investigations revealed that 4.3 and 1.5 V are critical voltages for the management of the degradation electrolytes in positrodes and negatrodes of $\frac{Li^+}{Li}$ EDLCs, respectively [150]. Some studies investigated the combinations of activated carbon (AC) negatrodes, with many materials that battery positrodes are fabricated out of, such as $LiMn_2O_4$, $LiFePO_4$, and $LiCoO_2$ [51, 151–153].

8.3.3 Solid-State Electrolytes

State-of-art electrochemical capacitors require reliable, safe, and bend-resistive solutions for small and smart electronic devices. The fact that common electrolytes are liquids makes achieving the mentioned goals in electronic devices hard, as abnormal conditions may lead to excessive pressure build-up in the capacitor or even an explosive reaction. Using solid-state materials as an electrolyte was proposed to address these issues. A set of water-based polymer electrolytes have been developed, which are conductors of protons, hydroxide ions, or neutral salt ions [154]. Mixing polyacrylamide and ionic species, such as tetraethylammonium hydroxide, silicotungstic acid, or lithium chloride, in polymer electrolyte systems provides a reliable substitute for liquid electrolytes and showed promising performance in conditions where they were bent and stretched (Figure 8.5). The manufacturing process of such mixtures usually involves the external preparation of an ionic gel, which includes water and different salts as contents, and then a membrane will be placed in the cell. Using ionic gels is advantageous because, based on the requested cell's design and the membrane's shape or size, the produced gel can be easily scaled (Figure 8.5). By the way, dry cells perform poorly because of relatively mediocre electrical conductivity and moderate capacitance. The reason for this is that solid or highly viscous electrolytes cannot fully penetrate inside the electrode material to maintain good conductivity and charge propagation. The moderate conductivity of ion gels can be fixed by a rational design (i.e., changing the thickness of electrodes) or an *in situ* synthesis/gelation, but controlling the thickness of the membrane may be challenging in cases where the *in-situ* gelation method

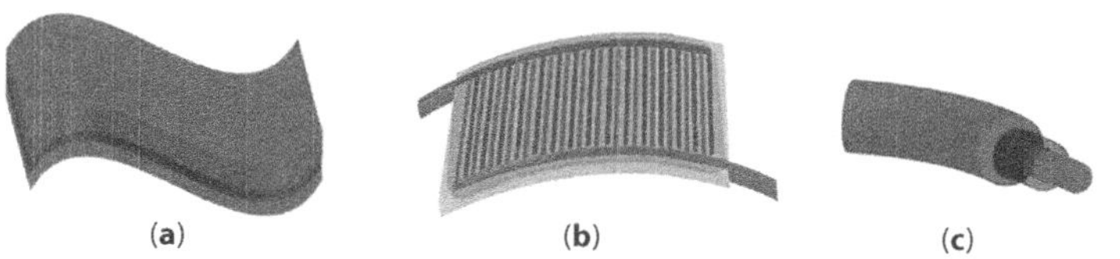

Figure 8.5 Different arrangements of polymer electrolytes; (a) flexible 3D sandwich; (b) 2D interdigitated; (c) 1D fiber (reproduced from [163], Copyright ©, 2021, Elsevier).

is chosen to be used [12]. Poly(vinyl alcohol) [155], poly(ethylene oxide) [156], and many other polymers can be used to make solid electrolytes. Biopolymers are stable and sustainable materials for producing solid electrolytes. They are primarily fit for aqueous [157] and IL electrolytes [12]. Chitosan-based solid ion conductors (CS) can be used as hosts of neutral aqueous electrolytes [158]. CS can form gel-like electrolytes, which has been repeatedly demonstrated in systems that use chitosan/$LiClO_4$/ethylene carbonate/propylene carbonate [159]. This plasticized mixture includes chitosan starch and $LiClO_4$ [160], a mixture of ionic liquids and chitosan [161], or a photopolymerized mixture of chitosan, hydroxyethyl methacrylate, and ILs [162].

Based on their properties, Ishikawa *et al.* [164] proposed some practical uses for chitosan-based gel electrolytes, including an aqueous medium and an IL. Some of these uses are shown in Figure 8.6.

Properties of cellulose, chitin, and biopolymers can be used to prepare acidic chitin-cellulose hybrid gels for capacitors [165, 166]. These hybrid gels make the production of transparent electronic devices possible. Thus, these materials are foreseen to be utilized in personal electronic devices. These materials have been studied in several configurations using a mixture of sulfuric acid and ionic liquids. Studies showed that the devices that utilize these materials exhibit good performances in different temperatures and ranges of electrical current loads. Nanoparticles of cellulose are used to prepare imidazolium-based ILs [167]. Imidazolium-based ILs are transparent and flexible solid-state electrolytes. Some studies suggested using biodegradable gel electrolytes based on agarose and $NaCl$ [168]. Starch is another abundant and biodegradable material derived from corn, consisting of amylose and amylopectin. Investigation showed that corn starch could be used for preparing gels of polymer electrolytes and can be doped with lithium acetate [169]. However, it performs poorly due to the sluggish release of ions during charging and discharging, and

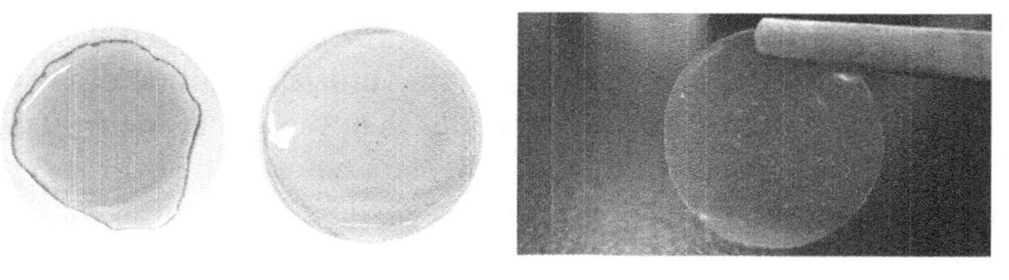

Figure 8.6 A Chitosan and IL gel electrolyte (left); A combination of powdered activated carbon and chitosan combined with a 3% solution of acetic acid in water (middle) and the same mixture that includes an aqueous solution of poly(ethylene oxide) (right) (reprinted from [164], Copyright © 2013, Elsevier).

its performance decreases after some cycles [12]. Another gelating agent in polymer electrolytes is carboxymethylcellulose, which has also been used to prepare thin mixtures of flexible solid-state materials in symmetric capacitors [170].

Solid-state electrolytes can be utilized in electrochemical capacitors, but they poorly perform when applied to high currents. This poor performance is due to sluggish ion diffusion inside the polymer. Since such electrolytes are not introduced to industries yet, predicting whether solid-state electrolytes derived from biomaterials can turn into a commercially important option or be a green substitute for PVA-based electrolytes is difficult. However, starch is already a promising additive in the battery manufacturing industry. Hence, natural derivatives cannot be excluded from future applications [12].

8.4 Performance of Sustainable Electrolytes Compared to Fluorinated Electrolytes

Electrolytes are known to affect the performance of ESs. Thus, an evaluation of the performance of different electrolyte materials is necessary. Regardless of the material used in the manufacture of electrodes, the cell's voltage in acidic and alkaline electrolytes is limited to 1.3 volts [80]. The highest reported cell voltage for neutral electrolytes is 2.2 volts [53]. ESs that use aqueous electrolytes are only functional in temperatures below the boiling and above the freezing points of the aqueous solutions. Although numerous researchers have studied aqueous electrolytes, currently, organic electrolytes reign the market, thanks to being able to operate in different voltages from 2.5 to 2.8 volts. As previously described, the cell's voltage can be raised to improve power and energy densities significantly. Organic electrolytes can make using cheaper materials (such as Aluminum) in current collectors and packages feasible [80].

8.4.1 Strongly Acidic Electrolytes

Sulfuric acid (H_2SO_4) is used chiefly as an acidic electrolyte in water-based ESs, among various acidic electrolytes, thanks to being a good conductor of electricity (EC = 0.8 S/cm for a 1 M solution of H_2SO_4 at 25 °C). Since the EC strongly depends on the concentration of H_2SO_4, optimal concentrations for achieving maximum ionic conductivity at a specific temperature in H_2SO_4-based electrolytes have been comprehensively studied. Based on performed studies, the maximum ionic conductivity of electrolytes that use H_2SO_4 is obtained at a concentration of 1.0 M at 25 °C. Thus, in most studies, a 1.0 M solution of H_2SO_4 is used as an electrolyte, especially for those ESs that utilize carbon-based electrode materials. Most comparative studies indicated that EDLCs that utilize a solution of H_2SO_4 as an electrolyte have a higher amount of specific capacities than those that use neutral electrolytes [171–174]. The published articles in the last few years reported that the specific capacities of EDLCs that utilize strongly acidic electrolytes, such as H_2SO_4, are generally in a range that starts from 100 to 300 F/g, which is more than the amount of capacitance obtained in organic electrolytes. Many studies noted that the benefit of enhancing the electrode material to the specific capacitance is higher than enhancing the used electrolyte materials [7, 118, 175–177]. However, observations showed that H_2SO_4 electrolyte-based EDLCs have a specific capacitance higher than organic electrolyte-based capacitors. This phenomenon can be seen even in capacitors that use the same electrode materials, which can be due to interactions between electrodes and electrolytes caused by the usage of different electrolytes [80].

8.4.2 Strong Alkaline Electrolytes

Compared to strong acid electrolytes, relatively cheap metals can be used as electrical current collectors in alkaline electrolyte-based ESs. The most widely used electrolyte, among the various alkaline electrolytes, is *KOH*, thanks to its high EC ($0.6\frac{s}{cm}$ for a solution with a concentration of 6 *M* at 25 °C). Although, the usage of other alkaline electrolytes, such as *NaOH* and *LiOH*, has also been investigated. Reported values for energy density and specific capacitance in EDLCs that use aqueous electrolytes that contain potassium hydroxide are generally similar to those that are reported in EDLCs that use H_2SO_4 as an electrolyte. In addition to attempts to utilize strong acidic electrolytes in ESs, many studies have put their attention to improving the energy density of ESs by utilizing alkaline electrolytes [80].

8.4.3 Neutral Electrolytes

Neutral electrolytes have some advantages over other electrolytes, such as having more expansive operating potential windows, less corrosiveness, and being safer. Neutral electrolytes can consist of salts of lithium (like Li_2SO_4, $LiClO_4$, and *LiCl*), sodium (like Na_2SO_4, $NaNO_3$, and NaCl), potassium (like KCl, KNO_3, and K_2SO_4), calcium ($Ca(NO_3)_2$) and magnesium (like $MgSO_4$). The solution of Na_2SO_4 is chiefly used among various other neutral salts to prepare electrolytes [171–174]. Equivalent series resistance (ESR) and EC in ESs that utilize neutral electrolytes are lower than KOH or H_2SO_4 electrolyte-based ESs [171, 173, 174]. Higher voltages can be generated in carbon-based ESs that use neutral electrolytes, and this is because the electrochemical stable potential window (ESPW) in carbon-based ESs is higher for neutral electrolytes compared to alkaline and acidic aqueous electrolytes [53, 178, 179]. Since neutral electrolyte does not have a high concentration of H^+ and OH^- ions than alkaline and acidic electrolytes, it is expected to have a higher overpotential for oxygen and hydrogen evolution reactions. Thus, an increase in ESPW is expected in devices that utilize neutral electrolytes. Based on studies by Demarconnay *et al.*, an excellent 10,000 charge-discharge cycle was achieved for ESs that utilized symmetric activated carbon with a 0.5 M Na_2SO_4 electrolyte at 1.6 V [179]. Zhao *et al.* successfully raised the voltage in cells of ESs to 1.8 volts, and they managed to obtain energy densities above 29.1 W/kg at a power density equal to 450 W/kg with a 1 molar solution of Na_2SO_4 as the electrolyte by using Oxygen and Nitrogen Doping for Carbon Nanofiber Electrodes [80, 180].

8.4.4 Organic Electrolytes

There are issues to consider in organic electrolytes. In contrast to aqueous electrolytes, organic electrolytes are highly expensive and have lower specific capacitance and conductivity. Also, there are some safety concerns related to the toxicity, volatility, and flammability of organic electrolytes. Also, the utilization of organic electrolytes demands complicated purification procedures and be assembled in a well-controlled environment to avoid contaminating the electrolyte with residual impurities (such as water) that can cause severe degradation of the electrolyte and self-discharge problems [80, 181].

8.5 Final Remarks

The electrolytes used in modern electrochemical energy storage devices are flammable and mainly toxic, and most of these electrolytes are liquid, which means that these devices cannot be designed to be flexible. Various researchers have proposed using fluorinated electrolytes to address the problem of electrolytes' flammability and to increase EES methods' safety. Fluorinated electrolytes have very low flammability and a low flash-point and are very stable materials, and this is due to the strong bonds that fluorine atoms form. However, these electrolytes are potential environmental pollutants and, at the end of the life cycle of electronic devices, can leak into the environment through poorly managed e-wastes. Developing and utilizing sustainable, renewable, and environment-friendly electrolyte materials is indispensable for solving this problem. The introduced alternative materials can be classified into three categories: aqueous electrolytes, solid-state electrolytes, and organic electrolytes. Seawater is one of these sustainable aqueous electrolytes, but unfortunately, very little research has been done on its utilization in electrochemical supercapacitors. Another type of aqueous electrolyte proposed as a sustainable alternative to fluorinated electrolytes is redox-active electrolytes. The primary purpose of their introduction as an electrolyte in EDLCs is to increase the specific energy of supercapacitors. However, these solutions lose their efficiency over time due to aging. Several additives that have good stability have been proposed to solve this problem, of which iodides, quinones, bromides, thiocyanates, and ferric cyanides are noteworthy. However, some of these additives can intensify the self-discharge problem in EDLCs. Recently various research has been done on solid-state electrolytes. In addition to being very stable, these electrolytes are flexible, and polymer and plastic materials that they are fabricated out of can be easily obtained from nature and have a good performance. However, due to their very high viscosity, they have a very low ionic conductivity.

Although developing sustainable, renewable, and environment-friendly electrolytes with desirable properties such as flexibility and safety are vital, very little research has been done in this field. Most of the research on sustainable materials for ESs is about electrode materials. Thus, due to the necessity of this field, more research is needed to be done on developing new electrolyte materials.

References

1. Mahmood, N., Zhang, C., Yin, H., Hou, Y., Graphene-based nanocomposites for energy storage and conversion in lithium batteries, supercapacitors and fuel cells. *J. Mater. Chem. A*, 2, 1, 15–32, 2014.
2. Theerthagiri, J., Lee, S.J., Shanmugam, P., Choi, M.Y., Chapter 1 - Basic principles in energy conversion and storage, in: *Nanostructured, Functional, and Flexible Materials for Energy*

Conversion and Storage Systems, A. Pandikumar and P. Rameshkumar (Eds.), pp. 1–14, Elsevier, United States, 2020.

3. Mousavi, S.M., Low, F.W., Hashemi, S.A., Lai, C.W., Ghasemi, Y., Soroshnia, S., Savardashtaki, A., Babapoor, A., Pynadathu Rumjit, N., Goh, S.M., Development of graphene based nanocomposites towards medical and biological applications. *Artif. Cells Nanomed. Biotechnol.*, 48, 1, 1189–1205, 2020.
4. Ghadimi, L.S., Arsalani, N., Ahadzadeh, I., Hajalilou, A., Abouzari-Lotf, E., Effect of synthesis route on the electrochemical performance of CoMnFeO4 nanoparticles as a novel supercapacitor electrode material. *Appl. Surf. Sci.*, 494, 440–451, 2019.
5. Raghavendra, K.V.G., Vinoth, R., Zeb, K., Gopi, C.V.M., Sambasivam, S., Kummara, M.R., Obaidat, I.M., Kim, H.J., An intuitive review of supercapacitors with recent progress and novel device applications. *J. Energy Storage*, 31, 101652, 2020.
6. Ahmadi, S., Fazilati, M., Mousavi, S.M., Nazem, H., Anti-bacterial/fungal and anti-cancer performance of green synthesized Ag nanoparticles using summer savory extract. *J. Exp. Nanosci.*, 15, 1, 363–380, 2020.
7. Yan, J., Wang, Q., Wei, T., Fan, Z., Recent advances in design and fabrication of electrochemical supercapacitors with high energy densities. *Adv. Energy Mater.*, 4, 4, 1300816, 2014.
8. Mousavi, S.M., Hashemi, S.A., Zarei, M., Gholami, A., Lai, C.W., Chiang, W.H., Omidifar, N., Bahrani, S., Mazraedoost, S., Recent progress in chemical composition, production, and pharmaceutical effects of kombucha beverage: A complementary and alternative medicine. *Evid.-Based Complementary Altern. Med.*, 2020, 4397543, 2020.
9. Largeot, C., Portet, C., Chmiola, J., Taberna, P.-L., Gogotsi, Y., Simon, P., Relation between the ion size and pore size for an electric double-layer capacitor. *J. Am. Chem. Soc.*, 130, 9, 2730–2731, 2008.
10. Kandalkar, S., Dhawale, D., Kim, C.-K., Lokhande, C., Chemical synthesis of cobalt oxide thin film electrode for supercapacitor application. *Synth. Met.*, 160, 11–12, 1299–1302, 2010.
11. Mousavi, S.M., Hashemi, S.A., Zarei, M., Bahrani, S., Savardashtaki, A., Esmaeili, H., Lai, C.W., Mazraedoost, S., Abassi, M., Ramavandi, B., Data on cytotoxic and antibacterial activity of synthesized Fe3O4 nanoparticles using Malva sylvestris. *Data Brief*, 28, 104929, 2020.
12. Fic, K., Platek, A., Piwek, J., Frackowiak, E., Sustainable materials for electrochemical capacitors. *Mater. Today*, 21, 4, 437–454, 2018.
13. Goodenough, J.B. and Kim, Y., Challenges for rechargeable Li batteries. *Chem. Mater.*, 22, 3, 587–603, 2010.
14. Goodenough, J.B. and Park, K.-S., The Li-ion rechargeable battery: A perspective. *J. Am. Chem. Soc.*, 135, 4, 1167–1176, 2013.
15. Choi, N.S., Chen, Z., Freunberger, S.A., Ji, X., Sun, Y.K., Amine, K., Yushin, G., Nazar, L.F., Cho, J., Bruce, P.G., Challenges facing lithium batteries and electrical double-layer capacitors. *Angew. Chem. Int. Ed.*, 51, 40, 9994–10024, 2012.
16. Frackowiak, E. and Béguin, F., *Supercapacitors: Materials, systems and applications*, Wiley-VCH Verlag GmbH & Co, Poznan, 2013.
17. Mousavi, S., Arjmand, O., Hashemi, S., Banaei, N., Modification of the epoxy resin mechanical and thermal properties with silicon acrylate and montmorillonite nanoparticles. *Polym. Renewable Resour.*, 7, 3, 101–113, 2016.
18. Guo, Y.G., Hu, J.S., Wan, L.J., Nanostructured materials for electrochemical energy conversion and storage devices. *Adv. Mater.*, 20, 15, 2878–2887, 2008.
19. Zhang, Q., Uchaker, E., Candelaria, S.L., Cao, G., Nanomaterials for energy conversion and storage. *Chem. Soc. Rev.*, 42, 7, 3127–3171, 2013.

20. Liu, J., Zhang, J.G., Yang, Z., Lemmon, J.P., Imhoff, C., Graff, G.L., Li, L., Hu, J., Wang, C., Xiao, J., Materials science and materials chemistry for large scale electrochemical energy storage: From transportation to electrical grid. *Adv. Funct. Mater.*, 23, 8, 929–946, 2013.
21. Mousavi, S.M., Hashemi, S.A., Ramakrishna, S., Esmaeili, H., Bahrani, S., Koosha, M., Babapoor, A., Green synthesis of supermagnetic Fe3O4–MgO nanoparticles via Nutmeg essential oil toward superior anti-bacterial and anti-fungal performance. *J. Drug Delivery Sci. Technol.*, 54, 101352, 2019.
22. Wang, G., Zhang, L., Zhang, J., A review of electrode materials for electrochemical supercapacitors. *Chem. Soc. Rev.*, 41, 2, 797–828, 2012.
23. Kötz, R. and Carlen, M., Principles and applications of electrochemical capacitors. *Electrochim. Acta*, 45, 15–16, 2483–2498, 2000.
24. Armand, M. and Tarascon, J.-M., Building better batteries. *Nature*, 451, 7179, 652–657, 2008.
25. Yang, Z., Zhang, J., Kintner-Meyer, M.C., Lu, X., Choi, D., Lemmon, J.P., Liu, J., Electrochemical energy storage for green grid. *Chem. Rev.*, 111, 5, 3577–3613, 2011.
26. Rolison, D.R. and Nazar, L.F., Electrochemical energy storage to power the 21st century. *MRS Bull.*, 36, 7, 486–493, 2011.
27. Zhang, L., Liu, Z., Cui, G., Chen, L., Biomass-derived materials for electrochemical energy storages. *Prog. Polym. Sci.*, 43, 136–164, 2015.
28. Hashemi, S.A., Mousavi, S.M., Bahrani, S., Ramakrishna, S., Integrated polyaniline with graphene oxide-iron tungsten nitride nanoflakes as ultrasensitive electrochemical sensor for precise detection of 4-nitrophenol within aquatic media. *J. Electroanal. Chem.*, 873, 114406, 2020.
29. Lazzari, M., Mastragostino, M., Pandolfo, A., Ruiz, V., Soavi, F., Role of carbon porosity and ion size in the development of ionic liquid based supercapacitors. *J. Electrochem. Soc.*, 158, 1, A22, 2010.
30. Lazzari, M., Soavi, F., Mastragostino, M., Mesoporous carbon design for ionic liquid-based, double-layer supercapacitors. *Fuel Cells*, 10, 5, 840–847, 2010.
31. Arbizzani, C., Biso, M., Cericola, D., Lazzari, M., Soavi, F., Mastragostino, M., Safe, high-energy supercapacitors based on solvent-free ionic liquid electrolytes. *J. Power Sources*, 185, 2, 1575–1579, 2008.
32. Galiński, M., Lewandowski, A., Stępniak, I., Ionic liquids as electrolytes. *Electrochim. Acta*, 51, 26, 5567–5580, 2006.
33. Mousavi, S.M., Zarei, M., Hashemi, S.A., Ramakrishna, S., Chiang, W.-H., Lai, C.W., Gholami, A., Omidifar, N., Shokripour, M., Asymmetric membranes: A potential scaffold for wound healing applications. *Symmetry.*, 12, 7, 1100, 2020.
34. Francke, R., Cericola, D., Kötz, R., Weingarth, D., Waldvogel, S.R., Novel electrolytes for electrochemical double layer capacitors based on 1, 1, 1, 3, 3, 3-hexafluoropropan-2-ol. *Electrochim. Acta*, 62, 372–380, 2012.
35. Kirsch, P., *Modern fluoroorganic chemistry: Synthesis, reactivity, applications*, John Wiley & Sons, Germany, 2013.
36. Nakajima, T. and Groult, H., *Fluorinated materials for energy conversion*, Elsevier, Amsterdam, 2005.
37. Dyatkin, B., Presser, V., Heon, M., Lukatskaya, M.R., Beidaghi, M., Gogotsi, Y., Development of a green supercapacitor composed entirely of environmentally friendly materials. *ChemSusChem*, 6, 12, 2269–2280, 2013.
38. Mousavi, S., Esmaeili, H., Arjmand, O., Karimi, S., Hashemi, S., Biodegradation study of nanocomposites of phenol novolac epoxy/unsaturated polyester resin/egg shell nanoparticles using natural polymers. *J. Mater.*, 2015, 1–6, 2015.

39. Waldvogel, S.R., Malkowsky, I.M., Griesbach, U., Pütter, H., Fischer, A., Hahn, M., Kötz, R., Novel fluorine-free electrolyte system for supercapacitors. *Electrochem. Commun.*, 11, 6, 1237–1241, 2009.
40. Francke, R., Cericola, D., Kötz, R., Schnakenburg, G., Waldvogel, S.R., Bis (2, 2′-biphenoxy) borates for electrochemical double-layer capacitor electrolytes. *Chem.–Eur. J.*, 17, 11, 3082–3085, 2011.
41. Azhdari, R., Mousavi, S.M., Hashemi, S.A., Bahrani, S., Ramakrishna, S., Decorated graphene with aluminum fumarate metal organic framework as a superior non-toxic agent for efficient removal of Congo Red dye from wastewater. *J. Environ. Chem. Eng.*, 7, 6, 103437, 2019.
42. Nanbu, N., Suzuki, K., Yagi, N., Sugahara, M., Takehara, M., Ue, M., Sasaki, Y., Use of fluoroethylene carbonate as solvent for electric double-layer capacitors. *Electrochemistry*, 75, 8, 607–610, 2007.
43. Suzuki, K., Shin-ya, M., Ono, Y., Matsumoto, T., Nanbu, N., Takehara, M., Ue, M., Sasaki, Y., Physical and electrochemical properties of fluoroacetonitrile and its application to electric double-layer capacitors. *Electrochemistry*, 75, 8, 611–614, 2007.
44. Zhang, S. and Angell, C., A novel electrolyte solvent for rechargeable lithium and lithium-ion batteries. *J. Electrochem. Soc.*, 143, 12, 4047, 1996.
45. Vardar, G., Sleightholme, A.E., Naruse, J., Hiramatsu, H., Siegel, D.J., Monroe, C.W., Electrochemistry of magnesium electrolytes in ionic liquids for secondary batteries. *ACS Appl. Mater. Interfaces*, 6, 20, 18033–18039, 2014.
46. Monti, D., Ponrouch, A., Palacín, M.R., Johansson, P., Towards safer sodium-ion batteries via organic solvent/ionic liquid based hybrid electrolytes. *J. Power Sources*, 324, 712–721, 2016.
47. Bruce, P.G., Freunberger, S.A., Hardwick, L.J., Tarascon, J.-M., Li–O2 and Li–S batteries with high energy storage. *Nat. Mater.*, 11, 1, 19–29, 2012.
48. Kozen, A.C., Lin, C.-F., Pearse, A.J., Schroeder, M.A., Han, X., Hu, L., Lee, S.-B., Rubloff, G.W., Noked, M., Next-generation lithium metal anode engineering via atomic layer deposition. *ACS Nano*, 9, 6, 5884–5892, 2015.
49. Balducci, A., Bardi, U., Caporali, S., Mastragostino, M., Soavi, F., Ionic liquids for hybrid supercapacitors. *Electrochem. Commun.*, 6, 6, 566–570, 2004.
50. Mousavi, S., Aghili, A., Hashemi, S., Goudarzian, N., Bakhoda, Z., Baseri, S., Improved morphology and properties of nanocomposites, linear low density polyethylene, ethylene-co-vinyl acetate and nano clay particles by electron beam. *Polym. Renew. Resour.*, 7, 4, 135–153, 2016.
51. Xia, L., Yu, L., Hu, D., Chen, G.Z., Electrolytes for electrochemical energy storage. *Mater. Chem. Front.*, 1, 4, 584–618, 2017.
52. Mousavi, S.M., Hashemi, S.A., Amani, A.M., Saed, H., Jahandideh, S., Mojoudi, F., Polyethylene terephthalate/acryl butadiene styrene copolymer incorporated with oak shell, potassium sorbate and egg shell nanoparticles for food packaging applications: Control of bacteria growth, physical and mechanical properties. *Polym. Renew. Resour.*, 8, 4, 177–196, 2017.
53. Fic, K., Lota, G., Meller, M., Frackowiak, E., Novel insight into neutral medium as electrolyte for high-voltage supercapacitors. *Energy Environ. Sci.*, 5, 2, 5842–5850, 2012.
54. Choi, J.W. and Aurbach, D., Promise and reality of post-lithium-ion batteries with high energy densities. *Nat. Rev. Mater.*, 1, 4, 1–16, 2016.
55. Orikasa, Y., Masese, T., Koyama, Y., Mori, T., Hattori, M., Yamamoto, K., Okado, T., Huang, Z.-D., Minato, T., Tassel, C., High energy density rechargeable magnesium battery using earth-abundant and non-toxic elements. *Sci. Rep.*, 4, 1, 1–6, 2014.
56. Seyed, M.M., Unsaturated polyester resins modified with cresol novolac epoxy and silica nanoparticles: Processing and mechanical properties. *Int. J. Chem. Pet. Sci. (IJCPS)*, 5, 1, 13–26, 2016.

57. Kita, Y., Tohma, H., Inagaki, M., Hatanaka, K., Yakura, T., A novel oxidative azidation of aromatic compounds with hypervalent iodine reagent, phenyliodine (III) bis (trifluoroacetate) (PIFA) and trimethylsilyl azide. *Tetrahedron Lett.*, 32, 34, 4321–4324, 1991.
58. Kita, Y., Tohma, H., Hatanaka, K., Takada, T., Fujita, S., Mitoh, S., Sakurai, H., Oka, S., Hypervalent iodine-induced nucleophilic substitution of para-substituted phenol ethers. Generation of cation radicals as reactive intermediates. *J. Am. Chem. Soc.*, 116, 9, 3684–3691, 1994.
59. Steenken, S. and McClelland, R.A., 248-nm Laser flash photoprotonation of mesitylene, hexamethylbenzene, and 1, 3, 5-trimethoxybenzene in 1, 1, 1, 3, 3, 3-hexafluoroisopropyl alcohol. Formation and electrophilic reactivities of the cyclohexadienyl cations. *J. Am. Chem. Soc.*, 112, 26, 9648–9649, 1990.
60. Shuklov, I.A., Dubrovina, N.V., Boerner, A., Fluorinated alcohols as solvents, cosolvents and additives in homogeneous catalysis. *Synthesis*, 2007, 19, 2925–2943, 2007.
61. Mousavi, S.M., Hashemi, S.A., Esmaeili, H., Amani, A.M., Mojoudi, F., Synthesis of Fe3O4 nanoparticles modified by oak shell for treatment of wastewater containing Ni (II). *Acta Chim. Slov.*, 65, 3, 750–756, 2018.
62. Carre, B. and Devynck, J., The acidity functions of trifluoroethanol and hexafluoroisopropanol, and their mixtures with water. *Anal. Chim. Acta*, 131, 141–147, 1981.
63. Begue, J.-P., Bonnet-Delpon, D., Crousse, B., Fluorinated alcohols: A new medium for selective and clean reaction. *Synlett.*, 2004, 01, 18–29, 2004.
64. Krossing, I., The facile preparation of weakly coordinating anions: Structure and characterisation of silverpolyfluoroalkoxyaluminates AgAl (ORF) 4, calculation of the alkoxide ion affinity. *Chem.-Eur. J.*, 7, 2, 490–502, 2001.
65. Raabe, I., Wagner, K., Guttsche, K., Wang, M., Grätzel, M., Santiso-Quiñones, G., Krossing, I., Tetraalkylammonium salts of weakly coordinating aluminates: Ionic liquids, materials for electrochemical applications and useful compounds for anion investigation. *Chem.-Eur. J.*, 15, 8, 1966–1976, 2009.
66. Ivanova, S.M., Nolan, B.G., Kobayashi, Y., Miller, S.M., Anderson, O.P., Strauss, S.H., Relative Lewis basicities of six Al (ORF) 4– superweak anions and the structures of LiAl {OCH (CF3) 2} 4 and [1-Et-3-Me-1, 3-C3H3N2][Li {Al {OCH (CF3) 2} 4} 2]. *Chem.-Eur. J.*, 7, 2, 503–510, 2001.
67. Gholami, A., Hashemi, S.A., Yousefi, K., Mousavi, S.M., Chiang, W.-H., Ramakrishna, S., Mazraedoost, S., Alizadeh, A., Omidifar, N., Behbudi, G., 3D nanostructures for tissue engineering, cancer therapy, and gene delivery. *J. Nanomater.*, 2020, 1852946, 2020.
68. Kirste, A., Nieger, M., Malkowsky, I.M., Stecker, F., Fischer, A., Waldvogel, S.R., ortho-Selective phenol-coupling reaction by anodic treatment on boron-doped diamond electrode using fluorinated alcohols. *Chem.-Eur. J.*, 15, 10, 2273–2277, 2009.
69. Kirste, A., Schnakenburg, G., Stecker, F., Fischer, A., Waldvogel, S.R., Anodic phenol–arene cross-coupling reaction on boron-doped diamond electrodes. *Angew. Chem. Int. Ed.*, 49, 5, 971–975, 2010.
70. Lee, H., Yang, X., Xiang, C., McBreen, J., Choi, L., The synthesis of a new family of boron-based anion receptors and the study of their effect on ion pair dissociation and conductivity of lithium salts in nonaqueous solutions. *J. Electrochem. Soc.*, 145, 8, 2813, 1998.
71. Li, L., Lee, H., Li, H., Yang, X., Nam, K., Yoon, W., McBreen, J., Huang, X., New electrolytes for lithium ion batteries using LiF salt and boron based anion receptors. *J. Power Sources*, 184, 2, 517–521, 2008.
72. Lee, H., Ma, Z., Yang, X., Sun, X., McBreen, J., Synthesis of a series of fluorinated boronate compounds and their use as additives in lithium battery electrolytes. *J. Electrochem. Soc.*, 151, 9, A1429, 2004.

73. Nagasubramanian, G. and Sanchez, B., A new chemical approach to improving discharge capacity of Li/(CFx) n cells. *J. Power Sources*, 165, 2, 630–634, 2007.
74. Hashemi, S.A. and Mousavi, S.M., Effect of bubble based degradation on the physical properties of Single Wall Carbon Nanotube/Epoxy Resin composite and new approach in bubbles reduction. *Compos. Part A: Appl. Sci. Manuf.*, 90, 457–469, 2016.
75. Landolt, H., Börnstein, R., Mitsui, T., Hellwege, K.-H., Hellwege, A., Numerical data and functional relationships in science and technology: New series. Crystal and solid state physics. *Ferroelectrics and Related Substances: Revised, Updated and Extended Edition of Volumes III/3 and III/9.* Non-oxides/T. Mitsui [and Others], K.H. Hellwege and A.M. Hellwege (Eds.), Springer, Berlin, 1982.
76. Ue, M., Ida, K., Mori, S., Electrochemical properties of organic liquid electrolytes based on quaternary onium salts for electrical double-layer capacitors. *J. Electrochem. Soc.*, 141, 11, 2989, 1994.
77. Dura, H., Perry, J., Lecacou, T., Markoulidis, F., Lei, C., Khalil, S., Decker, M., Weil, M., Cost analysis of supercapacitor cell production, in: *2013 International Conference on Clean Electrical Power (ICCEP)*, IEEE, 2013.
78. Amani, A.M., Hashemi, S.A., Mousavi, S.M., Abrishamifar, S.M., Vojood, A., Electric field induced alignment of carbon nanotubes: Methodology and outcomes, in: *Carbon Nanotubes-Recent Progress*, IntechOpen, London, 2017.
79. Lee, H. and Goodenough, J., Supercapacitor behavior with KCl electrolyte. *J. Solid State Chem.*, 144, 1, 220–223, 1999.
80. Zhong, C., Deng, Y., Hu, W., Qiao, J., Zhang, L., Zhang, J., A review of electrolyte materials and compositions for electrochemical supercapacitors. *Chem. Soc. Rev.*, 44, 21, 7484–7539, 2015.
81. Wang, Y., Chen, B., Zhang, Y., Fu, L., Zhu, Y., Zhang, L., Wu, Y., ZIF-8@ MWCNT-derived carbon composite as electrode of high performance for supercapacitor. *Electrochim. Acta*, 213, 260–269, 2016.
82. Gao, P., Grätzel, M., Nazeeruddin, M.K., Organohalide lead perovskites for photovoltaic applications. *Energy Environ. Sci.*, 7, 8, 2448–2463, 2014.
83. Manikandan, R., Raj, C.J., Rajesh, M., Kim, B.C., Sim, J.Y., Yu, K.H., Electrochemical behaviour of lithium, sodium and potassium ion electrolytes in a Na0. 33V2O5 symmetric pseudocapacitor with high performance and high cyclic stability. *ChemElectroChem*, 5, 1, 101–111, 2018.
84. Zhao, C. and Zheng, W., A review for aqueous electrochemical supercapacitors. *Front. Energy Res.*, 3, 23, 2015.
85. Challagulla, N.V., Vijayakumar, M., Sri Rohita, D., Elsa, G., Bharathi Sankar, A., Narasinga Rao, T., Karthik, M., Hierarchical activated carbon fibers as a sustainable electrode and natural seawater as a sustainable electrolyte for high-performance supercapacitor. *Energy Technol.*, 8, 9, 2000417, 2020.
86. Mousavi, S.M., Hashemi, S.A., Jahandideh, S., Baseri, S., Zarei, M., Azadi, S., Modification of phenol novolac epoxy resin and unsaturated polyester using sasobit and silica nanoparticles. *Polym. Renewable Resour.*, 8, 3, 117–132, 2017.
87. Hilbertz, W., Electrodeposition of minerals in sea water: Experiments and applications. *IEEE J. Oceanic Eng.*, 4, 3, 94–113, 1979.
88. Hahn, R., Mainert, J., Glaw, F., Lang, K.-D., Sea water magnesium fuel cell power supply. *J. Power Sources*, 288, 26–35, 2015.
89. Millero, F.J., Feistel, R., Wright, D.G., McDougall, T.J., The composition of standard seawater and the definition of the reference-composition salinity scale. *Deep Sea Res. Part I: Oceanogr. Res. Pap.*, 55, 1, 50–72, 2008.
90. Kim, J.-K., Mueller, F., Kim, H., Bresser, D., Park, J.-S., Lim, D.-H., Kim, G.-T., Passerini, S., Kim, Y., Rechargeable-hybrid-seawater fuel cell. *NPG Asia Mater.*, 6, 11, e144–e144, 2014.

91. Kim, J.K., Lee, E., Kim, H., Johnson, C., Cho, J., Kim, Y., Rechargeable seawater battery and its electrochemical mechanism. *ChemElectroChem*, 2, 3, 328–332, 2015.
92. Hasvold, Ø., Henriksen, H., Melv, E., Citi, G., Johansen, B.Ø., Kjønigsen, T., Galetti, R., Seawater battery for subsea control systems. *J. Power Sources*, 65, 1–2, 253–261, 1997.
93. Zhao, P., Yao, M., Zhang, Q., Wang, N., Hu, W., Komarneni, S., Electrochemical behavior of representative electrode materials in artificial seawater for fabricating supercapacitors. *Electrochim. Acta*, 318, 211–219, 2019.
94. Mousavi, S., Zarei, M., Hashemi, S., Polydopamine for biomedical application and drug delivery system. *Med. Chem. (Los Angeles)*, 8, 218–29, 2018.
95. Mousavi, S.M., Hashemi, S.A., Kalashgrani, M.Y., Omidifar, N., Bahrani, S., Vijayakameswara Rao, N., Babapoor, A., Gholami, A., Chiang, W.-H., Bioactive graphene quantum dots based polymer composite for biomedical applications. *Polymers*, 14, 3, 617, 2022.
96. Frackowiak, E., Meller, M., Menzel, J., Gastol, D., Fic, K., Redox-active electrolyte for supercapacitor application. *Faraday Discuss.*, 172, 179–198, 2014.
97. Gastol, D., Walkowiak, J., Fic, K., Frackowiak, E., Enhancement of the carbon electrode capacitance by brominated hydroquinones. *J. Power Sources*, 326, 587–594, 2016.
98. Menzel, J., Fic, K., Meller, M., Frackowiak, E., The effect of halide ion concentration on capacitor performance. *J. Appl. Electrochem.*, 44, 4, 439–445, 2014.
99. Przygocki, P., Abbas, Q., Babuchowska, P., Béguin, F., Confinement of iodides in carbon porosity to prevent from positive electrode oxidation in high voltage aqueous hybrid electrochemical capacitors. *Carbon*, 125, 391–400, 2017.
100. Senthilkumar, S., Selvan, R.K., Lee, Y., Melo, J., Electric double layer capacitor and its improved specific capacitance using redox additive electrolyte. *J. Mater. Chem. A*, 1, 4, 1086–1095, 2013.
101. Gao, Z., Zhang, L., Chang, J., Wang, Z., Wu, D., Xu, F., Guo, Y., Jiang, K., Catalytic electrode-redox electrolyte supercapacitor system with enhanced capacitive performance. *Chem. Eng. J.*, 335, 590–599, 2018.
102. Menzel, J., Fic, K., Frackowiak, E., Hybrid aqueous capacitors with improved energy/power performance. *Prog. Nat. Sci.: Mater. Int.*, 25, 6, 642–649, 2015.
103. Roldán, S., Blanco, C., Granda, M., Menéndez, R., Santamaría, R., Towards a further generation of high-energy carbon-based capacitors by using redox-active electrolytes. *Angew. Chem. Int. Ed.*, 50, 7, 1699–1701, 2011.
104. Roldan, S., Granda, M., Menendez, R., Santamaría, R., Blanco, C., Mechanisms of energy storage in carbon-based supercapacitors modified with a quinoid redox-active electrolyte. *J. Phys. Chem. C*, 115, 35, 17606–17611, 2011.
105. Ren, L., Zhang, G., Lei, J., Wang, Y., Hu, D., Novel layered polyaniline-poly (hydroquinone)/graphene film as supercapacitor electrode with enhanced rate performance and cycling stability. *J. Colloid Interface Sci.*, 512, 300–307, 2018.
106. Vonlanthen, D., Lazarev, P., See, K.A., Wudl, F., Heeger, A.J., A stable polyaniline-benzoquinone-hydroquinone supercapacitor. *Adv. Mater.*, 26, 30, 5095–5100, 2014.
107. Navarro-Suárez, A.M., Casado, N., Carretero-González, J., Mecerreyes, D., Rojo, T., Full-cell quinone/hydroquinone supercapacitors based on partially reduced graphite oxide and lignin/PEDOT electrodes. *J. Mater. Chem. A*, 5, 15, 7137–7143, 2017.
108. Kalashgarani, M.Y. and Babapoor, A., Application of nano-antibiotics in the diagnosis and treatment of infectious diseases. *Adv. Appl. NanoBio-Tech.*, 3, 1, 22–35, 2022.
109. Senthilkumar, S., Selvan, R.K., Ponpandian, N., Melo, J., Redox additive aqueous polymer gel electrolyte for an electric double layer capacitor. *RSC Adv.*, 2, 24, 8937–8940, 2012.
110. Dubal, D.P., Nagar, B., Suarez-Guevara, J., Tonti, D., Enciso, E., Palomino, P., Gomez-Romero, P., Ultrahigh energy density supercapacitors through a double hybrid strategy. *Mater. Today Energy*, 5, 58–65, 2017.

111. Kim, D., Lee, G., Kim, D., Yun, J., Lee, S.-S., Ha, J.S., High performance flexible double-sided micro-supercapacitors with an organic gel electrolyte containing a redox-active additive. *Nanoscale*, 8, 34, 15611–15620, 2016.
112. Hwang, J., Knapik, G.G., Dufour, J.S., Best, T.M., Khan, S.N., Mendel, E., Marras, W.S., Validation of a personalized curved muscle model of the lumbar spine during complex dynamic exertions. *J. Electromyogr. Kinesiol.*, 33, 1–9, 2017.
113. Lee, J., Choudhury, S., Weingarth, D., Kim, D., Presser, V., High performance hybrid energy storage with potassium ferricyanide redox electrolyte. *ACS Appl. Mater. Interfaces*, 8, 36, 23676–23687, 2016.
114. Alipour, A. and Kalashgarani, M.Y., Nano protein and peptides for drug delivery and anticancer agents. *Adv. Appl. NanoBio-Technol.*, 3, 1, 60–64, 2022.
115. Fic, K., Meller, M., Frackowiak, E., Strategies for enhancing the performance of carbon/carbon supercapacitors in aqueous electrolytes. *Electrochim. Acta*, 128, 210–217, 2014.
116. Gorska, B., Bujewska, P., Fic, K., Thiocyanates as attractive redox-active electrolytes for high-energy and environmentally-friendly electrochemical capacitors. *Phys. Chem. Chem. Phys.*, 19, 11, 7923–7935, 2017.
117. Lewandowski, A., Olejniczak, A., Galinski, M., Stepniak, I., Performance of carbon–carbon supercapacitors based on organic, aqueous and ionic liquid electrolytes. *J. Power Sources*, 195, 17, 5814–5819, 2010.
118. Béguin, F., Presser, V., Balducci, A., Frackowiak, E., Carbons and electrolytes for advanced supercapacitors. *Adv. Mater.*, 26, 14, 2219–2251, 2014.
119. Iwama, E., Taberna, P.-L., Azais, P., Brégeon, L., Simon, P., Characterization of commercial supercapacitors for low temperature applications. *J. Power Sources*, 219, 235–239, 2012.
120. Jiang, D.-e., Jin, Z., Henderson, D., Wu, J., Solvent effect on the pore-size dependence of an organic electrolyte supercapacitor. *J. Phys. Chem. Lett.*, 3, 13, 1727–1731, 2012.
121. Liu, P., Verbrugge, M., Soukiazian, S., Influence of temperature and electrolyte on the performance of activated-carbon supercapacitors. *J. Power Sources*, 156, 2, 712–718, 2006.
122. Erik Brandon, M.S. and West, W., Supercapacitor electrolyte solvents with liquid range below −80 °C. *NASA Tech Briefs*, 34, 21, 2010.
123. Ue, M., Mobility and ionic association of lithium and quaternary ammonium salts in propylene carbonate and γ-butyrolactone. *J. Electrochem. Soc.*, 141, 12, 3336, 1994.
124. Ue, M., Conductivities and ion association of quaternary ammonium tetrafluoroborates in propylene carbonate. *Electrochim. Acta*, 39, 13, 2083–2087, 1994.
125. Ue, M., Takeda, M., Takehara, M., Mori, S., Electrochemical properties of quaternary ammonium salts for electrochemical capacitors. *J. Electrochem. Soc.*, 144, 8, 2684, 1997.
126. S. Mori, K. Ida, M. Ue, Mitsubishi Chemical Co. Ltd, US Pat. 4892944, 1990.
127. Ue, M., Chemical capacitors and quaternary ammonium salts. *Electrochemistry*, 75, 8, 565–572, 2007.
128. UE, M., Takehara, M., Oura, Y., Toriumi, A., Takeda, M., High sensitive detection of impurities in nonaqueous electrolyte solutions by a rotating disk electrode. *Electrochemistry*, 69, 6, 458–461, 2001.
129. UE, M., Takehara, M., Takeda, M., Triethylmethylammonium tetrafluoroborate as a highly soluble supporting electrolyte salt for electrochemical capacitors. *Denki Kagaku oyobi Kogyo Butsuri Kagaku*, 65, 11, 969–971, 1997.
130. Chiba, K. and Futagami, A., Enhanced bending stability of carbon-nanotube-reinforced indium tin oxide films on flexible plastic substrates. *Appl. Phys. Lett.*, 93, 1, 013114, 2008.
131. Simon, P. and Gogotsi, Y., Materials for electrochemical capacitors, in: *Nanoscience and Technology: A Collection of Reviews from Nature Journals*, pp. 320–329, World Scientific, Singapore, London, 2010.

132. Kurzweil, P. and Chwistek, M., Electrochemical stability of organic electrolytes in supercapacitors: Spectroscopy and gas analysis of decomposition products. *J. Power Sources*, 176, 2, 555–567, 2008.
133. Koh, A.R., Hwang, B., Roh, K.C., Kim, K., The effect of the ionic size of small quaternary ammonium BF 4 salts on electrochemical double layer capacitors. *Phys. Chem. Chem. Phys.*, 16, 29, 15146–15151, 2014.
134. Jänes, A., Kurig, H., Romann, T., Lust, E., Novel doubly charged cation based electrolytes for non-aqueous supercapacitors. *Electrochem. Commun.*, 12, 4, 535–539, 2010.
135. Yokoyama, Y., Shimosaka, N., Matsumoto, H., Yoshio, M., Ishihara, T., Effects of supporting electrolyte on the storage capacity of hybrid capacitors using graphitic and activated carbon. *Electrochem. Solidstate Lett.*, 11, 5, A72, 2008.
136. Xu, K., Ding, M.S., Jow, T.R., Quaternary onium salts as nonaqueous electrolytes for electrochemical capacitors. *J. Electrochem. Soc.*, 148, 3, A267, 2001.
137. Kurig, H., Jänes, A., Lust, E., Substituted phosphonium cation based electrolytes for nonaqueous electrical double-layer capacitors. *J. Mater. Res.*, 25, 8, 1447–1450, 2010.
138. Krause, A. and Balducci, A., High voltage electrochemical double layer capacitor containing mixtures of ionic liquids and organic carbonate as electrolytes. *Electrochem. Commun.*, 13, 8, 814–817, 2011.
139. Pohlmann, S. and Balducci, A., A new conducting salt for high voltage propylene carbonate-based electrochemical double layer capacitors. *Electrochim. Acta*, 110, 221–227, 2013.
140. Yu, X., Ruan, D., Wu, C., Wang, J., Shi, Z., Spiro-(1, 1′)-bipyrrolidinium tetrafluoroborate salt as high voltage electrolyte for electric double layer capacitors. *J. Power Sources*, 265, 309–316, 2014.
141. Hu, C., Qu, W., Rajagopalan, R., Randall, C., Factors influencing high voltage performance of coconut char derived carbon based electrical double layer capacitor made using acetonitrile and propylene carbonate based electrolytes. *J. Power Sources*, 272, 90–99, 2014.
142. Naoi, K., Ishimoto, S., Miyamoto, J.-i., Naoi, W., Second generation 'nanohybrid supercapacitor': Evolution of capacitive energy storage devices. *Energy Environ. Sci.*, 5, 11, 9363–9373, 2012.
143. He, L., Li, J., Gao, F., Wang, X., Ye, F., Yang, J., Pressure evolution and analysis of aged electrodes at high temperature in electrochemical double layer capacitors. *Electrochemistry*, 79, 12, 934–940, 2011.
144. Ruch, P., Cericola, D., Foelske-Schmitz, A., Kötz, R., Wokaun, A., Aging of electrochemical double layer capacitors with acetonitrile-based electrolyte at elevated voltages. *Electrochim. Acta*, 55, 15, 4412–4420, 2010.
145. Ishimoto, S., Asakawa, Y., Shinya, M., Naoi, K., Degradation responses of activated-carbon-based EDLCs for higher voltage operation and their factors. *J. Electrochem. Soc.*, 156, 7, A563, 2009.
146. Hahn, M., Würsig, A., Gallay, R., Novák, P., Kötz, R., Gas evolution in activated carbon/propylene carbonate based double-layer capacitors. *Electrochem. Commun.*, 7, 9, 925–930, 2005.
147. Chiba, K., Ueda, T., Yamaguchi, Y., Oki, Y., Shimodate, F., Naoi, K., Electrolyte systems for high withstand voltage and durability I. Linear sulfones for electric double-layer capacitors. *J. Electrochem. Soc.*, 158, 8, A872, 2011.
148. Chiba, K., Ueda, T., Yamaguchi, Y., Oki, Y., Saiki, F., Naoi, K., Electrolyte systems for high withstand voltage and durability II. Alkylated cyclic carbonates for electric double-layer capacitors. *J. Electrochem. Soc.*, 158, 12, A1320, 2011.
149. Brandt, A., Isken, P., Lex-Balducci, A., Balducci, A., Adiponitrile-based electrochemical double layer capacitor. *J. Power Sources*, 204, 213–219, 2012.

150. Naoi, K., 'Nanohybrid capacitor': The next generation electrochemical capacitors. *Fuel Cells*, 10, 5, 825–833, 2010.
151. Du Pasquier, A., Plitz, I., Menocal, S., Amatucci, G., A comparative study of Li-ion battery, supercapacitor and nonaqueous asymmetric hybrid devices for automotive applications. *J. Power Sources*, 115, 1, 171–178, 2003.
152. Cericola, D., Novák, P., Wokaun, A., Kötz, R., Hybridization of electrochemical capacitors and rechargeable batteries: An experimental analysis of the different possible approaches utilizing activated carbon, Li4Ti5O12 and LiMn2O4. *J. Power Sources*, 196, 23, 10305–10313, 2011.
153. Krause, A., Kossyrev, P., Oljaca, M., Passerini, S., Winter, M., Balducci, A., Electrochemical double layer capacitor and lithium-ion capacitor based on carbon black. *J. Power Sources*, 196, 20, 8836–8842, 2011.
154. Manyala, N., Bello, A., Barzegar, F., Khaleed, A.A., Momodu, D.Y., Dangbegnon, J.K., Coniferous pine biomass: A novel insight into sustainable carbon materials for supercapacitors electrode. *Mater. Chem. Phys.*, 182, 139–147, 2016.
155. Yu, H., Wu, J., Fan, L., Xu, K., Zhong, X., Lin, Y., Lin, J., Improvement of the performance for quasi-solidstate supercapacitor by using PVA–KOH–KI polymer gel electrolyte. *Electrochim. Acta*, 56, 20, 6881–6886, 2011.
156. Lewandowski, A., Zajder, M., Frąckowiak, E., Beguin, F., Supercapacitor based on activated carbon and polyethylene oxide–KOH–H2O polymer electrolyte. *Electrochim. Acta*, 46, 18, 2777–2780, 2001.
157. Pérez-Madrigal, M.M., Estrany, F., Armelin, E., Díaz, D.D., Alemán, C., Towards sustainable solidstate supercapacitors: Electroactive conducting polymers combined with biohydrogels. *J. Mater. Chem. A*, 4, 5, 1792–1805, 2016.
158. He, X., Ling, P., Qiu, J., Yu, M., Zhang, X., Yu, C., Zheng, M., Efficient preparation of biomass-based mesoporous carbons for supercapacitors with both high energy density and high power density. *J. Power Sources*, 240, 109–113, 2013.
159. Kumar, M.S. and Bhat, D.K., LiClO4-doped plasticized chitosan as biodegradable polymer gel electrolyte for supercapacitors. *J. Appl. Polym. Sci.*, 114, 4, 2445–2454, 2009.
160. Sudhakar, Y. and Selvakumar, M., Lithium perchlorate doped plasticized chitosan and starch blend as biodegradable polymer electrolyte for supercapacitors. *Electrochim. Acta*, 78, 398–405, 2012.
161. Soeda, K., Yamagata, M., Yamazaki, S., Ishikawa, M., Application of chitosan-based gel electrolytes with ionic liquids for high-performance and safe electric double layer capacitors. *Electrochemistry*, 81, 10, 867–872, 2013.
162. Liu, X., Wu, D., Wang, H., Wang, Q., Self-recovering tough gel electrolyte with adjustable supercapacitor performance. *Adv. Mater.*, 26, 25, 4370–4375, 2014.
163. Virya, A. and Lian, K., A review of neutral pH polymer electrolytes for electrochemical capacitors: Transitioning from liquid to solid devices. *Mater. Rep.: Energy*, 1, 1, 100005, 2021.
164. Yamagata, M., Soeda, K., Ikebe, S., Yamazaki, S., Ishikawa, M., Chitosan-based gel electrolyte containing an ionic liquid for high-performance nonaqueous supercapacitors. *Electrochim. Acta*, 100, 275–280, 2013.
165. Yamazaki, S., Takegawa, A., Kaneko, Y., Kadokawa, J.-i., Yamagata, M., Ishikawa, M., An acidic cellulose–chitin hybrid gel as novel electrolyte for an electric double layer capacitor. *Electrochem. Commun.*, 11, 1, 68–70, 2009.
166. Yamazaki, S., Takegawa, A., Kaneko, Y., Kadokawa, J.-i., Yamagata, M., Ishikawa, M., High/low temperature operation of electric double layer capacitor utilizing acidic cellulose–chitin hybrid gel electrolyte. *J. Power Sources*, 195, 18, 6245–6249, 2010.

167. Wei, D., Wakeham, S.J., Ng, T.W., Thwaites, M.J., Brown, H., Beecher, P., Transparent, flexible and solidstate supercapacitors based on room temperature ionic liquid gel. *Electrochem. Commun.*, 11, 12, 2285–2287, 2009.
168. Moon, W.G., Kim, G.-P., Lee, M., Song, H.D., Yi, J., A biodegradable gel electrolyte for use in high-performance flexible supercapacitors. *ACS Appl. Mater. Interfaces*, 7, 6, 3503–3511, 2015.
169. Shukur, M., Ithnin, R., Kadir, M., Electrical characterization of corn starch-LiOAc electrolytes and application in electrochemical double layer capacitor. *Electrochim. Acta*, 136, 204–216, 2014.
170. Chodankar, N.R., Dubal, D.P., Gund, G.S., Lokhande, C.D., Flexible all-solidstate MnO2 thin films based symmetric supercapacitors. *Electrochim. Acta*, 165, 338–347, 2015.
171. Jiménez-Cordero, D., Heras, F., Gilarranz, M.A., Raymundo-Piñero, E., Grape seed carbons for studying the influence of texture on supercapacitor behaviour in aqueous electrolytes. *Carbon*, 71, 127–138, 2014.
172. Torchała, K., Kierzek, K., Machnikowski, J., Capacitance behavior of KOH activated mesocarbon microbeads in different aqueous electrolytes. *Electrochim. Acta*, 86, 260–267, 2012.
173. Wu, H., Wang, X., Jiang, L., Wu, C., Zhao, Q., Liu, X., Yi, L., The effects of electrolyte on the supercapacitive performance of activated calcium carbide-derived carbon. *J. Power Sources*, 226, 202–209, 2013.
174. Zhang, X., Wang, X., Jiang, L., Wu, H., Wu, C., Su, J., Effect of aqueous electrolytes on the electrochemical behaviors of supercapacitors based on hierarchically porous carbons. *J. Power Sources*, 216, 290–296, 2012.
175. Zhai, Y., Dou, Y., Zhao, D., Fulvio, P.F., Mayes, R.T., Dai, S., Carbon materials for chemical capacitive energy storage. *Adv. Mater.*, 23, 42, 4828–4850, 2011.
176. Zhang, L.L. and Zhao, X., Carbon-based materials as supercapacitor electrodes. *Chem. Soc. Rev.*, 38, 9, 2520–2531, 2009.
177. Lee, S.W., Gallant, B.M., Byon, H.R., Hammond, P.T., Shao-Horn, Y., Nanostructured carbon-based electrodes: Bridging the gap between thin-film lithium-ion batteries and electrochemical capacitors. *Energy Environ. Sci.*, 4, 6, 1972–1985, 2011.
178. Bichat, M., Raymundo-Piñero, E., Béguin, F., High voltage supercapacitor built with seaweed carbons in neutral aqueous electrolyte. *Carbon*, 48, 15, 4351–4361, 2010.
179. Demarconnay, L., Raymundo-Piñero, E., Béguin, F., A symmetric carbon/carbon supercapacitor operating at 1.6 V by using a neutral aqueous solution. *Electrochem. Commun.*, 12, 10, 1275–1278, 2010.
180. Zhao, L., Qiu, Y., Yu, J., Deng, X., Dai, C., Bai, X., Carbon nanofibers with radially grown graphene sheets derived from electrospinning for aqueous supercapacitors with high working voltage and energy density. *Nanoscale*, 5, 11, 4902–4909, 2013.
181. Conway, B., *I.E. Supercapacitors, Scientific fundamentals and technological applications*, p. 528, KluwerAcademic/Plenum Publishers, New York, 1999.

9

Aqueous Redox-Active Electrolytes

Ranganatha S.

Dept. of Chemistry, Presidency University, Rajanakunte, Bengaluru, India

Abstract

It is necessary to improve the supercapacitors' energy particulars due to low energy storage capacity and charge/discharge tolerance. Electrolytes are considered to be important part in the output delivery of supercapacitors which are electrochemical in nature. Organic electrolytes are being used in the majority of commercially marketed electrochemical supercapacitors, which has some advantages such as a high operating voltage. Organic alternatives, on the other hand, are more expensive than aqueous ones. Electrolytes are costly, combustible, and hazardous in few instances. It's a safe bet to say that. Despite the fact that aqueous electrolytes are subjected to a low workable voltage, they exhibit far quicker charge carrier rates than non-aqueous electrolytes and do improve the efficiency of supercapacitors. As a result, research works are being done concentrating voltage and energy density in supercapacitor devices by expanding the workable potential range of aqua based electrolyte. This book chapter summarizes the most recent developments related to the concerned interest and gives overview of aqua based electrolyte.

***Keywords*:** Electrolytes, aqueous electrolytes, supercapacitors, acid electrolytes, alkaline electrolytes, neutral electrolytes, non-aqueous electrolytes, ionic conductance

9.1 Introduction

Global warming is accelerating, and the world's access to fossil fuels is diminishing. Society must adapt to sustainable resources which are reusable in nature. Because of this, advancement in sustainable energy technologies has been seen as a significant concern around the world. Most of these reusable energy resources are reliant on weather conditions around the country. Electrochemical supercapacitors/ultracapacitors, fuel cells and batteries are the most competent and widely employed among various energy storage technologies [1]. The greatest common feature with these systems is the phenomenon taking place at electrode/electrolyte interfaces in these devices.

Supercapacitors have gained much of interest in academia and industry because of the appreciable rate of energy delivery along with high cyclability to the par with fuel cells and batteries [2]. Electrochemical supercapacitors are grouped into three kinds with respect to

Email: kamath.ranganath@gmail.com

Inamuddin, Tariq Altalhi and Sayed Mohammed Adnan (eds.) Sustainable Materials for Electrochemical Capacitors, (247–260) © 2023 Scrivener Publishing LLC

their charge storage process: i. EDLC or electric double layer, ii. pseudo, and iii. hybrid [3]. Activated carbon which behaves as EDLC is used in practically all commercially available supercapacitors because to their technical maturity. Compared to rechargeable fuel cells and battery systems, supercapacitors face the most critical obstacle of low values for storage capacity (10 watt hour per kg) [4]. As a result, massive investigations made to improve energy density of these systems [5].

The charge storage of supercapacitors is directly proportional to capacitance and square of operating voltage. As a result, increasing capacitance and working potential are seen as possible strategies in improving high charge storage supercapacitors. By selecting high capacity materials as electrodes and an electrolyte possessing a high working voltage, a high energy density can be achieved. As voltage (square of) and energy stored are directly proportional enhancing the workable voltage range rather than improving the specific capacitance seems to be successful approach in obtaining higher charge storage. Therefore, establishing a novel electrolyte with wide voltage range is the primary parameter to consider that of searching for novel negative and positive electrodes.

9.2 Effect of the Electrolyte on Supercapacitor Performance

With respect to supercapacitors, liquid or solid electrolytic materials are one of most important parts, since they possess current carrying capacity, allowing for charge balancing during the electrochemical process [6, 7]. Electrolyte's ionic conductivity, which can affect the electrochemical series resistance of the supercapacitors, and stability range (electrochemical), which determines the operating voltage of the supercapacitors, are the two key features of electrolytes for supercapacitors. Therefore, as mentioned in the previous sections, both energy storage and rate of energy delivery will rise. As a result, for improved capacitive performance in supercapacitors, electrolytes with a broader workable and stable voltage range and higher ionic conductance will be preferred. The quality of electrolyte which constitutes, radius and nature of ion, strength and ion-solvent behavior, all have a substantial impact on supercapacitor performance. Also, action of electrolyte ions and electrodes has a substantial impact on supercapacitor capacitive performance. The mismatch in electrolytic ion size and the carbonaceous electrode's porosity, for example, imparts significant impact on charge storage ability in the case of supercapacitors.

Electrolytic solutions in supercapacitors aid in the stabilization of other critical supercapacitor properties. Voltage drop, rate capability, temperature stability, number of charge/discharges, discharge when no load, and environmental benignity are all factors to consider when using supercapacitors in practice. Figure 9.1 depicts the influence of electrolyte on different supercapacitor characteristics, which affect the device's strength.

In the representation, first circle contains supercapacitors with electrolytes; second circle contains electrolytes and electrochemical supercapacitors; and circle 3 contains electrolytes and electrochemical supercapacitors. The parameters that strongly influence functioning of supercapacitor are found in second circular region. Electrolyte factors influencing the parameters in third circular are shown (circle 2). Charge carriage and movement of electrolyte ions and the resistance they have for the mobility, for example, impact directly influence supercapacitor electrochemical series resistance (ESR) (see Figure 9.1). The parameters within each circle, on the other hand, have an effect on each other. As depicted in second

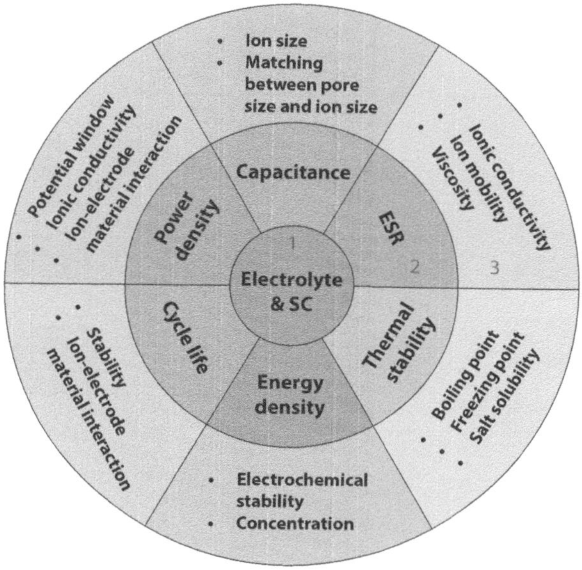

Figure 9.1 Influence of electrolyte nature on various supercapacitor characteristics (reused with the permission [8]).

circle, more the series resistance lower will be the rate of energy delivery. It demonstrates that qualities of electrolytic solutions specified in third circular region are critical in determining supercapacitor performance in one way or another.

Different electrolytes are designed and published by the scientists for Supercapacitors. Based on the physical state it is primarily liquid and solid state. Based on aqua, organic, or ionic nature, liquid electrolytes can be classified. Electrolytes based on polymers, gel polymer and inorganic based solid are three types of solid or quasi-solid electrolytes [8, 9]. The schematic in Figure 9.2 depicts the general classification of electrolytes.

So far, no single electrolyte has just been found to meet all of the requirements for supercapacitors. Between the two crucial electrolyte properties of conductance and stability range, there is always a trade-off. Supercapacitors with aqueous electrolytes, for example, have a significant conductance and charge storage in comparison. However, its stability range is restricted by electrolyte's small breakdown (One-volt limit), ending up with the supercapacitors with small value for energy storage. Supercapacitors with electrolytes sourced from organic or ionic sources, on the other hand, can be operated at greater voltages and have a significant energy storage than those aqua based. Furthermore, solid state has the benefit of not having the outflow problems associated with liquid electrolytes, although they often have lesser ionic conductivity. As a result, each electrolyte possesses its inherent benefits and limitations.

Therefore, many methodologies are used to formulate novel and effective electrolytes with broader working voltage range, better conductance with a wider temperature range of operation. This will allow researchers to take advantage of the beneficial impacts of specific electrolytes on supercapacitor performance metrics like energy storage, rate of energy delivery etc. Detailed outlook on typical electrolytic forms for supercapacitors, like aqua and non-aqua based electrolytes, will be presented in the following subsections.

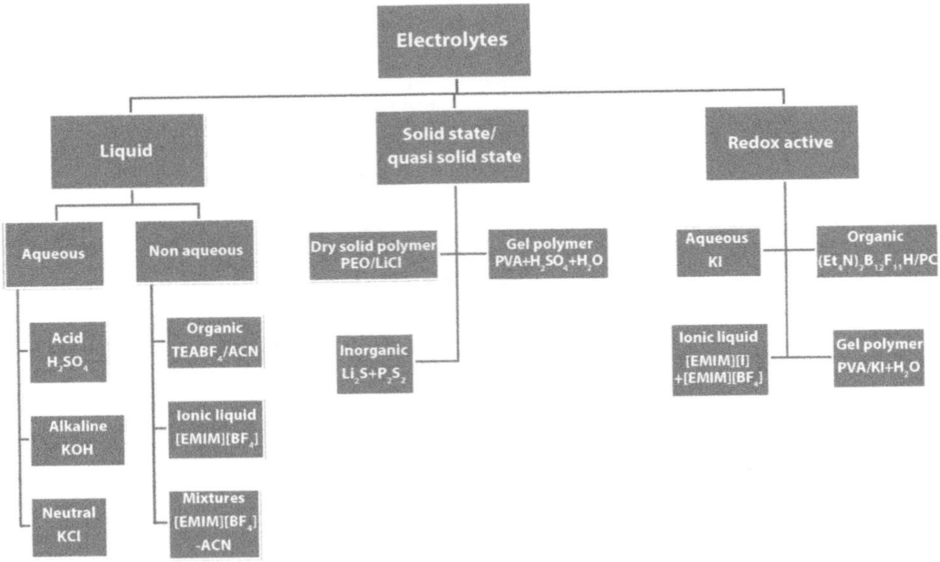

Figure 9.2 Classification of electrolytes for supercapacitors (reused with the permission [8]).

9.3 Aqueous Electrolytes

Because of their narrow voltage window, aqueous electrolytes are usually not a great choice for practical supercapacitor devices when it comes to energy density studies. This could explain why most of the practical supercapacitors employ non aqueous (organic) electrolyte rather than inorganic electrolyte. Electrolytes in aqueous solution are less effective than electrolytes in aqueous solution. Despite the fact that aqueous electrolytes have insufficient cell voltage, they have been extensively used in the literature since 1997 as they appear to be less expensive and may be easily managed in the laboratory.

To keep moisture out of ionic liquids and organic electrolytes, extensive purification processes in a highly restricted environment are required. These characteristics of aqua based electrolytes help to understand the design and function of the supercapacitors. Aqueous electrolytes, in general, have a very high conductivity of higher magnitude than organic or ionic liquid electrolytes. The aqueous electrolyte's greater conductivity makes it ideal for lowering solution resistance leads to supercapacitors with much higher power density.

Some typical factors should be considered when evaluating the comprehensive performance of aqua based electrolytes, like original size of the ions and hydrated counterparts, ionic movement modifying the electrolytic conductance, and charge storage. Acidic, basic and neutral are the three types of aqueous electrolytes. KOH, H_2SO_4, and sodium sulphate are commonly used aqueous electrolytes, respectively.

9.4 Acidic Electrolytes

Because of extraordinarily significant conductance (0.8 cm^{-1} at room temperature for 1 molar sulphuric acid), sulphuric acid is very commonly utilized aqueous acid electrolyte in supercapacitors [10]. Because of direct relation between concentration and the ionic conductance of the electrolytic solution, appropriate molar concentrations for achieving the appreciable conductance at a given temperature have already been researched. When the concentration of the electrolyte high or low, conductance drops quickly. As a result, the majority of research investigations, particularly for carbon-based supercapacitors, use a 1 M H_2SO_4 solution.

9.4.1 Sulfuric Acid Electrolyte-Based EDLC and Pseudocapacitors

Many researchers recently observed the specific capacitance found with the electrolyte sulphuric acid is substantially greater compared to that of specific capacitance obtained in a neutral for carbon-based capacitors (EDLCs). Furthermore, the equivalent series resistance is substantially lesser compared to neutral one because of high ionic conductivity of H_2SO_4. A group investigated electric double layer capacitor behavior w.r.t. activated carbon in the electrolyte of sulphuric acid and discovered higher electrolyte conductivity improving the specific capacitance [11]. Many publications over the last few years have claimed that carbon-based materials can give capacity in the scale of 100–300 Farads per gram. When compared to the commercial electrolytes, certain reported values are frequently greater. Changes in specific capacitance could mimic the unique interactions between electrode materials and ions of electrolyte generated by different electrolytes. As previously stated, charge storage of EDLCs is severely decreased due to limited voltage range of aqua electrolyte. So, massive exercises are considered so as to improve the electrochemical supercapacitor storage capacity by studying alternative approaches, such as exploiting pseudocapacitive behavior in electrodes.

The sulphuric acid when employed as electrolyte contributes the pseudocapacitance owing to redox active reactions associated with carbon surface's functionalities, adding to usual double layer capacitance. Adding various heteroatoms to the carbon materials surface, such as nitrogen, oxygen, and boron, could improve the pseudocapacitance contributions [12]. Heteroatom doping not only boosts charge storage capacity with improving cycleability in aqueous electrolytes, at higher potentials too. In an aqua electrolyte, redox type electrodes such as oxides and sulphides of metal, and polymers which electrically conduct have substantially greater expected capacity values than carbonaceous materials [13]. Because metallic compounds are extremely responsive to the environment and acidity/basicity, they are often unstable in electrolytes lower pH value below 7, limiting their usage in commercial goods. Ruthenium oxide (RuO_2), as is widely known, is one among the superior electrodes that has been thoroughly investigated for its pseudocapacitance wherein sulphuric acid is as the electrolyte. Because of the ease with which the proton can be inserted into the amorphous structure, non-crystalline Ruthenium oxide produces significant capacitance 10^3 Farads/gram in the electrolytes of low pH. Regrettably, because of higher price and less availability Ru sources, their practical application has been limited.

9.4.2 H_2SO_4 Electrolyte-Based Hybrid Supercapacitors

Constructing a hybrid supercapacitor is the very efficient and reliable technique to boost the energy storage in the electrolytes of water base. The highest symmetric supercapacitor EMF employing the same electrodes in an water based electrolyte is limited due to the gas evolution reactions. Asymmetric supercapacitors combine two alternate electrodes in distinct working potentials, resulting in a broader functional voltage range in an aqueous electrolyte. There have been many studies on performances of symmetric supercapacitors in the electrolytes of the type strongly acidic nature, as like combination of carbonaceous material along with metal oxides of lead and ruthernium, and carbon with varied weight and characteristics in the electrodes. In the acidic electrolyte, the carbon/PbO_2 asymmetric supercapacitor supplied an energy density 25–30 Watt hour per kilo gram, showing significantly higher value compared to symmetric carbonaceous capacitors (3–6 Wh/kg). Polymers which are electrically conducting in nature and carbonaceous asymmetric supercapacitors, like metal oxide, have provided enough energy density for commercial use. In the case of polyaniline employed as a +ve, carbonaceous material as –ve one at wider voltage window (negative 1 to positive 1 V), storage capacity may reach up to 11.46 Wh/kg [14]. Polyaniline has found restricted application in real world because of weak stability, despite having a larger energy density than symmetric carbon capacitors. The utilization of a sulphuric acid in the case of polymer systems can increase stability. In contrast to polymer in solid state, liquid H_2SO_4 has a great thermal stability. So, stability and energy density of supercapacitors in liquid-based electrolytes may be increased. The utilization of a sulphuric acid as electrolyte in case of polymer-based electrodes can increase stability. Considering all these characteristics, there is a scope to enhance charge storage of supercapacitors when liquid based electrolytes are employed. In the case of ruthenium oxide and carbon based symmetric electrochemical supercapacitors, sulphuric acid given storage capacity of roughly 20 Watt hour per kilogram, which is double compared to the electrolyte related to polymer gel low pH [15].

9.5 Alkaline Electrolytes

Electrolytes of alkaline nature have been used widely by the research community because acidic electrolytes are typically not suited for various economical metallic and metal compounds (Ni, Co, Eu, etc.). Due to much higher conductance of about 0.6 Siemen per centimeter, potassium hydroxide (KOH) is the much often utilized aqueous alkaline electrolyte.

9.5.1 Alkaline Electrolyte-Based EDLC and Pseudocapacitors

In aqueous KOH, charge storage capacity of carbonaceous material based devices is comparable with sulphuric acid electrolytes. As a result, significant effort has been made to improve EDLC material energy density in basic electrolyte by raising capacity and increasing working potential. Few of the approaches include:

(1) Doping some pseudocapacitance functional groups into carbon materials to increase specific capacitance,
(2) Finding out highly capacitive electrode materials of pseudocapacitive in nature,

(3) developing novel hybrid materials which can contribute both EDLC and pseudocapacitance, and
(4) Designing high performance asymmetric supercapacitors possessing wide voltage range.

As stated earlier, redox reactions associated with functionalized carbon materials take place as a result of electrolyte ions interacting with the grafted reactive groups. According to reports, potassium hydroxide was found to be helpful for the doping or functioning of carbon based electrodes improvise pseudocapacitance. Possession of reactive moiety with oxygen w.r.t. carbonaceous materials can improve the overall capacitance of the electrode by increasing the pseudocapacitance contribution in potassium hydroxide electrolyte [16]. It is widely understood that the interaction of ions of electrolyte and electrode plays an important part in pseudocapacitance contribution as a result of this. Some oxides of metal, sulphides, and hydroxides, like Co_3O_4, MnO_2, NiO, $NiCo_2O_4$, CoS, nickel hydroxide etc., have extraordinarily high predicted energy storage, and are studied for supercapacitor usage in potassium hydroxide, e.g., in the electrolyte of 2 M potassium hydroxide, charge storage of Co_3O_4 nanosized films attained extraordinarily to greater values of about one thousand and four hundred Farads per gram [17]. In the same way, while metal sulphides related to cobalt, manganese, nickel etc., possess lesser value for faradaic capacitance in potassium hydroxide, electrochemical conversion of sulphides into various electroactive products like nickel hydroxide, cobalt hydroxide may increase the capacitance of electrodes. The operating temperature and supercapacitor working is possible to be influenced by parameters related to electrolyte including type of ion, ionic size, and concentrations. Because the pH of potassium hydroxide electrolyte will be obviously high, corrosion of surface of the electrode may happen during the usage of high electrolyte concentration is utilized. As a result, optimizing strength of electrolyte in order in minimizing current collector corrosion. Overall performance of electrochemical supercapacitor is sensitive for ionic nature and concentration in the electrolyte. Insertion and disinsertion of the electrolytic ion can determine the extent to which material's pseudocapacitance. Because of Li^+ ions' small ionic radius, Misnon *et al.* [18] were able to produce a greater energy characteristics of manganese oxide in lithium hydroxide compared to potassium hydroxide as the electrolyte. Smaller the ion, easy is the intercalation and de-intercalation which leads to result in good faradaic processes. Moreover, influence of different electrolytes of water base on $NiCo_2O_4$ nanomaterials in wider temperature range were examined, in potassium hydroxide electrolyte obtaining the maximum energy density when compared to sodium hydroxide and lithium hydroxide. The ion mobility in KOH increased dramatically due to the reduced size of Potassium ions, facilitating quick redox reactions of electrode and electrolyte [19, 20]. In addition to cations, anions pose a crucial role in electrochemical supercapacitor function. A research group investigated effect of negative ions on charge storage behavior of potassium-based alkaline (potassium hydroxide) and inert (potassium chloride) in depth [21]. They observed less bulky hydroxide ions could improve the faradaic processes of sulphide of cobalt and graphene composites with potassium hydroxide. There are fewer reports accessible for comparative investigations of variety of alkaline electrolytic phenomenon and therefore mechanisms stay unclear.

Pseudocapacitive materials have low cyclic stability in alkaline electrolytes in contrast with EDLC materials, which could be due to the electrolyte's recurrent faradaic reactions.

A previous study looked into using carbon materials to extend the cycle life in redox type electrodes (composites). Composites expanded operating potential range of alkaline electrolytes while also improving electrode stability. Fe_2O_3 nanodots/N-doped graphene hybrid, for example, exhibited significantly high cycle life stable up to of 75.3 percent over 1 lakh charge/discharges in potassium hydroxide [22], which could be owing to the pyrolic Nitrogen atoms in graphene promoting the composite's transfer of electron processes.

9.5.2 Alkaline Electrolyte-Based Hybrid Supercapacitors

For hybrid supercapacitors, series of solutions which are alkaline in nature has been researched to boost charge storage over long voltage window. When asymmetric electrodes are utilized, applied potential of potassium hydroxide can be increased by till 1.7 V. In the KOH, asymmetric carbon/metal hydroxides (carbon/Nickel hydroxide 1.7 V and carbon/ Cobalt hydroxide) 1.6 Volts) and C/Ruthenium oxide (1.4 Volts) capacitors are seen to provide remarkable capacity storage. Most hybrid systems could produce greater capacity storage in the range of 20 and 40 Watt-hour per kilogram as a result of their long voltage range and strong redox reactions, which was close to the values related to that of lithium ion systems. The performance and operating voltage of an asymmetric device were improved by in situ synthesis of electroactive colloids of vanadium with high cation utilization [23]. As a result, building an asymmetric hybrid supercapacitor is an economical solution to enhance the working potential, and it might be utilized in KOH electrolyte applications.

9.6 Neutral Electrolyte

Neutral electrolytes have also been thoroughly studied as compared to alkaline and acid electrolytes because of wider working potential range and lower corrosive properties. Variety of inert electrolytic solutions like lithium chloride, lithium sulphite, sodium sulfate, sodium chloride, potassium chloride, potassium sulfate and others, have been employed in supercapacitor research to date. Na_2SO_4 neutral electrolyte, in particular, has demonstrated favorable electrochemical responses for electrodes, particularly pseudocapacitance materials.

9.6.1 Neutral Electrolyte-Based EDLC and Pseudocapacitors

Compared to alkaline electrolyte, smaller H^+ and OH^- concentrations in neutral ones, allow the voltage of water splitting reactions to get pushed to a different voltage range, enabling working voltage range to get extended. At higher cell EMF of 1.6 V in 0.5 molar sodium sulfate, Demarconnay *et al.* showed outstanding cyclic stability w.r.t. symmetric carbonaceous device even after 10,000 charged-discharge cycles [24]. Because of smaller ionic size in Li_2SO_4 electrolyte, EMF of the device should be able to increase till 2.2 V for carbonaceous system (Table 9.1).

Because Li^+ ions are smaller, they can speed up the charge storage process in the interface of carbon electrode/electrolyte and prevent gases like oxygen and hydrogen from escaping at higher potentials. In comparison to H_2SO_4 and KOH electrolytes, the reported working potential of inert electrolyte w.r.t. carbon based electrode systems is high with less harsh

Table 9.1 Ionic sizes and their conductance values.

Ion type	Bare ion size (Å)	Hydrated ion size (Å)	Gibbs energy (kcal/mol)	Ionic conductance (S cm² mol⁻¹)
H^+	1.15	2.80		350.1
Li^+	0.60	3.82	138.4	38.69
Na^+	0.95	3.58	162.3	50.11
K	1.33	3.31	179.9	73.5
NH_4^+	1.48	3.31	–	73.7
Mg_2^+	0.72	4.28	–	106.12
Ca_2^+	1.00	4.12	–	119
Ba_2^+	1.35	4.04	–	127.8
Cl^-	1.81	3.32	–	76.31
NO_3^-	2.64	3.35	–	71.42
SO_4^{2-}	2.90	3.79	–	160.0
OH^-	1.76	3.00	–	198
ClO_4^-	2.92	3.38	–	67.3
PO_4^{3-}	2.23	3.39	–	207
CO_3^{2-}	2.66	3.94	–	138.6

chemically. In a neutral electrolyte, symmetric carbon supercapacitors have been found as a possible contender for devices with increased charge storage and fewer environmental consequence. There have been some similar investigations on the systems with inert electrolytes of different metals including lithium, potassium, and sodium, to see how ions affect capacitive performance. Several research [25, 26] investigated the carbon based compounds in inert electrolytes and their performance is in following sequence lithium sulfate > sodium sulfate > potassium sulfate.

Electrochemical supercapacitors' function is influenced not only by nature of electrolyte used, also by the preparation of electrode and conditions at which analysis carried out, such as rate of scanning, temperature, and potential range. Accordingly, more investigation is necessary so as to analyze effect of electrolyte on electrochemical supercapacitor functioning.

In neutral electrolytes, faradaic materials like RuO_2, MnO_2 and vanadium oxide demonstrated good electrochemical supercapacitor performance. The most prevalent pseudocapacitive compound is MnO_2, which has been extensively researched in inert electrolytic solution. In neutral electrolytes, Li *et al.* [27] emphasized the sequence of charge storage and capacitance of porous manganese oxide electrodes is as follows: Li_2SO_4 > Na_2SO_4 > K_2SO_4

Figure 9.3 Ionic dimension and size in water based electrolytes as obtained by modeling (reused with the permission [28]).

is the order of importance. This behavior seems to be linked to sizes of alkali metal both in normal and hydrated condition, Lithium Sodium, and Potassium ions, implying smaller the ionic size better is functioning. Fic *et al.* demonstrated the same behavior in the case of ruthenium oxide in various inert electrolytes. It is found that ionic motion increases in sequence of Li^+ Na^+ K^+. Higher movement of ionic Lithium is advantageous to improvise altogether functioning of the faradaic electrodes, according to shape and size of the ion (sourced from modeling) [25], as illustrated in Figure 9.3.

9.6.2 Neutral Electrolyte-Based Hybrid Supercapacitors

Neutral electrolytes were used in devices of different positive and negative electrodes to provide larger variety of possible windows for achieving enhanced charge storage. Devices with Manganese oxide electrodes electrochemical supercapacitors provided sufficient specific capacitance, their working potential and total charge storage are limited. In many situations, a device potential of around 1 Volt was used. The MnO_2 cell potential can be greatly improved by using carbon materials (activated carbon and CNT) as the negative electrode [29]. Symmetric devices with manganese phosphate electrodes, on the other hand, may be operated at an extraordinarily high voltage (1.8 V) in the Na_2SO_4 electrolyte, resulting in a high charge storage of about 19 Watt-hour per kilogram [30]. Devices with same and different types of positive and negative electrodes have generated significant charge storage in inert electrolytic solution and not in the alkaline. In neutral electrolyte, rationale in the case of $Mn_3(PO_4)_2$'s improved performance and great cycle stability is still unknown. Asymmetric capacitors in inert electrolytic solutions were studied using a variety of cathodes and anodes, some of them are included in Table 9.2. In the 1.5–2.0 V range, voltage range of this set of asymmetric supercapacitors is wider than that of electrolytes of non-neutral pH. Benefit of inert water based electrolytes that they not only solve supercapacitor device corrosion concerns, but they also provide cheap and benign option for improving output of supercapacitors with higher working potential and charge storage. The charge/discharge stability of supercapacitors w.r.t. inert aqua electrolytic solutions is currently a developing field.

Table 9.2 Few significant water based electrolytic supercapacitor devices and their output.

Water based electrolyte/ concentration	Electrode	Cs (F/g)	E (W h kg^{-1})	Voltage range (V)	Ref.
Sulfuric acid/1 M	RuO_2-carbon nano-onion	570	10.62	1	[21]
Sulfuric acid/0.5 Molar	RuO_2-graphene	479	20.28	1.2	[31]
H_2SO_4/1 M	Graphene-mesoporous PANI	510	133	1	[32]
Potassium hydroxide/6 M	Graphene	303	6.5	1	[33]
Potassium hydroxide/2 M	$NiCo_2O_4$	1647.6	38	0.41	[34]
Potassium hydroxide/1 M	$Ni(OH)_2$/graphene	160	48	1.5	[35]
Potassium hydroxide/6 M	Iron nanosheets/ graphene	720	140	1.2	[36]
KCl/3 M	Ni $(OH)_2$	718	–	0.5	[37]
NaOH/1 M	Fe_2O_3	178	–	0.5	[38]
Na_2SO_4/1 M	CNT/V_2O_5 nanocomposite/ MnO_2/C	–	16	1.6	[39]
Na_2SO_4/1 M	Hydrous RuO_2	56.66	18.77	1.6	[40]

9.7 Conclusion and Future Research Directions

To summarize, this contribution provides a thorough examination of evolution and current practices in water based electrolytic solutions for Electrochemical Supercapacitors. Impact of these electrolytic solutions parameters such as conductance, size of ion, workable voltage range, and other factors on supercapacitor output specifications such as cyclic stability, energy density, specific capacitance and temperature stability is thoroughly examined. The effect of interplay between electrodes and electrolytes has been examined in particular. Current discussion demonstrates electrochemical supercapacitors depending on neutral aqueous electrolytes give various advantages over other electrolyte systems, including low cost, environmental friendliness, and safety. Despite significant progress in this field, numerous challenges remain, like the lesser charge storage and working potential associated with water based electrolytes.

To address above issues, future exploration should be in the following path:

1. Extending electrolyte stable potential window values (ESPW): The ESPW values of aqueous electrolytes are determined by the ions which it is composed with. As a result, discovering novel water based conducting electrolytic salt along with optimization could help to improve the electrolyte stable potential window values.
2. Increasing the effective area of surface of the faradaic electrode materials: Because energy density is mostly determined by capacitance, the performance of supercapacitors might be improved by increasing the electrode material's charge capacity. The symmetric and asymmetric pseudocapacitive material-based supercapacitors bring better energy density in aqueous electrolyte than EDLC.
3. Building water based electrolytic solution purity: In few situations, impurity associated with electrolyte has a detrimental impact on working voltage range which results in automatous discharging. As a result, developing purest form of electrolytic solution to increase supercapacitor output is highly considered.

References

1. Ramachandran, R., Saranya, M., Grace, A.N., Wang, F., MnS nanocomposites based on doped graphene: Simple synthesis by a wet chemical route and improved electrochemical properties as an electrode material for supercapacitors. *RSC Adv.*, 7, 2249, 2017.
2. Yu, G., Xie, X., Pan, L., Bao, Z., Cui, Y., Hybrid nanostructured materials for high-performance electrochemical capacitors. *Nano Energy*, 2, 213, 2013.
3. Ramachandran, R., Saranya, M., Kollu, P., Ragupathy, B.P.C., Jeong, S.K., Grace, A.N., Solvothermal synthesis of zinc sulfide decorated graphene (ZnS/G) nanocomposites for novel supercapacitor electrodes. *Electrochim. Acta*, 178, 647, 2005.
4. Conway, B.E., *Electrochemical supercapacitors: Scientific fundamentals and technological applications*, Kluwer Academic/Plenum, New York, 1999.
5. Xia, L., Yu, L., Chen, G.Z., Electrolytes for electrochemical energy storage. *Mater. Chem. Front.*, 1, 584, 2017.
6. Matthes, F.C., Hermann, H., Zimmer, W., *The vision scenario for the European Union 2011 update for the EU-27*, Berlin, 2011.
7. Moseley, P.T. and Garche, J., *Electrochemical energy storage for renewable sources and grid balancing*, Elsevier, Amsterdam, 2014.
8. Zhong, C., Deng, Y., Hu, W., Qiao, J., Zhang, L., Zhang, J., A review of electrolyte materials and compositions for electrochemical supercapacitors. *Chem. Soc. Rev.*, 44, 7484, 2015.
9. Shaplov, A.S., Marcilla, R., Mecerreyes, D., New family of highly conductive and low viscous ionic liquids with asymmetric 2,2,2-trifluoromethylsulfonyl-N-cyanoamide anion. *Electrochim. Acta*, 175, 18, 2015.
10. Galinski, M., Lewandowski, A., Pniak, I.S., Ionic liquids as electrolytes. *Electrochim. Acta*, 51, 5567, 2006.
11. Torchala, K., Kierzek, K., Machnikowski, J., Capacitance behavior of KOH activated mesocarbon microbeads in different aqueous electrolytes. *Electrochim. Acta*, 86, 260, 2012.

12. Pognon, G., Brousse, T., Demarconnay, L., Belanger, D., Performance and stability of electrochemical capacitor based on anthraquinone modified activated carbon. *J. Power Sources*, 196, 4117, 2011.
13. Augustyn, V., Simon, P., Dunn, B., Pseudocapacitive oxide materials for high-rate electrochemical energy storage. *Energy Environ. Sci.*, 7, 1597, 2014.
14. Khomenko, V., Pinero, E.R., Frackowiak, E., Beguin, F., High-voltage asymmetric supercapacitors operating in aqueous electrolyte. *Appl. Phys. A*, 82, 567, 2006.
15. Muniraj, V.K.A., Kamaja, C.K., Shelke, M.V., RuO_2·nHO nanoparticles anchored on carbon nano-onions: An efficient electrode for solid state flexible electrochemical supercapacitor. *ACS Sustainable Chem. Eng.*, 4, 2528, 2016.
16. Ramachandran, R., Saranya, M., Velmurugan, V., Ragupathy, B.P.C., Jeong, S.K., Grace, A.N., Effect of reducing agent on graphene synthesis and its influence on charge storage towards supercapacitor applications. *Appl. Energy*, 153, 22, 2015.
17. Feng, C., Zhang, J.F., He, Y., Zhong, C., Hu, W.B., Liu, L., Deng, Y.D., Sub-3 nm Co_3O_4 nanofilms with enhanced supercapacitor properties. *ACS Nano*, 9, 1730, 2015.
18. Misnon, I.I., Aziz, R.A., Zain, N.K., Vidhyadharan, B., Krishnan, S.K., Jose, R., High performance MnO_2 nanoflower electrode and the relationship between solvated ion size and specific capacitance in highly conductive electrolytes. *Mater. Res. Bull.*, 57, 221, 2014.
19. Gupta, R.K., Candler, J., Palchoudhury, S., Ramasamy, K., Gupta, B.K., Flexible and high performance supercapacitors based on $NiCo_2O_4$ for wide temperature range applications. *Sci. Rep.*, 5, 15265, 2015.
20. Barzegar, F., Momodu, D.F., Fashedemi, O.O., Bello, A., Dangbegnon, J.K., Manyala, N., Investigation of different aqueous electrolytes on the electrochemical performance of activated carbon-based supercapacitors. *RSC Adv.*, 5, 107482, 2015.
21. Ramachandran, R., Grace, A.N., Subramaniam, C.K., Cobalt sulfide-graphene (CoSG) composite based electrochemical double layer capacitors. *MRS Proc.*, 1786, 19, 2015.
22. Liu, L., Lang, J., Zhang, P., Hu, B., Yan, X., Facile synthesis of Fe_2O_3 nano-dots@nitrogen-doped graphene for supercapacitor electrode with ultralong cycle life in KOH electrolyte. *ACS Appl. Mater. Interfaces*, 8, 9335, 2016.
23. Chen, K. and Xue, D., High energy density hybrid supercapacitor: *In-situ* functionalization of vanadium-based colloidal cathode. *ACS Appl. Mater. Interfaces*, 8, 29522–29528, 2016.
24. Demarconnay, L., Pinero, E.R., Beguin, F., A symmetric carbon/carbon supercapacitor operating at 1.6 V by using a neutral aqueous solution. *Electrochem. Commun.*, 12, 1275–1278, 2010.
25. Fic, K., Lota, G., Meller, M., Frackowiak, E., Novel insight into neutral medium as electrolyte for high-voltage supercapacitors. *Energy Environ. Sci.*, 5, 5842, 2012.
26. Chae, J.H. and Chen, G.H., Influences of ions and temperature on performance of carbon nanoparticulates in supercapacitors with neutral aqueous electrolytes. *Particuology*, 15, 9–17, 2014.
27. Li, S., Qi, L., Lu, L., Wang, H., Facile preparation and performance of mesoporous manganese oxide for supercapacitors utilizing neutral aqueous electrolytes. *RSC Adv.*, 2, 3298, 2012.
28. Pal, B., Yang, S., Ramesh, S., Thangadurai, V., Jose, R., Electrolyte selection for supercapacitive devices:a critical review. *Nanoscale Adv.*, 1, 3807, 2019.
29. Long, J.W., Belanger, D., Brousse, T., Sugimoto, W., Sassin, M.B., Crosnier, O., Asymmetric electrochemical capacitors—Stretching the limits of aqueous electrolytes. *MRS Bull.*, 36, 513, 2011.
30. Ma, X.J., Zhang, W.B., Kong, L.B., Luo, Y.C., Kang, L., Electrochemical performance in alkaline and neutral electrolytes of a manganese phosphate material possessing a broad potential window. *RSC Adv.*, 6, 40077, 2016.
31. Deng, L.J., Wang, J.F., Zhu, G., Kang, L.P., Hao, Z.P., Lei, Z.B., Yang, Z.P., Liu, Z.H., RuO_2/graphene hybrid material for high performance electrochemical capacitor. *J. Power Sources*, 248, 407, 2014.

32. Wang, Q., Yan, J., Fan, Z.J., Wei, T., Zhang, M.L., Jing, X.Y., Mesoporous polyaniline film on ultra-thin graphene sheets for high performance supercapacitors. *J. Power Sources*, 247, 197, 2014.
33. Wang, H.J., Sun, X.X., Liu, Z.H., Lei, Z.B., Creation of nanopores on graphene planes with MgO template for preparing high-performance supercapacitor electrodes. *Nanoscale*, 6, 6577, 2014.
34. Li, L., Peng, S., Cheah, Y., Teh, P., Wang, J., Wee, G., Ko, Y., Woung, C., Srinivasan, M., Electrospun porous $NiCo_2O_4$ nanotubes as advanced electrodes for electrochemical capacitors. *Chem. Eur. J.*, 19, 5892, 2013.
35. Wang, H., Liang, Y., Mirfakhrai, T., Chen, Z., Casalomgue, H.S., Dai, H., Advanced asymmetrical supercapacitors based on graphene hybrid materials. *Nano Res.*, 4, 8, 729, 2011.
36. Long, C., Wei, T., Yan, J., Jiang, L., Fan, Z., Supercapacitors based on graphene-supported iron nanosheets as negative electrode materials. *ACS Nano*, 7, 12, 11325, 2013.
37. Lee, J.W., Ahn, T., Kim, J.H., Ko, J.M., Kim, J.D., Nanosheets based mesoporous NiO microspherical structures via facile and template-free method for high performance super- capacitors. *Electrochim. Acta*, 56, 4849, 2011.
38. Kulal, P.M., Dubal, D.P., Lokhande, C.D., Fulari, V.J., Chemical synthesis of Fe_2O_3 thin films for supercapacitor application. *J. Alloys Compd.*, 509, 2567, 2011.
39. Q, Q.T., Shi, Y., Li, L.L., Guo, W.L., Wu, Y.P., Zhang, H.P., Guan, S.Y., Holze, R., $V_2O_5 \cdot 0._6H_2O$ nanoribbons as cathode material for asymmetric supercapacitor in K_2SO_4 solution. *Electrochem. Commun.*, 11, 1325, 2009.
40. Xia, H., Meng, Y.S., Yuan, G., Cui, C., Lu, L., A symmetric RuO_2/RuO_2 supercapacitor operating at 1.6 V by using a neutral aqueous electrolyte. *Electrochem. Solid-State Lett.*, 15, A60, 2012.

10

Biodegradable Electrolytes

Tuba Saleem[1], Ijaz Rasul[1], Habibullah Nadeem[1], Sanora Sehar[1] and Arfaa Sajid[2*]

[1]*Department of Bioinformatics and Biotechnology, Government College University Faisalabad, Faisalabad, Pakistan*

[2]*Department of Chemistry, The University of Lahore, Lahore, Pakistan*

Abstract

Fossil fuels are consumed at a large amount to meet the needs of energy demand and this consumption of fossil fuels has a negative impact on environment and human activities. Biodegradable/biopolymer electrolytes are the materials composed of salt and solvent having different properties like good ionic conductivity, cost effective, safe to use and good chemical stability. A number of bio-based polymers are used as host polymer in the formation of biodegradable electrolytes. Biodegradable polymer electrolytes are simply prepared by the solution casting method. Free charge carriers and their mobility, glass transition temperature and dielectric constant highly affect their ionic conductivity. Biodegradable electrolytes have been used due to their renewability, environmental stability, and safety. They can meet the energy requirements and hence are renewable made from biomaterials, not consuming fossil fuels and are cost effective. Biodegradable polymer electrolytes have different applications, like fuel cells, batteries, supercapacitors and dye sensitized solar cells etc.

Keywords: Solid polymer electrolytes, gel polymer electrolytes, composite polymer electrolytes, fuel cells, batteries, supercapacitors

10.1 Introduction

Fossil fuels are consumed at a large amount to meet the needs of energy demand and this consumption of fossil fuels has a negative impact on environment and human activities. It is very harmful for the environment to have such a huge waste and to dispose it in a manner which requires time, energy, and money. So, it is mandatory to have a suitable disposal way of waste and to not harm the environment. Therefore, biodegradable polymers are coming fast in environment to slower the rate of waste and to replace the non-biodegradable polymers. Biodegradable polymers are of great importance due to their tremendous advantages like non-toxicity, biodegradability, environment friendly, and biocompatible. They can serve as good replacement of some hazardous materials in environment. To maintain the energy sources and made it friendly, biodegradable electrolytes have been introduced which are extensively studied due to their good environmental effects, eliminating

**Corresponding author*: arfaa.sajid@chem.uol.edu.pk

Inamuddin, Tariq Altalhi and Sayed Mohammed Adnan (eds.) Sustainable Materials for Electrochemical Capacitors, (261–278) © 2023 Scrivener Publishing LLC

pollution and non-dependency of fossil fuels. Most famous of them are lithium ion-based polymer electrolytes. These are better than lead-acid batteries due to their safety and life span duration used in electric vehicles based on solid state batteries.

To form environment friendly biodegradable polymer electrolytes, biodegradable polymers are extensively studied. Biodegradable polymers can be used in solar cells, supercapacitors, batteries, and electrolytes. They can be used in lithium batteries, dye sensitized solar cells and fuel cells [1]. Biodegradable polymer electrolytes are the materials composed of salt and solvent having different properties like good ionic conductivity, cost effective, safe to use and good chemical stability [2].

Bio-based polymer electrolytes in solid state are one of the electrolytes having high ionic conductivity. They are better in sense of mechanical strength, good electrolyte-electrode contact, shape, flexibility, cost, dimensional and electrochemical stability, safety use and life span [3].

A number of bio-based polymers are used as host polymer in the formation of biodegradable electrolytes. They may be categorized as polysaccharides, polyanhydrides and polyesters (Figure 10.1) [4, 5]. Some of them are alginate, chitosan, pectin, carrageenan, agarose, cellulose, and starch based. The best-known bio-based polymer is alginate extracted naturally from brown seaweeds. It has several characteristics like good adhesive ability, low cost, semi-crystalline, low toxicity and biocompatible [6].

Some starch based and rice starch-based polymer electrolytes' aspects like conductivity, thermal and electrical properties, and rheological effects have been studied. Starch is one of the low-cost, abundantly found biodegradable polymers with unique thermal properties and steel adhesion characteristics. One of the biggest hurdles is to get high conductivity and some efforts have been made to do that. Kobayashi *et al.* reported that lithium salt and its concentration can affect the ion conductivity [7]. One method to improve the conductivity is the use of room temperature ionic liquids (RTILs), plasticizers and fillers [6]. RTILs are a mixture of molten organic salts formed by organic cations and organic or inorganic anions [8, 9]. RTILs have unique inherent chemical and physical characteristics like good chemical stability, high solubility, low level of vapor pressure, non-combustible and high ionic

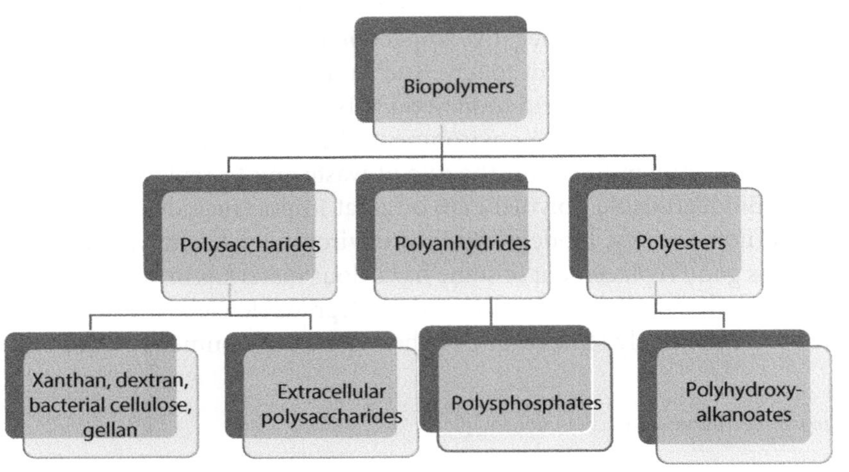

Figure 10.1 Biopolymers used for biodegradable electrolytes.

conductivity. It seems that RTILs can also be recycled and could be the substitute for many traditional solvents. They are assumed to be a green alternative [10].

There is also a cost-effective replacement of RTILs commonly known as deep-eutectic solvents (DES). This type of ionic liquids is prepared by mixing different salts (ammonium, phosphate, and analogs) with hydrogen bond donors (HBDs). These HBDs can be acids, amines, or alcohols. They form eutectic which have low melting point than their components [11, 12]. Most commonly used biopolymers are categorized in Figure 10.1.

10.2 Classification of Biodegradable Electrolytes

Biodegradable electrolytes took a long time to become what they are. The progress and advancement of polymer electrolytes took three stages to fully develop. These three stages are now the three classes of biodegradable electrolytes. They can fulfill the requirements of electrochemical industry at majority level and have applications in energy storage devices.

The classification of polymer electrolytes is as follows.

1. Solid polymer electrolytes (SPEs)
2. Gel polymer electrolytes (GPEs)
3. Composite polymer electrolytes (CPEs)

10.2.1 Solid Polymer Electrolytes

Electrolytes are the necessary part of energy storage devices for their accurate performance. SPEs have astonishing properties like process ability, flexibility, good thermal stability, and electrochemical stability. Electrochemical functioning is determined by thermal and potential strength frame of SPEs [13]. The ionic conductivity of SPEs is carried out by the lone pair of electrons present on the polymer chain and the salt's cation. These polar groups make a complex and cause ionic conductivity. Inter- and intra-molecular interactions are dependent on the distance among coordinating groups which can maximize the interaction force of polymer with cation and the polymer should have the capacity of adopting low-energy conformations [14]. It indicates that the host polymer of SPEs should be amorphous and of low molecular weight. Polyethylene oxide (PEO) is one of the mostly used polymers due to its extraordinary ability of solvating towards salt. The research being carried out in SPEs is with lithium as ionic salt and PEO as polymer host. The Lithium-ion systems are investigated largely due to their excellent uses in electrochemical devices and in primary or secondary batteries. These systems are not harmful for environment, are environmentally compatible.

Ionic liquids (ILs) are the replacement of carbonates' electrolytes which are flammable and volatile. It is the class of organic salts used as solvents or additives of bio-polymers and are molten salts at room temperature. They are good in conductivity, have stable electrochemical frame and strongly biocompatible when used in electrolytes [15]. They own broad temperature range and have low melting point when used at below room temperature. These remarkable properties make them a suitable option to be used in high-performance electrochemical devices instead of organic electrolytes.

Although SPEs have a lot of benefits but at the same time they have some drawbacks that restrict their use in commercial devices. These flaws are their interfacial resistance and strong crystallizing tendency due to polymer materials [16, 17]. But the high ionic conductivity of SPEs as compared to liquid electrolytes has made them a main focus of researchers [18]. Chatterjee *et al.* [19] reported the formation of nanocomposites based SPEs including biopolymer starch/Poly vinyl alcohol (PVA) and nano fumed silica, glycerol and glutaraldehyde was used as plasticizer and cross linker respectively. It was seen that different ratios of silica enhanced the ionic conductivity when it forms the bond with starch or PVA (C-O-Si). FTIR results confirmed the structural changes happening during the variation of nano fumed silica content. Further FTIR and dielectric measurements revealed that the most suitable concentration of nano fumed silica was 3% (by mass) which showed maximum conductivity and three-dimensional network formation.

A study has revealed the formation of SPEs using methyl cellulose as host polymer and magnesium nitrate [$Mg(NO_3)_2.6H_2O$] salt carried by solution casting method. It was seen that glass transition temperature boosted by increase in salt intensity and the ionic conductivity was increased. The electrochemical stability of this SPE also expanded and got the increased ionic conductivity and considered as suitable replacement of synthetic electrolytes for energy storage devices. This biodegradable polymer electrolytes has a positive effect on environment as reducing the chemical waste [20].

Biodegradable electrolytes using lithium iodide (LiI) and biodegradable rice starch was prepared. The mobile ions of lithium affected the ionic conductivity and transport parameters. The ionic conductivity of the polymer is affected by the addition of LiI salt which increases the number and density of mobile ions [7].

A solid polymer electrolyte was designed through PEO/$LiCF_3SO_3$ (Lithium triflate) polymer electrolytes and 1-ethyl-3-methylmidazolium trifluoromethanesulfonate (EMITf) ionic liquid. The ionic liquid was added, and the glass transition temperature is lowered and the amorphous content is increased. The mobility of free ions was also dependent upon the concentration of ionic liquid [21]. Liew *et al.* [10] designed PVA-based SPE using solution casting technique which includes the PVA/(CH_3COONH_4) ammonium acetate as host polymers and 1-butyl-3-methylimidazolium chloride (BmImCl) as ionic liquid. The ionic conductivity of this SPE increased with change in temperature due to expansion effect of polymer.

10.2.2 Gel Polymer Electrolytes

It is a gel-based, third generation of polymer electrolytes that is suspended in an inert polymer host, a complex of lithium salt and low molecular weight compound. On the basis of preparation method gel polymer electrolytes have two classes. They are prepared when SPEs matrix is infused with plasticizer, liquid or any solvent.

1. Chemical gels that are formed when polymers' functional group form a chemical bond with cross-linked compound when cross linker is added to the system.
2. Physical gels are prepared when the polymer host and liquid electrolytes make complex without making any connection between polymer and solvent [13].

GPEs have drawbacks same as liquid electrolytes. But they have a number of advantages such as fast charging, high ambient ionic conductivity, lightweight, high-power density and cost effective. GPEs offered high ionic conductivity, good compatibility of electrode/electrolyte and huge electrochemical stability. GPEs has low mechanical stability which can be improved with the linkage of component and thermoset to GPEs [22]. GPEs retain properties of both liquid and solid electrolytes, cohesive as of solid and diffusive one as of liquid electrolytes [23]. The solid state matrix of GPEs in solvent is maintained by the polymer to support ions' movement and displays high ionic conductivity of 10^{-2} to 10^{-3} S/cm at ambient temperature [24].

Different host polymers are used for the synthesis of GPEs having conductivity of 10^{-4} to 10^{-3} S/cm at ambient temperature. These host polymers include PEO (polyethylene oxide), PMMA (polymethyl methacrylate), PAN (poly acrylonitrile), and PVdF (polyvinylidene fluoride). Plasticizers used in GPEs preparation are propylene carbonate (PC), diethyl carbonate (DEC), ethylene carbonate (EC), dimethyl carbonate (DMC), and dimethyl formamide (DMF) having low evaporation rates. These plasticizers retain the ability of enhancing the amorphous phase of polymer host.

Different GPEs have been studied include several lithium salts i.e., lithium perchlorate ($LiClO_4$), lithium triflate ($LiCF_3SO_3$) and lithium tetrafluoroborate ($LiBF_4$) and propylene carbonate (PC) as plasticizer. It was seen through a comparison study that PCL (Poly(ε-caprolactone) based GPEs had greater ionic conductivity than pure PCL. GPEs of PCL retained high degree of dissociation of lithium salt followed by the polymer-salt complex association and plasticizing agent. Addition of PC as plasticizer decreased the crystallinity of the polymer and a better thermal behavior was evaluated. Its high ionic conductivity and biodegradable nature make it a better candidate for green batteries [25].

A novel GPE has been prepared using carboxymethylated nanocellulose (CMNC) that was cross linked with epichlorohydrin (ECH). They were used in lithium-ion batteries due to their excellent flexibility and mechanical stability. The usage of CMNC resulted in dense structure of GPE which make them avoid short circuits. The CMNC immersed in liquid electrolyte and formed GCMNC (gel carboxymethylated nanocellulose) GPE with improved ionic conductivity, denser structure and number of lithium ions [26].

A novel compliant GPE consisting of copolymer methyl acrylate-acrylonitrile using polyvinyl alcohol as stabilizer P(MA-co-AN)/PVA was prepared through suspension polymerization with ionic conductivity of 0.98 mS/cm [27]. A study conducted by Jayathilaka *et al.* [28] revealed that EC/PC exists in two environments making the gel electrolyte solvate by both PAN and EC/PC. The GPE system PAN/EC/PC doped in Lithium bis-(trifluoromethanesulfonyl) imide (LiTFSI) showed greater ionic conductivity. A GPE was prepared using poly(vinylidene fluoride–hexafluoropropylene) (PVDF–HFP) and PMMA as host polymers, $LiClO_4$ as salt, PC/DEC as plasticizers. The electrical conductivity of 10^{-3} S/cm was achieved by this electrolyte system. It was noticed that electrical conductivity was dependent on morphology of GPE [29]. Another GPE was prepared using poly(vinylidene fluoride-hexafluoropropylene) (PVdF-HFP) and ($LiBF_4$) with EC/PC as plasticizers with 10^{-4} S/cm of ionic conductivity and showed wide electrochemical window, low crystallinity [30]. Prabakaran *et al.* [24] prepared a polymer electrolytes system consisting of PEO-PVDF-HFP/$LiClO_4$ (polyethylene oxide/poly vinylidene flouride-hexa flouro propylene with lithium perchlorate) plasticized with EC using solution casting method.

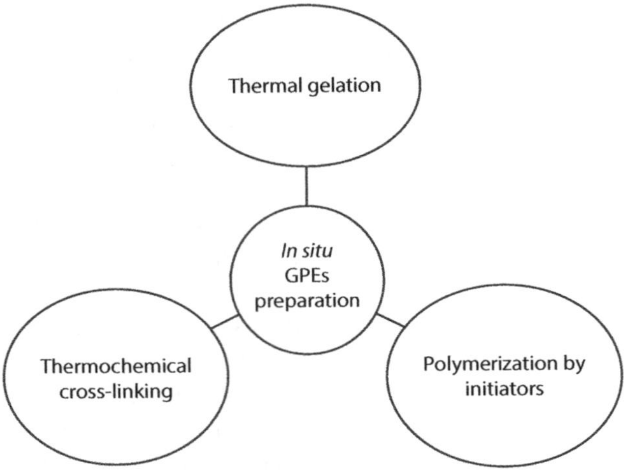

Figure 10.2 *In situ* preparation of GPEs.

A number of methods are used to prepare *in situ* gel polymer electrolytes (Figure 10.2). These methods are used to make GPEs' flexible, thermal resistant and inhibit the growth of dendrites. This *in situ* preparation is regulated and optimized by optimizing cross linking and polymerization to make GPEs function properly [31].

10.2.3 Composite Polymer Electrolytes

The composite polymer electrolytes are made up of three components, namely, salt, polymer matrix and an inorganic particle/filler. It remains the focus of study due to its quality of being divider and an electrolyte. It has been noticed that SPEs' mechanical and electrochemical stability can be enhanced by the addition of inorganic particle which are the home properties of CPEs and an advantage of CPEs to SPEs. The most widely used inert particles were silicon dioxide (SiO_2) [32], aluminum oxide (Al_2O_3) [33], titanium dioxide (TiO_2) [34], magnesium oxide (MgO) [35], and zinc oxide (ZnO) [36]. CPEs can enhance the performance of SPEs by improving properties like flexibility, process-ability, adhesion, ionic conductivity and mechanical properties. Its synergistic effects can improve the several parameters like safety, longevity, and energy density of the system.

The components of CPEs widely affect their performance; these could be the interaction of host polymer with inorganic particle or by the characteristics of inorganic filler. They show Lewis acid-base interactions and lower the crystallinity of polymer and make a Li^+ pathway for polymer-particle window. The conductivity of PEO has been improved by the incorporation of inorganic particles at room temperature by affecting the recrystallizing ability of PEO matrices. Matrices having crystalizing ability other than PEO can also get advantage from these particles [37]. Eriksson *et al.* [38] studied the poly(ε-caprolactone) (PCL)-based CPEs' consisting of LiTFSI and inactive particles TiO_2, Al_2O_3, and h-BN (hexagonal boron nitride). The crystallinity of the host polymer can be tackled by the addition of salt (lithium) and further can be reduced by the usage of inorganic particle. Thus, lower degree of crystallinity of PCL matrix increases the ionic conductivity at ambient temperature by one order of magnitude of CPEs. However, the addition of particle didn't affect the mechanical properties of CPE system and results in improved amorphous structure and free-standing film of CPEs. Croce *et al.* [39] studied the effect of nanoparticles coating on PEO electrolytes' ionic conductivity. Table 10.1 describes different polymer electrolytes with each component.

Ion conducting green polymer electrolytes was prepared to check the plasticizing effect on CPEs, when it is plasticized with PC. Lithium triflate ($LiCF_3SO_3$) was used as salt and the interaction between polymer and salt evaluated. CPEs' plasticization improved the

Table 10.1 A review of different polymer electrolytes including components.

Polymer electrolyte	Polymer host	Salt	Plasticizer	Filler	ILs	References
SPEs	PVA	CH_3COONH_4	-	-	BmImCl	[10]
	Starch-PVA nanocomposite electrolyte	LiBr	-	-	SiO_2	[19]
	PEO	$LiCF_3SO_3$	-	-	EMITf	[21]
GPEs	PEO-PVDF-HFP	$LiClO_4$	Ethylene carbonate	-	-	[24]
	P(MA-co-AN)-PVA	$LiPF_6$	Ethylene carbonate/ dimethyl carbonate/ diethyl carbonate	-	-	[27]
	PAN	LiTFSI	Ethylene carbonate/ propylene carbonate	-	-	[28]
	PVDF-HFP-PMMA	$LiClO_4$	Propylene carbonate/ diethyl carbonate	-	-	[29]
	PVDF-HFP	$LiBF_4$	Ethylene carbonate/ propylene carbonate	-	-	[30]
CPEs	PCL	LiTFSI	-	TiO_2, Al_2O_3	-	[38]
	PEO	Li_2SO_4	-	SiO_2	-	[42]
	PVC-PAN	$LiClO_4$	-	TiO_2	-	[43]

electrical conductivity, physical properties, and amorphous content. Spectroscopy analysis revealed that interaction of salt, polymer and plasticizer can enhance the ionic mobility by changing the matrix [40].

Biodegradable polymer layered silicate nanocomposite electrolytes (PLSNEs) were prepared using nano clay, dopant salt (zinc triflate) and host polymer poly(caprolactone). Octadecyl amine modified montmorillonite (ODAMMT) nano clay formed the complex with polymer and showed an increased ionic conductivity and decrease in crystallinity level by an increase in nano clay concentration. With different analysis technology, thermal and electrochemical stability was evaluated and considered them good enough to use these electrolytes in commercial systems [41]. An effort has been made by Yu *et al.* [42] to design CPE consisting of PEO as polymer and Li_2SO_4 as salt with SiO_2 as filler which showed excellent properties of flexibility and feasibility. The doping of salt and filler increased the lithium-ion conductivity. Composite polymer electrolytes film formation was studied in PVC-PAN polymers complexed with $LiClO_4$ salt with varying concentration of TiO_2 through solution casting technique. It was observed that ionic conductivity of the electrolyte film was improved by the addition of TiO_2 [43].

10.3 Preparation of Biodegradable Electrolytes

Biodegradable polymer electrolytes are simply prepared by the solution casting method. Its common schematic flowchart is shown in Figure 10.3. This technique involves the continuous stirring of salt and polymer in the solvent, which is then allowed to fully dissolute and make the viscous solution after 24h stirring. Then this solution becomes transparent and pored off in polypropylene dishes to evaporate the solvent and air dry. These polypropylene dishes can also be put in dry oven to get free standing films which are fully dried. This slow evaporation of solvent results in exceptional film which is flexible, transparent, and uniform. However, the plasticizers, fillers and ionic liquids added to the polymer to enhance conductivity are also formed by this method which are added to the starting solution and salt [44].

10.4 Some Defined Ways to Increase the Ionic Conductivity

The low level of ionic conductivity of biodegradable polymer electrolytes is a drawback and making them unsuitable to replace synthetic plastic was not cherishing. So, it was tried to enhance the ionic conduction of biodegradable polymer electrolytes. Some effective ways to achieve the good ionic conductivity of polymer electrolytes are described below.

10.4.1 Polymer Blending

Polymer blending includes a blend of two or more polymers which play their roles individually. One of the polymers works to enhance the ion transportation and the second one works to maintain the good mechanical stability. It is cost-effective as compared to polymer modification because the method to modify the polymer is costly as it needs the polymerization process [45, 46].

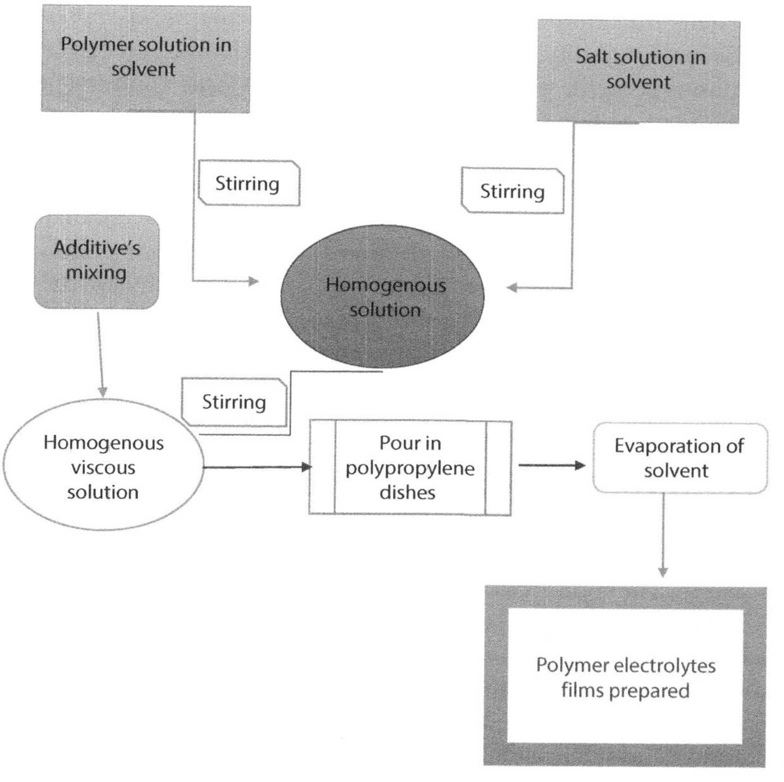

Figure 10.3 A schematic diagram for the preparation of different polymer electrolytes.

10.4.2 Incorporation of Additives

Different types of additives like plasticizers, nano fillers, and ionic liquids are used. Plasticizer reduces the crystallinity and increases the ionic conductivity, considers one of the best ionic conductivity enhancers. Some of the plasticizers are PC, EC, DEC and PEG (polyethylene glycol). The plasticizers have ability to decrease the transition temperature which leads to improved ionic conductivity [47, 48]. Ionic liquids are molten salts and works to enhance the electrochemical characteristics and permeate the low interfacial stability. They are also known as RTILs [49, 50]. Nano-fillers are added in small concentration and their affect is wide. They can enhance the mechanical and chemical window of polymer electrolyte as well as the ionic conductivity [51, 52].

10.5 Factors Affecting Ion Conduction of Biodegradable Polymer Electrolytes

Free charge carriers and their mobility highly affect the ionic conductivity of polymer electrolytes. The mobility of free charge carriers is represented by

$$\sigma = n \times q \times \mu$$

Here 'n' represents the number of charge carriers/unit volume; 'q' represents the charge of ions and 'μ' for ionic mobility. High mobility of ions and low degree of crystallinity gave increased amorphous content, which is essential for good ionic conductivity. Other factor may be the glass transition temperature (T_g) which is the transformation stage of temperature to rubbery state from glassy state. Ionic mobility is affected by T_g when it is in glass state, ionic mobility is restricted. Dielectric constant is also one of the factors affecting ion conduction. It regulates the ion movement and restricts the ion accumulation and maintains the ion separation properly [44]. So, the given factors should be highly maintained to get biodegradable polymer electrolytes of improved ionic conductivity. The mobile ions in the polymer electrolyte is the detrimental factor of ionic conductivity and this is boosted up with the addition of salt to host polymers to support the mobile ions movement and their number [4].

10.6 Properties of Ideal Biodegradable Electrolyte System

The ideal biodegradable electrolytes system should have high ionic conductivity with high mechanical stability. The ionic conductivity at room temperature plays a key role in the electrochemical performance of electrolyte system. It has less electronic conductivity which leads to decrease in self-discharging property of electrolyte system. Energy density and high transference number of cations in the electrolyte system enhances the performance of the battery. Thermal and mechanical stability of the electrolyte system make it non-flammable in nature, safe to use and non-brittle with good stress-strain property respectively [53, 54]. More precisely the properties common to all types of biodegradable electrolytes have been shown in Figure 10.4.

10.7 Applications of Biodegradable Electrolytes

Biodegradable electrolytes have been used due to their renewability, environmental stability, and safety. They can meet the energy requirements and hence are renewable made from biomaterials, not consuming fossil fuels and are cost effective. SPEs and GPEs are the biodegradable polymer electrolytes used widely to replace the conventional liquid electrolytes. Lithium salt is being used with biomaterials for efficient formation of polymer electrolytes

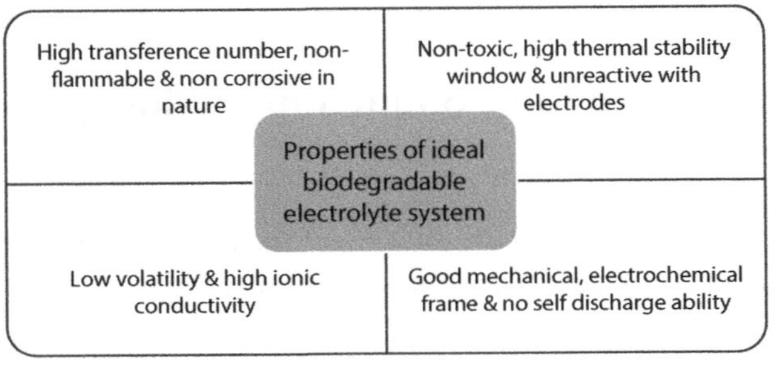

Figure 10.4 Common properties of an ideal electrolyte system.

due to their good solubility and mechanical properties. There is a dire need of biodegradable electrolytes due to the increasing demand of electrical consumables. So, the need is to meet the energy demand in a focused and mannered way by making it easy, cheaper, long lasting, and high-performance energy resources. Biodegradable electrolytes (solid, gel or composite electrolytes) had demolished the safety issues concerned to conventional liquid electrolytes. They are considered as advancement in the era of energy storage devices [55].

10.7.1 Biodegradable Electrolytes in Fuel Cells

Fuel cells have been remarkable replacement of synthetic power generation devices formed by fossil fuels. These fuel cells are sustainable and environment friendly. They offer high level electrochemical performance and can meet the energy demands efficiently being cost-effective. Portable power, stationary power plants and transportation are some of the departments using fuel cells as energy storage devices [56]. Fuel cells are the devices which work on the principle of electrochemical reaction by converting chemical energy into electrical energy; proton exchange membrane fuel cells (PEMFCs) are one of them. PEMFCs have two parts, one is electrolyte which is proton-conducting membrane, and the second is electrode catalyst layers (cathode or anode). Both of these electrodes make the electrolyte sandwiched between them [57]. The electricity in these fuel cells is generated through the electrochemical bonding of oxygen (air) with hydrogenated fuel. The fuel cells got recharged when the supply of fuel is cut off. The both electrodes are feed with fuel and oxygenated air and are disbursed in that form of ions and electrons, when these electrons pass over the external circuit it produces power and ions move towards the opposite electrode through electrolyte [58]. This type of storage offers energy of 10 magnitudes high/unit weight as compared to lead-acid batteries.

Gadim *et al.* [59] has reported the manufacturing of membrane electrode assembly (MEA) consisting of Pt/C electrodes and Nafion-BC (bacterial cellulose) electrolytes on the carbon gas diffusion layer. The study revealed that the open circuit voltage was 0.9 V, and the highest power density was 16 mW/cm^2 at current density of 40 mA/cm^2. The membrane of this fuel cell assembly found to be degraded after testing. Later with some modifications, the BC-PSSA (poly(4-styrene sulfonic acid) MEA showed no significant degradation of membrane after testing and noted that the voltage of open-circuit and current density had also improved.

Buraidah *et al.* [60] prepared the fuel cells though solution casting method consisting of chitosan-PVA indulge in ammonium iodide salt. Electrical impended spectroscopy (EIS) was used to check the impedance and their conductivity.

Other studies were carried out for the preparation of Nafion-BC blends and PSS-BC membrane. The power density and electrical conductivity was evaluated. The electrochemical performance of PSS-BS membrane was checked by single-chamber microbial fuel cells (MFCs) [61, 62].

10.7.2 Biodegradable Electrolytes and Batteries

Batteries are considered the most suitable and reliable energy storage electrical systems due to their performance and advantages. The lithium-ion batteries were introduced in 1991 and advanced the era of portable energy devices. The secondary batteries have safety issues and environmentally hazardous that needs to be sought out. It urged the researchers to revolutionize the electrical devices with some better performing devices being environment friendly and sustainable. The all typical batteries include cathode, anode and a separator which is used to prevent the physical contact or damage between electrolyte and electrodes [63]. Electrolyte is an important part of the battery that maintains the battery's efficiency in terms of charging and discharging, performance/functioning cycle and current density. Ions travel to anode from cathode during charging via electrolytes and move back to cathode from anode when it is discharged [64]. Several studies were conducted to examine the biodegradable electrolytes' applications in metal-ion batteries [65, 66].

Zhange *et al.* [67] studied the Zn-air batteries using electrolytes of BC hydrogel doped in KI and KOH, and anode of Zn foil and the cathode was the carbon cloth loaded with a catalyst. This battery offered a great conductivity, flexibility, and power density. Further, Zhang *et al.* [67] prepared the more accurate and higher performing Zn-ion batteries using xanthan-based electrolytes and electrodes. Electrolyte was socked in $ZnSO_4$/$MnSO_4$, and the Zn anode with MnO_2/CNT (manganese dioxide/carbon nanotube) cathode. Xanthan-based batteries showed a reduction in self-corrosion and zinc dendrites [68]. Zinc dendrites cause the internal short circuits which is their drawback. Xu *et al.* [69] prepared the Li-ion battery using Li anode and lithium iron phosphate ($LIPOFe_4$) cathode with BC-polymer cross linked electrolyte. Cross linking of the polymer increased the mechanical stability and decreased the formation of zinc dendrite in gel polymer electrolytes [70].

10.7.3 Supercapacitors in Terms of Biodegradable Electrolytes

Supercapacitors are also one of the energy storage devices, possessing high power density, highest capacitance, and energy charge density than other energy storage devices. These capacitors are categorized into three classes; redox supercapacitors, hybrid capacitors and electrical double-layer capacitors (EDLC) based on their charge storage and electrode material system. The use of bio-polymers in the formation of supercapacitors help to make better capacitors in terms of performance [71]. A study conducted by Arof *et al.* [72] has revealed the potential of carbonaceous material to be used as electrode in EDLCs. They are alternative to secondary batteries and conventional capacitors. EDLCs have a vast range of applications due to their remarkable properties like easy to manage, high specific capacity, non-hazardous and long-life cycle. The use of liquid electrolytes in supercapacitors is not good due to some drawbacks of leakage, corrosiveness, and self-discharging. But the use of SPEs and GPEs has improved the performance of capacitors and solved the issues of discharging and leakage and make their usage safe and environment friendly. The formation of these capacitors is followed by the insertion of biodegradable electrolytes between two carbon-based electrodes which assemble current. The cations in electrolyte will move towards the cathode (negatively charged) and anions towards the positively charged anode. The ions will be collected on the electrodes until the charging completes [71]. The energy is stored by the accumulation of ions at the interface between electrolyte and electrodes making the process electrostatic and non-faradic [73]. Kotatha *et al.* [74] also developed the EDLCs using BC gel doped in chitosan and alginate electrolytes and carbon-based electrodes absorbed in $EMImBF_4$.

10.7.4 Biodegradable Electrolytes in Dye Sensitized Solar Cells

Dye sensitized solar cells (DSSCs) are one of the less investigated energy storage devices. They convert sunlight into electrical energy, manufactured easily on the principle of photovoltaic technology and are versatile, cost-effective and flexible [75]. Typical DSSCs are composed of a platinum counter electrode, a photoelectrode of nanocrystalline TiO_2, an electrolyte's solution sandwiching the electrodes and a mesoporous layer sensitized with a dye. Their remarkable characteristics due to nanocrystalline film formation of TiO_2 has attracted the interest of researchers and commercial market [18]. The electrolyte in this system serves as charge transportation system for dye regeneration by redox couple [76]. A study showed that the biodegradable electrolytes based on biopolymers are efficient alternatives to develop DSSCs for high conductivity and efficiency making the environment clean [77]. Galliano *et al.* [78] developed the DSSCs based on xanthan gum gel that showed the short circuit current density of 6.58 mA/cm^2 and open circuit voltage of 0.62 V.

10.8 Conclusion

Development of biodegradable electrolyte is the advancement in the field of electronics or energy storage devices with a number of key advantage and remarkable properties. They are safe to use, environment friendly, being non-hazardous, non-toxic, and biodegradable. They have no issue of leakage as liquid electrolytes. The biodegradable polymer electrolytes have high ionic conductivity and better performance due to good thermal and mechanical stability and electrochemical properties. Different methods have been investigated to make the biodegradable electrolytes more efficient and cost-effective. A number of bacterial polymers are used in the synthesis of polymer electrolytes to make them biocompatible and biodegradable. It has revolutionized the line of research in the formation of energy storage devices through biodegradable electrolytes. However, there are some aspects (manufacturing cost, high ionic conductivity, long-lasting performance) to be addressed properly to commercialize these biodegradable electrolytes at a wide range.

References

1. Sudhakar, Y.N. and Selvakumar, M., Lithium perchlorate doped plasticized chitosan and starch blend as biodegradable polymer electrolyte for supercapacitors. *Electrochim. Acta*, 78, 398–405, 2012.
2. Tang, W., Ogo, Y., Minamimoto, N., Takeoka, M., Preparation of a lithium ion conductor through polyethylene glycol and $LiSO_3CF_3$ sorption in zeolite. *Chem. Lett.*, 35, 674–675, 2006.
3. Putri, R.M., Sundari, C.D.D., Floweri, O., Mayangsari, T.R., Ivansyah, A.L., Santosa, S.P., Arcana, I.M., Iskandar, F., PEO/PVA/LiOH solid polymer electrolyte prepared via ultrasound-assisted solution cast method. *J. Non-Cryst. Solids*, 556, 120549, 2021.
4. Torres, F.G., De-la-Torre, G.E., Gonzales, K.N., Troncoso, O.P., Bacterial-polymer-based electrolytes: Recent progress and applications. *ACS Appl. Energy Mater.*, 3, 11500–11515, 2020.
5. Torres, F.G., Arroyo, J.J., Troncoso, O.P., Bacterial cellulose nanocomposites: An all-nano type of material. *Mater. Sci. Eng. C*, 98, 1277–1293, 2019.

6. Fuzlin, A.F., Misnon, I.I., Nagao, Y., Samsudin, A.S., Study on ionic conduction of alginate bio-based polymer electrolytes by incorporating ionic liquid. *Mater. Today Proc.*, 51, 1455–1459, 2022.
7. Khanmirzaei, M.H. and Ramesh, S., Ionic transport and FTIR properties of lithium iodide doped biodegradable rice starch based polymer electrolytes. *Int. J. Electrochem. Sci.*, 8, 9977–9991, 2013.
8. Shamsipur, M., Beigi, A.A.M., Teymouri, M., Pourmortazavi, S.M., Irandoust, M., Physical and electrochemical properties of ionic liquids 1-ethyl-3-methylimidazolium tetrafluoroborate, 1-butyl-3-methylimidazolium trifluoromethanesulfonate and 1-butyl-1-methylpyrrolidinium bis (trifluoromethylsulfonyl) imide. *J. Mol. Liq.*, 157, 43–50, 2010.
9. Ramesh, S., Liew, C.-W., Ramesh, K., Evaluation and investigation on the effect of ionic liquid onto PMMA-PVC gel polymer blend electrolytes. *J. Non-Cryst. Solids*, 357, 2132–2138, 2011.
10. Liew, C.-W., Ramesh, S., Ramesh, K., Arof, A.K., Preparation and characterization of lithium ion conducting ionic liquid-based biodegradable corn starch polymer electrolytes. *J. Solid State Electrochem.*, 16, 1869–1875, 2012.
11. Fernandes, P.M.V., Campiña, J.M., Pereira, N.M., Pereira, C.M., Silva, F., Biodegradable deep-eutectic mixtures as electrolytes for the electrochemical synthesis of conducting polymers. *J. Appl. Electrochem.*, 42, 997–1003, 2012.
12. Shahbaz, K., Mjalli, F.S., Hashim, M.A., AlNashef, I.M., Prediction of deep eutectic solvents densities at different temperatures. *Thermochim. Acta*, 515, 67–72, 2011.
13. Arya, A. and Sharma, A.L., Polymer electrolytes for lithium ion batteries: A critical study. *Ionics*, 23, 497–540, 2017.
14. Silva, M.M., de Zea Bermudez, V., Pawlicka, A., Insight on polymer electrolytes for electrochemical devices applications, in: *Polymer Electrolytes*, pp. 113–136, John Wiley & Sons, Ltd, Elsevier, Amsterdam, 2020.
15. Jia, X., Wang, C., Ranganathan, V., Napier, B., Yu, C., Chao, Y., Forsyth, M., Omenetto, F.G., Macfarlane, D.R., Wallace, G.G., A biodegradable thin-film magnesium primary battery using silk fibroin-ionic liquid polymer electrolyte. *ACS Energy Lett.*, 2, 831–836, 2017.
16. Ngai, K.S., Ramesh, S., Ramesh, K., Juan, J.C., A review of polymer electrolytes: Fundamental, approaches and applications. *Ionics*, 22, 1259–1279, 2016.
17. Sequeira, C. and Santos, D. (Eds.), *Polymer electrolytes: Fundamentals and applications*, Elsevier, Amsterdam, 2010.
18. Ghazali, N.M. and Samsudin, A.S., Progress on biopolymer as an application in electrolytes system: A review study. *Mater. Today Proc.*, 49, 3668–3678, 2022.
19. Chatterjee, B., Kulshrestha, N., Gupta, P.N., Nano composite solid polymer electrolytes based on biodegradable polymers starch and poly vinyl alcohol. *Measurement*, 82, 490–499, 2016.
20. Koliyoor, J., Ismayil, Hegde, S., Vasachar, R., Sanjeev, G., Novel solid biopolymer electrolyte based on methyl cellulose with enhanced ion transport properties. *J. Appl. Polym. Sci.*, 139, 51826, 2022.
21. Karmakar, A. and Ghosh, A., Structure and ionic conductivity of ionic liquid embedded PEO-$LiCF_3SO_3$ polymer electrolyte. *AIP Adv.*, 4, 087112, 2014.
22. Pandey, G.P., Hashmi, S.A., Agrawal, R.C., Experimental investigations on a proton conducting nanocomposite polymer electrolyte. *J. Phys. Appl. Phys.*, 41, 055409, 2008.
23. Capiglia, C., Yang, J., Imanishi, N., Hirano, A., Takeda, Y., Yamamoto, O., Composite polymer electrolyte: The role of filler grain size. *Solid State Ionics*, 154–155, 7–14, 2002.
24. Prabakaran, P., Manimuthu, R.P., Gurusamy, S., Sebasthiyan, E., Plasticized polymer electrolyte membranes based on PEO/PVdF-HFP for use as an effective electrolyte in lithium-ion batteries. *Chin. J. Polym. Sci.*, 35, 407–421, 2017.

25. Fonseca, C.P., Cavalcante, F., Amaral, F.A., Souza, C.Z., Neves, S., Thermal and conduction properties of a PCL-biodegradable gel polymer electrolyte with $LiClO_4$, LiF_3CSO_3, and $LiBF_4$ salts. *Int. J. Electrochem. Sci.*, 2, 52, 2007.
26. Wang, J., Wang, C., Wang, W., Li, W., Lou, J., Carboxymethylated nanocellulose-based gel polymer electrolyte with a high lithium ion transfer number for flexible lithium-ion batteries application. *Chem. Eng. J.*, 428, 132604, 2022.
27. Ma, X., Huang, X., Gao, J., Zhang, S., Deng, Z., Suo, J., Compliant gel polymer electrolyte based on poly (methyl acrylate-co-acrylonitrile)/poly (vinyl alcohol) for flexible lithium-ion batteries. *Electrochim. Acta*, 115, 216–222, 2014.
28. Jayathilaka, P., Dissanayake, M., Albinsson, I., Mellander, B.-E., Dielectric relaxation, ionic conductivity and thermal studies of the gel polymer electrolyte system PAN/EC/PC/LiTFSI. *Solid State Ionics*, 156, 179–195, 2003.
29. Gohel, K. and Kanchan, D.K., Effect of PC: DEC plasticizers on structural and electrical properties of PVDF–HFP: PMMA based gel polymer electrolyte system. *J. Mater. Sci. Mater. Electron.*, 30, 12260–12268, 2019.
30. Song, S., Wang, J., Tang, J., Muchakayala, R., Ma, R., Preparation, properties, and Li-ion battery application of EC+ PC-modified PVdF-HFP gel polymer electrolyte films. *Ionics*, 23, 3365–3375, 2017.
31. Ma, C., Cui, W., Liu, X., Ding, Y., Wang, Y., *In situ* preparation of gel polymer electrolyte for lithium batteries: Progress and perspectives. *InfoMat*, 4, e12232, 2022.
32. Zhu, Y., Cao, J., Chen, H., Yu, Q., Li, B., High electrochemical stability of a 3D cross-linked network PEO@nano-SiO_2 composite polymer electrolyte for lithium metal batteries. *J. Mater. Chem. A*, 7, 6832–6839, 2019.
33. Masoud, E.M., El-Bellihi, A.-A., Bayoumy, W.A., Mousa, M.A., Organic–inorganic composite polymer electrolyte based on PEO–$LiClO_4$ and nano-Al_2O_3 filler for lithium polymer batteries: Dielectric and transport properties. *J. Alloys Compd.*, 575, 223–228, 2013.
34. Arya, A., Sadiq, M., Sharma, A.L., Effect of variation of different nanofillers on structural, electrical, dielectric, and transport properties of blend polymer nanocomposites. *Ionics*, 24, 2295–2319, 2018.
35. Masoud, E.M., El-Bellihi, A.-A., Bayoumy, W.A., Mohamed, E.A., Polymercomposite containing nano magnesium oxide filler and lithiumtriflate salt: An efficient polymer electrolyte for lithium ion batteries application. *J. Mol. Liq.*, 260, 237–244, 2018.
36. Dhatarwal, P., Choudhary, S., Sengwa, R.J., Electrochemical performance of Li^+-ion conducting solid polymer electrolytes based on PEO–PMMA blend matrix incorporated with various inorganic nanoparticles for the lithium ion batteries. *Compos. Commun.*, 10, 11–17, 2018.
37. Meng, N., Zhu, X., Lian, F., Particles in composite polymer electrolyte for solid-state lithium batteries: A review. *Particuology*, 60, 14–36, 2022.
38. Eriksson, T., Mindemark, J., Yue, M., Brandell, D., Effects of nanoparticle addition to poly (ε-caprolactone) electrolytes: Crystallinity, conductivity and ambient temperature battery cycling. *Electrochim. Acta*, 300, 489–496, 2019.
39. Croce, F., Persi, L., Scrosati, B., Serraino-Fiory, F., Plichta, E., Hendrickson, M.A., Role of the ceramic fillers in enhancing the transport properties of composite polymer electrolytes. *Electrochim. Acta*, 46, 2457–2461, 2001.
40. Aziz, S.B., Hamsan, M.H., Kadir, M.F., Karim, W.O., Abdullah, R.M., Development of polymer blend electrolyte membranes based on chitosan: Dextran with high ion transport properties for EDLC application. *Int. J. Mol. Sci.*, 20, 13, 3369, 2019.
41. Sownthari, K. and Suthanthiraraj, S.A., Preparation and properties of biodegradable polymer-layered silicate nanocomposite electrolytes for zinc based batteries. *Electrochim. Acta*, 174, 885–892, 2015.

42. Yu, J., Wang, C., Li, S., Liu, N., Zhu, J., Lu, Z., Li^+-Containing, continuous silica nanofibers for high Li^+ conductivity in composite polymer electrolyte. *Small*, 15, 1902729, 2019.
43. Rajendran, S., Shanker Babu, R., Sivakumar, P., Investigations on PVC/PAN composite polymer electrolytes. *J. Membr. Sci.*, 315, 67–73, 2008.
44. Bharti, V., Singh, P.K., Sharma, J.P., Development of polymer electrolyte membranes based on biodegradable polymer. *Mater. Today Proc.*, 34, 856–862, 2021.
45. Abdullah, O.G., Aziz, S.B., Rasheed, M.A., Incorporation of NH_4NO_3 into MC-PVA blend-based polymer to prepare proton-conducting polymer electrolyte films. *Ionics*, 24, 777–785, 2018.
46. Ahmed, H.T. and Abdullah, O.G., Structural and ionic conductivity characterization of PEO: MC-NH_4I proton-conducting polymer blend electrolytes based films. *Results Phys.*, 16, 102861, 2020.
47. Hamsan, M.H., Aziz, S.B., Shukur, M.F., Kadir, M.F.Z., Protonic cell performance employing electrolytes based on plasticized methylcellulose-potato starch-NH_4NO_3. *Ionics*, 25, 559–572, 2019.
48. Shaari, N. and Kamarudin, S.K., Recent advances in additive-enhanced polymer electrolyte membrane properties in fuel cell applications: An overview. *Int. J. Energy Res.*, 43, 2756–2794, 2019.
49. Kawazoe, T., Hashimoto, K., Kitazawa, Y., Kokubo, H., Watanabe, M., A polymer electrolyte containing solvate ionic liquid with increased mechanical strength formed by self-assembly of ABA-type ionomer triblock copolymer. *Electrochim. Acta*, 235, 287–294, 2017.
50. Kitazawa, Y., Iwata, K., Kido, R., Imaizumi, S., Tsuzuki, S., Shinoda, W., Ueno, K., Mandaai, T., Kokubo, H., Dokko, K., Polymer electrolytes containing solvate ionic liquids: A new approach to achieve high ionic conductivity, thermal stability, and a wide potential window. *Chem. Mater.*, 30, 252–261, 2018.
51. Fang, R., Xiao, W., Miao, C., Mei, P., Zhang, Y., Yan, X., Jiang, Y., Enhanced lithium storage performance of core-shell structural Si/ TiO_2/NC composite anode via facile sol-gel and *in situ* N-doped carbon coating processes. *Electrochim. Acta*, 317, 575–582, 2019.
52. Li, R., Xiao, W., Miao, C., Fang, R., Wang, Z., Zhang, M., Sphere-like SnO_2/TiO_2 composites as high-performance anodes for lithium ion batteries. *Ceram. Int.*, 45, 13530–13535, 2019.
53. Fong, K.D., Self, J., Diederichsen, K.M., Wood, B.M., McCloskey, B.D., Persson, K.A., Ion transport and the true transference number in nonaqueous polyelectrolyte solutions for lithium ion batteries. *ACS Cent. Sci.*, 5, 1250–1260, 2019.
54. Koduru, H.K., Marinov, Y.G., Scaramuzza, N., Review on microstructural and ion-conductivity properties of biodegradable starch-based solid polymer electrolyte membranes. *Starch-Stärke*, 74, 2100170, 2022.
55. Fu, X. and Zhong, W.-H., Biomaterials for high-energy lithium-based batteries: Strategies, challenges, and perspectives. *Adv. Energy Mater.*, 9, 1901774, 2019.
56. Cacciola, G., Antonucci, V., Freni, S., Technology up date and new strategies on fuel cells. *J. Power Sources*, 100, 67–79, 2001.
57. Breitwieser, M., Klingele, M., Vierrath, S., Zengerle, R., Thiele, S., Tailoring the membrane-electrode interface in PEM fuel cells: A review and perspective on novel engineering approaches. *Adv. Energy Mater.*, 8, 1701257, 2018.
58. Vaghari, H., Jafarizadeh-Malmiri, H., Berenjian, A., Anarjan, N., Recent advances in application of chitosan in fuel cells. *Sustain. Chem. Process.*, 1, 1–12, 2013.
59. Gadim, T.D., Loureiro, F.-J.A., Vilela, C., Rosero-Navarro, N., Silvestre, A.-J.D., Freir, C.-S.R., Figueiredo, F.-M.L., Protonic conductivity and fuel cell tests of nanocomposite membranes based on bacterial cellulose. *Electrochim. Acta*, 233, 52–61, 2017.

60. Buraidah, M.H. and Arof, A.K., Characterization of chitosan/PVA blended electrolyte doped with NH_4I. *J. Non-Cryst. Solids*, 357, 3261–3266, 2011.
61. Jiang, G., Zhang, J., Qiao, J.-L., Jiang, Y.-M., Zarrin, H., Chen, Z., Hong, F., Bacterial nanocellulose/Nafion composite membranes for low temperature polymer electrolyte fuel cells. *J. Power Sources*, 273, 697–706, 2015.
62. Vilela, C., Cordeiro, D.M., Boas, J.V., Barbosa, P., Nolasco, M., Vaz, P.D., Rudić, S., Ribeiro-Claro, P., Silvestre, A.-J.D., Oliveira, V.B., Poly (4-styrene sulfonic acid)/bacterial cellulose membranes: Electrochemical performance in a single-chamber microbial fuel cell. *Bioresour. Technol. Rep.*, 9, 100376, 2020.
63. Mekhilef, S., Saidur, R., Kamalisarvestani, M., Effect of dust, humidity and air velocity on efficiency of photovoltaic cells. *Renew. Sustain. Energy Rev.*, 16, 2920–2925, 2012.
64. Rayung, M., Aung, M.-M., Azhar, S.C., Abdullah, L.C., Su'ait, M.S., Ahmad, A., Jamil, S. N. A.-Md., Bio-based polymer electrolytes for electrochemical devices: Insight into the ionic conductivity performance. *Materials*, 13, 838, 2020.
65. Li, W., An unpredictable hazard in lithium-ion batteries from transition metal ions: Dissolution from cathodes, deposition on anodes and elimination strategies. *J. Electrochem. Soc.*, 167, 090514, 2020.
66. Cho, J., Jeong, S., Kim, Y., Commercial and research battery technologies for electrical energy storage applications. *Prog. Energy Combust. Sci.*, 48, 84–101, 2015.
67. Zhang, Y., Chen, Y., Li, X., Alfred, M., Li, D., Huang, F., Wei, Q., Bacterial cellulose hydrogel: A promising electrolyte for flexible zinc-air batteries. *J. Power Sources*, 482, 228963, 2021.
68. Higashi, S., Lee, S.W., Lee, J.S., Takechi, K., Cui, Y., Avoiding short circuits from zinc metal dendrites in anode by backside-plating configuration. *Nat. Commun.*, 7, 1–6, 2016.
69. Xu, D., Wang, B., Wang, Q., Gu, S., Li, W., Jin, J., Chen, C., Wen, Z., High-strength internal cross-linking bacterial cellulose-network-based gel polymer electrolyte for dendrite-suppressing and high-rate lithium batteries. *ACS Appl. Mater. Interfaces*, 10, 17809–17819, 2018.
70. Shin, D.W., Guiver, M.D., Lee, Y.M., Hydrocarbon-based polymer electrolyte membranes: Importance of morphology on ion transport and membrane stability. *Chem. Rev.*, 117, 4759–4805, 2017.
71. Winie, T. and Arof, A.K., Biopolymer electrolytes for energy devices. *Nanostruct. Polym. Membr. Appl.*, 2, 311–356, 2016.
72. Arof, A.K., Kufian, M.Z., Syukur, M.F., Aziz, M.F., Abdelrahman, A.E., Majid, S.R., Electrical double layer capacitor using poly (methyl methacrylate)–C_4BO_8Li gel polymer electrolyte and carbonaceous material from shells of mata kucing (Dimocarpus longan) fruit. *Electrochim. Acta*, 74, 39–45, 2012.
73. Chen, T. and Dai, L., Carbon nanomaterials for high-performance supercapacitors. *Mater. Today*, 16, 272–280, 2013.
74. Kotatha, D., Morishima, K., Uchida, S., Ogino, M., Ishikawa, M., Furuike, T., Tamura, H., Preparation and characterization of gel electrolyte with bacterial cellulose coated with alternating layers of chitosan and alginate for electric double-layer capacitors. *Res. Chem. Intermed.*, 44, 4971–4987, 2018.
75. Teo, L.P., Buraidah, M.H., Arof, A.K., Polyacrylonitrile-based gel polymer electrolytes for dye-sensitized solar cells: A review. *Ionics*, 26, 4215–4238, 2020.
76. Tomar, N., Agrawal, A., Dhaka, V.S., Surolia, P.K., Ruthenium complexes based dye sensitized solar cells: Fundamentals and research trends. *Sol. Energy*, 207, 59–76, 2020.

77. Singh, R., Polu, A.R., Bhattacharya, B., Rhee, H.-W., Varlikli, C., Singh, P.K., Perspectives for solid biopolymer electrolytes in dye sensitized solar cell and battery application. *Renew. Sustain. Energy Rev.*, 65, 1098–1117, 2016.
78. Galliano, S., Bella, F., Bonomo, M., Viscardi, G., Gerbaldi, C., Boschloo, G., Barolo, C., Hydrogel electrolytes based on xanthan gum: Green route towards stable dye-sensitized solar cells. *Nanomaterials*, 10, 1585, 2020.

11

Supercapattery: An Electrochemical Energy Storage Device

Fiona Joyline Mascarenhas, Shreeganesh Subraya Hegde* and Badekai Ramachandra Bhat†

Catalysis and Materials Chemistry Laboratory, Department of Chemistry, National Institute of Technology Karnataka, Surathkal, Mangalore, Karnataka, India

Abstract

The emission of greenhouse gases into the environment due to the use of non-renewable energy sources has become one of the biggest challenges to the world research community. Hence, there is a necessity for the evolution of clean energy sources since the energy demand has exponentially increased globally because of the expansion in the world's population leading to modernization. Therefore, a substantial amount of research is being implemented to efficiently convert renewable energy and capture it in energy storage devices. One of the gravitating energy storage devices is a supercapattery device made from the battery and capacitor-type electrode materials with tremendous specific energy and power. The current chapter outlines the development and various aspects of supercapattery devices, including the device's working mechanism, the materials used, the recent advancement, challenges, and future perspectives.

***Keywords*:** Supercapattery, energy storage, electrode material, clean energy, sustainable energy, battery

11.1 Introduction

Considerable advancement has been witnessed in several areas: industrialization, and urbanization, which led to economic and social modernization [1]. Consequently, consumption of non-renewable energy sources like fossil fuels has surged to a greater degree [2]. The outcome of employing unreplenishable energy sources correlated with the emission of carbon dioxide and methane, which affects people's health. It poses severe implications on the world ecology and environment, leading to global warming. In addition to this, the inflation in the human population has triggered the global energy demand [1, 3, 4]. The world's present economic model is pushing beyond the threshold of the planet's ability to supply natural resources, instigating the exhaustion of fossil fuel reserves. However, the negative impact of carbon-intensive fossil fuels can be significantly limited by

**Corresponding author*: hegdeshreeganesh@gmail.com; ORCID(S): 0000-0003-2678-7481
†*Corresponding author*: ram@nitk.edu.in; ORCID(S): 0000-0002-8169-1055

Inamuddin, Tariq Altalhi and Sayed Mohammed Adnan (eds.) Sustainable Materials for Electrochemical Capacitors, (279–290) © 2023 Scrivener Publishing LLC

the utilization of alternative energy strategies which would ultimately replace and preserve fossil sources [5]. Therefore, a significant amount of attention has been taken to develop and execute environmentally friendly renewable energy roots including solar, bioenergy, wind, hydro, and geothermal to address the energy challenges. In this context, researchers have developed novel energy storage equipment and appliances that are the key to storing and supplying the energy procured from these renewable sources at affordable costs compared to conventional non-renewable options. Electrochemical energy storage and conversion systems consist of batteries, fuel cells, and electrochemical capacitors (ECs) [6]. Though the mechanisms of all these systems are different, they can store charges quickly and efficiently [7].

11.2 Batteries and Capacitors

The electrochemical energy storage systems are broadly classified into two principal divisions: batteries and capacitors. Batteries can store energy in the form of chemical reactants [8]. They get charged when the voltage is applied across the terminals, thereby instigating a chemical reaction that is reformed into electrical energy during discharging of the battery. Capacitors store electrical energy as an electrostatic charge that can charge and discharge almost spontaneously [9]. Batteries exhibit higher energy density; they can significantly accumulate greater energy per unit mass as compared to capacitors because of their storage mechanism. Batteries must be retrieved after several thousand charge-discharge cycles since the redox reactions perturb the stability by altering the molecular or crystalline nature of the electrode materials. Besides, capacitors can be charged and discharged up to a millionfold. Even though the energy density of capacitors is lower compared to batteries, they own higher power densities because of the quick charge-discharge processes employed in capacitors compared to the faradaic processes in batteries. Additionally, their operation time scales are pretty distinct. The rechargeable batteries demand several minutes to charge and hours or days to discharge; on the other hand, capacitors can be charged and discharged in seconds [8].

The supercapacitor is another form of a capacitor, also known as an ultracapacitor, whose power density is higher than that of batteries but is still lower than that of capacitors. Nevertheless, their specific energy is several orders of magnitude higher than capacitors [4]. Supercapacitors are interesting because they have advantages in their potent thermal operating range, excellent specific power, and generous numbers of charging-discharging cycles [1]. The performances of the different energy storage devices can be compared with the help of the chart, known as the Ragone plot is used. In the Ragone plot (Figure 11.1), the energy density (in Wh/kg) is plotted against power density (in W/kg). It is clear that capacitors and supercapacitors attain the highest power density with relatively low energy density when compared with other energy storage systems, whereas batteries show relatively low power characteristics [10]. Significant progress made in both capacitor and battery systems opened various applications depending on the performance needs. Still, these technologies alone cannot be considered the storage device for replacing fossil fuels. By overseeing the situation, researchers acknowledged the need to develop various hybrid devices that will change the direction of electrochemical energy storage (EES) technologies. The one that

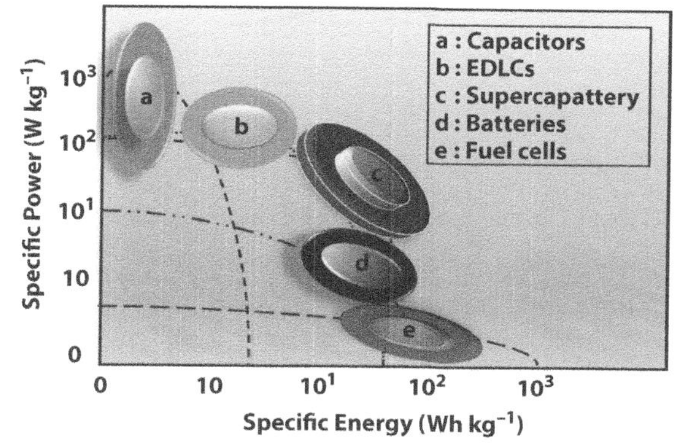

Figure 11.1 Ragone plot (reprinted with the permission from Elsevier [License Number:5294681187333]). [14].

procured ample consideration among the scientific community is the supercapattery device [11, 12].

11.3 Supercapattery Device and Electrode Materials

Supercapattery (supercapacitor + battery) has been suggested as a new frame to epitomize an extensive verity of devices that take advantage of both capacitive and battery-type materials, which meets the requirement of an ideal energy storage device, as shown in Figure 11.2. It is the amalgamation of high energy density and high-power density from the battery and capacitive materials, correspondingly. This supercapattery device includes both capacitive and faradaic charge storage mechanisms [3, 12, 13]. It consists of a battery-grade and capacitive electrode, as described in Figure 11.3. The positive terminal will be a battery-grade electrode, and the negative terminal will be a capacitive electrode. The capacitive material

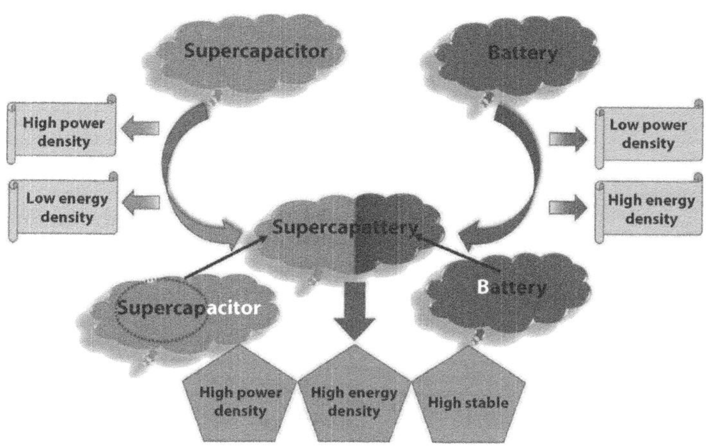

Figure 11.2 Overview of supercapattery (Reprinted with the permission from Elsevier [License Number: 5294681187333]) [14].

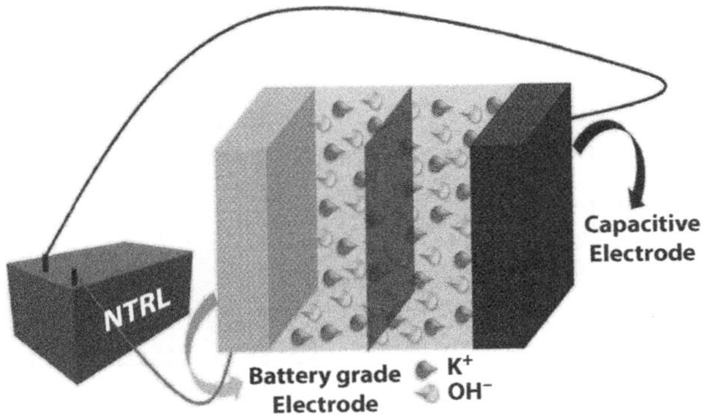

Figure 11.3 Schematic representation of a supercapattery device (Reprinted with the permission from Elsevier [License Number: 5294681187333]) [14].

represents power capability and strives to get maximum values, whereas the battery-grade material stands for the energy density of the supercapattery device [13]. Research on supercapattery persistently strives to obtain new battery and capacitive materials so that the high energy density can be coupled with high power density to construct an exceptionally good supercapattery device [14]. The battery-type materials and pseudocapacitive materials such as metal oxides, metal sulfides, metal phosphates, and conducting polymers such as PANI, and PPy are most commonly employed as positive electrode. Carbonaceous materials such as activated carbon, carbon nanotubes, carbon aerogel, and graphene are utilized in the fabrication of negative electrode [13]. Table 11.1 produces the list of electrode materials and their parameters used in the supercapattery device.

Table 11.1 Synthetic methods and performance evaluation of supercapattery electrode materials.

Material	Energy density ($Wh\ kg^{-1}$)	Power density ($W\ kg^{-1}$)	Synthetic method	Ref.
Co_3O_4	40	742	Sol-gel	[15]
Co_3O_4	38.5	8532.5	Hydrothermal	[16]
Co_3O_4	63	16,200	Wet chemical process	[17]
CuO/CoO	56.5	831.9	Hydrothermal	[18]
rGO/TiO_2	54.37	420.48	Hydrothermal	[20]
CoMnS	104	24,000	Electrodeposition	[21]
NiCoS-Se	39.6	1501	Hydrothermal	[22]
$NiS_2@NiV_2S_4$	19.4	140	Hydrothermal	[23]
$Co_{0.5}Mn_{0.5}S$	61.34	850	Sonochemical	[24]

(*Continued*)

Table 11.1 Synthetic methods and performance evaluation of supercapattery electrode materials. (*Continued*)

Material	Energy density (Wh kg^{-1})	Power density (W kg^{-1})	Synthetic method	Ref.
$Co_3(PO_4)_2$	62.6	425	Sonochemical	[25]
$Co_{0\cdot50}Cu_{0\cdot50}MnPO_4$	19	6400	Sonochemical	[26]
NiMn $(PO_4)_2$	63.8	11,892	Sonochemical	[27]
$Zn_{0.50}CoO_{0.50}Mn\ (PO_4)_2$	45.45	425	Sonochemical	[28]
SrO-PANI	24	2240	Physical blending	[29]
NiMn $(PO4)_2$-PANI	71.3	12,686	Sonochemical	[30]
$Zn_3(PO_4)_2$/PANI	25.16	3400	Hydrothermal	[31]
CeO_2	20	500	Soft template	[32]
$Co_3(PO_4)_2$@ PANI	53.2	432	Sonochemical	[34]
Strontium phosphide@PANI	28.9	1020	Sonochemical	[35]
NiMnP@MWCNT	78	698	Hydrothermal	[36]
ZnP- rGO	24.26	510	Hydrothermal	[37]
gCN/BVO	61	16.2	Hydrothermal	[39]
CNT-$CuCo_2O_4$@Ag	91	420	Condensation	[40]
Co_3O_4/Au@MWCNT	18.80	302	Hydrothermal	[41]

11.3.1 Metal-Based Materials and Their Composites

Metal oxides possess high specific capacitance, porous nature, and large surface area because of which they are used as supercapattery electrode material. Devi *et al.* synthesized nanostructured Co_3O_4 through a sol-gel technique. The highest energy density of 40Wh kg^{-1} and power density of 742 W kg^{-1} were obtained when studied for supercapattery applications where Co_3O_4 was coupled with reduced graphene oxide (rGO). The coulombic efficiency of the fabricated material was turned to be 98% [15]. Micro-bundle structure of Co_3O_4 was synthesized through hydrothermal and annealing technique by Chen *et al.* A high specific capacity of 282.3 C g^{-1} at 1 A g^{-1} was obtained when it was studied in a three-electrode setup. When this was studied with the two-electrode arrangement, an excellent specific capacity of 144.1 C g^{-1} at 1 A g^{-1}was exhibited and obtained a 126.4% of retention in capacity even succeeding 5000 charging-discharging cycles at 5 A g^{-1}. The high energy density and power density values were evaluated to be 38.5 Wh kg^{-1} and 8532.5 W kg^{-1}, respectively [16]. The three-electrode assembly setup was used by Iqbal *et al.* with the prepared Co_3O_4 electrode and procured the specific capacitance of 567 F g^{-1} at 3mV s^{-1}. The two-electrode assembly exhibits good performance by exhibiting a high energy density of 63 Wh kg^{-1} at

0.7 A g^{-1} and a power density of 16,200 W kg^{-1} at 13 A g^{-1}. Also, the capacitive retention of the system was found to be 82% even after 6000 cycles [17]. Gopi *et al.* investigated the electrochemical performances of copper oxide/cobalt oxide, synthesized through a hydrothermal technique followed by annealing. Three electrode assemblies reveals that the electrode has shown excellent specific capacitance. Later, setting up a two-electrode supercapattery device by coupling the synthesized electrode with G-ink, the energy and power densities were 56.5 Wh kg^{-1} and 831.9 W kg^{-1}, correspondingly. Even after 4000 GCD cycles, the device exhibited excellent stability retention of 98.8% [18]. Using a hydrothermal process, ZnO_2/MnO_2 nanostructure was prepared by yang *et al.* Initially, the electrochemical behavior was investigated using the three-electrode system and an aerial capacitance was found to be 271.1 $mFcm^{-1}$. A supercapattery device with two electrode assemblage was set up by pairing the prepared electrode with activated carbon, which revealed elevated energy and power density values of 610.2 mWh cm^{-1} and 6480 mW cm^{-2} correspondingly [19]. The hydrothermal method was utilized by Heng *et al.* to synthesize TiO_2 material along with rGO. Investigating the prepared electrode in a three-electrode system revealed 116.70 mAh g^{-1} of specific capacitance 0.2 A g^{-1}. Later on, two electrode assemblies was built where the rGO/TiO_2 material was coupled with activated carbon and proved that the material was apt for the supercapattery application since it showed an energy density of 54.37 Wh kg^{-1} at a power density of 420.48W kg^{-1}. Furthermore, excellent retention in the capacity of 92% was obtained even after 3000 cycles [20].

Metal sulfides have earned exquisite consideration due to their superior electrochemical performances. Since they exhibit high electrical conductivity, these materials have become superior materials for energy storage [14]. Haider *et al.* used the electrodeposition technique for the deposition of CoMnS on Ni-foam. A sample with a definite deposition cycle delivered a specific capacitance of 2291 F g^{-1}. Furthermore, the real supercapattery device was assessed where the prepared electrode was coupled with activated carbon. The two-electrode system revealed a phenomenal energy density of 104 Wh kg^{-1} and supreme power density of 24,000 W kg^{-1}. Meanwhile, after 2500 cycles, 94% capacitance retention was obtained [21]. Lin *et al.* used the hydrothermal and selenylation reaction to dope selenium with nickel-cobalt sulfide. The composition $NiCo_2S_{2.2}Se_{1.8,}$ when tested in a three-electrode cell, revealed 870 C g^{-1} at 2.5 A g^{-1} specific capacitance. A supercapattery device was formulated by considering activated carbon as a cathode to review its electrochemical performance. The energy density was found to be 39.6 Wh kg^{-1}, and the power density achieved was 1501 Wh kg^{-1} [22]. A composite of nickel sulfide was prepared with nickel vanadium sulfide ($NiS_2@NiV_2S_4$) by Manikandan *et al.* using the hydrothermal method. The highest specific capacitance of 520 C g^{-1} was delivered at 1 A g^{-1} when the prepared electrode was investigated in a three-electrode setup. A two-electrode assemblage was depicted for the real device measurements, along with activated carbon showed an energy density of 19.4 Wh kg^{-1} and a power density of 140 W kg^{-1}. After completing 10,000 cycles, 90% of its capacity was sustained [23]. Binary metal sulfide $Co_{0.5}Mn_{0.5}S$ was synthesized by Iqbal *et al.* through the sonochemical route. The electrode cell configuration of the prepared electrode exhibited superior electrochemical performance, which is 893.3 F g^{-1} specific capacitance at 1.0 A g^{-1}. To look into the device assessment, a two-electrode cell was depicted by coupling with activated carbon, which showed excellent electrochemical values. The specific energy and specific power attained were 61.34 Wh kg^{-1} of 850 W kg^{-1} at 1 A g^{-1}, respectively. When the fabricated material was made to run 2000 cycles, it retained 90.19% of its capacity [24].

Metal phosphates are an important material in supercapatteries because of the faradic reactions taking place in GCD processes [14]. $Co_3(PO^4)_2$ were developed using the sonochemical method by Iqbal *et al.* The specific capacity of 403.1 C g^{-1} was attained when tested the prepared electrode in a three-electrode configuration. In a two-electrode assembly, the device has showed an energy density of 62.6 Wh kg^{-1} and power density of 425.0 W kg^{-1} at 0.5 A g^{-1}. Also, over 5000 cycles, the capacity retention was found to be 92.9% [25]. Cobalt-manganese phosphate doped with copper was synthesized using the sonochemical method by Alzaid *et al.* The prepared sample showed a tremendous specific capacity value of 340 C g^{-1} at 0.5 A g^{-1}. Two electrode assembly $Co_{0·50}Cu_{0·50}Mn\ (PO_4)_2$/AC was fabricated to assess the supercapattery device, and it delivered high specific energy of 19 Wh kg^{-1} and specific power of 6400 W kg^{-1}. Also, the material showed 90% capacitance retention [26]. The binary composite nickel manganese phosphates were synthesized by Alam *et al.* using the sonochemical method. Primarily, a specific capacity of 678 C g^{-1} was attained at 0.4 A g^{-1} from the three-electrode arrangement. Later on, to assess the electrochemical performance, the composite was coupled with activated carbon, and extraordinary results were obtained for energy density, and power density is 63.8 Wh kg^{-1} and 11,892 W kg^{-1}, correspondingly [27]. Mixed metal phosphate nanomaterial $Zn_{0.50}CoO_{0.50}Mn\ (PO_4)_2$ was synthesized through a sonochemical approach by Iqbal *et al.* It was found from the electrochemical measurement that the material revealed a high specific capacitance of 1704.21 F g^{-1} at 1.2 A g^{-1}. For the real device assessment, a two-electrode configuration was built by coupling the nanomaterial with activated carbon delivering specific energy of 45.45 W h kg^{-1} with the specific power of 425 W kg^{-1} at 0.5 A g^{-1} [28].

11.3.2 Polymers and their Composites

Conducting polymers are studied for their unique properties, such as high capacitance, good redox capability, and large surface area. Also, they exhibit better conductivity and low equivalent series resistance [5]. Iqbal *et al.* synthesized a composite of strontium oxide/polyaniline using the physical blending technique. The composite revealed a high specific capacitance of 258 C g^{-1} when the cyclic study was performed using a three-electrode assembly at 0.8 A g^{-1}. The SrO-PANI composite was examined in a two-electrode assemblage, with the composite being the positive and activated carbon being the negative electrode, respectively. Furthermore, a supercapattery device delivered the highest power density of 2240 W kg^{-1} and energy density of 24 Wh kg^{-1} [29]. Alam *et al.* synthesized a ternary composite of $NiMn\ (PO_4)_2$-PANI using the sonochemical method. The three-electrode system was built to know the electrochemical performance and the composite with 30% wt obtained the specific capacity of 847 C g^{-1} at 0.5 A g^{-1}. Later on, with the help of a two-electrode configuration, the supercapattery device was investigated by coupling the composite with activated carbon, and outstanding performance was obtained for power density and energy density which are 12,686 W kg^{-1} and 71.3 Wh kg^{-1} [30]. A nontoxic composite of zinc phosphate/polyaniline was synthesized through the hydrothermal method by Fisal *et al.* When the tested in a three-electrode cell, the specific capacity of the best composition $Zn_3(PO_4)_2$/PANI (50/50%) was attained to be 118 C g^{-1} (CV) and 130 C g^{-1} (GCD). The supercapattery device was fabricated where the composite was coupled with activated carbon for the two-electrode device measurement. The device presented a tremendous energy density of 25.16 Wh kg^{-1} and a power density of 3400 W kg^{-1} [31]. Junais *et al.* studied the polymer soft template technique for the preparation of Cerium oxide nanoparticles. At the current density of 1 A g^{-1}, the electrode delivered a specific capacity of 400 C g^{-1}. The two-electrode assembly was constructed by pairing with the reduced graphene oxide, revealing specific energy and power of 20 W h kg^{-1} and 500 W kg^{-1}, respectively [32]. Iqbal *et al.* synthesized a composite Ag/Co_3O_4@PPy where polypyrrole was engulfed with cobalt oxide nanograin and silver nanoparticles via a hydrothermal method. The assembled electrode revealed outstanding electrochemical properties in comparison with the individual materials. The two electrode supercapattery device was set up by coupling nanocomposite with activated carbon. The satisfactory values of energy density and power density were obtained at 0.7 A g^{-1}. Also, the fabricated supercapattery device revealed massive retention in capacity, which is 153.67% over 3000 cycles [33]. Alam *et al.* synthesized a binary composite of $Co_3(PO_4)_2$ with PANI at a different weight %. A specific capacity of 70:30 wt.% of the composite was found to be 638 C g^{-1} at 0.8 A g^{-1} when tested in a cell consisting of three electrodes. A two-electrode device is assembled to examine the charge storage by coupling the composite with activated carbon. The energy density and power density of the device were found to be 53.2 Wh $kg^{-1}$and 432 W kg^{-1}, respectively, at 0.5 A g^{-1}. Moreover, the device retained 97.6% of cyclic stability following 5000 charge-discharge cycles [34]. A binary composite of strontium phosphide with polyaniline composite was examined through the sonochemical method by Iqbal *et al.* The composite with particular composition indicates the highest specific capacity of 192 C g^{-1} (CV at 10 mV s) and 196 C g^{-1} (GCD at 0.4 A g^{-1}) The real supercapattery device exhibited the maximum energy and power density to be 28.9 Wh kg^{-1} and 5100 W kg^{-1}, respectively. Furthermore, the device was made to undergo 2000 cycles to investigate its longevity [35].

11.3.3 Carbon Materials and Their Composites

Carbon materials are employed as electrode material in supercapatteries due to their high surface area, high electrical conductivity, and low cost. These materials are also useful since they are thermally, chemically, and electrochemically stable and eco-friendly [5]. A composite of NiMnP with MWCNTs was prepared through a hydrothermal process by Sharmila *et al.* A three-electrode assembly was built to analyze the composite, and it showed a high capacitance of 812 F g^{-1} at 2 A g^{-1}. Meanwhile, a two-electrode assembly is designed where the material was optimized with an activated carbon resulting in maximum capacitance of 380 C g^{-1} at 1 A g^{-1}. Furthermore, 698 W kg^{-1} of maximum power density at the energy density of 78 Wh kg^{-1} were attained [36]. The electrode material comprising zinc phosphate and rGO powder was prepared hydrothermally. The 70:30% composition of ZnP and rGO revealed excellent magnificent capacitance of 278.6 F g^{-1} at 0.6 A g^{-1}. Two electrode supercapattery device was built, and the synthesized material was employed along with reduced graphene oxide and procured specific energy of 24.26 Wh kg^{-1}, specific power of 510 W kg^{-1} at 0.6 A g^{-1}. Also, the device is made to undergo 2000 charge-discharge cycles to probe its durability, and it retained 71% capacity retention after 2000 cycles [37]. $CoSe_2$/GO was synthesized through the hydrothermal method by Wang *et al.* The prepared electrode was assessed in a three-electrode system, and the specific capacity was achieved to be 108.31 mAh g^{-1} at 1 A g^{-1}. Furthermore, capacity retention was found to be 91.2% after 5,000 cycles [38]. Murugan *et al.* synthesized Nanocomposite of Graphitic Carbon Nitride-Bismuth Vanadate (gCN/BVO) via a hydrothermal process in an aqueous electrolyte.

The electrochemical measurements show that 06 wt % of nanocomposite exhibit a high specific capacity of 2171 C g^{-1} at a current density of 2 A g^{-1}. For the real device measurement, two electrode assembly is set up and showed a high energy density of 61 Wh kg^{-1} and power densities of 16.2 kW kg^{-1}. The device, after undergoing 20000 cycles at 20 A/g, presented 98.8% of coulombic efficiency [39]. Vadiyar *et al.* studied the synthesis of the composite CNT-$CuCo_2O_4$@Ag where Ag dendrites are grown on carbon nanotubes wrapped $CuCo_2O_4$ nanoflowers. The electrochemical performance of the prepared electrode obtained an extraordinary capacity of 590 mAh g^{-1} at 0.5 A g^{-1} when tested in a three-electrode configuration. Subsequently, the supercapattery device has been investigated by setting up a two-electrode cell by coupling the composite with the activated carbon. The energy density was found to be 91 Wh kg^{-1} at 0.42 kW kg^{-1} power density. The device delivered excellent cyclic stability of 98% succeeding 20,000 cycles [40]. A composite Co_3O_4/Au@MWCNT was synthesized via a single-step hydrothermal route by Iqbal *et al.* The composite was utilized in a three-electrode assembly and disclosed that the composite delivers outstanding electrochemical performance in comparison with the metal oxide without MWCNT. To examine the supercapattery device, two electrode assembly was built by coupling Co_3O_4/Au@MWCNT with activated carbon attaining a power density of 302.00 W kg^{-1} and energy density of 18.80 W h kg^{-1}. Moreover, the device retained 91.90% capacity even after 3500 cycles [41].

11.4 Advantages and Challenges of Supercapatteries

Supercapatteries have marked a noteworthy position in this field of research because of their extraordinary characteristics. Also, an appreciable number of publications has arisen since it has been recognized among various types of electrochemical energy storage devices [42]. It is found from the reported research that supercapatteries are high-performance devices that take advantage of both batteries and capacitors [43]. Exploring new and advanced battery and capacitor electrode materials also succor the supercapatteries to prosper [42]. Hence, Various electrode materials, from metal oxides to conducting polymers, are used to provide a greater probability and excellent prospects of assembling a supercapattery device. Despite all the advantages, supercapattery technology also owns its constraints. Occasionally it fails to satisfy the energy demand due to lower specific energy [14]. Also, this technology lacks commercialization of the electrode material on a larger scale [42]. In addition to this the challenges faced by these materials, include inadequate capacitance, lesser rate capability, and lower electrical conductivity, which retards their application and can be decoded by profuse possibilities [14].

11.5 Conclusions

Supercapattery device is highly accepted and favored in this research era since they can profit from both battery type and capacitive type electrode materials. To satisfy the energy demand, future perspectives not only aim for the optimized energy and power density but also to obtain significant cyclic stability after several charge-discharge cycles as well. Further, economic fabrication methods, efficient packing techniques, and compactness of the device are crucial while designing the device [13]. But sometimes it is highly challenging and laborious to implement a device with outstanding power and energy densities. Therefore, more effort and recognition must be put into bringing out practically and commercially applicable high-energy supercapattery devices with ample power. Meanwhile, new electrode materials must be developed that provide high surface area and porosity so that higher capacitance is achieved.

References

1. Vangari, M., Pryor, T., Jiang, L., Supercapacitors: Review of materials and fabrication methods. *J. Energy Eng.*, 139, 72, 2013.
2. Li, K. and Lin, B., Impacts of urbanization and industrialization on energy consumption/CO2 emissions: Does the level of development matter? *Renewable Sustainable Energy Rev.*, 52, 1107, 2015.
3. Yu, L. and Chen, G.Z., Redox electrode materials for supercapattery. *J. Power Sources*, 326, 604, 2016.
4. González, A., Goikolea, E., Barrena, J.A., Mysyk, R., Review on supercapacitors: Technologies and materials. *Renewable Sustainable Energy Rev.*, 58, 1189, 2016.
5. Forouzandeh, P., Kumaravel, V., Pillai, S.C., Electrode materials for supercapacitors: A review of recent advances. *Catalysts*, 10, 969, 2020.
6. Krishnan, S.K., Kandasamy, S., Subbiah, K., Fabrication of microbial fuel cells with nanoelectrodes for enhanced bioenergy production, R. Praveen Kumar, B. Bharathiraja (eds.), Nanomaterials, Academic Press, 2021, ISBN 9780128224014, https://doi.org/10.1016/B978-0-12-822401-4.00003-9. (https://www.sciencedirect.com/science/article/pii/B9780128224014000039)
7. Winter, M. and Brodd, R.J., What are batteries, fuel cells, and supercapacitors? *Chem. Rev.*, 104, 4245, 2004.
8. Abruña, H.D., Kiya, Y., Henderson, J.C., Batteries and electrochemical capacitors. *Phys. Today*, 61, 43, 2008.
9. Elliman, R., Gould, C., Al-Tai, M., Review of current and future electrical energy storage devices, in: *2015 50th International Universities Power Engineering Conference (UPEC), IEEE, Stoke On Trent,* United Kingdom, p. 1, 2015.
10. Yao, F., Pham, D.T., Lee, Y.H., Carbon-based materials for lithium-ion batteries, electrochemical capacitors, and their hybrid devices. *ChemSusChem*, 8, 2284, 2015.
11. Iqbal, M.Z., Faisal, M.M., Ali, S.R., Farid, S., Afzal, A.M., Co-MOF/polyaniline-based electrode material for high performance supercapattery devices. *Electrochim. Acta*, 346, 136039, 2020.
12. Das, A., Raj, B., Mohapatra, M., Andersen, S.M., Basu, S., Performance and future directions of transition metal sulfide-based electrode materials towards supercapacitor/supercapattery. *WIREs Energy Environ.*, 11, 2022. https://wires.onlinelibrary.wiley.com/doi/epdf/10.1002/wene.414
13. Balasubramaniam, S., Mohanty, A., Balasingam, S.K., Kim, S.J., Ramadoss, A., Comprehensive insight into the mechanism, material selection and performance evaluation of supercapattery. *Nano-Micro Lett.*, 12, 85, 2020.
14. Iqbal, M.Z. and Aziz, U., Supercapattery: Merging of battery-supercapacitor electrodes for hybrid energy storage devices. *J. Energy Storage*, 46, 103823, 2022.
15. Sankar Devi, V., Athika, M., Duraisamy, E., Prasath, A., Selva Sharma, A., Elumalai, P., Facile sol-gel derived nanostructured spinel Co3O4 as electrode material for high-performance supercapattery and lithium-ion storage. *J. Energy Storage*, 25, 100815, 2019.

16. Chen, H., Du, X., Sun, J., Wu, R., Wang, Y., Xu, C., Template-free synthesis of novel Co3O4 micro-bundles assembled with flakes for high-performance hybrid supercapacitors. *Ceram. Int.*, 47, 716, 2021.
17. Iqbal, M.Z., Haider, S.S., Zakar, S., Alzaid, M., Afzal, A.M., Aftab, S., Cobalt-oxide/carbon composites for asymmetric solid-state supercapacitors. *Mater. Res. Bull.*, 131, 110974, 2020.
18. Muralee Gopi, C.V.V., Vinodh, R., Sambasivam, S., Obaidat, I.M., Naidu Kalla, R.M., Kim, H.-J., One-pot synthesis of copper oxide–cobalt oxide core–shell nanocactus-like heterostructures as binder-free electrode materials for high-rate hybrid supercapacitors. *Mater. Today Energy*, 14, 100358, 2019.
19. Yang, M., Li, K., Xiao, L., Zinc oxide/manganese oxide hybrid nanostructure for electrode and asymmetric supercapacitor with long-term cyclic life. *Mater. Sci. Eng.: B*, 269, 115173, 2021.
20. Heng, I., Low, F.W., Lai, C.W., Juan, J.C., Amin, N., Tiong, S.K., High performance supercapattery with rGO/TiO2 nanocomposites anode and activated carbon cathode. *J. Alloys Compd.*, 796, 13, 2019.
21. Haider, S.S., Iqbal, M.Z., Zakar, S., Afzal, A.M., Yaqoob, K., Aftab, S., Superior performance of electrodeposited CoMnS as novel electrode material for supercapattery devices. *J. Energy Storage*, 39, 102608, 2021.
22. Lin, J., Zhong, Z., Wang, H., Zheng, X., Wang, Y., Qi, J., Cao, J., Fei, W., Huang, Y., Feng, J., Rational constructing free-standing Se doped nickel-cobalt sulfides nanotubes as battery-type electrode for high-performance supercapattery. *J. Power Sources*, 407, 6, 2018.
23. Manikandan, R., Raj, C.J., Yu, K.H., Kim, B.C., Self-coupled nickel sulfide @ nickel vanadium sulfide nanostructure as a novel high capacity electrode material for supercapattery. *Appl. Surf. Sci.*, 497, 143778, 2019.
24. Iqbal, M.Z. and Khan, J., Optimization of cobalt-manganese binary sulfide for high performance supercapattery devices. *Electrochim. Acta*, 368, 137529, 2021.
25. Iqbal, M.Z., Khan, J., Alam, S., Ali, R., Iqbal, M.J., Afzal, A.M., Aftab, S., Enhanced electrochemical performance of battery-grade cobalt phosphate via magnetron sputtered copper interfacial layer for potential supercapattery applications. *Int. J. Energy Res.*, 45, 18658, 2021.
26. Alzaid, M., Iqbal, M.Z., Siddique, S., Hadia, N.M.A., Exploring the electrochemical performance of copper-doped cobalt–manganese phosphates for potential supercapattery applications. *RSC Adv.*, 11, 28042, 2021.
27. Alam, S. and Iqbal, M.Z., Nickel-manganese phosphate: An efficient battery-grade electrode for supercapattery devices. *Ceram. Int.*, 47, 11220, 2021.
28. Iqbal, M.Z., Khan, J., Awan, H.T.A., Alzaid, M., Afzal, A.M., Aftab, S., Cobalt–manganese-zinc ternary phosphate for high performance supercapattery devices. *Dalton Trans.*, 49, 16715, 2020.
29. Iqbal, M.Z., Alam, S., Afzal, A.M., Iqbal, M.J., Yaqoob, K., Kamran, M.A., Karim, M.R.A., Alherbi, T., Binary composites of strontium oxide/polyaniline for high performance supercapattery devices. *Solid State Ionics*, 347, 115276, 2020.
30. Alam, S., Iqbal, M.Z., Khan, J., Green synthesis of nickel-manganese/polyaniline-based ternary composites for high-performance supercapattery devices. *Int. J. Energy Res.*, 45, 11109, 2021.
31. Faisal, M.M., Ali, S.R., Iqbal, M.Z., Iqbal, M.W., Numan, A., Sanal, K.C., Effect of polyaniline on the performance of zinc phosphate as a battery-grade material for supercapattery. *J. Energy Storage*, 44, 103329, 2021.
32. Junais, P.M., Athika, M., Govindaraj, G., Elumalai, P., Supercapattery performances of nanostructured cerium oxide synthesized using polymer soft-template. *J. Energy Storage*, 28, 101241, 2020.
33. Iqbal, J., Numan, A., Ansari, M.O., Jagadish, P.R., Jafer, R., Bashir, S., Mohamad, S., Ramesh, K., Ramesh, S., Facile synthesis of ternary nanocomposite of polypyrrole incorporated with cobalt

oxide and silver nanoparticles for high performance supercapattery. *Electrochim. Acta*, 348, 136313, 2020.

34. Alam, F., Iqbal, M.Z., Alam, S., Alhassan, N., Binary composites of sonochemically synthesized cobalt phosphates/polyaniline for supercapattery devices. *J. Energy Storage*, 42, 103150, 2021.
35. Iqbal, M.Z., Faisal, M.M., Ali, S.R., Afzal, A.M., Abdul Karim, M.R., Kamran, M.A., Alharbi, T., Strontium phosphide-polyaniline composites for high performance supercapattery devices. *Ceram. Int.*, 46, 10203, 2020.
36. Sharmila, V., Packiaraj, R., Nallamuthu, N., Parthibavarman, M., Fabrication of MWCNTs wrapped nickel manganese phosphate asymmetric capacitor as a supercapattery electrode for energy storage applications. *Inorg. Chem. Commun.*, 121, 108194, 2020.
37. Iqbal, M.Z., Faisal, M.M., Ali, S.R., Afzal, A.M., Hydrothermally synthesized zinc phosphate-rGO composites for supercapattery devices. *J. Electroanal. Chem.*, 871, 114299, 2020.
38. Wang, C., Zheng, W., Wang, Z., Yin, Z.-Z., Qin, Y., Kong, Y., Synthesis of graphene oxide supported CoSe2 as high-performance supercapattery electrodes. *J. Electroanal. Chem.*, 901, 115759, 2021.
39. Murugan, C., Subramani, K., Subash, R., Sathish, M., Pandikumar, A., High-performance high-voltage symmetric supercapattery based on a graphitic carbon nitride/bismuth vanadate nanocomposite. *Energy Fuels*, 34, 16858, 2020.
40. Vadiyar, M.M., Liu, X., Ye, Z., Highly porous silver dendrites on carbon nanotube wrapped copper cobaltite nano-flowers for boosting energy density and cycle stability of asymmetric supercapattery. *J. Power Sources*, 415, 154, 2019.
41. Iqbal, J., Li, L., Numan, A., Rafique, S., Jafer, R., Mohamad, S., Khalid, M., Ramesh, K., Ramesh, S., Density functional theory simulation of cobalt oxide aggregation and facile synthesis of a cobalt oxide, gold and multiwalled carbon nanotube based ternary composite for a high performance supercapattery. *New J. Chem.*, 43, 13183, 2019.
42. Yu, L. and Chen, G.Z., Supercapatteries as high-performance electrochemical energy storage devices. *Electrochem. Energy Rev.*, 3, 271, 2020.
43. Guan, L., Chen, G.Z., Croft, A.K., Grant, D.M., Perspective—Redox ionic liquid electrolytes for supercapattery. *J. Electrochem. Soc.*, 169, 030529, 2022.

12

Ceramic Multilayers and Films for High-Performance Supercapacitors

Sonali Verma, Bhavya Padha and Sandeep Arya*

Department of Physics, University of Jammu, Jammu, India

Abstract

Supercapacitors are the electrochemical devices used to store and release energy via reversible adsorption and desorption of ions at the interface of active material and electrolyte interface. This energy storage device achieved high popularity due to its superior storage capability, better cyclic performance, high power density, and eco-friendly nature. Nanomaterials have been the subject of extensive research to develop low-cost, highly efficient, and stable electrodes. Ceramic materials exhibit the potential to be employed as supercapacitor electrode materials. On the other hand, the low specific surface area and low surface activity of ceramic materials significantly limit the performance of supercapacitors. The fundamental for knowledge of ceramics, the optimization approach, and the recent research in this field are critical for enabling superior electrical conductivity and good electron transportation properties and achieving economically practical ceramic electrodes in industry. This chapter discusses various advances in ceramic-based supercapacitor electrodes, mainly metal oxides, multi-elemental oxides, metal sulphides, and other unique ceramics (MXene), as well as multilayered ceramics, have been discussed. Additionally, the current challenges, as well as the perspective of ceramic-based materials for supercapacitors, are also covered in this chapter.

Keywords: Ceramic material, specific capacitance, supercapacitors, energy storage

12.1 Introduction

Technology advanced rapidly in several nations with the advent of the Industrial Revolution, resulting in an economic boom [1]. Mobile communications and electronic healthcare equipment are among technical advancements as prominent fields in a modern community. Unfortunately, society suffered a high cost for such developments, which included severe implications for the environment and adverse effects on employment and health due to the extensive utilization of fossil fuels (oil, coal and gas) [2]. By 2035, global fossil fuel consumption is anticipated to rise to 18.3 billion tons of oil equivalent (BTOE), around 13.7 BTOE in 2012 [3]. One of its detrimental repercussions is the yearly global carbon dioxide increase (in 2011, total emission was 1.09 billion tons which rose to nearly 33.5 times in 2018) [4].

**Corresponding author*: snp09arya@gmail.com

Inamuddin, Tariq Altalhi and Sayed Mohammed Adnan (eds.) Sustainable Materials for Electrochemical Capacitors, (291–304) © 2023 Scrivener Publishing LLC

The development of associated energy generation, storage, and conservation technologies is driven by the demand for renewable and clean energy alternatives, for instance, biofuels, tidal, geothermal, solar, and wind [5, 6]. Numerous technologies for storing electrical energy have been reported so far, including electrochemical energy storage (batteries and pseudocapacitors), electrical energy storage (electric double-layer capacitors and physical capacitors), mechanical energy storage (flywheels), thermochemical and thermal energy storage (solar hydrogen and heat storage), and chemical energy conversion and storage (fuel cells and water splitting) [7–9]. Amongst these storage devices, supercapacitor has a high power density, excellent reversibility, extended cycle stability, environmental friendliness, simple operation mode, and higher energy density [10, 11]. Supercapacitors can cover the energy-power gap among fuel cells/batteries and capacitors because of the above features [12]. Furthermore, because of an uncomplicated charge storing mechanism, supercapacitors emit a lesser amount of thermo-chemical heat [13], allowing them to be used in a broader range of applications like memory backup systems, consumer electronics, as well as in industrial power [14, 15].

Different materials have been investigated for supercapacitor electrodes, including MXene, MOFs, nitrides, carbides, sulphides, hydroxides, oxides, and carbon-based materials (conducting polymers, GO, CNTs and activated carbon) [16–26]. In general, capacitance, stability, and operating voltage are employed to evaluate supercapacitor electrodes' performance [27, 28]. Electrode materials, electrolytes, and potential windows are the primary determinants of its electrochemical performance [29, 30]. Nevertheless, most of the latest studies have focused on the design optimization and structure of the electrode materials. This leads to a strong conductivity, large pores, higher specific surface area, ion diffusion, and electron transportation [31].

As electrode materials for supercapacitors, a diverse range of materials have recently been explored, including COFs, MOFs, ceramics, polymers, and metals [32–34]. Despite their low surface activity and specific surface area, thermal shock, radiation, high temperature, and corrosion resistance render these ceramic materials appealing for supercapacitor electrode [35–38]. In this chapter, recent advances in the field of ceramics for supercapacitors have been summarized comprehensively.

12.2 Different Types of Ceramic Materials

Ceramic is a solid and hard material with high stability and significant optical, magnetic, and electrical properties, making it employable in supercapacitors (SCs), batteries, solar cells, fuel cells [36–42] and catalysis [43–48]. Ceramic materials have a variety of uses, and they are becoming essential in energy applications. Scientists are creating new materials to save energy and reduce emissions while adhering to sustainable progress. Among these, the evolution of ceramics is one.

Ceramics have high corrosion, temperature, radiation, and thermal shock resistance, making them viable in SCs. In addition, many studies illustrate novel approaches for optimizing ceramic materials' size, shape, and composition for attaining high specific surface area and high porosity. Henceforth, numerous ceramic-based materials as electrodes for SCs have been investigated, including oxides of metals, multi-elemental oxides, and other unique ceramics.

12.2.1 Metal Oxides

Metal oxides are used in energy storage and conversion applications like SCs as they are abundant in nature, inexpensive, and environmentally friendly [49–51]. Transition metal oxides have gained importance because of their minimal cost and environmental affability [52–55]. The change in electrovalency of metal ions via metal ion intercalation-deintercalation [56], surface adsorption-desorption [57], and surface reduction-oxidation processes with negative ions [58] lead to charge storage for the electrochemical behavior of metal oxides-based electrodes (such as NiO, Co_3O_4 and MnO_2) in aqueous electrolytes.

12.2.2 Multi-Elemental Oxides

Multi-elemental oxides have been used for fabricating SC electrodes. Two of them are discussed below.

12.2.2.1 Spinel Oxides

Such materials include two metals, resulting in more complex chemistry than single-component oxides [59]. However, single-component oxides have higher electric conductivity and better electrochemical performance compared to binary oxides. For instance, spinel ferrites (MFe_2O_4) were employed as SC electrodes.

12.2.2.2 Barium Titanate ($BaTiO_3$)

Polymer ceramics have been employed as polymer separators because polymers contain different alkaline earth metal-based titanites. For instance, phase transition through precipitation of solvent using a combination of polypropylene (PP), polyvinylidene fluoride (PVDF), as well as lithium chloride (LiCl) as a precursor was accompanied via heat treatment and subsequent thermal annealing to produce a porous, sponge-type polymer $BaTiO_3$ film [60]. To enhance the material's porosity, LiCl was added, resulting in high charge transfer rates. Gu *et al.* fabricated a barium titanate/polymer/aluminum (BPA) composite film through roll deposition and thermal annealing [61].

12.2.2.3 Other Unique Ceramics for Supercapacitors

Besides these, some other ceramic materials like $(Sm_{0.02}Ag_{0.94})(Nb_{0.9}Ta_{0.1})O_3$ ceramic [62], $Li_2SeP_2S_5$ glass-ceramic electrolyte [63], and MXene ceramic [64] show exceptional energy storage performance. Han *et al.* [62] employed a solid-state reaction to generate doped $AgNbO_3$ lead-free ceramic.

12.3 Multilayer Structure

Multilayer structures are created by layering two or more different layers. In case the hetero-stacked layers exhibit the cyclical characteristic, giving arise to superlattice structure.

The thickness of the periodic layers was calculated using an epitaxial growth model and an expanded perspective of the out-of-plane q–2q multilayer. The formula used for calculating thickness:

$$L = \frac{\lambda}{2(\sin\theta_{(n+1)} - \sin\theta_n)} \tag{12.1}$$

where λ, L, n, and (n+1) are the X-ray wavelength, the layer thickness, and the orders of the satellite peak, respectively. In most cases, varying the thickness and no. of cycles is one of the effective techniques to change the characteristics and performance of multilayer dielectrics. Furthermore, stacking layers with various characteristics is frequently given precedence for exploiting hybridization and coupling. More significantly, the multilayer structure always displays an apparent increase in functional characteristics than the monolayer or some unexpected physical processes absent in its parent due to the interconnection of spin, charge, orbital, and strain degrees of freedom at the interfaces of various layers. Multilayer systems containing N equals six exhibit good thermal stability. As a result, different capacitors with a multilayered configuration having diverse arrangements of parent dielectrics in both practical and theoretical fields have been researched in recent decades [65].

12.4 Supercapacitors Based on Ceramic Materials

12.4.1 Metal Oxide Ceramics

Lang *et al.* reported MnO_2 electrode with a nanoporous V_2O_3 skeleton that can give excellent electric power of about 422 W cm^{-3} [66]. The electrode's conductivity was significantly improved due to the metallization of its electron-related structure. Additionally, under room conditions, the structure demonstrated an increase in the pseudocapacitive characteristics and rate capability of MnO_2 by facilitating the transfer of charges at the interface of V_2O_3/MnO_2.

Ceramics such as Al_2O_3 and SnO_2 could also be used in supercapacitors. With alumina ceramic plates, Li *et al.* reported a simple and quick method for making planar micro-supercapacitors [67].

Initially, selective laser etching was performed on the plates' surfaces, resulting in SH and SHL patterns. Then, based on different wettability properties of the etched surface, graphene ink is introduced that quickly covers up the patterned interdigital region of the electrode. Finally, the PVA-H_2SO_4 was applied to the interdigital area as a gel electrolyte. The developed supercapacitor has a 5.5 mF cm^{-2} area capacitance and exhibits excellent capacitive retention. Alkaline earth stannates (AESs), such as strontium and calcium, play the role of sensitizer on SnO_2 ceramic [68].

The synthesized sample's crystalline surface was coated by a very thin barrier layer of AESs, which successfully elevated the SnO_2 flat-band potential and lowered the electron trapping-detrapping rates. The system's efficiency can be tweaked more to provide a range of sensitive sensors.

Transition metal oxide ceramics were also synthesized using the electrophoretic deposition technique. Ata *et al.* [69] used this approach to create a porous MnO_2 films-based nanostructure. The applied electric field and stirring during electrophoretic deposition may cause nanoparticle aggregation. The utilization of Catechol type anionic dispersants, on the other hand, can effectively disperse MnO_2 nanoparticles, making the obtained porous nanostructure a potential material for energy storage in the field of supercapacitors.

12.4.2 Multi-Elemental Oxide Ceramics

Rani *et al.* reported $Co_xZn_{0.04-x}Fe_2O_4$ (x = 0, 0.01, 0.02) ceramic by making use of a less toxic co-precipitation technique [70]. Big clusters and numerous spherical grain nanoparticles made up the cubic phase $Co_xZn_{0.04-x}Fe_2O_4$. At a scan rate of 10 mV s^{-1}, the cobalt substitution boosted the electrode's capacitance to around 377 F g^{-1}.

Further, the synthesis of $NiCo_2O_4$ as one of the potential electrode materials for supercapacitors was described by Kovalenko *et al.* [71]. This nanocrystalline material with a 25–30% open porosity was formed by combining a nitrate solution's crystallization and thermal and ultrasonic treatment. The variable valence of Ni and Co, like other spinel oxides, improved the electrode's electrical conductivity and pseudocapacity, making it a potential ceramic-based electrode for storing energy. Anandhan and colleagues used sol-gel-aided electrospinning followed by calcination to make mesoporous spinel $ZnMn_2O_4$ nanofibers (NFs) [72]. The material was thermally stable and consisted of hexagonal $ZnMnO_3$ and MnO phases in $ZnMn_2O_4$ NFs. It was reported that after calcination, the synthesized ceramic had four band gaps and a large specific surface area of 79.51 m^2 g^{-1}, indicating that ceramic nanofibers might be used as an electrode material.

$BaTiO_3$ nanowires with a high aspect ratio in a polymer matrix have also been synthesized and investigated for energy storage applications [73]. In place of nanowire morphology, the shape of $BaTiO_3$ during the transition would be cubic, snowflakes or starfish-like, depending on the hydrothermal reaction parameters. In addition, amine groups were introduced to the $BaTiO_3$ nanowires' surfaces to improve the filler-matrix compatibility. As a result of the low volume fraction (17.5%) of $BaTiO_3$, the reported ceramic had 10.48 J cm^{-3} energy density at a 300 MV m^{-1} electric field.

Zhao *et al.* [74] reported pinecone-like porous nanospheres of $MnCoS_5$ (Figures 12.1 (a, b)) that offer adequate space for expansion and shrinkage of the electrode throughout the charging-discharging cycles. $MnCoS_5$ nanospheres' mixed phase, high porosity, as well as good specific surface area, allow numerous reduction-oxidation processes, have good electrical conductivity, as well as transferable electrons and ions pathways, leading to high supercapacitor performance with a specific capacitance of 992 F g^{-1} and retained 102% of capacitance after 5500 cycles at 10 A g^{-1} (Figures 12.1 (c, d)) [75].

Liu *et al.* deposited Cu_9S_8 on carbon fibers and then electrochemically covered it using polypyrrole (PPy) [76]. The open-pore configuration and good electric conductivity of the reported PPy/Cu_9S_8@C-CC composite resulted in high electrochemical performance (270 F g^{-1} at 10 mVs^{-1}) and high (80%) capacitive retention at the end of 3000 cycles.

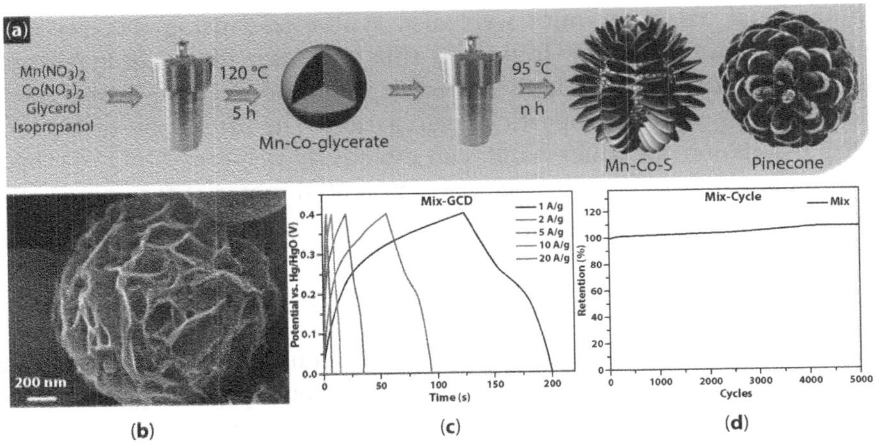

Figure 12.1 (a) Schemetic of the synthesis process, (b) SEM results, (c) GCD curves, and (d) Cycling test of Mn-Co-S nanospheres. Reproduced from Ref. [75] under creative commons license (CC-BY 3.0) (Copyright 2021, Elsevier).

12.4.3 Other Special Ceramics

MXene is a 2D transition metal nitride/carbide ceramic material made via selective etching of a layer of the MAX phase. $M_{n+1}XT_x$ is the general molecular formula for these ceramics [77]. In this formula, M represents transition metals, X denotes nitrogen/carbon, and T represents active terminal groups. MXene is frequently utilized as potential active material in supercapacitors [78] for storing energy due to the excellent conductivity as well as layered configuration.

Gogotsi *et al.* demonstrated that 2D MXene might operate at speeds faster than electrochemical double-layer capacitors (EDLCs), delivering superior values of capacitance in comparison to carbonaceous materials, polymers, and oxides of transition metals [79]. To boost ion accessibility to redox-active regions, this team used two distinct designs on the MXene architecture. At 10 V s^{-1}, the macroporous $Ti_3C_2T_x$ sheet offers a specific capacitance of about 210 F g^{-1}, outperforming the finest carbon-based electrodes for supercapacitors. MXene hydrogel, on the other hand, might have a capacitance of 1500 F cm^{-3} with respect to the volume.

Tian *et al.* [80] combined $Ti_3C_2T_x$ Mxene with 1D cellulose nanofibrils (CNFs) to create a highly efficient composite. The nanofibril structures between neighboring layers were visible in the freestanding nanopapers with 10% CNF. $Ti_3C_2T_x$/10% CNF nanopaper shows a good specific capacitance value (324 F g^{-1} at 2 mV s^{-1}). Additionally, RuO_2 particles were combined with $Ti_3C_2T_x$ MXene to create an asymmetrical device for energy storage [81].

MXene exhibited hydrophilicity, metallic conductivity, and surface redox reactivity, making it a suitable pseudocapacitor electrode material. The titanium carbide MXene ($Ti_3C_2T_x$) was initially produced using an intensive layer delamination process, followed by a dropped-cast on conductive carbon fabric as a negative electrode. RuO_2 was formed via hydrothermal process on the carbon fabric (CF) to act as a positive electrode. The asymmetric device demonstrated capacitance of 93 F g^{-1} at 50 mV s^{-1} within a potential range of 1.5 V, whereas the composite electrode with evenly dispersed RuO_2 and $Ti_3C_2T_x$ over the CF

Table 12.1 Synthesis methods and performance evaluation of multilayered ceramic material-based supercapacitors.

Ceramic materials	Synthesis methods	Specific capacitances	Ref.
PPy/$Al_5Y_3O_{12}$	Electropolymerization	254 F g^{-1}	[82]
NiO/YSZ	Mixing and calcination	272 F g^{-1}	[83]
$CaCu_3Ti_4O_{12}$	Molten salt method	1185 mF cm^{-2}	[84]
PANI/$CaCu_3Ti_4O_{12}$	Chemical oxidative polymerization	610 C g^{-1}	[85]
$Li_{3x}La_{2/3-x}TiO_3$	Solid-state reaction	0.28 mF g^{-1}	[86]
CuO/$La_{1-x}Sr_xCoO_{3-\delta}$	Solid-state reaction	545 F g^{-1}	[87]
NiO/$La_{0.7}Sr_{0.3}CoO_{3-\delta}$	Solid-state reaction	1064.1 F g^{-1}	[88]
$Ca_8H_2(PO_4)_6{\cdot}5H_2O$	Solution precipitation	6 mAh g^{-1}	[89]
$Li_3V_2(PO_4)_3$	Heat treatment	116.4 F g^{-1}	[90]
$Li_{1.3}Al_{0.3}Ti_{1.7}P_3O_{12}$	Solid-state reaction	0.13 F cm^{-1}	[91]

shows 401 F g^{-1} at 10 mV s^{-1}. Table 12.1 shows synthesis methods and performance evaluation of multilayered ceramic material-based supercapacitors.

12.5 Challenges and Prospects

Despite the wide range of SCs applications for ceramic multilayers and thin films, some limitations prohibit their usage in the commercial sector. We have highlighted a few areas that we feel should be given attention.

- **Procedures used for molding and sintering:** Ceramic materials' high sintering temperatures, as well as prolonged sintering and molding times, would result in significant energy consumption, severely limiting their usage of renewable energy. Moreover, the limitation of grains at high-temperature sintering of ceramic materials is typically complex to manage using traditional molding/sintering procedures, resulting in poor stability of the formed ceramics. More dangerous, the high density of high-temperature sintered ceramic materials frequently results in limited surface area and less porous structure that significantly limits ionic transport in an electrochemical reaction, leading to low performance. Advanced molding and sintering processes, like cold sintering and 3D printing, will enable exceedingly complex designs and low shrinkage ceramic materials to be produced quickly [81]. These approaches not only save money and energy, but they also increase ceramics' electrochemical performance. As a result, better molding and sintering methods are desperately required.

- **Poor flexibility and portability:** Micro and flexible electronics' lightweight and convenience open up new opportunities for folding displays, intelligent e-skins, and implantable medical systems. Flexible supercapacitors undergo complete charging in seconds, resulting in ultra-high areal power density (10 mW cm^{-2}) as well as excellent cyclic stability (10,000 cycles) [92]. On the other hand, ceramics are more complex to create due to their great hardness and poor tenacity, making them ideal for ultra-thin and flexible systems. There are several ways to increase the hardness of ceramics now available, including introducing secondary phases and creating laminate or net-like structures to increase the robustness of ceramics.
- **Poor efficiency of ceramic SC electrodes:** Ceramics' electrochemical characteristics are still improving. The key factors impacting performance are electrolyte, electrode, and device assembly. Various tactics have been efficiently utilized to increase ceramic materials' electrochemical properties, including nanostructure designing, interface alteration, heteroatom doping, and defect manufacture. Self-supporting, binder-free, large area ceramic-based materials will become popular in the future since they make the device much more efficient. Simultaneously, stimulating the ceramic electrode's surface is a highly successful method. This technique can eliminate the non-conducting and non-electrochemically active passivation layer, allowing the electrolyte ions to reach the electrode more efficiently [93]. The voltage range for the aqueous electrolytes is restricted because of the production of oxygen and hydrogen in the electrolyte.

On the other hand, the ionic liquid electrolytes allow for a larger output voltage. Gel electrolytes are also preferred because of their durability. Anhydrous gel electrolytes have a wide potential window (3×10^5 V) and are relatively safe [94, 95]. The electrochemical performance of an SC is affected by the current collector, the distance among the electrodes, and the SC fabrication.

12.6 Conclusion

The latest accomplishments of different ceramic-based materials for supercapacitors, such as metal oxides ceramics, multi-elemental oxides ceramics, and others, have been discussed in this chapter. Specific examples of different ceramic materials used as SC electrodes are included, focusing on the relationship between electrochemical performance and composition. For attaining high specific surface area, ample pores, quick electron transportation with effectual ionic diffusions, developing potential ways to modify ceramics' size, shape, internal configuration, and compositions, novel ceramic materials and knowledge of their energy-storing processes are being developed. Currently, ceramic electrodes are mainly focused on a few materials, and new ceramic electrode development, fundamental electrochemistry understanding of new materials, is inadequate. Intercalative and extrinsic capacitance are predicted to be provided by the new ceramic electrodes. A ceramic-based battery and capacitor electrodes hybrid system can also be investigated. Furthermore, the energy-storing strategy of novel ceramic materials, like the MXenes ceramic electrode, has not been thoroughly investigated.

References

1. Raza, W., Ali, F., Raza, N., Luo, Y., Kim, K.H., Yang, J., Kumar, S., Mehmood, A., Kwon, E.E., Recent advancements in supercapacitor technology. *Nano Energy*, 52, 441, 2018.
2. Chu, S., Cui, Y., Liu, N., The path towards sustainable energy. *Nat. Mater.*, 16, 16, 2017.
3. Singh, S., Jain, S., Venkateswaran, P.S., Tiwari, A.K., Nouni, M.R., Pandey, J.K., Goel, S., Hydrogen: A sustainable fuel for future of the transport sector. *Renewable Sustainable Energy Rev.*, 51, 623, 2015.
4. Azcarate, C., Mallor, F., Mateo, P., Tactical and operational management of wind energy systems with storage using a probabilistic forecast of the energy resource. *Renewable Energy*, 102, 445, 2017.
5. McKone, J.R., DiSalvo, F.J., Abruña, H.D., Solar energy conversion, storage, and release using an integrated solar-driven redox flow battery. *J. Mater. Chem. A*, 5, 5362, 2017.
6. Chen, H., Cong, T.N., Yang, W., Tan, C., Li, Y., Ding, Y., Progress in electrical energy storage system: A critical review. *Prog. Nat. Sci.*, 19, 291, 22009.
7. Noori, A., El-Kady, M.F., Rahmanifar, M.S., Kaner, R.B., Mousavi, M.F., Towards establishing standard performance metrics for batteries, supercapacitors and beyond. *Chem. Soc. Rev.*, 48, 1272, 2019.
8. Zhang, L.L. and Zhao, X.S., Carbon-based materials as supercapacitor electrodes. *Chem. Soc. Rev.*, 38, 2520, 2009.
9. Zeng, X., Yang, B., Li, X., Yu, R., Three-dimensional hollow CoS2 nanoframes fabricated by anion replacement and their enhanced pseudocapacitive performances. *Electrochim. Acta*, 240, 341, 2017.
10. Verma, S., Gupta, V., Khosla, A., Kumar, S., Arya, S., High performance asymmetric supercapacitor based on vertical nanowire arrays of a novel Ni@Co–Fe LDH core@shell as negative and $Ni(OH)_2$ as positive electrode. *Nanotechnology*, 31, 245401, 2020.
11. Kurzweil, P. and Chwistek, M., Electrochemical stability of organic electrolytes in supercapacitors: Spectroscopy and gas analysis of decomposition products. *J. Power Sources*, 176, 555, 2008.
12. Verma, S., Khosla, A., Arya, S., Performance of electrochemically synthesized nickel-zinc and nickel-iron (Ni–Zn//Ni–Fe) nanowires as battery type supercapacitor. *J. Electrochem. Soc.*, 167, 120527, 2020.
13. Xue, Q., Sun, J., Huang, Y., Zhu, M., Pei, Z., Li, H., Wang, Y., Li, N., Zhang, H., Zhi, C., Recent progress on flexible and wearable supercapacitors. *Small*, 13, 1701827, 2017.
14. Miller, E.E., Hua, Y., Tezel, F.H., Materials for energy storage: Review of electrode materials and methods of increasing capacitance for supercapacitors. *J. Energy Storage*, 20, 30, 2018.
15. Najib, S. and Erdem, E., Current progress achieved in novel materials for supercapacitor electrodes: Mini review. *Nanoscale Adv.*, 1, 2817, 2019.
16. Verma, S., Padha, B., Singh, A., Khajuria, S., Sharma, A., Mahajan, P., Singh, B., Arya, S., Sol-gel synthesized carbon nanoparticles as supercapacitor electrodes with ultralong cycling stability. *Fuller. Nanotub. Carbon Nanostructures*, 29, 1045, 2021.
17. Gonçalves, J.M., da Silva, M.I., Toma, H.E., Angnes, L., Martins, P.R., Araki, K., Trimetallic oxides/hydroxides as hybrid supercapacitor electrode materials: A review. *J. Mater. Chem. A*, 8, 10534, 2020.
18. Verma, S., Arya, S., Gupta, V., Khosla, A., Highly stable self-charging piezoelectric (Rochelle salt) driven supercapacitor based on Ni nanowires. *Chem. Eng. J.*, 424, 130567, 2021.
19. Zhao, Y., Shi, Z., Li, H., Wang, C.A., Designing pinecone-like and hierarchical manganese cobalt sulfides for advanced supercapacitor electrodes. *J. Mater. Chem. A*, 6, 12782, 2018.
20. Chen, T., Li, M., Song, S., Kim, P., Bae, J., Biotemplate preparation of multilayered TiC nanoflakes for high performance symmetric supercapacitor. *Nano Energy*, 71, 104549, 2020.

21. Liu, Y., Wu, Q., Liu, L., Manasa, P., Kang, L., Ran, F., Vanadium nitride for aqueous supercapacitors: A topic review. *J. Mater. Chem. A*, 8, 8218, 2020.
22. Wang, H., Li, J., Li, K., Lin, Y., Chen, J., Gao, L., Nicolosi, V., Xiao, X., Lee, J.M., Transition metal nitrides for electrochemical energy applications. *Chem. Soc. Rev.*, 5, 1354, 2021.
23. Li, Q., Dai, Z., Wu, J., Liu, W., Di, T., Jiang, R., Zheng, X., Wang, W., Ji, X., Li, P., Xu, Z., Fabrication of ordered macro-microporous single-crystalline MOF and its derivative carbon material for supercapacitor. *Adv. Energy Mater.*, 10, 1903750, 2020.
24. Hu, M., Zhang, H., Hu, T., Fan, B., Wang, X., Li, Z., Emerging 2D MXenes for supercapacitors: Status, challenges and prospects. *Chem. Soc. Rev.*, 49, 6666, 2020.
25. Allagui, A., Freeborn, T.J., Elwakil, A.S., Fouda, M.E., Maundy, B.J., Radwan, A.G., Said, Z., Abdelkareem, M.A., Review of fractional-order electrical characterization of supercapacitors. *J. Power Sources*, 400, 457, 2018.
26. Mohanty, A., Jaihindh, D., Fu, Y.P., Senanayak, S.P., Mende, L.S., Ramadoss, A., An extensive review on three dimension architectural Metal-Organic Frameworks towards supercapacitor application. *J. Power Sources*, 488, 229444, 2021.
27. González, A., Goikolea, E., Barrena, J.A., Mysyk, R., Review on supercapacitors: Technologies and materials. *Renewable Sustainable Energy Rev.*, 58, 1189, 2016.
28. Zhao, H., Liu, L., Vellacheri, R., Lei, Y., Recent advances in designing and fabricating self-supported nanoelectrodes for supercapacitors. *Adv. Sci.*, 4, 1700188, 2017.
29. Wang, F., Wu, X., Yuan, X., Liu, Z., Zhang, Y., Fu, L., Zhu, Y., Zhou, Q., Wu, Y., Huang, W., Latest advances in supercapacitors: From new electrode materials to novel device designs. *Chem. Soc. Rev.*, 46, 6816, 2017.
30. Zhang, Y.Z., Wang, Y., Cheng, T., Yao, L.Q., Li, X., Lai, W.Y., Huang, W., Printed supercapacitors: Materials, printing and applications. *Chem. Soc. Rev.*, 48, 3229, 2019.
31. Zhou, J. and Wang, B., Emerging crystalline porous materials as a multifunctional platform for electrochemical energy storage. *Chem. Soc. Rev.*, 46, 6927, 2017.
32. Choudhary, N., Li, C., Moore, J., Nagaiah, N., Zhai, L., Jung, Y., Thomas, J., Asymmetric supercapacitor electrodes and devices. *Adv. Mater.*, 29, 1605336, 2017.
33. Shen, Z.Y., Wang, Y., Tang, Y., Yu, Y., Luo, W.Q., Wang, X., Li, Y., Wang, Z., Song, F., Glass modified barium strontium titanate ceramics for energy storage capacitor at elevated temperatures. *J. Materiomics*, 5, 641, 2019.
34. Colombo, P., Conventional and novel processing methods for cellular ceramics. *Philos. Trans. Math. Phys. Eng. Sci.*, 364, 109, 2006.
35. Studart, A.R., Gonzenbach, U.T., Tervoort, E., Gauckler, L.J., Processing routes to macroporous ceramics: A review. *J. Am. Ceram. Soc.*, 89, 1771, 2006.
36. Li, D.X., Shen, Z.Y., Li, Z.P., Luo, W.Q., Song, F.S., Wang, X.C., Wang, Z.M., Li, Y.M., Optimization of polarization behavior in (1-x)BSBNT-xNN ceramics for pulsed power capacitors. *J. Mater. Chem. C*, 8, 7650, 2020.
37. Dutta, P., Sikdar, A., Majumdar, A., Borah, M., Padma, N., Ghosh, S., Mait, U.N., Graphene aided gelation of MXene with oxidation protected surface for supercapacitor electrodes with excellent gravimetric performance. *Carbon*, 169, 225, 2020.
38. Wang, X., Zhai, H.W., Qie, B.Y., Cheng, Q., Li, A.J., Borovilas, J., Xu, B.Q., Shi, C.M., Jin, T.W., Liao, X.B., Li, Y.B., He, X.D., Du, S.Y., Fu, Y.K., Dontigny, M., Zaghib, K., Yang, Y., Rechargeable solid-state lithium metal batteries with vertically aligned ceramic nanoparticle/polymer composite electrolyte. *Nano Energy*, 60, 205, 2019.
39. Zhao, Y., Yan, J., Cai, W., Lai, Y., Song, J., Yu, J., Ding, B., Elastic and well-aligned ceramic LLZO nanofiber-based electrolytes for solid-state lithium batteries. *Energy Storage Mater.*, 23, 306, 2019.
40. Chen, G., Chen, J., Pei, W., Lu, Y., Zhang, Q., Zhang, Q., He, Y., Bismuth ferrite materials for solar cells: Current status and prospects. *Mater. Res. Bull.*, 110, 39, 2019.

41. Li, D., Shen, Z.Y., Li, Z., Luo, W., Wang, X., Wang, Z., Song, F., Li, Y., P-E hysteresis loop going slim in Ba0.3Sr0.7TiO3-modified Bi0.5Na0.5TiO3 ceramics for energy storage applications. *J. Adv. Ceram.*, 9, 183, 2020.
42. Bae, K., Kim, D.H., Choi, H.J., Son, J.W., Shim, J.H., High-performance protonic ceramic fuel cells with 1 mm thick Y: Ba(Ce, Zr)O3 electrolytes. *Adv. Energy Mater.*, 8, 1801315, 2018.
43. Zeng, X., Shui, J., Liu, X., Liu, Q., Li, Y., Shang, J., Zheng, L., Yu, R., Single-Atom to single-atom grafting of Pt1 onto Fe-N4 center: Pt1@Fe-N-C multifunctional electrocatalyst with significantly enhanced properties. *Adv. Energy Mater.*, 8, 1701345, 2018.
44. Liu, X., Xiao, L., Zhang, Y., Sun, H., Significantly enhanced piezo-photocatalytic capability in BaTiO3 nanowires for degrading organic dye. *J. Materiomics*, 6, 256, 2020.
45. Singh, P., Kaur, G., Singh, K., Singh, B., Kaur, M., Kaur, M., Krishnan, U., Kumar, M., Bala, R., Kumar, A., Specially designed B4C/SnO2 nanocomposite for photocatalysis: Traditional ceramic with unique properties. *Appl. Nanosci.*, 8, 1, 2018.
46. Celebanska, A. and Opallo, M., Layer-by-layer gold-ceramic nanoparticulate electrodes for electrocatalysis. *ChemElectroChem*, 3, 1629, 2016.
47. Verma, S. and Sinha-Ray, S., Electrospun CNF supported ceramics as electrochemical catalysts for water splitting and fuel cell: A review. *Polymers*, 12, 238, 2020.
48. Singh, P., Kaur, G., Singh, K., Singh, B., Kaur, M., Kaur, M., Krishnan, U., Kumar, M., Bala, R., Kumar, A., Porous ceramic nanofibers as new catalysts toward heterogeneous reactions. *Compos. Commun.*, 15, 168, 2019.
49. Kuang, M., Han, P., Huang, L., Cao, N., Qian, L., Zheng, G., Electronic tuning of Co, Ni-based nanostructured (hydr) oxides for aqueous electrocatalysis. *Adv. Funct. Mater.*, 28, 1804886, 2018.
50. Kwon, H.J., Shin, K., Soh, M., Chang, H., Kim, J., Lee, J., Ko, G., Kim, B.H., Kim, D., Hyeon, T., Large-scale synthesis and medical applications of uniform-sized metal oxide nanoparticles. *Adv. Mater.*, 30, 1704290, 2018.
51. Hu, M., Yang, W., Tan, H., Jin, L., Zhang, L., Kerns, P., Dang, Y., Dissanayake, S., Schaefer, S., Liu, B., Zhu, Y., Template-free synthesis of mesoporous and crystalline transition metal oxide nanoplates with abundant surface defects. *Matter*, 2, 1244, 2020.
52. Osgood, H., Devaguptapu, S.V., Xu, H., Cho, J., Wu, G., Transition metal (Fe, Co, Ni, and Mn) oxides for oxygen reduction and evolution bifunctional catalysts in alkaline media. *Nano Today*, 11, 601, 2016.
53. Liu, Y., Tian, D., Biswas, A.N., Xie, Z., Hwang, S., Lee, J.H., Meng, H., Chen, J.G., Transition metal nitrides as promising catalyst supports for tuning CO/H2 syngas production from electrochemical CO2 reduction. *Angew. Chem. Int. Ed.*, 59, 11345, 2020.
54. Wang, C., Zeng, X., Jiang, G., Chen, M., Zhu, L., Yu, R., General self-template synthesis of transition-metal oxide microspheres and their excellent charge storage properties. *Electrochim. Acta*, 283, 190, 2018.
55. Liu, Y., Li, X., Shen, W., Dai, Y., Kou, W., Zheng, W., Jiang, X., He, G., Multishelled transition metal-based microspheres: Synthesis and applications for batteries and supercapacitors. *Small*, 15, 1804737, 2019.
56. Li, J.F., Che, D., Wu, Q.S., α-Fe2O3 cased carbon composite as pure negative electrode for application as supercapacitor. *Eur. J. Inorg. Chem.*, 2019, 1301, 2019.
57. Sugimoto, W., Yokoshima, K., Murakami, Y., Takasu, Y., Charge storage mechanism of nanostructured anhydrous and hydrous ruthenium-based oxides. *Electrochim. Acta*, 52, 1742, 2006.
58. Casella, I.G. and Gatta, M., Study of the electrochemical deposition and properties of cobalt oxide species in citrate alkaline solutions. *J. Electroanal. Chem.*, 534, 31, 2002.
59. Zhang, Y., Guo, Z.H., Han, Z.Y., Xiao, X.Y., Effect of rare earth oxides doping on MgAl2O4 spinel obtained by sintering of secondary aluminium dross. *J. Alloys Compd.*, 735, 2597, 2018.

60. Alvarez-Sanchez, C.O., Lasalde-Ramírez, J.A., Ortiz-Quiles, E.O., Massó-Ferret, R., Nicolau, E., Polymer-MTiO3 (M ¼ Ca, Sr, Ba) composites as facile and scalable supercapacitor separators. *Energy Sci. Eng.*, 7, 730, 2019.
61. Gu, L.L., Li, T., Xu, Y.J., Sun, C.H., Yang, Z.Y., Zhu, D.L., Chen, D.L., Effects of the particle size of BaTiO3 fillers on fabrication and dielectric properties of BaTiO3 polymer/Al films for capacitor energy-storage application. *Materials*, 12, 439, 2019.
62. Han, K., Luo, N., Mao, S., Zhuo, F., Liu, L., Peng, B., Chen, X., Hu, C., Zhou, H., Wei, Y., Ultrahigh energy-storage density in A-/B-site co-doped AgNbO3 lead-free antiferroelectric ceramics: Insight into the origin of antiferroelectricity. *J. Mater. Chem. A*, 7, 26293, 2019.
63. Francisco, B.E., Jones, C.M., Lee, S.H., Stoldt, C.R., Nanostructured all-solid-state supercapacitor based on Li2S-P2S5 glass-ceramic electrolytes. *Appl. Phys. Lett.*, 100, 103902, 2012.
64. Naguib, M., Kurtoglu, M., Presser, V., Lu, J., Niu, J., Heon, M., Hultman, L., Gogotsi, Y., Barsoum, M.W., Two-dimensional nanocrystals produced by exfoliation of Ti3AlC2. *Adv. Mater.*, 23, 4248, 2011.
65. Fan, Z., Li, L., Mei, X., Zhao, F., Li, H., Zhuo, X., Zhang, X., Lu, Y., Zhang, L., Liu, M., Multilayer ceramic film capacitors for high-performance energy storage: Progress and outlook. *J. Mater. Chem. A*, 9, 9462, 2021.
66. Lang, X.Y., Liu, B.T., Shi, X.M., Li, Y.Q., Wen, Z., Jiang, Q., Ultrahigh-power pseudocapacitors based on ordered porous heterostructures of electron-correlated oxides. *Adv. Sci.*, 3, 1500319, 2016.
67. Li, T., Cao, Y., Xue, W., Sun, B., Zhu, D., Self-assembly of graphene-based planar micro-supercapacitor with selective laser etching-induced superhydrophobic/superhydrophilic pattern. *SN Appl. Sci.*, 2, 1, 2020.
68. Tennakone, K., Apsonsu, G.L.M.P., Ariyasinghe, Y.P.Y.P., Buchanan, R.C., Perera, V.P.S., Tennakone, H., Wijayarathna, T.R.C.K., Dye-sensitized solar cells based on nanostructured semiconductor oxide ceramics with ultra-thin barrier layers. *Integr. Ferroelectr.*, 115, 120, 2010.
69. Ata, M.S. and Zhitomirsky, I., Electrophoretic nanotechnology of ceramic films. *Adv. Appl. Ceram.*, 111, 345, 2012.
70. Rani, B.J., Ravi, G., Yuvakkumar, R., Ganesh, V., Ravichandran, S., Thambidurai, M., Rajalakshmi, A.P., Sakunthala, A., Pure and cobalt-substituted zinc-ferrite magnetic ceramics for supercapacitor applications. *Appl. Phys. A*, 124, 1, 2018.
71. Kovalenko, A.S., Shilova, O.A., Morozova, L.V., Kalinina, M.V., Drozdova, I.A., Arsent'ev, M.Y., Features of the synthesis and the study of nanocrystalline cobalt-nickel spinel. *Glass Phys. Chem.*, 40, 106, 2014.
72. Shamitha, C., Senthil, T., Wu, L., Kumar, B.S., Anandhan, S., Sol–gel electrospun mesoporous ZnMn2O4 nanofibers with superior specific surface area. *J. Mater. Sci.: Mater. Electron.*, 28, 15846, 2017.
73. Tang, H., Lin, Y., Sodano, H.A., Synthesis of high aspect ratio BaTiO3 nanowires for high energy density nanocomposite capacitors. *Adv. Energy Mater.*, 3, 451, 2013.
74. Zhao, Y., Shi, Z., Li, H., Wang, C.A., Designing pinecone-like and hierarchical manganese cobalt sulfides for advanced supercapacitor electrodes. *J. Mater. Chem. A*, 6, 12782, 2018.
75. Zeng, X., Song, H., Shen, Z.Y., Moskovits, M., Progress and challenges of ceramics for supercapacitors. *J. Materiomics*, 7, 1198, 2021.
76. Liu, Y.P., Qi, X.H., Li, L., Zhang, S.H., Bi, T., MOF-derived PPy/carbon-coated copper sulfide ceramic nanocomposite as high-performance electrode for supercapacitor. *Ceram. Int.*, 45, 17216, 2019.
77. Zhao, S., Zhang, H.B., Luo, J.Q., Wang, Q.W., Xu, B., Hong, S., Yu, Z.Z., Highly electrically conductive three-dimensional Ti3C2T x MXene/reduced graphene oxide hybrid aerogels with excellent electromagnetic interference shielding performances. *ACS Nano*, 12, 11193, 2018.

78. Zhang, M., Cao, J., Wang, Y., Song, J., Jiang, T., Zhang, Y., Si, W., Li, X., Meng, B., Wen, G., Electrolyte-mediated dense integration of graphene-MXene films for high volumetric capacitance flexible supercapacitors. *Nano Res.*, 14, 699, 2021.
79. Lukatskaya, M.R., Kota, S., Lin, Z., Zhao, M.Q., Shpigel, N., Levi, M.D., Halim, J., Taberna, P.L., Barsoum, M.W., Simon, P., Gogotsi, Y., Ultra-high-rate pseudocapacitive energy storage in two-dimensional transition metal carbides. *Nat. Energy*, 2, 1, 2017.
80. Tian, W., Vahid Mohammadi, A., Reid, M.S., Wang, Z., Ouyang, L., Erlandsson, J., Pettersson, T., Wågberg, L., Beidaghi, M., Hamedi, M.M., Multifunctional nanocomposites with high strength and capacitance using 2D MXene and 1D nanocellulose. *Adv. Mater.*, 31, 1902977, 2019.
81. Jiang, Q., Kurra, N., Alhabeb, M., Gogotsi, Y., Alshareef, H.N., All pseudocapacitive MXene-RuO2 asymmetric supercapacitors. *Adv. Energy Mater.*, 8, 1703043, 2018.
82. Shiri, H.M., Ehsani, A., Shayeh, J.S., Synthesis and highly efficient supercapacitor behavior of a novel poly pyrrole/ceramic oxide nanocomposite film. *RSC Adv.*, 5, 91062, 2015.
83. Zhang, X., He, B.L., Zhao, Y.Y., Tang, Q.W., A porous ceramic membrane tailored high-temperature supercapacitor. *J. Power Sources*, 379, 60, 2018.
84. Maity, S., Samanta, M., Sen, A., Chattopadhyay, K.K., Investigation of electrochemical performances of ceramic oxide CaCu3Ti4O12 nanostructures. *J. Solid State Chem.*, 269, 600, 2019.
85. Padmini, M., Elumalai, P., Thomas, P., Symmetric supercapacitor performances of CaCu3Ti4O12 decorated polyaniline nanocomposite. *Electrochim. Acta*, 292, 558, 2018.
86. Lu, D.L., Zhao, R.R., Wu, J.L., Ma, J.M., Huang, M.L., Yao, Y.B., Tao, T., Liang, B., Zhai, J.W., Lu, S.G., Investigations on the properties of Li3xLa2/3-xTiO3 based all-solid-state supercapacitor: Relationships between the capacitance, ionic conductivity, and temperature. *J. Eur. Ceram. Soc.*, 40, 2396, 2020.
87. Chang, P., Mei, H., Zhao, Y., Huang, W.Z., Zhou, S.X., Cheng, L.F., 3D structural strengthening urchin-like Cu(OH)2-based symmetric supercapacitors with adjustable capacitance. *Adv. Funct. Mater.*, 29, 1903588, 2019.
88. Liu, P.P., Liu, Z.J., Wu, P., Qu, X., Zhang, Y.P., Cai, W.Z., Yu, F.Y., Ni, M., Cheng, S., Liu, M.L., Liu, J., Enhanced capacitive performance of nickel oxide on porous La0.7Sr0.3CoO3-d ceramic substrate for electrochemical capacitors. *Int. J. Hydrogen Energy*, 43, 19589, 2018.
89. Tuncer, M., Bakan, F., Gocmez, H., Erdem, E., Capacitive behaviour of nanocrystalline octacalcium phosphate (OCP) (Ca8H2(PO4)6,5H2O) as an electrode material for supercapacitors: Biosupercaps. *Nanoscale*, 11, 18375, 2019.
90. El-Desoky, M.M., Al-Syadi, A.M., Al-Assiri, M.S., Hassan, H.M.A., El Enany, G., Electrochemical performance of novel Li3V2(PO4)3 glass-ceramic nanocomposites as electrodes for energy storage devices. *J. Solid State Electrochem.*, 20, 2663, 2016.
91. Hu, X., Chen, Y.L., Hu, Z.C., Li, Y., Ling, Z.Y., All-solid-state supercapacitors based on a carbon-filled porous/dense/porous layered ceramic electrolyte. *J. Electrochem. Soc.*, 165, 1269, 2018.
92. Yang, L.L., Zeng, X.J., Zhang, Y., 3D printing of alumina ceramic parts by heat induced solidification with carrageenan. *Mater. Lett.*, 255, 126564, 2019.
93. Liu, N.S. and Gao, Y.H., Recent progress in micro-supercapacitors with in-plane interdigital electrode architecture. *Small*, 13, 1701989, 2017.
94. Dinh, T.M., Achour, A., Vizireanu, S., Dinescu, G., Nistor, L., Armstrong, K., Guay, D., Pech, D., Hydrous RuO2/carbon nanowalls hierarchical structures for all-solid-state ultrahigh-energy-density micro-supercapacitors. *Nano Energy*, 10, 288, 2014.
95. Kitazawa, Y., Iwata, K., Kido, R., Imaizumi, S., Tsuzuki, S., Shinoda, W., Ueno, K., Mandai, T., Kokubo, H., Dokko, K., Watanabe, M., Polymer electrolytes containing solvate ionic liquids: A new approach to achieve high ionic conductivity, thermal stability, and a wide potential window. *Chem. Mater.*, 30, 252, 2018.

13

Potential Applications in Sustainable Supercapacitors

Pitchaimani Veerakumar[1,2]

[1]*Department of Chemistry, National Taiwan University, Roosevelt Road, Taipei, Taiwan, ROC*
[2]*Institute of Atomic and Molecular Sciences, Academia Sinica, Roosevelt Road, Taipei, Taiwan, ROC*

Abstract

The next-generation supercapacitor (SC) ought to be deliver excellent high energy density (ED), power density (PD) with outstanding cycling stability using sustainable doping materials. Heteroatoms like N, S, B, and P doped and co-doped with carbon materials are examples of sustainable functional materials for next-generation SCs. In the case of SCs, heteroatom doping with carbon has produced interesting effects, which we will discuss in this chapter. Although there is a wealth of literature on N-doped composites study for energy storage applications, there has been a scarcity of reviews on other doped composites. As a result, the objective of this chapter is to present critical insights and an up-to-date outline of current advancements, future research prospects, and the processes used to synthesize these doped functional materials. Following that, the benefits and drawbacks of using sustainable heteroatom-doped carbon materials will be discussed. The electrochemical (EC) performance of mono or multi-doped carbon composites as electrodes in EC capacitors is critically evaluated.

Keywords: Carbon, nitrogen, sulfur, boron, phosphorus, supercapacitor, power density, energy density

Abbreviations

ASC	Asymmetric supercapacitor
CV	Cyclic voltammetry
CD	Current density
C_{sp}	Specific capacitance
GCD	Galvanostatic charge/discharge
EC	Electrochemical
ED	Energy density
EDLC	Electrical double layer capacitance
ESS	Energy storage systems
IL	Ionic liquids

Email: spveerakumar@gmail.com

Inamuddin, Tariq Altalhi and Sayed Mohammed Adnan (eds.) Sustainable Materials for Electrochemical Capacitors, (305–338) © 2023 Scrivener Publishing LLC

PD Power density
SC Supercapacitor

13.1 Introduction

The world faces an energy crisis consequently of increasing depletion of resources, rising global population, rapid advancements in portable electronics, and electric cars [1]. Because solar energy is intermittent, many applications require effective energy storage systems to assure a consistent supply of energy. Renewable energy generation from the sun and wind is also increasing dramatically, demanding the progress of efficient energy storage/conversion equipment [2]. Regenerative braking in automobiles, elevators, and cranes necessitates energy storage devices that can charge and discharge 10 times quicker than current lithium ion batteries (LIBs), the greatest energy density (ED) in commercially accessible batteries [3]. Batteries, fuel cells, and electrochemical (EC) energy storage are some of the most prevalent energy storage systems (ESS) is also called as supercapacitors (SCs) or ultracapacitors, feature a large specific capacitance (C_{sp}), a high power density (PD), quick charging, lengthy charge-discharge cycles, a simple operating principle, and a wide operating temperature range [4]. Various types of EES devices are anticipated to have a comprehensive performance consisting of high PD/ED, stable cycle-ability, and superior safety, which is mostly dependent on electrode materials. In order to achieve efficient energy storage, substantial research on improved nanostructured materials has been conducted to obtain modified electrodes with high specific surface area (SSA), rapid electron, ion movement, and long-term structural stability [5, 6]. Consumer gadgets, memory backup devices, public transportation, military gear, and emergency doors on Airbus A380 planes have all used SCs in recent years [7]. However, as related to fuel cells and batteries, they have a lesser ED as well as higher production cost. As a result, much research being conducted to improve the energy storage and minimize the cost of SCs. They are environmentally friendly and renewable energy systems that meet the energy demands of modern living [8, 9]. SCs have a high PD but a poor ED, whereas batteries have a both. Such devices should ideally have high PD and ED, fast charging/discharging times, and an extended life cycles. The structure and properties of the nanomaterials employed in these energy systems have a significant impact on their efficiency.

Carbon nanotubes stand out among the many energy storage nanomaterials because of their availability, superior conductivity and stability, and low environmental pollution [10–12]. There have been several papers on the synthesis and solicitation of several classes of carbon materials in the field of EES devices (e.g. graphene [13], activated carbons (AC) [14], nanotubes [15], nanofibers [16], aerogels [17]). Furthermore, coal-based activated carbons [18], metal organic frameworks [19], and polymer-based carbons [20] have also undergone extensive research. Because of its customizable packing assembly, effective doping control, and higher electrical conductivity; hence the carbon-based composites are preferred as promising electrode platform for ultra-stable SCs than transition metals and conductive polymers [21].

An increasing the C_{sp} of ECs is another way to improve the ED of an EC capacitor cell. Whereas, an improved C_{sp} can be attained by leading pseudo-capacitive entities on its surface or inside the structure of carbons, such as transition metal oxides/polymers [22], heteroatoms like N, S, B, and P, etc. A carbon-based nanostructured material exhibits EDLC and pseudocapacitors (PC), while PC was caused by a doping of metal oxides, polymers, and heteroelements. Owing to their high-cost, less conductivity, and narrow cyclic stability, thus, the metal oxide-based composites are rarely used in practical applications [23]. Notably, heteroatom with carbons has shown better capacitance owing their pseudocapacitive impact via a quick and totally reversible reaction [24], lack of sacrificing high PD or cycle life. Many studies have been conducted to investigate the contribution of N, S, B, and P atoms in the field of SC applications [25–28]. In recent years, more than a single atom can increase the capacitance of carbon composites and adjusting its features, such as increasing wettability, creating pseudo-capacitance, and lowering resistance [29]. When utilized an electrode in energy storage devices, these heteroatoms are included in several compositions to adjust carbon nanomaterials in a desirable mode [30].

13.2 Fundamentals and Components of SCs

13.2.1 Conventional Capacitor

A conventional capacitor involves of two electrode plates divided by an insulator containing of glass, paper, ceramic, plastic, or aluminum oxide. Two electrode plates of the capacitor are joined to an exterior power source during charging, resulting create a potential difference [31]. On one plate, positive charges are collected, while on the other plate, negative charges are gathered, charges continue even the exterior power cut [32]. It is a capacitor's charged condition. The capacitor discharges its stored energy by releasing both positive and negative charges were linked resistive load. The difference in potential between these plates can produce an electric field in the dielectric layer, which collects identical amounts charges in the two plates.

Capacitance (*C*) is the measurement of a capacitor's ability to energy storage over this preparation of charges, which was computed using the measurement of unit Farad (*F*).

$$C = A * \epsilon d \tag{13.1}$$

where, A, ε, and d were denotes the surface area of the two plates, permittivity of dielectric composite, and distance among the two plates, respectively. Commercial capacitors have capacitances ranging from pico- to microfarads. Because of their poor ED, ordinary capacitors have restricted applications. The low ED is caused by the plates' small surface area and short distance between them. Because of this flaw in traditional capacitors, a large effort has been made to build a new kind of energy storage device was known as a supercapacitor (SC) [33]. A SC involves anode and cathode electrodes divided by an electrolyte, and a separator allows ions to migrate between the electrodes while electrically isolating them. Unlike traditional capacitors, the SC's electrodes are usually constructed of porous materials with a large SSA, and separator in contact with electrolyte. The electrode/electrolyte interface, which has a greater surface area than dielectric, forms the electric double layer [33]. As a result, SCs can store more charges than standard capacitors. It has been discovered that a SC's improved structure allows for better energy storage than ordinary capacitors. The SCs

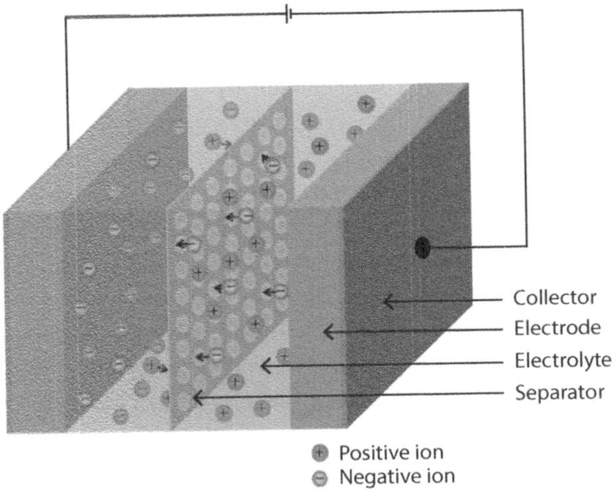

Figure 13.1 A graphic design of SC [34]. *Source:* Copyright 2020, Elsevier.

have an extended cycle life than batteries since the chemical phase shifts in their electrodes are far less than those in batteries during continuous charging/discharging process. Also, these are particularly appealing and adaptable as high-powered energy storages, as seen in Figure 13.1 [34].

To build flexible SCs for use in electronic devices, several researchers has created unique carbon and polymer nanomaterials [35–38]. Besides, substantial work has been put towards hybridizing the electrode in order to increase its capacitance. In this approach, battery and SC electrodes were combined towards produce hybrid SCs with improved ED. The results demonstrated that using hybrid SCs, an increase in ED may be achieved without sacrificing high PD [39, 40]. However, such improvements must continuously be adjusted compared to the loss of cycle stability and its performance.

13.2.2 Specific Capacitance

The specific capacitance (C_{sp}) was one of the most critical performance criteria for SCs, which was considered to relate the performance of different electrode masses. The C_{sp} of a capacitor was calculated using cyclic voltammetry (CV) and galvanostatic charge/discharge (GCD) curves. Also, CV delivers the potential of working electrode (WE) with the stable potential of the reference electrode (RE), and it is linearly swept between the two predetermined potentials [41, 42]. The plot of obtained current (I) against the potential affords, which was used to calculate the C_{sp}. GCD is an extra way for calculating a material's C_{sp}. A constant current density (CD) (A/g = ampere per gram) is used in this method, and the responsive potential is recorded throughout time. The WE was typically charged/discharge process in a predetermined potential, then to calculate C_{sp} using the following equation (13.2) for a three-electrode configuration:

$$C_{sp} = \frac{\int I dv}{V \times m \times \Delta V} \tag{13.2}$$

where, C_{sp} = WE system (F g^{-1}), $\int Idv$ = area of CV curve, V = potential scan rate (mV s^{-1}), and m = weight of active nanomaterials (g). Following equation (13.2) was used to calculate the C_{sp} from the charge/discharge profiles for three electrode setup.

$$C_{sp} = \frac{I \times \Delta t}{m \times \Delta V} \tag{13.3}$$

where I = discharge current (A), Δt = discharge time (seconds), m = weight of the electrode material at WE (g), and ΔV = potential drop during discharge (volts). Remarkably, unit of the C_{sp} is F g^{-1} where F is the Faraday.

13.2.3 Specific Energy and Power Density

Understanding performance of SCs in real-time applications requires an understanding of their energy and PD, both are calculated from CV and GCD. The ED was calculated using equation (13.4)

$$ED = \frac{\frac{1}{2} C_{sp} \Delta V^2}{3.6} (Whkg^{-1}) \tag{13.4}$$

The PD was calculated using equation (13.5)

$$PD = E \times \frac{3600}{\Delta t} (W\ kg^{-1}) \tag{13.5}$$

where Δt = discharging time.

ED denotes the amount of energy contained in a specific mass or volume. PD refers to the quantity of power contained in a particular mass. A high ED indicates that a device can store a large quantity of energy in a little quantity of mass. Moreover, a high ED does not automatically imply a high PD. A device can store a high ED on the other hand low PD as a function for an extended long time.

13.2.4 Electrolytes

SCs depend greatly on electrolytes to function properly. They have a direct impact on the cell's voltage range and resistance. The square voltage window was directly related to ED, but ionic resistance was inversely proportional to cell's power capability [43]. As a result, identifying the appropriate electrolyte for high-performance SCs was essential. Therefore, an aqueous, organic, and ionic liquids electrolytes are now used in SCs, because of their low cost and abundance, aqueous electrolytes are commonly used in research. The usual aqueous electrolytes are aqueous solutions of sulphuric acid (H_2SO_4), potassium hydroxide (KOH), and potassium chloride (KCl), which are easy to handle in an open surroundings and have a less ionic resistance. For instance, exchanging as of H_2SO_4 in to KOH in nickel oxide (NiO) electrode, raised after C_{sp} 16 F g^{-1} to 155 F g^{-1}, since NiO undergoes a Faradaic reaction in an hydroxide (OH) ions [44]. Because of their narrow voltage window, aqueous electrolytes are not suitable for commercial use (only 1.2 V). Water decomposes into oxygen (O_2) and hydrogen (H_2) molecules as the voltage is increased, creating pressure in the system and causing the cell to burst. Furthermore, metals can be corroded by strong acids or bases (e.g., nickel and aluminum). These metals can also be tampered with by chloride ions from KCl. Organic electrolytes like acetonitrile and propylene carbonate are commonly employed in commercially available SCs due to their high operating voltage (0.227 V). The wide voltage range boosts the ED to suit commercial requirements. Acetonitrile was preferred over PC because its ionic resistivity was one-third that of PC, but it was toxic and flammable. When carbon coated electrodes are exposed to organic electrolytes, they degrade slowly. The pores of the electrodes are limited by organic electrolyte breakdown, decreasing capacitive performance and cyclic stability [45].

Ionic liquid (IL) electrolytes are non-toxic, non-flammable, and they offer broadest voltage range (0–3.5 V) while avoiding heating and less chemical stability. They may also be heated to 300°C without vaporizing [46]. However, ILs have a lower ionic conductivity than the other two forms of electrolytes, which is a big disadvantage. However, ILs heating up to 125°C improves conductivity, however this introduces new strategy problems. El-Kady *et al.* [36] used aqueous, organic, and IL electrolytes to analyze laser-scribed graphene SCs. Moreover, ILs' performance is promising, but commercial implementations confront a number of challenges, including high costs.

13.2.5 Separators

A well-built separator is required, as a poorly constructed separator might cause additional resistances in the cell. Nonconductive (to impede electron transfer between electrodes), permeable to electrolyte ions with low ionic resistance, non-reactive with the electrolytes, mechanically strong, and easily wetted by the electrolytes are all requirements for an effective separator. Initially, glass, cellulose paper, and ceramics were used as separators. Polymer-based separators are now the most often used separators. Shulga *et al.* [47] used GO films as separators in SC applications with great success.

13.2.6 Current Collectors

The most popular of energy saving technologies involve current collectors to carry current from electrodes to external supply. As a result, it needs both electrically inert as well as electrolyte is conductive. collectors include aluminum, nickel foam, nickel foils, carbon fabric, steel, and iron. Furthermore, placing active nanocomposites directly onto current collectors, which can produce a strong chemical bonding connection, improving performance by reducing their interfacial resistance. Unfortunately, considerable resistance may result from the wide cycling of SCs and the physical awkwardness of nanomaterials. To prevent active components from dislocating from current collectors, some polymeric binding agents (such as Nafion and polytetrafluoroethylene) are used. Additionally, current collectors are required to disperse heat created within the cell [48]. As a result, aluminum was typically used, although carbon sheets were used in this investigation. Figure 13.2 shows a plot of PD vs. ED for diverse energy storage systems, dubbed the Ragone plot [49].

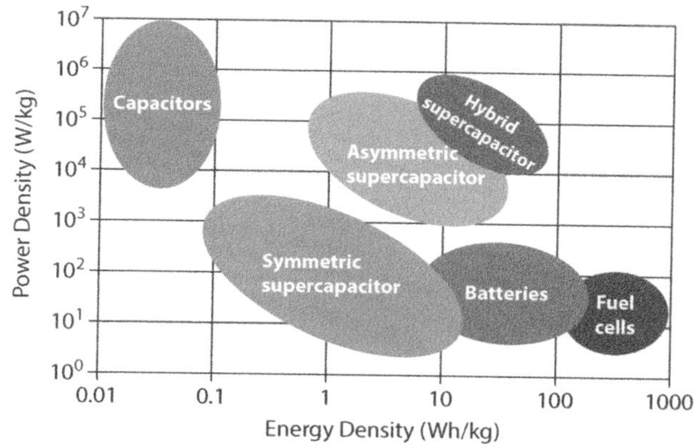

Figure 13.2 A Ragone plot for various energy storage technologies [49]. *Source:* Copyright 2017, Wiley.

SCs maintain a unique position, as shown in Figure 13.2. Their PD and ED are greater than both LIBs and traditional capacitors. As a result, they outperform alternative energy storage devices. In contrast, SCs have a high ED of about 510 Wh kg^{-1}. It was higher than dielectric capacitors, which have a low ED, but above batteries, such as LIBs (120-170 Wh kg^{-1}), which have a higher ED. It has been claimed that combining SCs with batteries can increase ED and provide energy for a longer amount of time [50]. However, increasing the ED near batteries is a difficult issue. When compared to batteries, fuel cells, and ASCs, it is clear that asymmetric SCs (ASCs) give much higher PD. Furthermore, the similar ED of asymmetric devices to LIBs predicts that they will be widely used in next-generation electronics and energy storage devices. Flexible electronics, e-textiles, and transportation are some of the application areas where ASCs are being used more frequently. Because of the restricted operating-voltage window, EC symmetric SCs (SSCs) produced with carbon-based electrodes had low ED in aqueous electrolytes. Consequently, all commercial SCs were used organic electrolytes only with 2.5–2.85 V and a high ED up to 5–10 Wh kg^{-1}. ASCs with aqueous electrolytes, on the other hand, may achieve cell voltages of more than 2.0 V [51] and are anticipated to supplant SSCs with organic electrolytes. Furthermore, because non-aqueous or IL electrolytes have a larger dissociation voltage, ASCs can achieve enormous operating voltages with extremely high EDs. However, there are just a few investigations on ASCs in organic electrolytes. Of explain the purpose of capacitors, a brief introduction to conventional capacitors is provided below.

13.3 Sustainable Nanomaterials in SCs

Faradic reactions occur on their electrode in pseudocapacitors (PC), hence C_{sp} was expected by the electrode materials' surface area. Accordingly, the following are some of the benefits of using nano electrode materials in super capacitors: (i) nanomaterials able to rise the area of interaction among the electrode/electrolyte in a given mass, allowing for a greater number of counter ions from electrolyte to reach the charged electrode surface; (ii) diffusion of electrons in the electrode/electrolyte towards electrode surface was significantly reduced.

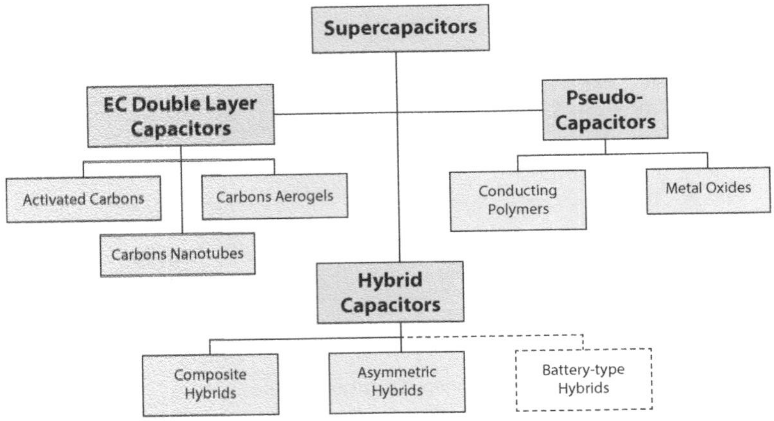

Figure 13.3 Types of SCs [53]. *Source:* Copyright 2009, Elsevier.

It has the ability to charge and discharge at a high rate, (iii) nanostructured materials have a higher tolerance to strain and structural distortion, allowing them to be cycled many times at a high rate, and (iv) nanoscale electrode materials have an exciting combination of properties, including low weight, transparency, flexibility, and biodegradability [52]. SCs are classified as follows: (i) electrical double-layer capacitors (EDLC), (ii) pseudocapacitors (PC), and (iii) asymmetric supercapacitors (ASC), depends on their energy storage mechanism [53] (Figure 13.3).

13.3.1 Electrical Double-Layer Capacitors (EDLCs)

An electrostatic interaction between the electrode charges and adsorbed ions of the electrolyte stores the charges in EDLCs, resulting in the development of two charged layers (double layer). The essential requirements for EDLC electrodes are their large SSA, quick charge/discharge rates, and good conductance. Carbon-modified electrodes are commonly used in EDLCs because they have a large SSA, are nontoxic, controllable porosity, and good electronic conductivity, which are available in different forms (tubes, powders, composites, sheets, aerogels) [54]. A potential difference between the EDLCs causes electrodes with different polarities and electrolyte ions to migrate to the electrodes' micropores. The double layer thickness at the electrode/electrolyte contact determines capacitance of EDLC (see Figure 13.4). Because the double layer's thickness is significantly less than the separator, however, EDLC basically larger capacitance than typical capacitor. A substantially bigger region of electrode/electrolyte interface than traditional capacitor able to give EDLC enhanced capacitance in a shorter separation distance. Furthermore, the electrolyte's superior electrical conductivity can lower electrode internal resistance, while the electrolyte's good wettability allows for greater ion mobility into electrode pores, enhancing the capacitive performance of EDLCs even further [34].

Normally, a specific type of AC was commonly used in commercial EDLCs. NaOH activation of polyacrylonitrile fibers yielded activated carbon fibers (ACFs) has large SSA and extremely micro/mesoporous structure for EDLCs. It has a high C_{sp} 371 F g^{-1} for aqueous 6 M KOH, 213 F g^{-1} for non-aqueous 1.0 M $LiClO_4$/PC, and 188 F g^{-1} for an IL containing

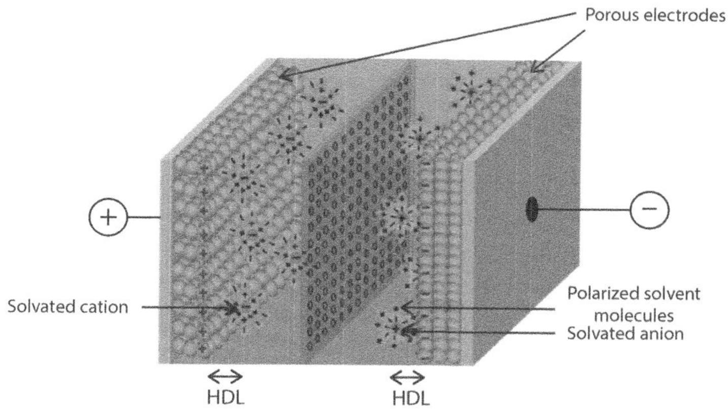

Figure 13.4 Charge storage mechanism of EDLC [34]. *Source:* Copyright 2020, Elsevier.

lithium bis(trifluoromethane sulfonyl)imide, and 2-oxazolidinone, signifying the ACF working as a favorable electrodes for EDLCs [55]. Three-electrode cell with $ZnCl_2/CO_2$-treated AC in aqueous 1.0 M H_2SO_4 electrolytes the achieved C_{sp} values of 118, 109.5, 101.7, 93.6, and 91.2 F g^{-1} at 1.0, 2.0, 4.0, 8.0, and 12 A g^{-1}, respectively [56].

Carbon nanotubes (CNTs), a sophisticated carbon structure, were investigated to enhance the capacitance of AC. CNTs are a type of carbon that has been developed to have better qualities than AC. However, nanotubes are created of graphitic walls which are arranged in one-dimensional cylinders (1-D). It can be categorized as single-walled carbon nanotubes (SWCNTs), multi-walled carbon nanotubes (MWCNTs) depends its thickness of tube structure. It has great mechanical qualities, prevents electron dispersion, and has electrical conductivity that is higher than AC. Highly organized CNT arrays feature a large porous region for electron storage and have a big surface area. CNTs have a C_{sp} of 15–200 F g^{-1} [57], which varies depends on shape and clarity. Nevertheless, one of the greatest obstacles to CNTs' possible use in commercial SCs is their high production cost.

13.3.2 Pseudocapacitors (PC)

PCs are also used in the current study. Reversible faradaic redox reactions between electrode materials produce capacitance in PCs. The electrode materials are mixtures of electrically conducting polymers and transition metal oxides. PCs have a higher ED than EDLCs, but their lifetime and PD are limited by segment changes inside the electrode caused by the Faradaic reaction. Energy storage mechanisms in EDLC and PCs can happen at the same time in some electrode materials, thus there's a probability of a synergistic impact. Some electrochemically active redox materials able to casted with AC to improve the C_{sp} of a PC. A quick reversible oxidation/reduction reaction(s) will occur in this manner, resulting in a capacitance of 10-100 fold greater than bare carbon-constructed EDLC [58]. Ruthenium oxide (RuO_2) is most commonly investigated materials for PCs. In proton-rich conditions (e.g., H_2SO_4), it exhibits numerous redox stages, supplying additional electron exchanges for high capacitance [59]. For example, RuO_2 nanotubes yielded C_{sp} ~1300 F g^{-1} in H_2SO_4 [54]. Unfortunately, despite its excellent performance, the toxicity, high cost, and scarcity of this RuO_2 electrode material limit its use in SCs. As a result, research and development has switched to less poisonous and less expensive alternatives, such as metal oxides like cobalt oxide (CoO_3), nickel oxide (NiO), as well as conducting polymers. For example, CoO_3 nanowires had a C_{sp} range of 570–700 F g^{-1} [60], whereas NiO nanoflowers shows C_{sp} of 600–760 F g^{-1}.

Polypyrrole, a conductive polymer, achieved C_{sp} of 150–400 F g^{-1} [61]. Though these choices may have interesting properties, their poor conductivity (in comparison to carbon nanomaterials) and shortage of long-cycle stability appear to exclude their use in PCs. Carbonaceous additions were investigated to integrate into the electrode for increased enactment to solve these disadvantages [62]. In recent years, graphene is a two-dimensional (2D) nanosheet with sp^2 bonded carbon atoms (i.e., an atomic layer thick) has sparked some research interest. It has a huge SSA of around 2600 m^2 g^{-1} with C_{sp} = 20 F cm^2, strong chemical stability with wide surface-to-volume ratio, a wide EC window, as well as making it appealing for LIBs ECs [54, 63, 64].

13.3.3 Asymmetric Supercapacitor

An asymmetric supercapacitor (ASC) has a battery-like Faradaic electrode, the unique strategy may modify the operating voltage window and boost ED owing to the EC redox course of reaction [50]. However, it is critical to optimize both electrodes over suspicious strategy. When a potential was given to electrode in a PC material, quick and reversible faradaic reactions (redox reactions) occur, and charge passes across the electrode/electrolyte contact. In theory, PCs have a greater ED than EDLCs, particularly in systems that can access several oxidation states. The C_{sp} of a PC can be 10–100 folds that of an EDLC, according to Conway *et al.* [65], but it has a lower PD and instability. This is due to the relatively slow Faradaic reactions and the physical changes occur during cycling process. A super capacitor's entire cell able to thought of two capacitors are connected. Because entire capacitance of the two electrodes was half of either electrode's capacitance, so-called a ASC. The comparable SC was dubbed ASC (see Figure 13.5) when anode and cathode electrode materials are different [49].

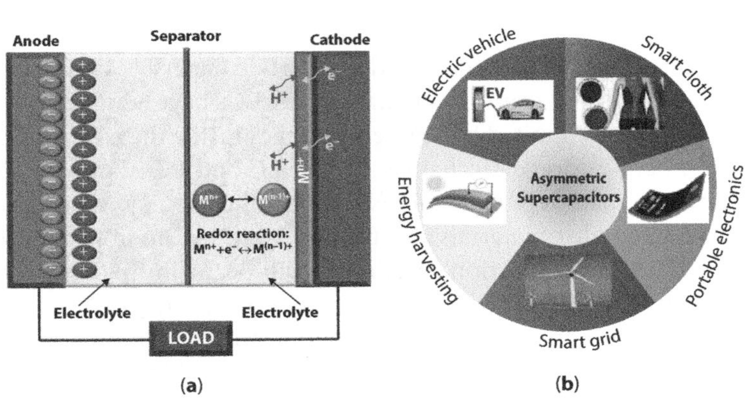

Figure 13.5 (a) Typical construction of ASC. (b) Some possible application areas of ASC devices [49]. *Source:* Copyright 2017, Wiley.

The capacitance and stored charge are mostly determined by the fabricated nanocomposites. A cell voltage was determined by electrode materials, which was controlled by thermodynamic stability of the electrolyte. ESR was influenced by the electrode matrix, electrolyte solution's intrinsic electrical properties, the mass transfer resistance, and contact resistance between the current collector and electrode [66]. To raise the overall cell capacitance, both electrode capacitances must be increased. For a SC to work well, it must have a big C_{sp}, high working cell voltage, and lower ESR.

13.4 Sustainable Carbon Nanomaterials for Energy Storage

Carbon-based nanomaterials are used as an ideal electrode in SC applications, due to its diverse structures and customizable surface properties, their low cost, abundant nature, nontoxic nature, high conductivity, mechanical stability, and so on [67]. Activated carbon, graphene, carbon nanotubes, onion-like carbons, and templated carbons have piqued attention as EDLC-type electrodes for SCs due to its high electrical conductance, huge SSA, ease of treatment, great chemical/thermal stability, and outstanding mechanical properties [68]. A comparison of various kind of carbons were frequently utilized as EDLC, as shown in Table 13.1.

13.4.1 Activated Carbon

Among them the activated (AC) distinguishes because of low cost, has a great theoretical surface area (3000 $m^2\ g^{-1}$) with wide pore size distribution ranging from >50-2 nm [69, 70]. The capacitance of activated carbons with numerous pore sizes has highly depends on the type of electrolyte used; for example, carbons with pore sizes (0.4-0.7 nm) are best suited for electro adsorption for aqueous electrolytes, however carbons have pore size should be ~0.8 nm, which was most appropriate for most organic electrolytes [71]. It was also shown that carbons with hole sizes equivalent to electrolyte ions had the highest electrical double layer capacitance, whereas pores that are larger than the ion size cause significant capacitance loss. Furthermore, surface functionalization of ACs has a significant impact on its capacitive performance. Inserting of heteroatoms (O, N, B, and S) into AC scaffold and make them hydrophilic in nature, allowing for improved ion adsorption [72]. For example, mild thermal oxidation of ACs has been shown to increase the C_{sp} by 37% [65]. N-doped ACs are more stable than oxygen functionalized ACs, and they can achieve a considerable pseudocapacitance reaction occur in both electrolytes, when introduced S-modified ACs for SCs, which resulted in a nearly 40% increase in C_{sp} [73].

13.4.2 Nitrogen-Doped Carbons

Doping N atoms on the carbon surface can increase capacitance performance by controlling an electron donor–acceptor, functional, and hydrophilic properties [74, 75]. The N-containing chemicals such as urea, nitric acid, ammonia, melamine, polyacrylonitrile, polyvinylpyridine, and glucosamine [76–80] were used to heat treatment or carbonization of crude carbons under inert atmosphere. Where N was introduced to an active material's surface or carbon structure. These N-doped carbons are commonly employed for energy

Table 13.1 Different types of carbons used in EDLCs[a].

Material	**Carbon onions**	**Carbon nanotubes**	**Graphene**	**Activated carbon**	**Carbide derived carbon**	**Templated carbon**
Dimensionality	0-D	1-D	2-D	3-D	3-D	3-D
Conductivity	High	High	High	Low	Moderate	Low
Volumetric Capitance	Low	Low	Moderate	High	High	Low
Cost	High	High	Moderate	Low	Moderate	High
Structure	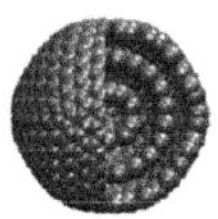	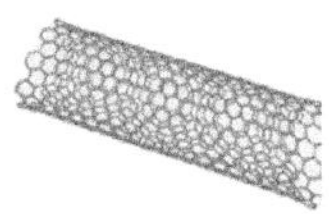	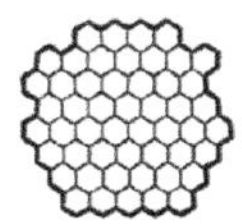	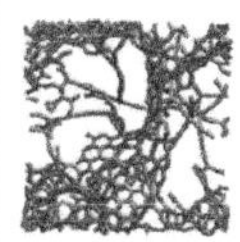	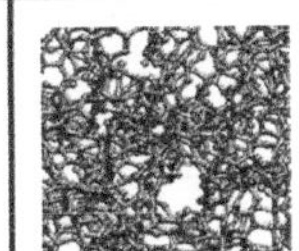	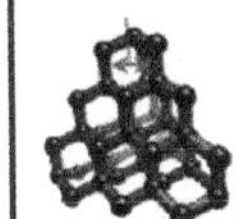

[a]Reproduced from Ref. [67] Source: Copyright 2013, The American Chemical Society.

storage in SCs. The C_{sp} of N-doped nanocomposite was summation of EDLC owing to its physical phenomenon arising at the interface of electrode/electrolyte as well as AC. Fast and fully revisable Faradic reaction combined with electronic transfer due to an electron donor of N [81], by Equations (13.6 and 13.7).

$$-C = NH + 2e^{-} \leftrightarrow -CH - NH_2 \tag{13.6}$$

$$-C = NHOH + 2e^{-} + 2H^{+} \leftrightarrow -C - NH_2 + H_2O \tag{13.7}$$

Han *et al.* [82] synthesized a *pueraria*-based carbon that was then N-doped using a thermal treatment of *pueraria* powder and melamine, as represented in Figure 13.6. Remarkably, N-doped carbon exhibited a larger C_{sp} of 250 F g^{-1} as related to 44 F g^{-1} for undoped carbon at CD of 0.5 Ag^{-1} in 6.0 M KOH with 92% of capacitance retention even after 10000 cycles. Additionally, an ED of 8.46 Wh kg^{-1} at PD of 123 W kg^{-1} were achieved with NPC-modified electrode in a SC, and it still reserved a specific ED and PD of 7.52 Wh kg^{-1} and 2417 W kg^{-1}, respectively.

Mao *et al.* [83] found that porous N-doped carbon (PNC) exhibit an improved EC performance, displaying an excellent areal capacitance was more than twice (C_{sp} = 683 mF cm^{-2} at 2 mA cm^{-2}) for N-doped carbon, whereas, un-doped carbon showed 330 mF cm^{-2} with excellent long term stability of more than 96% after 10000 cycles. N-doping able to use to boost their EDs, allowing them to be used more widely in practical applications. After the introduction of N functionality, the ED improved to 66.7 Wh kg^{-1} compared to 5.9 Wh kg^{-1}, demonstrating that N-doping is an effective technique to enhancing the ED's of SC electrodes, and that an increased ED's will lead to commercial solicitations [84]. When Dai and coworkers [85] used N-doped porous graphene acting as WE and $nBMIMBF_4$ electrolyte with a high C_{sp} (~130 F g^{-1}, 1.0 A g^{-1}) to benefit of ~3.5 V, they achieved an exceptional high

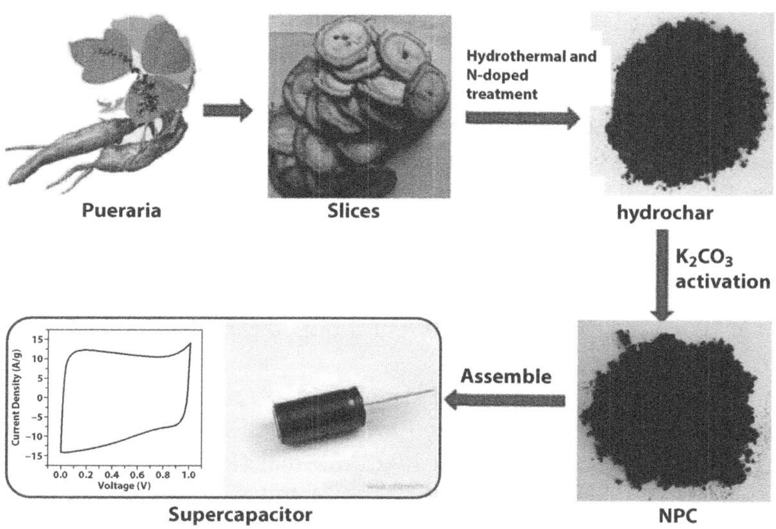

Figure 13.6 The representation of the synthesis route of NPC and application in SCs [82]. *Source:* Copyright 2018, Elsevier.

ED of 55 Wh kg^{-1} at a PD of 1800 W kg^{-1} with superior life cycling life (retaining 96.6% of the initial capacitance after 20,000 cycles). As a result, N-doping increases the electrolyte's wetting behavior, resulting in better electrode/electrolyte interface, lower solution resistance, and can improve the overall performance.

13.4.3 Sulphur-Doped Carbons

In past years, an introducing S heteroatoms to carbon surfaces have been increasing interest in order to improve capacitance, cyclability, corrosion resistance, and other features of the material [86]. Porous carbons with S-doping are still uncommon compared to those with N, O, or B, indicating a promising research potential in the area of long-term SCs. Because S changes the charge migration within the AC or its surface due to its electron acceptor/donor properties, incorporating S-containing groups within the AC scaffold or on surface can improve an electronic reactivity of active material, subsequent increased in electrode polarization, C_{sp} via a fast and fully reversible faradaic process [87, 88]. There are two methods such as direct thermal treatment and co-carbonization of carbon with S precursors, producing S-functionalized active carbons [89, 90]. The S could affect SC performance in a variety of ways. Carbonaceous materials have a higher electron density at the surface due to a synergetic effect between conjugated carbon and the electron-rich element S. Some of the redox processes that can occur in KOH electrolytes [91] are shown in the following equations (13.8 and 13.9).

$$> S - OH + OH^- \leftrightarrow > S = O + 2e^- + H_2O \tag{13.8}$$

$$> S = O + 2OH^- \leftrightarrow > SOO + 2e^- + H_2O \tag{13.9}$$

Doping of both N and S functional groups with the carbon electrode materials, which improves conductive performance and electrode/electrolyte wettability. Ma and colleagues discovered that S doping results in better conductive performance than N doping [92]. Because S-doping provides an excellent electronic conductivity, increased C_{sp} is achieved as a result of pseudo-capacitive contribution and EDCL resulting from S functionalities and porous characteristics of the active material. Furthermore, due to its better conductivity, S-doping improves the cell's ED and good PD.

Carbonizing sodium lignosulfonate resulted in large porous S-doped AC with SSA of 1592 m^2 g^{-1} and pore structure. AC material with 5.2 wt% of S was prepared, resulting in the large C_{sp} of 320 F g^{-1} at 0.2 A $g^{-1}$with a high ED of 8.2 Wh kg^{-1} at PD of 50 W kg^{-1} [93]. Similarly, the capacitive behavior of un-doped and S-doped graphene was improved from 145–160 F g^{-1} at 10 mV s^{-1}. Using 6.0 M KOH for doped carbon, a large ED of 160 Wh kg^{-1} at a PD of 5161 W kg^{-1} were achieved.

When N and S-doped AC was utilized as an active electrode material, which improves wetting behavior and capacitive behavior. Elm flower (EL)-derived biomass used to produce AC, through a simple and low cost hydrothermal (HT) method using NaOH activation [94]. Figure 13.7a depicts the synthesis of N and S co-doped EL-derived activated carbon (ELAC) and its contents were ca. 2.21 and 6.06 at%, respectively. In addition, ELAC have a large SSA (2048.6 m^2 g^{-1}) and a moderate pore volume of 0.88 cm^3 g^{-1}. TEM image

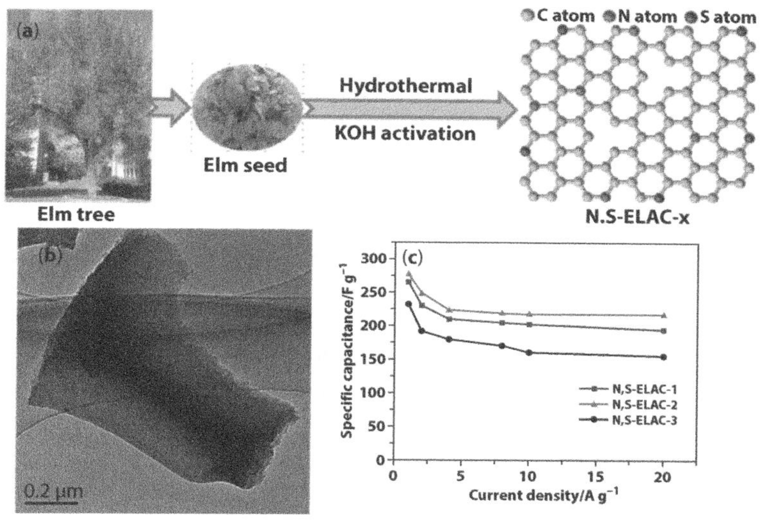

Figure 13.7 (a) The synthesis procedure for N, S-codoped ELAC-*x*, (b) TEM image, and (c) C_{sp} determined from GCD profiles [94]. *Source:* Copyright 2018, The American Chemical Society.

(Figure 13.7b) clearly identified rich micropore structures, which was consistent with the BET reports. ELAC achieved a C_{sp} of 275 F g^{-1} at 1.0 A g^{-1} and maintained C_{sp} of 216 F g^{-1} at 20 A g^{-1} in 6.0 M KOH due to their high SSA and N/S doping effect (see Figure 13.7c). Furthermore, an ASC electrode provided C_{sp} of 62 F g^{-1} at 10 A g^{-1}, with extreme ED and PD's of 16.8 Wh kg^{-1} and 600 W kg^{-1} with the capacitance retention was 87.2% at 4 A g^{-1} after 5000 cycles.

13.4.4 Boron-Doped Carbons

In the past years, boron is a one-of-a-kind element that has been explored like a carbon or diamond analogue to increase oxidation resistance behavior, Li-ion insertion, and electrochemical performance [95]. It enters the carbon lattice by exchanging carbon at trigonal locations, and because it has three valence electrons, so it acts as an electron acceptor nature, also shifting a Fermi level to the conducting band and altering the electrical configuration of B-doped carbon [96]. EDLC can be affected by changes in the electrical structure of carbon electrode materials. Most notably, low-level B doping has a electrocatalytic influence on chemisorption of O_2 on the AC surface, making redox processes possible on its surface, thus, the EC capacitance of carbons, which involves EDLC and pseudocapacitance, can be altered [97].

By adding boron to the carbon matrix, the electrical configuration of active carbon material can be changed. Because N and B have similar electrical configurations, sizes to carbon atoms, and are easier to dope with either N or B [98, 99]. The introduction of B into the carbon lattice, able to enhance charge transfer between neighboring carbon atoms, resulting in an uneven distribution of charges. Because of the pseudocapacitive contribution resulting from this electronic transfer (Faradic reaction) [100], this charge transfer improves electrochemical performance. Boron functionalization can be achieved utilizing a variety of synthetic processes, including laser ablation [101], arc discharge method [102, 103], HT reaction [104], boron oxide (B_2O_3) substitutional reaction [105–107], or chemical vapor deposition approach [108–110]. The HT reaction is the most prevalent method for producing B-doped active materials, and when B-doped graphene was synthesized using a thermal reaction, the C_{sp} was enhanced to 173 F g^{-1}. When 4.7% B was injected into the bulk of graphene, the optimum level of B-doping was discovered, resulting in a capacitance about 80% greater than un-doped ACs. As demonstrated in Figure 13.9(a,b), the EC cell had a greater ED of 3.86 Wh kg^{-1} at a PD of 125 W kg^{-1} and achieved to maintain the ED of 2.92 Wh kg^{-1} which was significantly greater PD of 5006 kW kg^{-1} with outstanding cycling stability of about 97% after 5000 cycles [111]. Typically, the template/nanocasting method have considered as beneficial method that aids in precisely handling the SSA, pore size, and pore shape, resulting a good effect on the EC cell's performance. Not only does B-doping improve capacitive behavior, but also it improves the wettability of electrode/electrolyte and lowering solution resistance.

Gao and coworkers used a hard template technique to synthesize B-doped meso-porous carbon with controlled porosity and found that after B-doping, the C_{sp} was 268 F g^{-1}, which was greater than un-doped carbon 221 F g^{-1} at 5 mV s^{-1}. Enhanced wettability with the inclusion of B functional groups also contributed to the extremely less solution resistance (R_S) of 1.05 U [112, 113]. Other advantages of improving the surface properties of electrode following B-doping include improved conductivity. As illustrated in Figure 13.8, thermally reduced graphene oxide with boron doping (BT-rGO) was created by simply heat annealing GO/B_2O_3. Because B-doping effects in an enhanced conductance of the GO, the extraordinarily high C_{sp} of 448 F g^{-1} was attained lacking of any conductance enhancers like as carbon black [114].

Huang *et al.* developed a pyrolysis method to preparing the B-modified phenol-formaldehyde resins staring material for B-doped carbon spheres (B-CSs), which was used

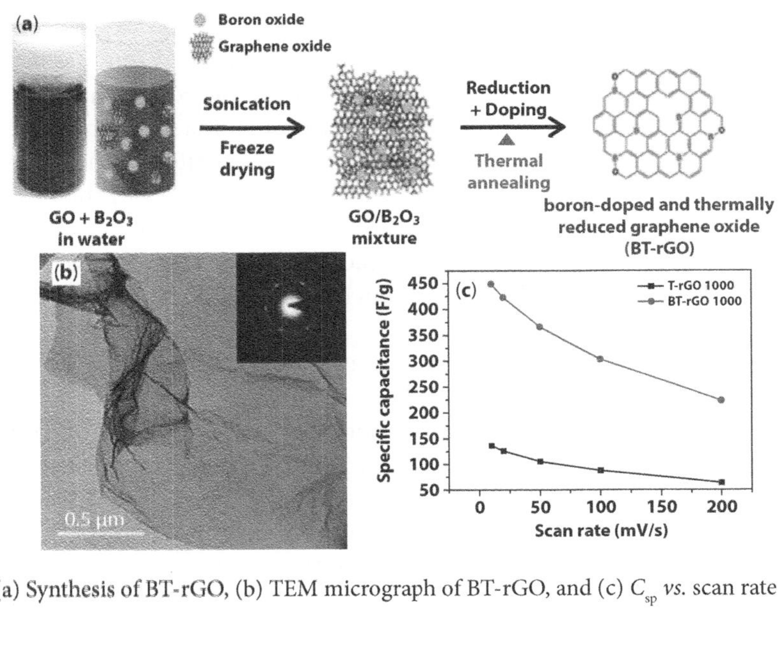

Figure 13.8 (a) Synthesis of BT-rGO, (b) TEM micrograph of BT-rGO, and (c) C_{sp} *vs.* scan rate [114].

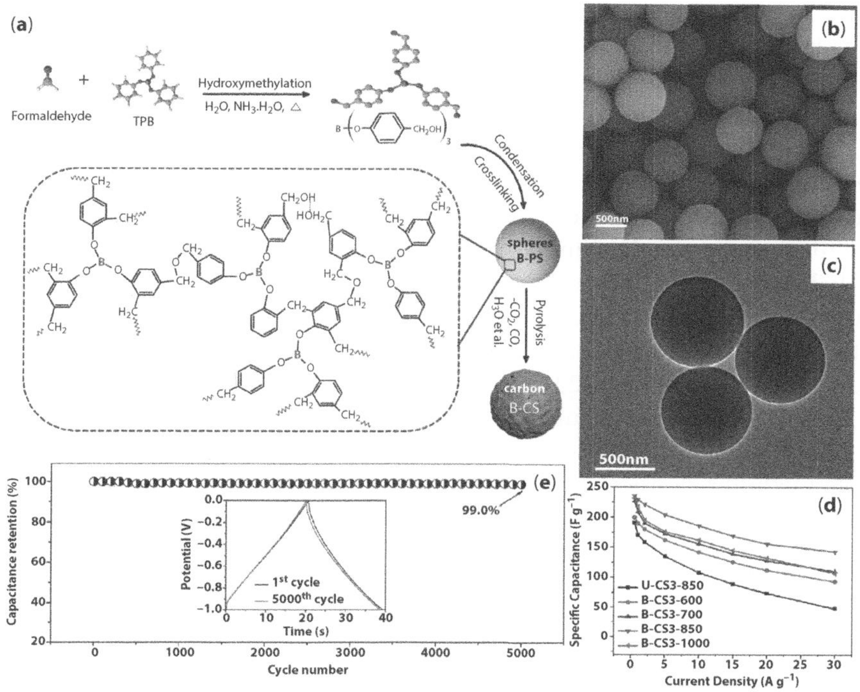

Figure 13.9 (a) Construction process of B-PS and B-CS, (b, c) SEM and TEM micrographs of B-CS3-850, (d) calculated C_{sp}, and (e) cycles stability at 10 A g^{-1} [115]. *Source:* Copyright 2022, Elsevier.

for capacitance performances [115]. B-CSs were synthesized using the HT pyrolysis process using triphenyl borate (TPB) as the B and carbon co-sources, as shown in Figure 13.9a. The microporous structure is obvious in SEM and TEM, and it has a smooth spherical shape and good homogeneity with a particle size of 780 nm (see Figures 13.9b,c). Also, the as prepared B-CSs possess a huge SSA of 560 m^2 g^{-1}, rich content of B (1.38 at%) with uniform scattering, and extraordinary surface polarity, they are contributes to pseudocapacitance development and calculated C_{sp} was shown in Figure 13.9d. By modifying the experimental conditions, the pore size, SSA, and B contents of B-CSs can be easily controlled. When compared to different pyrolyzed B-doped carbons, the acquired maximum C_{sp} of 235 F g^{-1} at the CD of 0.5 A g^{-1}, was obtained. Furthermore, later 5000 consecutive cycles, B-CS3-850 retains 99.0% of its capacity (Figure 13.9e). The introduction of active B atoms, which significantly improve conductivity and boost charge transmission, and the increase in surface area are the main reasons for this improvement. Moreover, we can believe this B-doped carbon material would provide a flexible and potential applications in sustainable SCs.

13.4.5 Phosphorus-Doped Carbons

Phosphorus exhibits chemical properties similar to nitrogen, because it has the equivalent amount of valence electrons as nitrogen. Conversely, its stronger electron-donating ability as well as bigger atomic radius, thus it was preferred for dopant in carbons [116]. Phosphorus-containing regents were used to thermally treat carbon during the carbonization and activation stages [117–119], resulting in the introduction of P on the carbon surface, while P species able to doped interior part of carbon matrix at higher temperature carbonization [120, 121]. At ease to create P-doped carbons using the first method, whereas the second method can be synthesized with precision control of the P concentration. P-doped carbons are a very intriguing idea for a diverse range of applications, like electrocatalysis, energy storage, EC sensors and so on. However, there has been limited application of P-doped AC employed as an electrode in EC capacitors, causing in a lack of accepting of its impact on physiochemical properties, limiting its potential as an active material and therefore on the complete behavior of SC [122]. When P is used as an active material, resulting the developed charge storage owing to addition of a pseudocapacitive component together with the electric double layer, electron-donor characteristics, high electrical conductivity, as well as enhanced transport capability [123].

Yi *et al.* [117] have successfully synthesized a 3D carbon framework-based on the NaH_2PO_4-$ZnCl_2$-treated cellulose-derived carbon (PZCC), which was readily prepared by a simple method schematically as shown in Figures 13.10a,b. Obviously, the HR-TEM image of the PZCC sample showing large mesopores, numerous micropores, and graphene-like layers at the pore edge are observed for the PZCC sample. The amorphous feature of the PZCC sample was established by the corresponding SAED pattern included (Figure 13.10c). Good capacitance retention of about 98%t after 10000 cycles was attained with 133 F g^{-1} at high CD of 10 A g^{-1} (Figure 13.10d).

According to Ma *et al.* [118], due to the additional capacitance resulting from the reversible redox reaction, P-doping not only improves capacitive behavior but it also prevents the formation of unstable functional groups, resulting at 3.0 V when used pure carbon (~2.5 V), resulting in an exceptionally of high ED of 38.65 Wh kg^{-1} at PD of 1500 W kg^{-1}.

Another study phytic acid (PA) utilized to treat a rich P-doped wood-derived carbon for SC, which able to establish hydrogen bonds with cellulose molecules [124]. P-doped carbon

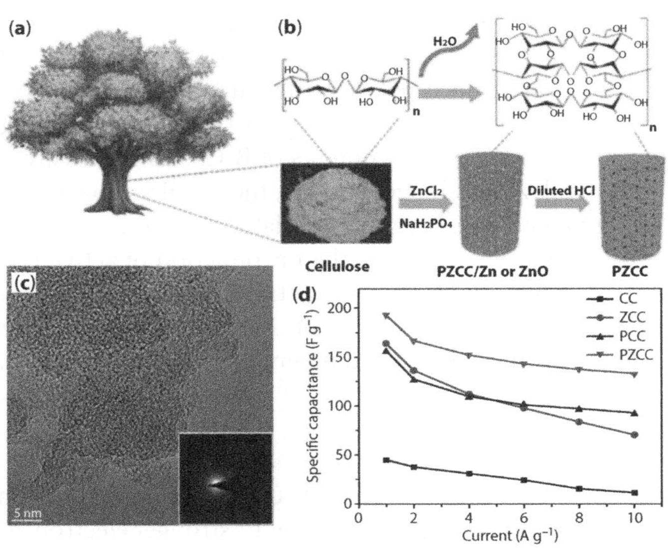

Figure 13.10 (a) The synthesis process for PZCC, (b) After carbonization, (c) HR-TEM micrographs, and (d) CD vs. capacitance for different electrodes [117]. *Source:* Copyright 2017, Elsevier.

hierarchical carbon wood (CW) construction with high P doping (9.24%) developed by employing PA-treated porous wood (as a carbon source). Negatively charged six phosphate groups in PA deliver many cross-linking spots, allowing rice levels of P-doping to be achieved, resulting in significantly higher capacitance. Figure 13.11a shows the construction of a P-doped wood-derived carbon electrode with a hierarchically porous structure. Indeed, CW possess macro-, meso-, and micropores, could help for transporting both ion and electrons, resulting in excellent EC performance. Notably, P-doping not only improved capacitive behavior, enlarged the cell's operating voltage window, resulting in increased ED. In addition, the C_{sp} of various electrodes was calculated, as shown in Figure 13.11b. Knowingly, the symmetrical SC device shown high areal, C_{sp} of 4.7 F cm^{-2} and 206.5 F g^{-1} at 1.0 mA cm^{-2} with significant retention of 90.5% after 20000 cycles at 20 mA cm^{-2}, as well as 0.94 mW h cm^{-2} (41.2 Wh kg^{-1}) at PD of 0.6 mW cm^{-2} (26.3 W kg^{-1}). It has a volumetric capacitance of 29.3 F cm^{-3} and an ED of 5.8 Wh cm^{-3}. In addition, LED (1.8 V, 1.0 W) able to lighted for 20 minutes in two series (Figure 13.11c inset), demonstrating that it has practical application potential. An excellent performance could be ascribed to hierarchical structure, low tortuosity, and rich content of P-doping, as well as the superb hierarchical structure, low tortuosity, and quick accessible pore channels at high CD. Our findings also suggest that the CW-P electrode could be used in SC devices, and they inspire future study into dense electrodes.

Surface oxygen functionalities like carboxylic groups and quinones are active component which have been shown to be electrochemically active and able to improve the overall behavior of the cell [125]. On the other hand, S-doping is inherently unstable and can result in a decrease in capacitive performance [126]. When P is incorporated into the carbon structure, it acts as an oxidation protector, preventing oxygen species from combusting, resulting in improved cell performance as well as an obstruction in the creation of electrophilic oxygen species [127]. Notably, P-doping can help achieve higher capacitive performance while also providing other benefits like enhanced conductance and lower charge

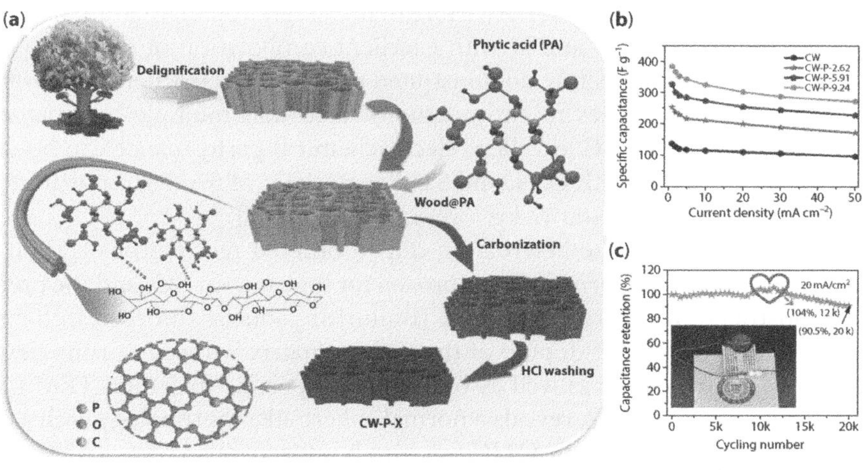

Figure 13.11 (a) Construction of CW-P-X electrodes, (b) C_{sp} of all modified electrodes, (c) The SSC device was tested at 20 mA cm^{-2} for 20000 cycles, and LED (1.8 V, 1.0 W) in two series [124]. *Source:* Copyright 2021, Elsevier.

transfer resistance. Nonetheless, much study has required to understand the essential reasons for these advances in order to commercialize P-doped carbons as electrodes for EC capacitors. Compared to carbon-containing B, N, P, and S-doping or functionalization is a simple, robust, and representing favorable way for creating novel active sites with improved capacitive performance [91].

13.4.6 Co-Doping of Carbons

To improve the performance of SCs, scientists have utilized a multiple doping of elements (N, S, B, and P). However, there is still probability to enhance the capacitance of these EC devices, which able to accomplished by co-doped carbons [128, 129] with a different composition (N/B, N/S, or more than two heteroatoms) has been employed [130–132]. In recent years, scientists have attempted to investigate the influence of multiple doping and the performance in energy storage [133, 134]. By altering aspects including wetting behavior toward the electrolyte, adding pseudocapacitive species, and lowering charge transfer resistance, more than one element able to increase the capacitive behavior of carbon [135]. Carbon materials are tuned in a desirable manner using N, B, P, and S heteroatoms in various combinations to improve the behavior of energy storage devices [136, 137].

According to Wang *et al.* [138] reported capacitance behavior of N and S co-doped carbon examined the synergetic pseudocapacitive contribution which was provided by N and S elements. The N, S, and N/S co-doped materials were utilized in SC cells in 6 M KOH and obtained the C_{sp} values of 371, 282, and 566 F g^{-1}, respectively. The C_{sp} of 272 F g^{-1} was achieved for IL electrolyte at a high CD of 0.1 A g^{-1} using N and S co-doped hollow cellular carbon nano-capsules [139]. When used as an electrode, N and S co-doped graphene aerogel (NS-GA) demonstrated a great ED of 100.7 Wh kg^{-1} at PD of 0.94 kW kg^{-1}, once the largest values even achieved for this kind of material. However, the IL ($EMIMBF_4$) used as an cell electrolyte, it deliver a high C_{sp} of 203 F g^{-1} and CD at 1.0 A g^{-1} [140].

Likewise, Chen *et al.* demonstrated the N and P co-doping effects shows an extraordinary C_{sp} of 337 F g^{-1} at 0.5 A g^{-1}, with EDs ranging from 23.1 to 12.4 Wh kg^{-1} and PDs ranging from 720.4 to 13950 W kg^{-1} [141]. More than a single dopant to promote the behavior of carbon electrode using B and N with excellent combination of these two elements and synergistic effects have recently demonstrated excellent electrochemical performance [142–145]. Zhao and colleagues recently demonstrated that multiple heteroatoms were introduced to highly porous AC, excellent electrochemical performance can be obtained. When carbon was co-doped with O, N, and S elements, a C_{sp} of 576 F g^{-1} and an extraordinary ED of 107 Wh kg^{-1} at PD 900 W kg^{-1} were obtained [146].

Du *et al.* [147] developed the heteroatom-doped ordered mesoporous carbon spheres (N/B-OMC) via simple ionothermal (IT) approach for high energy SCs and the preparation process was illustrated in Figure 13.12a. The $[Bmim]BF_4$ able to absorbed in the pores of PF spheres, resulting in N/B co-doping of the carbon matrix without causing destruction to the sphere-like morphology as well as ordered mesoporous structure. TEM picture of OMC, as shown in Figure 13.12b, reveals a normal sphere-like morphology with a diameter of ~90 nm. The morphology of N/B-OMC was also homogeneous sphere-like morphology with the identical diameter as shown in TEM, demonstrating that the morphology was not disrupted following IT treatment (Figure 13.12c). As SC electrode materials, N/B-OMC exhibits an obvious improvement in EC behavior with a high C_{sp} of 272 F g^{-1} at 0.5 A g^{-1}

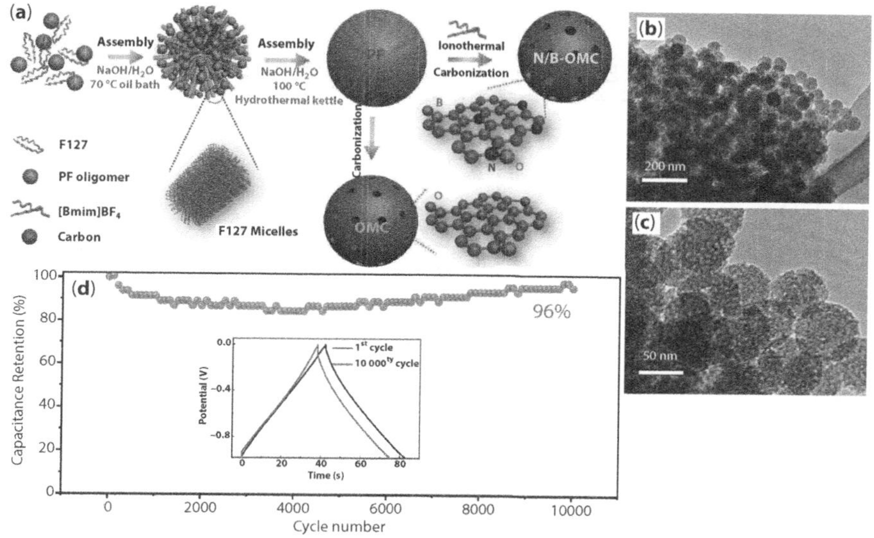

Figure 13.12 (a) Schematic illustration for synthesis of N/B-OMC by IT approach, (b,c) TEM images of OMC and N/B-OMC, (d) Cycle stability [147]. *Source:* Copyright 2021, Elsevier.

and a high rate capacity of 80.8% from 0.5–10 A g^{-1} and outstanding cyclic stability (4% capacitance) loss after 10000 cycles (Figure 13.12c). These findings suggest that N/B-OMC is a superior electrode material in terms of electrochemical stability. As a consequence, N/B-dual-heteroatom OMC's doping, high SSA, and increased mesopore size can improve EC performance, demonstrating its potential as a novel electrode material. Moreover, this IT approach may also afford a novel concept for multiple element doping in carbonaceous materials to increase pore size and SSA for improved EC performance.

Researchers have only recently attempted to estimate the effect of multiple doping are presented and the overall EC behavior was calculated [148–150]. In particular, N and S are thought to be an expected combination for maximal cell output, although a great compact of research is still needed to fine-tune composition of other dopants to enhance material conductivity. The influence of surface functional groups other than N in various combinations when used together with non-aqueous electrolytes to obtain high value of ED's remains a wide area of study investigation.

In summary, we compiled the reports on some heteroatom doped carbon-based materials which can play the role of electrodes in SCs, delivering enhanced electrochemical performances. Te related research results are briefly demonstrated in Table 13.2.

13.5 Conclusions

Despite the fact that N-doped carbons have been widely explored in EC SCs, in this study we exploring S, P, and B with physio/chemical properties that are ideal for greater EC cell outputs. The performance of an EC cell can be significantly developed by using these developing functional materials as electrodes. Because an incorporation of N to carbons, which results in better wettability behavior, benefits to sustain cell's low resistance, and maintaining

Table 13.2 Electrochemical performances of carbon-based sustainable catalysts.

Electrode	Method	Electrolyte	C_{sp}[a] ($F\,g^{-1}$), CD[b] ($A\,g^{-1}$)	ED[c] ($Wh\,kg^{-1}$), PD[d] ($W\,kg^{-1}$)	[c]CR (%), [d]CN	Ref.
N-doped porous activated carbon (NPC)	Carbonization	6.0 M KOH	250, 0.5	8.46, 123	92, 10000	[82]
Porous N-doped carbon (PNC)	Carbonization	1.0 M KOH	683 mF cm^{-2}, 2 mA cm^{-2}	55 , 1800	96, 10000	[83]
N-doped reduced graphene oxide (N-rGO)	Hydrothermal	$BMIMBF_4$	130, 1.0	55 , 1800	96.6, 20000	[85]
Olive mill wastewater (OMW)	Solvothermal	1.0 M H_2SO_4	325, 0.125	37, 274	99.3, 12500	[86]
Bamboo-like carbon nanotubes containing sulfur (BCNT-S)	Carbonization	6.0 M KOH	259, 1	–	97.7, 1000	[87]
S-doped graphene aerogel (GATUF)	Lyophilization	1.0 M KOH	1089, 1	43, 38	76, 10000	[89]
S-doped mesoporous carbons (SMCs)	Carbonization	6.0 M KOH	150, 1	–	95, 500	[92]
S-doped porous carbon (S-PC)	Carbonization	6.0 M KOH	320, 0.2	160, 5161	99, 20000	[93]
N,S-doped elm flower-derived activated carbon (N,S-ELAC)	Hydrothermal	6.0 M KOH	275, 1	16.8, 600	87.2, 5000	[94]
Hierarchically porous B-doped carbon (B-HPC)	Carbonization	1.0 M Na_2SO_4	379.9, 1	25.9, 447.2	96.3, 10000	[97]
B-doped graphene (BG-900)	Carbonization	6.0 M KOH	172.5, 0.5	3.86, 125	96.5, 5000	[111]
B-doped mesoporous carbons (BOMCs)	Nanocasting	6.0 M KOH	268, 5 mV s^{-1}	–	92, 10000	[112]
Ordered mesoporous carbons (OMCs)	Nanocasting	6.0 M KOH	221.8, 1	6.53, 5000	90, 10000	[113]

(Continued)

Table 13.2 Electrochemical performances of carbon-based sustainable catalysts. (*Continued*)

Electrode	Method	Electrolyte	C_{sp} [a] ($F\ g^{-1}$), CD[b] ($A\ g^{-1}$)	ED[c] ($Wh\ kg^{-1}$), PD[d] ($W\ kg^{-1}$)	[c]CR (%), [d]CN	Ref.
Thermally reduced graphene oxide with boron doping (BT-rGO)	Carbonization	6.0 M KOH	448, 10 mV S^{-1}	–	50, 3000	[114]
B-doped carbon spheres (B-CSs)	Hydrothermal	6.0 M KOH	235, 0.5	7.1, 250	99, 5000	[115]
NaH_2PO_4-$ZnCl_2$-treated cellulose-derived carbon (PZCC)	Carbonization	6.0 M KOH	133, 10	4.7, 833	97.9, 10000	[117]
KOH activated carbon (CK)	Carbonization	1.0 M Et_4NBF_4/PC	121, 1	38.65,· 1500	90.2, 20000	[118]
P-doped three-dimensional hierarchical porous carbons (P-3DHPCs)	Carbonization	6.0 M KOH	367, 0.3	8.5–11.11	88, 10000	[120]
Hierarchical porous N-doped carbon (HPNC)	Carbonization	EMI-BF_4	312, 1	44.7, 73.1	98, 200000	[123]
[f]Carbonized wood with P-doping (CW-P-9.24)	Carbonization	6.0 M KOH	206.5, 1.0 mA cm^{-2}	41.2, 26.3	90.5, 20000	[124]
N-doped active carbon (NAC)	Carbonization	6.0 M KOH	417, 0.5	–	99, 5000	[128]
Hierarchical multi-heteroatoms co-doped PC (MHPC)	Carbonization	6.0 M KOH	170, 0.5	8.2, 118.5	94.8, 14000	[129]
Tar-glucose carbon (TGC)	Hydrothermal	6.0 M KOH	199 ± 8.2, 10	6.1, 100	92, 5000	[133]
B,N co-doped porous carbons (BNCs)	Carbonization	6.0 M KOH	268, 0.1	3.0, 201	72, –	[136]
N,S-doped graphene hydrogels (NS-G)	Hydrothermal	6.0 M KOH	566, 0.5	29.4, 10	95, 2000	[138]

(*Continued*)

Table 13.2 Electrochemical performances of carbon-based sustainable catalysts. (*Continued*)

Electrode	Method	Electrolyte	C_{sp}[a] ($F\ g^{-1}$), CD[b] ($A\ g^{-1}$)	ED[c] ($Wh\ kg^{-1}$), PD[d] ($W\ kg^{-1}$)	[c]CR (%), [d]CN	Ref.
N,S co-doped hollow cellular carbon nanocapsules (h-CNST)	Carbonization	1.0 M H_2SO_4	272, 0.1	8.5, 2620	90.3, 2000	[139]
N,S co-doped graphene aerogel (NS-GA)	Carbonization	$EMIMBF_4$	203, 1.0	100.7, 0.94	90, 3000	[140]
N/P co-doped hierarchical porous carbon (CNP)	Carbonization	6.0 M KOH	337, 0.5	23.1, 720.4	97.8, 5000	[141]
B,N co-doped carbon materials (CNBs)	Carbonization	1.0 M H_2SO_4	299.4, 0.2	–	100, 1200	[142]
Activated hydrogel-derived heteroatom-doped (B,N co-doping) porous carbon networks (a-HHPC)	Carbonization	6.0 M KOH	214, 0.2	–	89, 3000	143]
B,N co-doped carbon (BNDC)	Annealing	1.0 M H_2SO_4	504, 1.0	22.7, 0.2	104.8, 5000	[144]
B/N co-doped 2D porous carbon nanosheets (BNHCs)	Carbonization	6.0 M KOH	311, 0.2	–	94.8, 10000	[145]
O-N-S co-doped hierarchical porous carbons (HPCs)	Carbonization	$EMIMBF_4$	576, 1.0	107, 900	95, 10000	[146]
N/B-OMC	Ionothermal	$[Bmim]BF_4$	272, 0.5	7.3, 214	96, 10000	[147]

[a]Specific capacitance.
[b]Current density.
[c]Capacitance retention.
[d]Cycle number.

the cell's high PD. When compared to N-doping, doping carbons with P resulted as better physio/chemical characteristics. Other advantages of utilizing P-doped carbons include a rise in the SC cell's working potential, which can improve its ED. Likewise, S-doping able to effective in raising the electrical reactivity of carbons, causing in a greater pseudocapacitive involvement, while compared to un-doped carbons.

However, recent focus has shifted to the use of dual dopants, where the synergistic impact on combined elements result in even better behavior. Despite the fact that N has been thoroughly studied and yielded promising results, a massive research effort is still required to investigate additional functional materials, as this topic is still in its infancy and has received little attention. Although these doping carbons have already revealed enormous potential, energy storage experts will be interested to watch how developments in doped carbons will begin to revolutionize the field of ACs for prospective energy uses in common, and applicability in SCs.

References

1. Wu, M., Liao, J., Yu, L., Lv, R., Li, P., Sun, W., Tan, R., Duan, X., Zhang, L., Li, F., Kim, J., Roadmap on carbon materials for energy storage and conversion. *Chem.–Asian J.*, 15, 995, 2020.
2. Wang, J., Nie, P., Ding, B., Dong, S., Hao, X., Dou, H., Zhang, X., Biomass derived carbon for energy storage devices. *J. Mater. Chem. A*, 5, 2411, 2017.
3. Sun, J., Luo, B., Li, H., A Review on the conventional capacitors, supercapacitors, and emerging hybrid ion capacitors: Past, present, and future. *Adv. Energy. Sustainability Res.*, 3, 2100191, 2022.
4. Liu, J., Wang, J., Xu, C., Jiang, H., Li, C., Zhang, L., Lin, J., Shen, Z.X., Advanced energy storage devices: Basic principles, analytical methods, and rational materials design. *Adv. Sci.*, 5, 1700322, 2018.
5. He, Y., Chen, W., Gao, C., Zhou, J., Li, X., Xie, E., An overview of carbon materials for flexible electrochemical capacitors. *Nanoscale*, 5, 8799, 2013.
6. Guo, Y.G., Hu, J.S., Wan, L.J., Nanostructured materials for electrochemical energy conversion and storage devices. *Adv. Mater.*, 20, 2878, 2008.
7. Yang, Y., Han, Y., Jiang, W., Zhang, Y., Xu, Y., Ahmed, A.M., Application of the supercapacitor for energy storage in China: Role and strategy. *Appl. Sci.*, 12, 354, 2021.
8. Wang, L., Xie, X., Dinh, K.N., Yan, Q., Ma, J., Synthesis, characterizations, and utilization of oxygen-deficient metal oxides for lithium/sodium-ion batteries and supercapacitors. *Coord. Chem. Rev.*, 397, 138, 2019.
9. Dong, Y., Feng, Y., Deng, J., He, P., Ma, J., Electrospun Sb_2Se_3@C nanofibers with excellent lithium storage properties. *Chin. Chem. Lett.*, 31, 909, 2020.
10. Ni, J.F. and Li, Y., Carbon nanomaterials in different dimensions for electrochemical energy storage. *Adv. Energy Mater.*, 6, 1600278, 2016.
11. Kong, D., Gao, Y., Xiao, Z., Xu, X., Li, X., Zhi, L., Rational design of carbon-rich materials for energy storage and conversion. *Adv. Mater.*, 31, 1804973, 2019.
12. Hu, C., Lin, Y., Connell, J.W., Cheng, H.M., Gogotsi, Y., Titirici, M.M., Dai, L., Carbon-based metal-free catalysts for energy storage and environmental remediation. *Adv. Mater.*, 31, 1806128, 2019.

13. Yan, J., Wang, Q., Lin, C., Wei, T., Fan, Z., Interconnected frameworks with a sandwiched porous carbon layer/graphene hybrids for supercapacitors with high gravimetric and volumetric performances. *Adv. Energy Mater.*, 4, 1400500, 2014.
14. Veerakumar, P., Rajkumar, C., Chen, S.M., Thirumalraj, B., Lin, K.C., Activated porous carbon supported rhenium composites as electrode materials for electrocatalytic and supercapacitor applications. *Electrochim. Acta*, 271, 433, 2018.
15. Li, Z., Yang, Q., Fan, K., Xie, W., Qiao, W., Shao, M., Wei, M., Atom–economical construction of carbon nanotube architectures for flexible supercapacitors with ultrahigh areal and volumetric capacities. *J. Mater. Chem. A*, 6, 21287, 2018.
16. Chen, S., Qiu, L., Cheng, H.M., Carbon-based fibers for advanced electrochemical energy storage devices. *Chem. Rev.*, 120, 2811, 2020.
17. Gu, P., Liu, W., Hou, Q., Ni, Y., Lignocellulose-derived hydrogel/aerogel-based flexible quasi-solid-state supercapacitors with high-performance: A review. *J. Mater. Chem. A*, 9, 14233, 2021.
18. Dong, D., Zhang, Y., Xiao, Y., Wang, T., Wang, J., Romero, C.E., Pan, W.P., High performance aqueous supercapacitor based on nitrogen-doped coal-based activated carbon electrode materials. *J. Colloid Interface Sci.*, 580, 77, 2020.
19. Mohanty, A., Jaihindh, D., Fu, Y.P., Senanayak, S.P., Mende, L.S., Ramadoss, A., An extensive review on three-dimension architectural Metal-Organic Frameworks towards supercapacitor application. *J. Power Sources*, 488, 229444, 2021.
20. Belaineh, D., Brooke, R., Sani, N., Say, M.G., Håkansson, K.M., Engquist, I., Berggren, M., Edberg, J., Printable carbon-based supercapacitors reinforced with cellulose and conductive polymers. *J. Energy Storage*, 50, 104224, 2022.
21. Zhang, L.L. and Zhao, X.S., Carbon-based materials as supercapacitor electrodes. *Chem. Soc. Rev.*, 38, 2520, 2009.
22. Olabi, A.G., Abdelkareem, M.A., Energy storage systems towards 2050. *Energy*, 219, 119634, 2021.
23. Abbas, Q., Mirzaeian, M., Ogwu, A.A., Electrochemical performance of controlled porosity resorcinol/formaldehyde based carbons as electrode materials for supercapacitor applications. *Int. J. Hydrogen Energy*, 42, 25588, 2017.
24. Patil, A.M., Moon, S., Seo, Y., Roy, S.B., Jadhav, A.A., Dubal, D.P., Kang, K., Jun, S.C., Reconfiguring the electronic structure of heteroatom doped carbon supported bimetallic oxide@metal sulfide core–shell heterostructure via *in situ* Nb incorporation toward extrinsic pseudocapacitor. *Small*, 19, 2205491 2023.
25. Zhou, J., Hou, L., Lian, J., Cheng, W., Wangv, D., Gou, H., Gao, F., Nitrogen-doped highly dense but porous carbon microspheres with ultrahigh volumetric capacitance and rate capability for supercapacitors. *J. Mater. Chem. A*, 7, 476, 2019.
26. Wu, Z.S., Tan, Y.Z., Zheng, S., Wang, S., Parvez, K., Qin, J., Shi, X., Sun, C., Bao, X., Feng, X., Müllen, K., Bottom-up fabrication of sulfur-doped graphene films derived from sulfur-annulated nanographene for ultrahigh volumetric capacitance micro-supercapacitors. *J. Am. Chem. Soc.*, 139, 4506, 2017.
27. Wang, D., Wang, Z., Li, Y., Dong, K., Shao, J., Luo, S., Liu, Y., Qi, X., *In situ* double-template fabrication of boron-doped 3D hierarchical porous carbon network as anode materials for Li- and Na-ion batteries. *Appl. Surf. Sci.*, 464, 422, 2019.
28. Patiño, J., López-Salas, N., Gutiérrez, M.C., Carriazo, D., Ferrer, M.L., del Monte, F., Phosphorus-doped carbon–carbon nanotube hierarchical monoliths as true three-dimensional electrodes in supercapacitor cells. *J. Mater. Chem. A*, 4, 1251, 2016.
29. Li, Y., Huang, G., Geng, Q., Liu, Y., Li, X., Yao, Y., Liu, Y., Xing, B., Liu, Q., Jia, J., Zhang, C., Nickel salts-induced microstructure modification of B–N co-doped porous carbons for high-performance supercapacitor electrodes. *J. Alloys Compd.*, 895, 162652, 2022.

30. Zhang, X., Yan, G., Li, H., Li, Z., Chen, J., Wang, L., Wu, Y., Slit needleless electrospun heteroatoms-doped hollow porous carbon nanofibers for solid-state flexible supercapacitors. *J. Alloys. Compds*, 943, 169188, 2023.
31. Nikolaidis, P. and Poullikkas, A., A comparative review of electrical energy storage systems for better sustainability. *J. Power Technol.*, 97, 220, 2017.
32. Chen, H., Ngoc, T., Yang, W., Tan, C., Li, Y., Progress in electrical energy storage system: A critical review. *Prog. Nat. Sci.*, 19, 291, 2009.
33. Wagner, L., Overview of energy storage methods, *MORA Assoc.*, p 15, 2007.
34. Choudhary, R.B., Ansari, S., Purty, B., Robust electrochemical performance of polypyrrole (PPy) and polyindole (PIn) based hybrid electrode materials for supercapacitor application: A review. *J. Energy Storage*, 29, 101302, 2020.
35. Yuan, L., Lu, X.H., Xiao, X., Zhai, T., Dai, J., Zhang, F., Hu, B., Wang, X., Gong, L., Chen, J., Hu, C., Tong, Y., Zhou, J., Wang, Z.L., Flexible solid-state supercapacitors based on carbon nanoparticles/MnO_2 nanorods hybrid structure. *ACS Nano*, 6, 656, 2012.
36. El-Kady, M.F., Strong, V., Dubin, S., Kaner, R.B., Laser scribing of high-performance and flexible graphene-based electrochemical capacitors. *Science*, 335, 1326, 2012.
37. Davies, A., Audette, P., Farrow, B., Hassan, F., Chen, Z., Choi, J.-Y., Yu, A., Graphene-based flexible supercapacitors: Pulse-electropolymerization of polypyrrole on free-standing graphene films. *J. Phys. Chem. C*, 115, 17612, 2011.
38. Meng, C., Liu, C., Chen, L., Hu, C., Fan, S., Highly flexible and all-solid-state paperlike polymer supercapacitors. *Nano Lett.*, 2010, 10, 4025.
39. Wang, Y. and Xia, Y., Recent progress in supercapacitors: From materials design to system construction. *Adv. Mater.*, 25, 5336, 2013.
40. Lang, X., Hirata, A., Fujita, T., Chen, M., Nanoporous metal/oxide hybrid electrodes for electrochemical supercapacitors. *Nat. Nanotechnol.*, 6, 232, 2011.
41. Wang, G., Zhang, L., Zhang, J., A review of electrode materials for electrochemical supercapacitors. *Chem. Soc. Rev.*, 41, 797, 2012.
42. Veerakumar, P., Sangili, A., Manavalan, S., Thanasekaran, P., Lin, K.C., Research progress on porous carbon supported metal/metal oxide nanomaterials for supercapacitor electrode applications. *Ind. Eng. Chem. Res.*, 59, 6347, 2020.
43. Zhong, J.H., Wang, A.L., Li, G.R., Wang, J.W., Ou, Y.N., Tong, Y.X., Co_3O_4/$Ni(OH)_2$ composite mesoporous nanosheet networks as a promising electrode for supercapacitor applications. *J. Mater. Chem.*, 22, 5656, 2012.
44. Burke, A., R&D considerations for the performance and application of electrochemical capacitors. *Electrochim. Acta*, 53, 1083, 2007.
45. Srinivasan, V. and Weidner, J.W., Studies on the capacitance of nickel oxide films: Effect of heating temperature and electrolyte concentration. *J. Electrochim. Soc.*, 147, 880, 2000.
46. Azais, P., Duclaux, L., Florian, P., Causes of supercapacitors ageing in organic electrolyte. *J. Power Sources*, 171, 1046, 2007.
47. Shulga, Y.M., Baskakov, S.A., Smirnov, V.A., Graphene oxide films as separators of polyaniline-based supercapacitors. *J. Power Sources*, 245, 33, 2014.
48. Samantara, A.K. and Ratha, S., Components of supercapacitor, in: *Materials Development for Active/Passive Components of a Supercapacitor.* Springer Briefs in Materials, pp. 11–39, Springer, Singapore, 2018.
49. Choudhary, N., Li, C., Moore, J., Nagaiah, N., Zhai, L., Jung, Y., Thomas, J., Asymmetric supercapacitor electrodes and devices. *Adv. Mater.*, 29, 1605336, 2017.
50. Ghosh, A. and Lee, Y.H., Carbon-based electrochemical capacitors. *ChemSusChem*, 5, 480, 2012.

51. Jiang, H., Li, C., Sun, T., Ma, J.A., green and high energy density asymmetric supercapacitor based on ultrathin MnO_2 nanostructures and functional mesoporous carbon nanotube electrodes. *Nanoscale*, 4, 807, 2012.
52. Dsoke, S., Pfeifer, K., Zhao, Z., The role of nanomaterials for supercapacitors and hybrid devices, in: *Frontiers of Nanoscience*, pp. 99–136, Elsevier, 2021.
53. Hadjipaschalis, I., Poullikkas, A., Efthimiou, V., Overview of current and future energy storage technologies for electric power applications. *Renew. Sustainable Energy Rev.*, 13, 1513, 2009.
54. Iqbal, S., Khatoon, H., Pandit, A.H., Ahmad, S., Recent development of carbon based materials for energy storage devices. *Mater. Sci. Energy Technol.*, 2, 417, 2019.
55. Xu, B., Wu, F., Chen, R., Cao, G., Chen, S., Zhou, Z., Yang, Y., Highly mesoporous and high surface area carbon: A high capacitance electrode material for EDLCs with various electrolytes. *Electrochem. Commun.*, 10, 795, 2008.
56. Veerakumar, P., Maiyalagan, T., Raj, B.G.S., Guruprasad, K., Jiang, Z., Lin, K.C., Paper flower-derived porous carbons with high-capacitance by chemical and physical activation for sustainable applications. *Arab. J. Chem.*, 13, 2995, 2020.
57. Liu, C., Li, F., Ma, L.P., Cheng, H.M., Advanced materials for energy storage. *Adv. Mater.*, 22, E28, 2010.
58. Zhu, Y., Murali, S., Stoller, M.D., Ganesh, K.J., Cai, W., Ferreira, P.J., Pirkle, A., Wallace, R.M., Cychosz, K.A., Thommes, M., Su, D., Carbon-based supercapacitors produced by activation of graphene. *Science*, 332, 1537, 2011.
59. Conway, B.E., Transition from "supercapacitor" to "battery" behavior in electrochemical energy storage. *J. Electrochem. Soc.*, 138, 1539, 1991.
60. Hu, C.C., Chang, K.H., Lin, M.C., Wu, Y.T., Design and tailoring of the nanotubular arrayed architecture of hydrous RuO_2 for next generation supercapacitors. *Nano Lett.*, 6, 2690, 2006.
61. Gao, Y., Chen, S., Cao, D., Wang, G., Yin, J., Electrochemical capacitance of Co_3O_4 nanowire arrays supported on nickel foam. *J. Power Sources*, 195, 1757, 2010.
62. Yuan, C., Zhang, X., Su, L., Gao, B., Shen, L., Facile synthesis and self-assembly of hierarchical porous NiO nano/micro spherical superstructures for high performance supercapacitors. *J. Mater. Chem.*, 19, 5772, 2009.
63. Kim, B.K., Chabot, V., Yu, A., Carbon nanomaterials supported $Ni(OH)_2$/NiO hybrid flower structure for supercapacitor. *Electrochim. Acta*, 109, 370, 2013.
64. Zhang, J., Kong, L.B., Li, H., Luo, Y.C., Kang, L., Synthesis of polypyrrole film by pulse galvanostatic method and its application as supercapacitor electrode materials. *J. Mater. Sci.*, 45, 1947, 2010.
65. Conway, B.E., Birss, V., Wojtowicz, J., The role and utilization of pseudocapacitance for energy storage by supercapacitors. *J. Power Sources*, 66, 1, 1997.
66. Geim, A.K. and Novoselov, K.S., The rise of graphene. *Nat. Mater.*, 6, 183, 2007.
67. Simon, P. and Gogotsi, Y., Capacitive energy storage in nanostructured carbon–electrolyte systems. *Acc. Chem. Res.*, 46, 1094, 2013.
68. Li, K.K., Liu, G.Y., Zheng, L.S., Jia, J., Zhu, Y.Y., Zhang, Y.T., Coal-derived carbon nanomaterials for sustainable energy storage applications. *New Carbon Mater.*, 36, 133, 2021.
69. Allen, M.J., Tung, V.C., Kaner, R.B., Honeycomb carbon: A review of graphene. *Chem. Rev.*, 110, 132, 2010.
70. Xia, J., Chen, F., Li, J., Tao, N., Measurement of the quantum capacitance of graphene. *Nat. Nanotechnol.*, 4, 505, 2009.
71. Raymundo-Pinero, E., Kierzek, K., Machnikowski, J., Béguin, F., Relationship between the nanoporous texture of activated carbons and their capacitance properties in different electrolytes. *Carbon*, 44, 2498, 2006.

72. Abbas, Q., Raza, R., Shabbir, I., Olabi, A.G., Heteroatom doped high porosity carbon nanomaterials as electrodes for energy storage in electrochemical capacitors: A review. *J. Sci.: Adv. Mater. Dev.*, 4, 341, 2019.
73. Huang, Y., Candelaria, S.L., Li, Y., Li, Z., Tian, J., Zhang, L., Cao, G., Sulfurized activated carbon for high energy density supercapacitors. *J. Power Sources*, 252, 90, 2014.
74. Li, L., Zhong, Q., Kim, N.D., Ruan, G., Yang, Y., Gao, C., Fei, H., Li, Y., Ji, Y., Tour, J.M., Nitrogen-doped carbonized cotton for highly flexible supercapacitors. *Carbon*, 105, 260, 2016.
75. He, D., Wu, L., Yao, Y., Zhang, J., Huang, Z.H., Wang, M.X., A facile route to high nitrogen-containing porous carbon fiber sheets from biomass-flax for high-performance flexible supercapacitors. *Appl. Surf. Sci.*, 507, 145108, 2020.
76. Zou, K., Deng, Y., Chen, J., Qian, Y., Yang, Y., Li, Y., Chen, G., Hierarchically porous nitrogen-doped carbon derived from the activation of agriculture waste by potassium hydroxide and urea for high-performance supercapacitors. *J. Power Sources*, 378, 579, 2018.
77. Arango, D.I., Zapata-Benabithe, Z., Arenas, E.C., Perez-Osorno, J.C., Influence of surface modification with nitric acid on electrochemical performance of agro-industrial waste-based activated carbon. *J. Mater. Sci. Mater. Electron.*, 29, 15557, 2018.
78. Li, S. and Fan, Z., Nitrogen-doped carbon mesh from pyrolysis of cotton in ammonia as binder-free electrodes of supercapacitors. *Microporous Mesoporous Mater.*, 274, 313, 2019.
79. Zhao, L., Baccile, N., Gross, S., Zhang, Y., Wei, W., Sun, Y., Antonietti, M., Titirici, M.M., Sustainable nitrogen-doped carbonaceous materials from biomass derivatives. *Carbon*, 48, 3778, 2010.
80. Sun, J., Niu, J., Liu, M., Ji, J., Dou, M., Wang, F., Biomass-derived nitrogen-doped porous carbons with tailored hierarchical porosity and high specific surface area for high energy and power density supercapacitors. *Appl. Surf. Sci.*, 427, 807, 2018.
81. Frackowiak, E. and Beguin, F., Carbon materials for the electrochemical storage of energy in capacitors. *Carbon*, 39, 937, 2001.
82. Han, X., Jiang, H., Zhou, Y., Hong, W., Zhou, Y., Gao, P., Ding, R., Liu, E., A high performance nitrogen-doped porous activated carbon for supercapacitor derived from pueraria. *J. Alloys Compd.*, 744, 544, 2018.
83. Mao, Z., Zhao, S., Wang, J., Zeng, Y., Lu, X., Tong, Y., Facile synthesis of nitrogen-doped porous carbon as robust electrode for supercapacitors. *Mater. Res. Bull.*, 101, 140, 2018.
84. Yue, Y., Huang, Y.L., Bian, S.W., Nitrogen-doped hierarchical porous carbon films derived from metal–organic framework/cotton composite fabrics as freestanding electrodes for flexible supercapacitors. *ACS Appl. Electron. Mater.*, 3, 2178, 2021.
85. Dai, S., Liu, Z., Zhao, B., Zeng, J., Hu, H., Zhang, Q., Chen, D., Qu, C., Dang, D., Liu, M., A high-performance supercapacitor electrode based on N-doped porous graphene. *J. Power Sources*, 387, 43, 2018.
86. Elmouwahidi, A., Castelo-Quibén, J., Vivo-Vilches, J.F., Perez-Cadenas, A.F., Maldonado-Hódar, F.J., Carrasco-Marin, F., Activated carbons from agricultural waste solvothermally doped with sulphur as electrodes for supercapacitors. *Chem. Eng. J.*, 334, 1835, 2018.
87. Yang, Y., Liu, L., Tang, Y., Zhang, Y., Jia, D., Kong, L., Bamboo-like carbon nanotubes containing sulfur for high performance supercapacitors. *Electrochim. Acta*, 191, 846, 2016.
88. Shaheen Shah, S., Abu Nayem, S.M., Sultana, N., Saleh Ahammad, A.J., Abdul Aziz, M., Preparation of sulfur-doped carbon for supercapacitor applications: A review. *ChemSusChem*, 15, e202101282, 2022.
89. Lee, W.S.V., Leng, M., Li, M., Huang, X.L., Xue, J.M., Sulphur-functionalized graphene towards high performance supercapacitor. *Nano Energy*, 12, 250, 2015.
90. Ghosh, S., Barg, S., Jeong, S.M., Ostrikov, K., Heteroatom-doped and oxygen-functionalized nanocarbons for high-performance supercapacitors. *Adv. Energy Mater.*, 10, 2001239, 2020.

91. Zhao, X., Zhang, Q., Chen, C.M., Zhang, B., Reiche, S., Wang, A., Zhang, T., Schlögl, R., Su, D.S., Aromatic sulfide, sulfoxide, and sulfone mediated mesoporous carbon monolith for use in supercapacitor. *Nano Energy*, 1, 624, 2012.
92. Ma, X., Ning, G., Kan, Y., Ma, Y., Qi, C., Chen, B., Li, Y., Lan, X., Gao, J., Synthesis of S-doped mesoporous carbon fibres with ultrahigh S concentration and their application as high performance electrodes in supercapacitors. *Electrochim. Acta*, 150, 108, 2014.
93. Tian, J., Zhang, H., Liu, Z., Qin, G., Li, Z., One-step synthesis of 3D sulfur-doped porous carbon with multilevel pore structure for high-rate supercapacitors. *Int. J. Hydrogen Energy*, 43, 1596, 2018.
94. Chen, H., Yu, F., Wang, G., Chen, L., Dai, B., Peng, S., Nitrogen and sulfur self-doped activated carbon directly derived from elm flower for high-performance supercapacitors. *ACS Omega*, 3, 4724–4732, 2018.
95. Peng, J., Dai, X., Huang, J., Zeng, J., Zheng, L., Chen, H., High-yield preparation of B/N co-doped porous carbon nanosheets from a cross-linked boronate polymer for supercapacitor applications. *J. Energy Storage*, 59, 106498, 2023.
96. Joshi, A., Tomar, A.K., Singh, G., Sharma, R.K., Engineering oxygen defects in the boron nanosheet for stabilizing complex bonding structure: An approach for high-performance supercapacitor. *Chem. Eng. J.*, 407, 127122, 2021.
97. Du, P., Liu, L., Dong, Y., Li, W., Li, J., Liu, Z., Wang, X., Synthesis of hierarchically porous boron-doped carbon material with enhanced surface hydrophobicity and porosity for improved supercapacitor performance. *Electrochim. Acta*, 370, 137801, 2021.
98. Wen, Z., Wang, X., Mao, S., Bo, Z., Kim, H., Cui, S., Lu, G., Feng, X., Chen, J., Crumpled nitrogen-doped graphene nanosheets with ultrahigh pore volume for high-performance supercapacitor. *Adv. Mater.*, 24, 5610, 2012.
99. Jeong, H.M., Lee, J.W., Shin, W.H., Choi, Y.J., Shin, H.J., Kang, J.K., Choi, J.W., Nitrogen-doped graphene for high-performance ultracapacitors and the importance of nitrogen-doped sites at basal planes. *Nano Lett.*, 11, 2472, 2011.
100. Wang, D.-W., Li, F., Chen, Z.-G., Lu, G.Q., Cheng, H.-M., Synthesis and electrochemical property of boron-doped mesoporous carbon in supercapacitor. *Chem. Mater.*, 20, 7195, 2008.
101. Ayala, P., Reppert, J., Grobosch, M., Knupfer, M., Pichler, T., Rao, A., Evidence for substitutional boron in doped single-walled carbon nanotubes. *Appl. Phys. Lett.*, 96, 183110, 2010.
102. Robin, J. and David, R., Boron-doping effects in carbon nanotubes. *J. Mater. Chem.*, 10, 1425, 2000.
103. Panchakarla, L.S., Subrahmanyam, K.S., Saha, S.K., Govindaraj, A., Krishnamurthy, H.R., Waghmare, U.V., Rao, C.N.R., Synthesis, structure, and properties of boron- and nitrogen-doped graphene. *Adv. Mater.*, 21, 4726, 2009.
104. Lin, T., Huang, F., Liang, J., Wang, Y., A facile preparation route for boron-doped graphene, and its CdTe solar cell application. *Energy Environ. Sci.*, 4, 862, 2011.
105. Sheng, Z.-H., Gao, H.-L., Bao, W.-J., Wang, F.-B., Xia, X.-H., Synthesis of boron doped graphene for oxygen reduction reaction in fuel cells. *J. Mater. Chem.*, 22, 390, 2012.
106. Han, W., Bando, Y., Kurashima, K., Sato, T., Boron-doped carbon nanotubes prepared through a substitution reaction. *Chem. Phys. Lett.*, 299, 368, 1999.
107. Hwang, S.K., Lee, J.M., Kim, S., Park, J.S., Park, H.I., Ahn, C.W., Lee, K.J., Lee, T., Kim, S.O., Flexible multilevel resistive memory with controlled charge trap B-and N-doped carbon nanotubes. *Nano Lett.*, 12, 2217, 2012.
108. Li, X., Fan, L., Li, Z., Wang, K., Zhong, M., Wei, J., Wu, D., Zhu, H., Boron doping of graphene for graphene-silicon pen junction solar cells. *Adv. Energy Mater.*, 2, 425, 2012.
109. Tang, Y.B., Yin, L.C., Yang, Y., Bo, X.H., Cao, Y.L., Wang, H.E., Zhang, W.J., Bello, I., Lee, S.T., Cheng, H.M., Lee, C.S., Tunable band gaps and p-type transport properties of boron-doped

graphenes by controllable ion doping using reactive microwave plasma. *ACS Nano*, 6, 1970, 2012.
110. Wu, T., Shen, H., Sun, L., Cheng, B., Liu, B., Shen, J., Nitrogen and boron doped monolayer graphene by chemical vapor deposition using polystyrene, urea and boric acid. *New J. Chem.*, 36, 1385, 2012.
111. Niu, L., Li, Z., Hong, W., Sun, J., Wang, Z., Ma, L., Wang, J., Yang, S., Pyrolytic synthesis of boron-doped graphene and its application as electrode material for supercapacitors. *Electrochim. Acta*, 108, 666, 2013.
112. Gao, J., Wang, X., Zhang, Y., Liu, J., Lu, Q., Liu, M., Boron-doped ordered mesoporous carbons for the application of supercapacitors. *Electrochim. Acta*, 207, 266, 2016.
113. Gao, J., Wang, X., Zhao, Q., Zhang, Y., Liu, J., Synthesis and supercapacitive performance of three-dimensional cubic-ordered mesoporous carbons. *Electrochim. Acta*, 163, 223, 2015.
114. Yeom, D.Y., Jeon, W., Tu, N.D.K., Yeo, S.Y., Lee, S.S., Sung, B.J., Chang, H., Lim, J.A., Kim, H., High concentration boron doping of graphene nanoplatelets by simple thermal annealing and their supercapacitive properties. *Sci. Rep.*, 5, 09817, 2015.
115. Huang, S., Ma, D.D., Wang, X., Shi, Y., Xun, R., Chen, H., Guan, H., Tong, Y., A space-sacrificed pyrolysis strategy for boron-doped carbon spheres with high supercapacitor performance. *J. Colloid Interface Sci.*, 608, 334, 2022.
116. Jiang, G., Senthil, R.A., Sun, Y., Kumar, T.R., Pan, J., Recent progress on porous carbon and its derivatives from plants as advanced electrode materials for supercapacitors. *J. Power Sources*, 520, 230886, 2022.
117. Yi, J., Qing, Y., Wu, C., Zeng, Y., Wu, Y., Lu, X., Tong, Y., Lignocellulose-derived porous phosphorus-doped carbon as advanced electrode for supercapacitors. *J. Power Sources*, 351, 130, 2017.
118. Ma, W., Xie, L., Dai, L., Sun, G., Chen, J., Su, F., Cao, Y., Lei, H., Kong, Q., Chen, C.M., Influence of phosphorus doping on surface chemistry and capacitive behaviors of porous carbon electrode. *Electrochim. Acta*, 266, 420, 2018.
119. Wu, Y., Yuan, W., Xu, M., Bai, S., Chen, Y., Tang, Z., Wang, C., Yang, Y., Zhang, X., Yuan, Y., Chen, M., Two-dimensional black phosphorus: Properties, fabrication and application for flexible supercapacitors. *Chem. Eng. J.*, 412, 128744, 2021.
120. Yang, W., Yang, W., Kong, L., Song, A., Qin, X., Shao, G., Phosphorus-doped 3D hierarchical porous carbon for high-performance supercapacitors: A balanced strategy for pore structure and chemical composition. *Carbon*, 127, 557, 2018.
121. Shaikh, J.S., Shaikh, N.S., Sabale, S., Parveen, N., Patil, S.P., Mishra, Y.K., Kanjanaboos, P., Praserthdam, S., Lokhande, C.D., A phosphorus integrated strategy for supercapacitor: 2D black phosphorus–doped and phosphorus-doped materials. *Mater. Today Chem.*, 21, 100480, 2021.
122. Zu, L., Gao, X., Lian, H., Cai, X., Li, C., Zhong, Y., Hao, Y., Zhang, Y., Gong, Z., Liu, Y., Wang, X., High electrochemical performance phosphorus-oxide modified graphene electrode for redox supercapacitors prepared by one-step electrochemical exfoliation. *Nanomaterials*, 86, 417, 2018.
123. Zhang, L., You, T., Zhou, T., Zhou, X., Xu, F., Interconnected hierarchical porous carbon from lignin-derived byproducts of bioethanol production for ultrahigh performance supercapacitors. *ACS Appl. Mater. Interfaces*, 8, 13918, 2016.
124. Wang, F., Cheong, J.Y., He, Q., Duan, G., He, S., Zhang, L., Zhao, Y., Kim, I.D., Jiang, S., Phosphorus-doped thick carbon electrode for high-energy density and long-life supercapacitors. *Chem. Eng. J.*, 414, 128767, 2021.
125. Abbas, Q., Mirzaeian, M., Ogwu, A.A., Mazur, M., Gibson, D., Effect of physical activation/surface functional groups on wettability and electrochemical performance of carbon/activated

carbon aerogels based electrode materials for electrochemical capacitors. *Int. J. Hydrogen Energy*, 45, 13586, 2020.
126. Andreas, H.A. and Conway, B.E., Examination of the double-layer capacitance of an high specific-area C-cloth electrode as titrated from acidic to alkaline pHs. *Electrochim. Acta*, 51, 6510, 2006.
127. Puziy, A., Poddubnaya, O., Ziatdinov, A., On the chemical structure of phosphorus compounds in phosphoric acid-activated carbon. *Appl. Surf. Sci.*, 252, 8036, 2006.
128. Cai, X., Xiao, Y., Sun, W., Yang, F., Glucose-derived activated carbons for supercapacitors: Comparison between single O doping and N/O co-doping. *Electrochim. Acta*, 406, 13986, 2022.
129. Ma, Q., Xi, H., Cui, F., Zhang, J., Chen, P., Cui, T., Self-templating synthesis of hierarchical porous carbon with multi-heteroatom co-doping from tea waste for high-performance supercapacitor. *J. Energy Storage*, 45, 103509, 2022.
130. Liu, B., Liu, Y., Chen, H., Yang, M., Li, H., Oxygen and nitrogen co-doped porous carbon nanosheets derived from Perilla frutescens for high volumetric performance supercapacitors. *J. Power Sources*, 341, 309, 2017.
131. Peng, C., Zeng, T., Yu, Y., Li, Z., Kuai, Z., Zhao, W., Fluorine and oxygen codoped porous carbons derived from third-class red dates for high performance symmetrical supercapacitors. *J. Mater. Sci. Mater. Electron.*, 29, 18674, 2018.
132. Chen, H., Xiong, Y., Yu, T., Zhu, P., Yan, X., Wang, Z., Guan, S., Boron and nitrogen codoped porous carbon with a high concentration of boron and its superior capacitive behavior. *Carbon*, 113, 266, 2017.
133. Li, D., Guo, Y., Li, Y., Liu, Z., Chen, Z., Waste-biomass tar functionalized carbon spheres with N/P Co-doping and hierarchical pores as sustainable low-cost energy storage materials. *Renewable Energy*, 188, 61, 2022.
134. Wang, X., Yang, C., Li, J., Chen, X.A., Yang, K., Yu, X., Lin, D., Zhang, Q., Wang, S., Wang, J., Xia, Z., Insights of heteroatoms doping-enhanced bifunctionalities on carbon based energy storage and conversion. *Adv. Funct. Mater.*, 31, 2009109, 2021.
135. Fulvio, P.F., Lee, J.S., Mayes, R.T., Wang, X., Mahurin, S.M., Dai, S., Boron and nitrogen-rich carbons from ionic liquid precursors with tailorable surface properties. *Phys. Chem. Chem. Phys.*, 13, 13486, 2011.
136. Guo, H. and Gao, Q., Boron and nitrogen co-doped porous carbon and its enhanced properties as supercapacitor. *J. Power Sources*, 186, 551, 2009.
137. Wang, C., Zhou, Y., Sun, L., Wan, P., Zhang, X., Qiu, J., Sustainable synthesis of phosphorus- and nitrogen-co-doped porous carbons with tunable surface properties for supercapacitors. *J. Power Sources*, 239, 81, 2013.
138. Wang, T., Wang, L.-X., Wu, D.-L., Xia, W., Jia, D.-Z., Interaction between nitrogen and sulfur in co-doped graphene and synergetic effect in supercapacitor. *Sci. Rep.*, 5, 9591, 2015.
139. Zhu, Q.L., Pachfule, P., Strubel, P., Li, Z., Zou, R., Liu, Z., Kaskel, S., Xu, Q., Fabrication of nitrogen and sulfur co-doped hollow cellular carbon nanocapsules as efficient electrode materials for energy storage. *Energy Storage Mater.*, 13, 72, 2018.
140. Chen, Y., Liu, Z., Sun, L., Lu, Z., Zhuo, K., Nitrogen and sulfur co-doped porous graphene aerogel as an efficient electrode material for high performance supercapacitor in ionic liquid electrolyte. *J. Power Sources*, 390, 215, 2018.
141. Chen, J., Wei, H., Chen, H., Yao, W., Lin, H., Han, S., N/P co-doped hierarchical porous carbon materials for superior performance supercapacitors. *Electrochim. Acta*, 271, 49, 2018.
142. Chang, Y., Yuan, C., Liu, C., Mao, J., Li, Y., Wu, H., Wu, Y., Xu, Y., Zeng, B., Dai, L., B, N co-doped carbon from cross-linking induced self-organization of boronate polymer for supercapacitor and oxygen reduction reaction. *J. Power Sources*, 365, 354, 2017.

143. You, B., Kang, F., Yin, P., Zhang, Q., Hydrogel-derived heteroatom-doped porous carbon networks for supercapacitor and electrocatalytic oxygen reduction. *Carbon*, 103, 9, 2016.
144. Zhao, Z. and Xie, Y., Electrochemical supercapacitor performance of boron and nitrogen co-doped porous carbon nanowires. *J. Power Sources*, 400, 264, 2018.
145. Geng, Q., Huang, G., Liu, Y., Li, Y., Liu, L., Yang, X., Wang, Q., Zhang, C., Facile synthesis of B/N co-doped 2D porous carbon nanosheets derived from ammonium humate for supercapacitor electrodes. *Electrochim. Acta*, 298, 1, 2019.
146. Zhao, G., Chen, C., Yu, D., Sun, L., Yang, C., Zhang, H., Sun, Y., Besenbacher, F., Yu, M., One-step production of ONS co-doped three-dimensional hierarchical porous carbons for high-performance supercapacitors. *Nano Energy*, 47, 547, 2018.
147. Du, J., Zhang, Y., Lv, H., Chen, A., N/B-co-doped ordered mesoporous carbon spheres by ionothermal strategy for enhancing supercapacitor performance. *J. Colloid Interface Sci.*, 587, 780, 2021.
148. Huang, C., Puziy, A.M., Poddubnaya, O.I., Hulicova-Jurcakova, D., Sobiesiak, M., Gawdzik, B., Phosphorus, nitrogen and oxygen co-doped polymer-based core-shell carbon sphere for high-performance hybrid supercapacitors. *Electrochim. Acta*, 270, 339, 2018.
149. Liu, F., Wang, Z., Zhang, H., Jin, L., Chu, X., Gu, B., Huang, H., Yang, W., Nitrogen, oxygen and sulfur co-doped hierarchical porous carbons toward high-performance supercapacitors by direct pyrolysis of kraft lignin. *Carbon*, 149, 105, 2019.
150. Gunasekaran, S.S. and Badhulika, S., Divulging the electrochemical hydrogen storage of ternary BNP-doped carbon derived from biomass scaled to a pouch cell supercapacitor. *Int. J. Hydrogen Energy*, 46, 35149, 2021.

14

Wearable Supercapacitors

Preety Ahuja[1*], Sanjeev Kumar Ujjain[1†], M. Ramanand Singh[2], Neelu Dheer[3] and Rajni Kanojia[4]

[1]Center for Advanced Sensor Technology, University of Maryland Baltimore County, MD, USA
[2]Department of Chemistry, Kirori Mal College, University of Delhi, Delhi, India
[3]Department of Chemistry, Acharya Narendra Dev College, University of Delhi, Delhi, India
[4]Department of Chemistry, Shivaji College, University of Delhi, Delhi, India

Abstract
The rapid advancement in wearable electronics has emerged the development of wearable supercapacitors with remarkable features of miniaturization, flexibility, and easy integration. Substantial efforts have been devoted to different challenges associated with wearable supercapacitors towards practical application. The present chapter is a brief the overview of working principle along with the preparation strategies of the materials. The current state of art pertinent to the challenges associated with the device is discussed with relevant examples. Additionally, the survey of wearable supercapacitors as a power source in integrated devices towards futuristic applications is also presented.

Keywords: Supercapacitor, flexible, wearable, air-permeable, self-healable

14.1 Introduction

Developing markets for wearable electronics have stimulated increasing demand for continuous supply of the portable and reliable energy storage via supercapacitors and batteries [1]. However, commercialized energy devices are rigid as they mostly used coin cells, leading to physical discomfort and meager portability. In order to enhance the compatibility with the future wearable electronics for upcoming technological trend, flexible and portable energy storage devices should be focused. For their practical realization, good mechanical compliance and durability regarding stretching, compression and high energy density are the key components for powering next era electronics [2].

Flexible all solid-state supercapacitors proved potential candidate owing to their outstanding performance [3]. Notably, their ability to withstand high strain conditions without degradation in their performance made them standalone alternative. Up to now, considerable attempts have been made to fabricate these devices in multidimensions to cope up with multifunctional integrated arrangements and micro-scale devices [4]. Baughman *et al.*

**Corresponding author*: drpreetyahuja@gmail.com
†*Corresponding author*: drsanjeevkujjain@gmail.com

Inamuddin, Tariq Altalhi and Sayed Mohammed Adnan (eds.) Sustainable Materials for Electrochemical Capacitors, (339–360) © 2023 Scrivener Publishing LLC

studied fiber-based supercapacitor constituting carbon nanotubes (CNT) for the first time, which was later used to develop fabric [5]. Later, these wearable devices made great progress with regard to their basic building blocks: electrode materials, substrates and electrolytes [4].

In this chapter, we presented the achievements of wearable supercapacitors, highlighting different aspects including structural, evaluation parameters, and the working principle. Furthermore, our focus will be on the advancements and associated issues in wearable supercapacitors during their practical evaluation, comprising material selection, mechanical adaptability, environmental tolerance, air permeability, and integration in the various applications. Strategies to render devices integrate in multifunctional electronics are also reviewed along with challenges and prospects in this evolving field.

14.2 Working Principle

The supercapacitors structurally include current collector, electrodes, electrolyte and separator [6]. Hence, vast research efforts have been focused to the advancements in electrode materials by considering the effective ions and electrons kinetics. Their charge storage mechanism primarily depends on the employed electrode materials. It is broadly categorized into following: electrochemical double layer capacitance (EDLC) and pseudocapacitance [7]. The erstwhile involves double layer formation at the electrode/electrolyte interface and latter involves quick and reversible redox processes. EDLCs mainly utilize carbonaceous materials comprising activated carbons (AC), carbon nanotube (CNT), graphene etc. whereas pseudocapacitors involves transition metal oxides/hydroxides and conducting polymers [8]. As per the charge storage mechanism, EDLCs deliver high power density, long, and stable cycling life. However, the shape and pore size of carbonaceous electrodes limits their capacitance value. On the contrary, pseudocapacitors provide dominant specific capacitance owing to the valence change and charge transfer in Faradaic redox reactions [9]. Besides, the volumetric alterations during insertion and exertion of electrolytic ions, further limits their stability during long term charge-discharge process [10]. Interestingly, hybrid supercapacitors can be used by combining EDLC and pseudocapacitance along with extended working potential leading to high capacitance and energy density.

The wearable supercapacitor mostly involves a quasi-solid electrolyte with high mechanical stability for practical suppleness, which is a key parameter for providing mechanical performance or flexibility to the fabricated supercapacitor device. However, there are no practical established methods to determine the flexibility of the electrolyte, or the related assembled device. Generally, 90% retention in electrochemical stability during various deformation states is widely accepted as a stable behavior [11]. Generally used electrolytes in flexible supercapacitors are hydrogel electrolytes comprising of hydrophilic polymer framework, dispersion medium (organic/aqueous solvent) and supporting electrolytic salt/acid/alkali to obtain desired conductivity. In addition to physical flexibility, they offered desirable properties involving high ionic conductivity and mechanical integrity. Numerous polymers, namely poly(propylene oxide), polyethylene oxide (PEO) and polyvinyl alcohol (PVA) have been utilized for the same purpose [12, 13]. Among them, PVA has the potential owing to its low cost, non-toxicity, and chemical stability [14]. However, they are synthesized by mixing and physical cross-linking leading to inferior stability with respect to structural and mechanical aspect [15]. Additionally, they have restricted potential range limiting the energy density of the device [16]. Whereas the counterpart, organic gel electrolytes have high potential range but being volatile, cannot be used under harsh conditions of low and high temperatures [17]. Another gel electrolytes are good substitute for wearable supercapacitors with potential to address the hydrogel issues [18]. They offered high electronic conductivity, a wide potential window along with excellent mechanical stability. Mostly gel electrolytes used ionic liquids owing to their outstanding properties including high conductivity and thermal stability [19]. In one such work, Lee *et al.* reported fabrication of wearable supercapacitor on cotton T-shirts via the UV curing-assisted stencil printing [20]. They comprised of multiwalled carbon nanotube (MWCNT) and activated carbon (AC) electrodes and a gel polymer electrolyte ([EMIM][TFSI]/thiolene polymer network skeleton/SiO_2), which were integrated on a T-shirt in form of symbols or letters. Such printed supercapacitor T-shirts offered outstanding flexibility and thermal stability owing to synergistic interactions. Importantly, their size and shape along with the electrochemical performance are maintained upon folding, laundering, and even ironing. Recently, another work involves MWCNT and reduced graphene oxide (RGO) based electrode along with propylene carbonate (PC) based gel electrolyte to assemble the supercapacitor [21]. The stability in a wide range of temperature offered by the electrolyte further facilitates the supercapacitor assembly with no capacitance decay when subjected to deformation involving bending and stretching.

It is further important to analyze the key parameters for instance cell capacitance, equivalent series resistance (ESR), energy density, and power density to evaluate the performance of wearable supercapacitor device [22]. These parameters are evaluated by different techniques namely galvanostatic charge discharge (GCD), electrochemical impedance spectroscopy (EIS) and cyclic voltammetry (CV) [23]. Primary objective is to estimate the electrochemical performances of supercapacitor device from different viewpoints. For instance, in cyclic voltammetry, current is measured when applied voltage becomes higher that predicted by Nernst equation at a fixed scan rate as follows:

$$C_{sp} = \frac{S}{2m\Delta V}$$

Where C_{sp} is specific capacitance, S represents the integral area of the curve associated in the CV, m represents electrodes mass, and ΔV represents change in potential of the device.

Galvanostatic charge/discharge (GCD) records the potential response relative to time and evaluate the electrochemical information associated with supercapacitor electrode material. It is most reliable technique for evaluating the electrochemical comportment. It measures in two phases involving the supercapacitor charge via a constant current, followed by discharging in a specific potential range or time. The GCD is used to estimate the capacitance value from the following:

$$C = \frac{I\Delta V}{m\Delta t}$$

where I and ΔV represent discharge current and voltage, respectively, m represents total electrodes mass, and Δt denotes discharging time of the device [24].

Furthermore, electrochemical impedance spectroscopy (EIS) can be used to evaluate the different processes contributing towards the overall performance of the device. It measures impedance or the capacitive attributes of a material by observing the impedance data at the open circuit potential in response to applied alternating potential over a large range of frequencies [25]. The result is expressed broadly in two plots involving Nyquist and Bode plot; from which various parameters can be calculated [25].

14.3 Design of Electrode Materials

The core of supercapacitor system i.e. low-cost and flexible electrode material should be regarded as the focus of supercapacitor research. These electrodes along with the other important characteristics including lightweight, high conductivity, and good electrochemical stability are required as a vital component for an efficient wearable supercapacitor device. The designing of the electrode can be distinguished based on their dimensionality, discussed in the following sections.

14.3.1 1D Yarn-Shaped Electrode

The yarn or fiber-shaped electrode possesses small volume and high flexibility owing to which they can be easily weaved and integrated to form wearable device. Different types of yarn can be employed for the same purpose including cotton fiber, carbon material-based fiber, Kevlar fiber metal wires and polymer nanofibers [11]. Using such fibers, supercapacitor electrodes can be formed via coating the fibers with the active materials using electrodeposition, chemical deposition, dip coating, or other deposition methods. These electrodes offer high conductivity, a crucial factor for supercapacitors with high performance. Further, reduced graphene oxide (RGO) coated textile electrode was fabricated by Liu *et al.* towards high volumetric power density (1400 mW cm^{-1}) [26]. Also, MnO_2 nanosheets was coated on carbon fibers for fabricating high capacitive electrodes for wearable supercapacitors [27]. It offered high capacitance of 87 Fg^{-1} along with good capacitance retention during charge discharge cycles. In other study, a conjugated microporous polymer (CMP) was grafted on CNT fibers (CNF@CMP) (Figure 14.1a) and demonstrated high areal specific capacitance [28]. Further, its symmetrical-twisted supercapacitor device with a gel electrolyte demonstrated wide working voltage of 1.4 V along with augmented electrochemical performance. Notably, they possess large mechanical strength (Figure 14.1b), maintaining capacitance retention (84.5%) even after 10,000 deformation (bending) cycles (Figure 14.1c). The fabricated supercapacitors are integrated in series and knitted in a watchband as demonstrated in Figure 14.1d. It powered a digital watch (Figure 14.1e) ascertaining its flexibility, electrochemical stability, and output energy for practical wearable applications.

Moreover, stretchability is of utmost importance in wearable devices which can be achieved by using different topologies in the device including parallel, twisted, helix and coaxial. Their electrodes can be mostly fabricated by using coiled/buckled fibers and elastic core fibers. One such study involves the fabrication of a stretchable CNT coiled yarn with deposition of MnO_2 involving a giant twist in the device to attain stretchability [29].

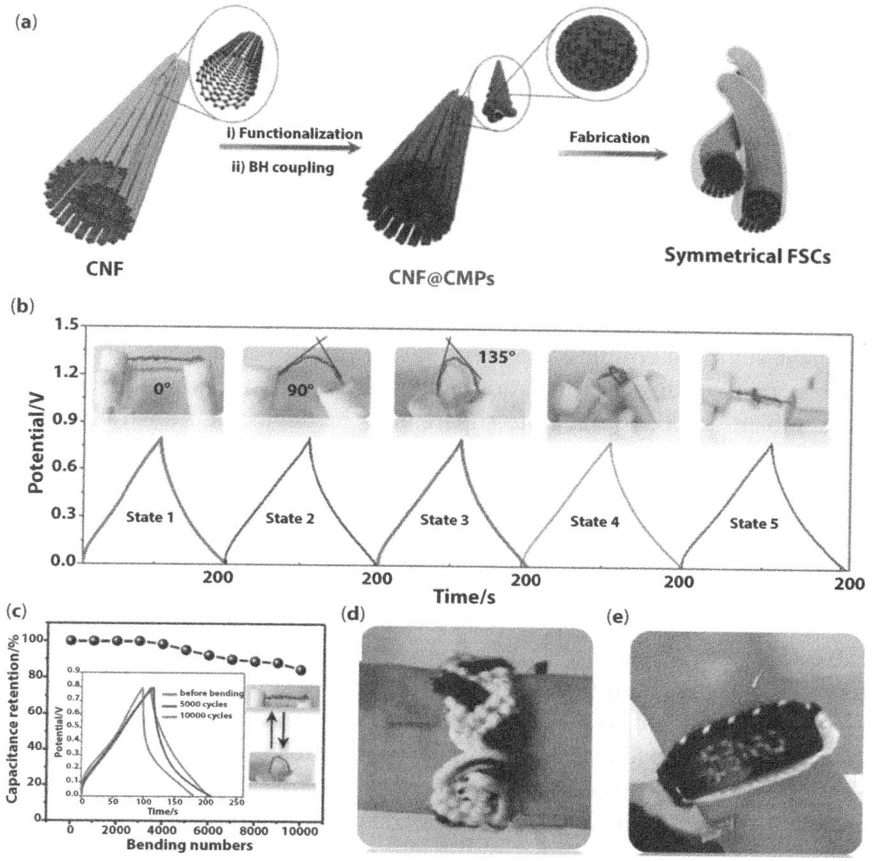

Figure 14.1 (a) Schematic demonstrating the fabrication of symmetrical-twisted supercapacitor device (FSCs). (b) GCD curves of supercapacitor in five different states at 0.68 mA cm^{-2} current density, revealing high mechanical strength of the fiber, (c) Capacitance retention of the supercapacitor with bending cycles at state 3. Fabricated fiber used in (d) Knitted watchband, and (e) Powering digital watch. Reproduced with permission from reference [28].

Their unique core-shell structure coiled electrode with PVA/LiCl gel electrolyte demonstrated high mechanical strength along with large volumetric capacitance. Further, the fabricated coiled supercapacitor displayed large capacitance retention during deformation and electrochemical evaluation. Other study involves overlapping CNT sheets around pre-strained elastic fibers and further treating with poly(3,4-ethylenedioxythiophene):poly(styrene sulfonate) (PEDOT:PSS) for achieving high electroactivity [30]. Two of these fibers were used as electrodes in twisted manner with hydrogel electrolyte to form the supercapacitor, which tolerated a stain of 350% with negligible change in the performance. Different configurations like helix or spring are also examined to render non-stretchable fiber with stretchability by using stretchable polymeric core as electrolyte. In a recent study, hybrid CNT/graphene/polyaniline (PANI) wires were twisted for constructing fiber supercapacitor, followed by coiling to form spring-like arrangement of the fabricated device [31].

The performance of the device was remarkably retained when subjected to large strain of 800%, endowing highly stretchable material for futuristic wearable devices.

14.3.2 2D-Shaped Electrodes

With the advancement in textile technology, several materials have been explored for preparing 2D flexible electrodes such as textiles, fabrics, and paper-based materials. Active materials are used with binder to form planar 2D electrode. However, they can be peeled off owing to the deformability of supercapacitors in wearable applications. Consequently, this urges for a binder free electrode with enhanced mechanical performance. This indeed demands for self-standing configuration with high flexibility and facile fabrication. In one demonstration, free standing carbon nanofiber/graphene self-standing films were fabricated [32]. The assembled solid-state flexible supercapacitor device demonstrated high energy and power densities. Cui *et al.* further informed about robust, self-standing $NiCo_2S_4$-carbon nanofibers electrode [33]. Such flexible carbon nanofiber electrodes are prepared by hybridization with Ag nanoparticles and graphene oxide by using electrospinning and carbonization process, followed by hydrothermal growth of $NiCo_2S_4$ nanotubes. Without using any mechanical support, it could be twisted and folded with achievement of high capacitance.

Further, stretchability can be successfully achieved by different structured (buckled/wavy) electrodes via pre-straining approach [34]. For this purpose, solid electrolytes are used preferentially in planar supercapacitor. Additionally, some stretchable substrates such as elastomers have been used with carbon materials or nanostructured metal for fabricating planar stretchable supercapacitors. In one such report, buckled structure electrode was prepared by CNT film with pre-stretched substrate of silicon rubber [35]. This could result in an omni-directionally stretchability to the fabricated supercapacitor device along with retention in its electrochemical performance under elongations of 200%. In other report, Ecoflex was used with CNT/PEDOT film and could undergo biaxial stretching [36]. Also, intrinsic stretchability of different substrates was successfully used to fabricate planar stretchable supercapacitor systems. For example, conductive textile was developed by employing textile with weft knitted structure by carbonizing [37]. They presented high electrical conductivity on subjection to 70% strain and could sustain 50% tensile strain along with high specific capacitance.

Stretchable microsupercapacitors are considered as a novel approach which could allow on-chip integration owing to their 2D configurations. For this purpose, different designs like serpentine and wavy have been implemented to sustain the applied tension strain on electrochemical performance of the device. Combining soft polydimethyl siloxane (PDMS)/Ecoflex as substrate and array of PDMS with polyethylene terephthalate (PET) to form micro-supercapacitors allow the accommodation of the strain in an effective manner [38]. Recently, patterned mold was produced by 3D-printing approach generating wavy structured PDMS electrode and MWCNT/PANI slurry acted as active materials. The fabricated microsupercapacitor could attain high areal specific capacitance along with stretchability of 40% [39]. Other microsupercapacitors are reported with tripod-structured PDMS and honey comb shaped PDMS to ensure their stability under deformation processes via associated intrinsic mechanism [40].

14.3.3 3D-Shaped Supercapacitor

Besides standard supercapacitor devices of 1D and 2D wearable supercapacitor, 3D wearable devices are recently upcoming candidates. They have high flexibility and elasticity owing to kirigami or pattern-based processing. This editing process involves the cutting of planar supercapacitor in linear patterns which can take honeycomb-like structure from the supercapacitor sheet under tensile stretching [41]. Such supercapacitors could accommodate 400% strain with remarkable electrochemical stability. Interestingly, kirigami, origami-based technology and other 3D stretchable patterns also come into play. Graphite-based isolated electrodes were utilized in developing origami-type supercapacitor. Cellular paper was used to coat with graphite based active material and then patterned gel electrolyte was used on the other side to combine electrodes. This allowed the supercapacitor to sustain wide deformability in large range of compression to stretching (-60% to 30%) [42].

Table 14.1 Different types of materials used in wearable supercapacitors with regard to their flexibility and wearability.

Types of materials	Specific examples	Associated challenges	Flexibility and wearability
Polymers	Stretchable Polymers or mixed polymers (PDMs, PVDF, PVA, PDMS+PI, PEDOT:PSS)	Degradation owing to • Environmental factors • Analytes	• Widely used as they have inherent stretchability • Capping materials can be used for protection against contaminants
Textiles	Different fibers/yarn knitted with polymers or grapheme or metal oxides	• Adhesion problems • Insufficient Structural integrity	• Pretreatment can be done to promote adhesion • Can be processed (heating, cooling and other related treatments)
2D Materials	Carbonaceous materials (Graphene) functionalized with metal nanowires/metal oxides/polymers.	• Brittleness	• Nanostructured materials can be easily utilized. • Besides ultrathin, they possess outstanding mechanical properties imparting high flexibility.
3D Materials	Carbonaceous Materials and Polymers with 3D stretchable patterns (E.g. Kirigami, origami)	• Wear and tear	• Pattern based processing impart high stretchability

The present status of 1D, 2D and 3D wearable supercapacitors demonstrated the flexibility and stretchability of the devices (Table 14.1) rather their practical applications which need to be discussed further. To conclude, 1D wearable supercapacitors hold instinctive advantage as they can be seamed to form textiles and can be used in regular clothing whereas 2D based supercapacitors are suitable for 'tattoo-like' devices. On the contrary, 3D devices are not so prolific in wearable energy applications owing to their bulkiness.

14.4 Wearable Supercapacitor

For developing highly efficient wearable supercapacitor device for practical applications, pre-requisite condition is to consider several important aspects. Among them, material selection should be on priority as stretchability is required for device fabrication. In this section, substrates for wearable supercapacitors and their essential features are summarized. Later, we will discuss wear and tear characteristic of the fabricated device along with their benignity with human body.

14.4.1 Material Selection

The pre-requisite condition of material selection involves the assortment of the electroactive material. As carbonaceous materials cannot deliver high capacitance, so they need to be hybridized with pseudocapacitive materials to augment the performance of the device. From the existing literature, it is concluded that conducting polymers are potential alternative as they can tolerate the stress induced performance degradation imparting high stretchability to the device; a critical issue for futuristic wearable devices. They can undergo shrinking and swelling in their intrinsic structure to accommodate the impact of strain and resulted in upholding their electrochemical performance. In one study, it is demonstrated that MnO_2 developed cracks in their structure along with peeling off at some region when the electrode is subjected to stretching [43]. This concern is relaxed by employing polypyrrole (PPy) which showed no considerable morphological variation on stretching/releasing cycles. Hence, the wrapping of MnO_2 by PPy proved advantageous in terms of electrochemical performance along with high flexibility in device fabrication. The concern with conducting polymers is their poor cycling stability, demonstrating remarkable degradation in 3000 cycles as compared to other counterpart electrode materials. Much efforts have been conducted in this field to show remarkable performance during consecutive cycling of polymeric electrodes. It is well reported in one of the literatures that electropolymerized PPy showed stable maintenance in capacitance with 98.5 % capacitive retention, in 1000 cycles as compared to 79 % retention in the composite without PPy [44].

Another important aspect involves the safety of the material for human body which involves air permeability, safety and their wear and tear characteristics towards environmental tolerance. For practical wearability, clothing comfort is essential, which has never been addressed. This necessitates air permeability of the fabric fabricated from the yarn. One strategy includes through-holes in the device, which poses requirement of robustness of the fabricated devices. In one such study, Dong *et al.* employed air-laid papers for flexible electrode fabrication with CNT and MnO_2 as electroactive material [45]. The device becomes highly air-permeable after making through-holes along with perpetuation of 94%

capacitance. Other technique involves the fabrication of yarn supercapacitors which later knitted to provide intrinsic air permeability. For this, conducting yarns are indispensable, including dip-coating with conducting active materials, metal deposition on the yarn and many others. However, they require multi-steps which are time consuming and hence proved expensive [46]. Prompted by this, direct carbonaceous materials are used into fiber and yarn fabrication for wearable applications [47]. Mao *et al.* informed the integration of carbon fiber yarn-based supercapacitor with PANI nanowires on carbon yarn as electrode and $EMIMBF_4$/PVDF/DMF gel electrolyte [48]. The assembled solid-state supercapacitors show high areal specific capacitance along with outstanding rate capability and electrochemical steadiness during charge-discharge cycles.

Concerning about the safety of human skin, with which the device will come in close contact, the best choice should be biocompatible materials. For example, natural bacterial cellulose was used as a flexible substrate for fabricating the supercapacitor device MnO_2-PANI as the electroactive material [49], maintaining high capacitance after 1000 cycles. In another report, cellulose fabric is used ensuring high porosity in the electrode material [50]. This fabric-based supercapacitor device is integrated with graphene as active material and dopamine as a binder and demonstrated ultrahigh areal specific capacitance. Other naturally occurring polymers say chitosan, gelatin and others also proved beneficial for the same purpose [51]. Regarding biocompatible electrolyte for wearable devices, human body sweat can be a good option owing to the presence of negative and positive ions in it. Manjakkal *et al.* reported this kind of biocompatible wearable supercapacitor with sweat as electrolyte with polyester cellulose fabric as electrode and PEDOT:PSS as the active material for the electrode [52]. They reported energy density and power density in presence of artificial sweat and real sweat where they diminished significantly, which may be attributed to low voltage window of the electrolyte.

After considering the components, safe packaging of wearable supercapacitors is another apprehension. Safe packaging is indispensable for biocompatible devices, as they can widen the lifespan of the device and ensure the efficient working of supercapacitor in hostile environment. We can avoid direct contact between human skin and electrolyte by preventing electrolyte leakage thereby extending the practically operable life of the device. Importantly, the post usage effect should be considered, such as the quick charging and discharging of the supercapacitor device which may increase the risk of temperature rise of the device. This may also lead to redox reaction on electrode surface accelerating the aging of the device or even an explosion, which should be taken into consideration.

14.4.2 Mechanical Adaptability

The mechanical adaptability with human skin is of utmost importance for hassle-free working of the device. The mechanical compliance between the device and human skin can be estimated by the device performance under different deformation conditions. Lu *et al.* designed mechanically compliant supercapacitor device by employing in situ growth of metal organic framework (MOF) on CNT surface [53]. He and coworkers reported the good electrochemical performance of the device owing to high surface area of MOF and flexibility from CNTs. Also, under bending conditions of the device from 0 to 180°, CV shape showed almost no considerable change implying the preservation of the device capacitance. Another study reported about the electrode fabricated from black phosphorous composite fiber based non-woven fabric via microfluid spinning process [54]. The as assembled supercapacitor device endows stable energy supply under large deformation in the device, 97.4% capacitance retention was achieved even after 1000 bending cycles of 180°. The compliant device can also be attained by employing stretchability in the device by using intrinsically elastic materials for electrode and stretchable structure in the device. For e.g., elastic fibers are used with PVA gel electrolyte and CNT sheet for fabricating supercapacitor device [55]. It can sustain 75% stretching with optimal electrochemical performance with these intrinsically elastic fibers. The main concern with these elastic materials is interfacial interactions of electrode and electrolyte owing to moduli mismatch, which become weak during stretching. Additionally, stainless steel fibers are coiled and nested together to form two-dimensional network for stretch recovery [56]. Further, hydrothermal method is employed to grow the active material for supercapacitor electrode. The decrement in capacitance is not considerable till 20% strain, but it declines to 23.6% on subjecting to 30% tensile strain. Similar to this, crumpled-graphene papers are utilized in fabricating highly stretchable supercapacitor [57]. Graphene paper was wrapped on an acquiescent substrate which can be crumpled into self-organized patterns; the crumpled patterns unfold on stretching the substrate. This consequently maintains invariable electrochemical performance during multiple cycles of large deformation of graphene paper. They can be subjected to larger strain from 300 to 800%, owing to its crumpled shape. In other study, wrinkle-structured material was explored to integrate porous stretchable supercapacitor [58]. They used wrinkled poly(3,4-ethylenedioxythiophene)-poly(styrenesulfonate) on a 3D porous reduced graphene oxide for coating on pre-strained elastic substrate. The resulted supercapacitor device maintained 85.1% capacitance at 300% strain along with good uphold after 500 stretching–releasing cycles.

Another study showed the performance of crumpled MXenes where they used crumpled MXenes on elastomeric substrate for formulation of the stretchable supercapacitor [59]. Firstly, their mechanical stability is enhanced by making their dispersions with polymer binder, sodium alginate (SA) followed by drop-casing on biaxially pre-stretched elastomer. They demonstrated excellent performance of the device at 0% and 80% strain evidenced by no considerable change in their capacitance. Their mechanical durability was demonstrated by 76% capacitance retention after enduring 1000 deformation cycles at 80% strain.

14.4.3 Self-Healable

Physical damage to the device can occur in response to external factors such as scratching, cutting, etc. Additionally, the devices in the market are not very cheap, so their self-regeneration or healing is indispensable. We will discuss self-healing process in context to materials first, followed by their applicability in the device. There are different self-healing mechanisms working in the chemical realm. For example, microencapsulation, the mostly studied mechanism involves polymer matrix with healing monomer which is microencapsulated along with a dispersed catalyst [60]. When crack reaches to microcapsule, healing monomer reached the crack and polymerizes itself owing to the presence of dispersed catalyst. This autonomous mechanism consequently inhibits further crack propagation. Another mechanism involves molecular interdiffusion at polymer-polymer interfaces [61]. As the name suggests, molecules diffuse at polymeric interfaces to entangle with each other and penetrate to ruptured site of the material. They can be observed in some of thermoplastic and thermosetting polymers including poly(methylmethacrylate) and polyurethane, respectively [62]. Different mechanisms of reversible covalent networking and reversible non-covalent crosslinking occurs in presence and absence of external stimulus such as heat or light, respectively [46]. They are mostly accounted in polymeric materials and endow futuristic materials for electronic device with self-healing ability.

On the contrary, self-healing in the device depends upon the self-healing materials and the alignment of the electrode. For fabricating this kind of supercapacitor, either a self-healing material is wrapped over the electrode or can be added in the middle as both electrode and electrolyte, which are not self-healable [63, 64]. These processes only allow 83% of capacitance retention in the device after only 5 cycles of cutting and healing. The electrochemically inactive self-healing material and the limited stretchability and flexibility of PVA-based electrolyte further limit the self-healability of the resulted device. There comes, hydrogel, which provide better alternative owing to high water content and stretchability. One of the example include poly acrylic acid (PAA), achieved by polymerization of acrylic acid monomers and vinyl group grafted silica nanoparticles [65]. The potential of these materials arises from their crosslinking capacity. On extending as a result of applied strain, the resulting stress was transferred by silica and chain breakage was re-joined via hydrogen bonds. Additionally, the water content can be controlled by immersion in phosphoric acid which correspondingly tuned the electrochemical activity. One literature reported the fabrication of self-healable wearable device based on rGO fiber spring supercapacitor wrapped in self-healable polyurethane hydrogel. It can successfully sustain 10% strain with 82.4% capacitance retention but can only retain 54.2% after 3 times self-repairing [66]. In other study, omni-healable and stretchable device was constructed by employing the PPy hydrogel electrodes (GCP@PP) [67]. The electrode possessed honeycombed structure, which was constructed by the growing PPy in the GCP framework followed by its chemical crosslinking with gold nanoparticles. The strong interaction between gold-sulphur coordination bonds reorganized the polymer chains when subjected to external stimulus of IR light imparting self-healability to the material. The large strain (2380%) tolerance and the healing with 94% efficiency proved the promising potential of the device in wearable electronics.

Importantly, a repetitively stretchable (500 times) hydrogel polyelectrolyte was fabricated by using azobenzeno-polyacrylamide (Azo-PAM) and α-cyclodextrin polymer (α-CDP) [68]. The host-guest interaction in this allows self-healing at room temperature without any induction. The electrochemical performance of fabricated supercapacitor using this hydrogel electrolyte sustain after several cycles of healing at the same bisected site. Further, they maintain the elastic performance of the device at the same time after repetitive strain application and consequent self-healing. Another quasi solid-state supercapacitor is reported by Vu *et al.* where they utilized PEDOT:PSS and MWCNT as electrode along with PVA/H_3PO_4 as gel electrolyte (Figure 14.2a) [69]. The electroactivity of the device was sustained to 94% even after 5th healing cycle via interfacial reconstruction (Figure 14.2b and 14.2c). Besides, good mechanical stability was also attained under various bending deformations of 180° bending as shown by capacitance retention under different bending conditions (Figure 14.2d). In this section, we have concise the performance of the fabricated devices in terms of mechanical endurance and healability to highlight the need in portable/wearable electronics.

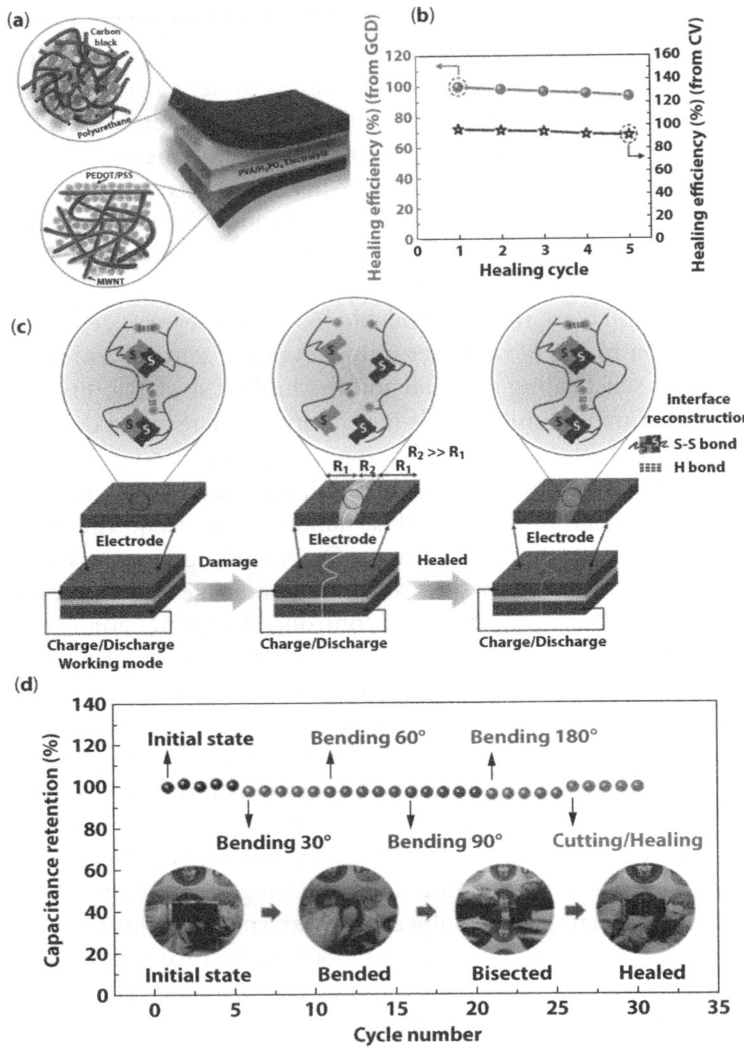

Figure 14.2 (a) Fabricated supercapacitor with PEDOT:PSS-MWCNT electrode coated with PU/CB and PVA/H_3PO_4 as gel electrolyte, (b) Healing efficiency of the device calculated from GCD and CV curves, (c) Schematic demonstrating the self-healing mechanism of the device and (d) Capacitance retention of the device at different bending deformations. The inset shows the device under different deformation conditions. Reproduced with permission from reference [69].

14.5 Integrated Application

Another important direction towards the development of wearable devices is to advance device-level integration technologies to enhance their versatility. For real time applications, it is of vital importance to integrate the wearable supercapacitors with other functional units to apprehend fully integrated portable systems. Till date, a series of integrated system have been developed, such as supercapacitor-sensor integration and self-driven integrated system (constituting nanogenerator, photovoltaic cell or wireless charge units), with strong perspective towards monitoring of health and artificial intelligence. Subsequent section will present the recent scenario of these portable and wearable integrated systems with highlighting the performance in multifunctional device constitutions and monitoring.

14.5.1 Supercapacitor with Sensing Applications

Conventional sensors require an external power source for their working which can be attained by connecting the power source such as supercapacitor in series with the sensing devices. Recently, a better alternative has been realized in which one single device can be used for multi-operations. For example, conductive textiles, being strain sensitive can be used for both the purposes. Huang *et al.* demonstrated this strategy by integrating sensing abilities in an energy storage device by using sensor-based electrodes [70]. Multifunctional material PPy is used as active material along with piezoresistive fabric geometry, facilitating supercapacitor and strain sensor functions. Other report suggested all-in-one stretchable coaxial-fiber based multifunctional system integrating strain detection and power supplying supercapacitor [71]. The supercapacitor is fabricated by electrodes comprised of manganese dioxide and PPy on CNT. Benefiting from this unique design, excellent flexibility is achieved at 200% strain with 85.1% retention in capacitance after 6000 stretching cycles. This self-sustained system under strain showed consistent performance with outstanding stability in this multifunctional device. Different from the above, compressible CNT array was fabricated with gradient cross-linked structure [72]. It can sustain large degree of compressive strain with reversible performance during several cycles. Later, with this as electrode, a supercapacitor with sensing application was fabricated which retained 94.5% capacitance after 3000 consecutive compression cycles at 60% strain.

Unlike the above systems, an all-in-one system was designed where planar microsupercapacitor, was also coupled with strain sensor [73]. For supercapacitor fabrication, manganese and vanadium based oxide was used with MWCNTs as an electrode with sulfone-based organic gel electrolyte, which offered large operational voltage to 2 V. A strain sensor was constructed by using graphene foams with PDMS film. This device can be attached on the skin via adhesive film and detect bio-signals of arterial pulse, brow frowning and many other gestures by using the power sourcing from the coupled microsupercapacitor.

14.5.2 Supercapacitor with Electrochromic Applications

Some electrodes materials undergo color change when they undergo electrochemical redox processes. Their optical behavior changes over reversibly on application of an external potential, and hence can be utilized as electrochromic materials and operated with the power source from supercapacitor. Broadly, these electrochromic materials have been divided in two categories involving organic and inorganic materials. The organic molecules having π-conjugation involve anthraquinone, triphenylamine, viologens, and the conducting polymers such as PEDOT, PPy, PANI, etc. The inorganic electrochromic materials include different transition metal oxides involving molybdenum oxide (MoO_3), tungsten oxide (WO_3) and many others [74]. One such report showed multiple color changing effect from deep red, green and gold in a set of electrochromic fibers [75]. The fiber-shaped supercapacitor devices are prepared by constructing dual helical electrodes structure involving a fiber-shaped plastic substrate deposited with tungsten oxide and poly(3-methylthiophene) along with $LiClO_4$/poly(methyl methacrylate) as solid electrolyte. Further, Nwanya *et al.* hybridized WO_3 in PANI to enhance electrochromic and electrochemical traits of PANI [76]. In addition to energy storage performance, the composite displayed reversible color changes on subjection to particular voltage i.e. from brownish green to light green and back to brownish green. Guo and co-workers designed an asymmetric supercapacitor device with electrochromic characteristics, by using TiO_2 based negative electrode and PEDOT as the positive electrodes [77]. The fabricated device offered electrochemically stability with 92% retention in capacitance after 5000 cycles and excellent electrochromic characteristics with reversible color transformation.

Another tactic to demonstrate the state-of-art of supercapacitor ability is further reported. Zhang *et al.* demonstrated in his report that the energy storage level of supercapacitor

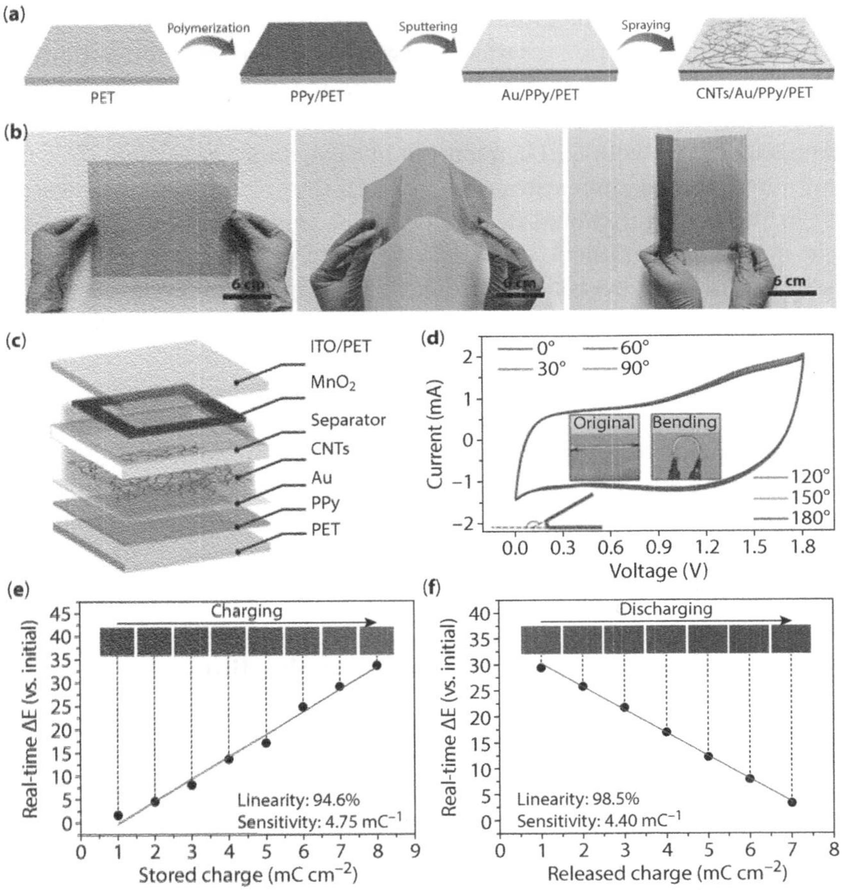

Figure 14.3 (a) Schematic illustrating the fabrication of CNTs/Au/PPy/PET electrode, (b) Digital photograph depicting excellent flexibility for the film, (c) Asymmetric supercapacitor fabrication process, (d) Cyclic voltammograms of the flexible asymmetric device under various deformations revealing high mechanical strength. The inset shows the device in original state and bending state (180°). Electrochromic contrast (ΔE, real-time) as a function of State of charge for device during the (e) Charge and (f) Discharge processes. The insets show the photographic images illustrating the colors of the device correlated to the real-time ΔE observed during charge and discharge of the device. Reproduced with permission from reference [79].

device can be noticed via optical changes observed through the electrochromism in the electrode material [78]. Recently another interesting feedback system is developed by using electrochromic properties of PPy based flexible electrode (Figure 14.3a and b) [79]. The as fabricated mechanically stable asymmetric supercapacitor device (Figure 14.3c and 14.3d) is coupled with electrochromic properties and a quantitative indicator has been developed. Benefiting from the inherent electrochromic properties of PPy, a linear dependence of real-time electrochromic contrast with state of charge of supercapacitor (Figure 14.3e and 14.3f), can be established endowing an optical energy-based monitoring system.

14.5.3 Supercapacitor with Shape-Memory Applications

As the name suggests, shape memory materials can recover their initial shape even from plastically deformed state when they are subjected to external stimuli such as heating. The shape memory property of these specific material can be utilized in supercapacitors as they encounter diverse irreversible deformations during practical usage. As a result, the potential physical fatigue of the supercapacitor device can be recovered to enhance their cycling stability.

Liu *et al.* developed a wrist band supercapacitor with innate shape memory properties. They reported asymmetric supercapacitor fabricated from Ti-Ni alloy with reduced graphene oxide and Ni-foil with MnO_2 as electrodes along with a gel electrolyte. It can sustain several dynamic shape memory cycles with high sweep speed [80]. Also, at 15°C this alloy demonstrates phase transition, thus human skin temperature can actuate their shape recovery and proved beneficial in the wearable electronics. Further, CNT sheet with polyurethane fiber coated with PVA based gel electrolyte supercapacitor was fabricated with good mechanical stability [81]. It was subjected to 20 deformation cycles and interestingly no considerable capacitance loss was observed owing to the shape memory function of the electrode material.

14.5.4 Supercapacitor with Energy Harvesting Applications

In addition to the above integrated systems with supercapacitors, energy harvesting has also been extensively explored. Energy from the environment in different forms can be converted into electric energy to improve energy efficiency. But they are intermittent and hence, their storage need is desired. In this section, we will showcase the supercapacitors and energy harvester integration comprising solar cells, tribo-generators, and thermo-electric devices. Research in this field is attracting much attention as they are considered as complementary technologies, extending the scope into promising applications of self-powering and wearable electronics.

Solar cells can work only in daytime, except rainy day and hence, their electric energy has to be stored. The integration with supercapacitor can be highly desirable for solar cells to work efficiently with no limitation. Du *et al.* demonstrated the coupling of a solid state supercapacitor with perovskite based solar cells which can be charged to 0.75 V along with lesser discharge time of 45 sec [82]. Further, Xu *et al.* showed such a powerful packet comprising a solar cell which is perovskite-based and supercapacitor which is PPy-based, connected in series [83]. It delivered a maximum energy storage efficiency 10% along with enhanced 1.45 V output voltage, when supercapacitor is fully charged.

Interestingly, device miniaturization and footprint area reduction prompted the researchers to fabricate photo-capacitor devices including shared electrode by solar cells and supercapacitors. For instance, Peng *et al.* showed a fiber based energy device having solar cell and supercapacitor at the two ends, with same Ti wire [84]. Although, such energy fiber demonstrates 0.82% storage efficiency, however, owing to their flexibility and weavability, they can form next generation smart fabric.

Additionally, mechanical motion of human body can be easily transformed to electrical energy by nanogenerators including piezoelectric, electromagnetic and triboelectric nanogenerators [85]. Among the three nanogenerators, the piezoelectric has good energy saving properties. Energy was harvested through human motion such as walking or running by mechanical pressure. For example, piezoelectric sensors were placed on the pressure centric parts i.e. heel and toe of the shoe, whereas piezoelectric generator is mounted inside the insole to analyze the coupling of this generator and supercapacitor [86]. Pressure is generated because of walk or run by a person, which is transformed into electrical power and used to charge the supercapacitor. Further, Liu *et al.* constructed nanowrinkle-patterned flexible triboelectric generator employing PDMS based films [87]. Systematic investigation delivered an instant high power density and a stable output voltage even after several consecutive cycles. Notably, this self-powered device can realize arm swing, elbow bending, and knee bending, with multi-adaptability and compatibility demonstrating the potential for energy harvesting and sensing applications without external energy.

Another important technology involves thermoelectric generator which harvests energy from heat of body. For example, electrical energy can be gathered from human sweat as in biofuels [85]. Also, the textile-based thermoelectric generators have been explored by exploiting the difference in human body temperature and the temperature of the surroundings. Owing to the Seebeck effect, the accumulated charge transporters on one side generate the voltage, called thermovoltage [88]. In one study a wearable thermoelectric nanogenerator was fabricated with n-type reduced graphene oxide and p-type (PEDOT:PSS) material [89]. It proved potential in transforming the body heat into electrical energy, between an external environment (20.0°C ± 5°C) and the human body (36.5°C) with generation of large open circuit voltage for other applications.

14.6 Conclusion

This chapter highlights the recent developments in wearable supercapacitors with desired characteristics of flexibility and stretchability for application in future electronics. It offers a comprehensive knowledge for designing of electrodes with pertinent to different existing challenges during material selection. Along with this, the recent techniques for designing different dimensional materials with enhanced electrolyte accessibility during device fabrication are reported. Despite the notable progress, several issues accompanying with the material selection, wearability, self healability, and mechanical adaptability during the construction of the device are addressed. Also, the last section has focused on the integration of the supercapacitor with other devices as power sources in multifunctional systems. With such continual efforts made by researchers, the development in wearable futuristic devices will prove to be a boon to our society.

References

1. Wang, Y., Zhang, L., Hou, H., Xu, W., Duan, G., He, S., Liu, K., Jiang, S., Recent progress in carbon-based materials for supercapacitor electrodes: A review. *J. Mater. Sci.*, **56**, 173, 2021.
2. Rani, S., Kumar, N., Sharma, Y., Recent progress and future perspectives for the development of micro-supercapacitors for portable/wearable electronics applications. *J. Phys. Energy*, **3**, 032017, 2021.
3. Asl, M.S., Hadi, R., Salehghadimi, L., Tabrizi, A.G., Farhoudian, S., Babapoor, A., Pahlevani, M., Flexible all-solid-state supercapacitors with high capacitance, long cycle life, and wide operational potential window: Recent progress and future perspectives. *J. Energy Storage*, **50**, 104223, 2022.
4. Fu, W., Turcheniuk, K., Naumov, O., Mysyk, R., Wang, F., Liu, M., Kim, D., Ren, X., Magasinski, A., Yu, M., Feng, X., Wang, Z.L., Yushin, G., Materials and technologies for multifunctional, flexible or integrated supercapacitors and batteries. *Mater. Today*, **48**, 176, 2021.
5. Dalton, A.B., Collins, S., Razal, J.M., Howard, V., Super-tough carbon-nanotube fibres. *Nature*, **423**, 703, 2003.
6. Ahuja, P., Ujjain, S.K., Kanojia, R., Electrochemical behaviour of manganese & ruthenium mixed oxide@reduced graphene oxide nanoribbon composite in symmetric and asymmetric supercapacitor. *Appl. Surf. Sci.*, **427**, 102, 2018.
7. Ahuja, P., Sharma, R.K., Singh, G., Solid-state, high-performance supercapacitor using graphene nanoribbons embedded with zinc manganite. *J. Mater. Chem. A*, **3**, 4931, 2015.
8. Ahuja, P., Sahu, V., Ujjain, S.K., Sharma, R.K., Singh, G., Performance evaluation of asymmetric supercapacitor based on cobalt manganite modified graphene nanoribbons. *Electrochim. Acta*, **146**, 429, 2014.
9. Ujjain, S.K., Ahuja, P., Sharma, R.K., Graphene nanoribbon wrapped cobalt manganite nanocubes for high performance all-solid-state flexible supercapacitors. *J. Mater. Chem. A*, **3**, 9925, 2015.
10. Ahuja, P., Ujjain, S.K., Sharma, R.K., Singh, G., Enhanced supercapacitor performance by incorporating nickel in manganese oxide. *RSC Adv.*, **4**, 57192, 2014.
11. Xue, Q., Sun, J., Huang, Y., Zhu, M., Pei, Z., Li, H., Wang, Y., Li, N., Zhang, H., Zhi, C., Recent progress on flexible and wearable supercapacitors. *Small*, **13**, 1701827, 2017.
12. Ahuja, P. and Ujjain, S.K., Graphene-based materials for flexible supercapacitors, in: *Self-Standing Substrates*, Inamuddin, R. Boddula, A.M. Asiri, (Eds.), pp. 297–326, Springer, Cham, 2020.
13. Ahuja, P. and Ujjain, S.K., Role of graphene in solid-state asymmetric supercapacitors, in: *Green Sustainable Process for Chemical and Environmental Engineering and Science*, Inamuddin, R. Boddula, A.M. Asiri, M.M. Rahman, (Eds.), pp. 123–148, Elsevier, Cambridge, MA, United States, 2021.
14. Wang, S., Liu, F., Gao, C., Wan, T., Wang, L., Wang, L., Wang, L., Enhancement of the thermoelectric property of nanostructured polyaniline/carbon nanotube composites by introducing pyrrole unit onto polyaniline backbone via a sustainable method. *Chem. Eng. J.*, **370**, 322, 2019.
15. Guo, L., Ma, W.B., Wang, Y., Song, X.Z., Ma, J., Han, X.D., Tao, X.Y., Guo, L.T., Fan, H.L., Liu, Z.S., Zhu, Y.B., Wei, X.Y., A chemically crosslinked hydrogel electrolyte based all-in-one flexible supercapacitor with superior performance. *J. Alloys Compd.*, **843**, 155895, 2020.
16. Ye, W., Wang, H., Ning, J., Zhong, Y., Hu, Y., New types of hybrid electrolytes for supercapacitors. *J. Energy Chem.*, **57**, 219, 2021.
17. Zhang, S., Liu, Y., Hao, J., Wallace, G.G., Beirne, S., Chen, J., 3D-printed wearable electrochemical energy devices. *Adv. Funct. Mater.*, **32**, 2103092, 2022.

18. Chen, N., Dai, Y., Xing, Y., Wang, L., Guo, C., Chen, R., Guo, S., Wu, F., Biomimetic ant-nest ionogel electrolyte boosts the performance of dendrite-free lithium batteries. *Energy Environ. Sci.*, **10**, 1660, 2017.
19. Ujjain, S.K., Ahuja, P., Bhatia, R., Attri, P., Printable multi-walled carbon nanotubes thin film for high performance all solid state flexible supercapacitors. *Mater. Res. Bull.*, **83**, 167, 2016.
20. Lee, S.S., Choi, K.H., Kim, S.H., Lee, S.Y., Wearable supercapacitors printed on garments. *Adv. Funct. Mater.*, **28**, 1705571, 2018.
21. Lee, H., Jung, G., Keum, K., Kim, J.W., Jeong, H., Lee, Y.H., Kim, D.S., Ha, J.S., A textile-based temperature-tolerant stretchable supercapacitor for wearable electronics. *Adv. Funct. Mater.*, **31**, 2106491, 2021.
22. Raza, W., Ali, F., Raza, N., Luo, Y., Kim, K.H., Yang, J., Kumar, S., Mehmood, A., Kwon, E.E., Recent advancements in supercapacitor technology. *Nano Energy*, **52**, 441, 2018.
23. Ujjain, S.K., Das, A., Srivastava, G., Ahuja, P., Roy, M., Arya, A., Bhargava, K., Sethy, N., Singh, S.K., Sharma, R.K., Das, M., Nanoceria based electrochemical sensor for hydrogen peroxide detection. *Biointerphases*, **9**, 031011, 2014.
24. Ujjain, S.K., Bhatia, R., Ahuja, P., Attri, P., Highly conductive aromatic functionalized multi-walled carbon nanotube for inkjet printable high performance supercapacitor electrodes. *PLoS ONE*, **10**, e0131475, 2015.
25. Ujjain, S.K., Ahuja, P., Sharma, R.K., Facile preparation of graphene nanoribbon/cobalt coordination polymer nanohybrid for non-enzymatic H_2O_2 sensing by dual transducer: Electrochemical as well as fluorescence. *J. Mater. Chem. B*, **3**, 7614, 2021.
26. Liu, L., Yu, Y., Yan, C., Li, K., Zheng, Z., Wearable energy-dense and power-dense supercapacitor yarns enabled by scalable graphene-metallic textile composite electrodes. *Nat. Commun.*, **6**, 7260, 2015.
27. Yu, N., Yin, H., Zhang, W., Liu, Y., Tang, Z., Zhu, M.Q., High-performance fiber-shaped all-solid-state asymmetric supercapacitors based on ultrathin MnO_2 nanosheet/carbon fiber cathodes for wearable electronics. *Adv. Energy Mater.*, **6**, 1501458, 2016.
28. Lyu, W., Zhang, W., Liu, H., Liu, Y., Zuo, H., Yan, C., Faul, C.F.J., Thomas, A., Zhu, M., Liao, Y., Conjugated microporous polymer network grafted carbon nanotube fibers with tunable redox activity for efficientf flexible wearable energy storage. *Chem. Mater.*, **32**, 8276, 2020.
29. Choi, C., Kim, K.M., Kim, K.J., Lepró, X., Spinks, G.M., Baughman, R.H., Kim, S.J., Improvement of system capacitance via weavable superelastic biscrolled yarn supercapacitors. *Nat. Commun.*, **7**, 13811, 2016.
30. Chen, T., Hao, R., Peng, H., Dai, L., High-performance, stretchable, wire-shaped supercapacitors. *Angew. Chem. Int. Ed.*, **54**, 618, 2015.
31. Lu, Z., Foroughi, J., Wang, C., Long, H., Wallace, G.G., Superelastic hybrid CNT/graphene fibers for wearable energy storage. *Adv. Energy Mater.*, **8**, 1702047, 2018.
32. Lee, D.G. and Kim, B.H., MnO_2 decorated on electrospun carbon nanofiber/graphene composites as supercapacitor electrode materials. *Synth. Met.*, **219**, 115, 2016.
33. Liu, Y., Jiang, G., Sun, S., Xu, B., Zhou, J., Zhang, Y., Yao, J., Growth of $NiCo_2S_4$ nanotubes on carbon nanofibers for high performance flexible supercapacitors. *J. Electroanal. Chem.*, **804**, 212, 2017.
34. Yu, Y., Fang, Z., Luo, Y., Wu, H., Li, Q., Fan, S., Wang, J., Ultra-stretchable supercapacitors based on biaxially pre-strained super-aligned carbon nanotube films. *Nanoscale*, **12**, 24259, 2020.
35. Yu, J., Lu, W., Pei, S., Gong, K., Wang, L., Meng, L., Huang, Y., Smith, J.P., Booksh, K.S., Li, Q., Byun, J.H., Oh, Y., Yan, Y., Chou, T.W., Omnidirectionally stretchable high-performance supercapacitor based on isotropic buckled carbon nanotube films. *ACS Nano*, **10**, 5204, 2016.
36. Kim, K.J., Lee, J.A., Lima, M.D., Baughman, R.H., Kim, S.J., Highly stretchable hybrid nanomembrane supercapacitors. *RSC Adv.*, **6**, 24756, 2016.

37. Wang, C., Zhang, M., Xia, K., Gong, X., Wang, H., Yin, Z., Guan, B., Zhang, Y., Intrinsically stretchable and conductive textile by a scalable process for elastic wearable electronics. *ACS Appl. Mater. Interfaces*, **9**, 13331, 2017.
38. Lim, Y., Yoon, J., Yun, J., Kim, D., Hong, S.Y., Lee, S.J., Zi, G., Ha, J.S., Biaxially stretchable, integrated array of high performance microsupercapacitors. *ACS Nano*, **8**, 11639, 2014.
39. Li, L., Lou, Z., Han, W., Chen, D., Jiang, K., Shen, G., Highly stretchable micro-supercapacitor arrays with hybrid MWCNT/PANI electrodes. *Adv. Mater. Technol.*, **2**, 1600282, 2017.
40. An, T. and Cheng, W., Recent progress on stretchable supercapacitors. *J. Mater. Chem. A*, **6**, 15478, 2018.
41. Lv, Z., Luo, Y., Tang, Y., Wei, J., Zhu, Z., Zhou, X., Li, W., Zeng, Y., Zhang, W., Zhang, Y., Qi, D., Pan, S., Loh, X.J., Chen, X., Editable supercapacitors with customizable stretchability based on mechanically strengthened ultralong MnO_2 nanowire composite. *Adv. Mater.*, **30**, 1704531, 2018.
42. Nam, I., Kim, G.P., Park, S., Han, J.W., Yi, J., All-solid-state, origami-type foldable supercapacitor chips with integrated series circuit analogues. *Energy Environ. Sci.*, **7**, 1095, 2014.
43. Tao, J., Liu, N., Ma, W., Ding, L., Li, L., Su, J., Gao, Y., Solid-state high performance flexible supercapacitors based on polypyrrole-MnO_2-carbon fiber hybrid structure. *Sci. Rep.*, **3**, 2286, 2013.
44. Zhou, J., Zhao, H., Mu, X., Chen, J., Zhang, P., Wang, Y., He, Y., Zhang, Z., Pan, X., Xie, E., Importance of polypyrrole in constructing 3D hierarchical carbon nanotube@MnO_2 perfect core-shell nanostructures for high performance flexible supercapacitors. *Nanoscale*, **7**, 14697, 2015.
45. Dong, L., Xu, C., Li, Y., Pan, Z., Liang, G., Zhou, E., Kang, F., Yang, Q.H., Breathable and wearable energy storage based on highly flexible paper electrodes. *Adv. Mater.*, **28**, 9313, 2016.
46. Huang, Y. and Zhi, C., Functional flexible and wearable supercapacitors. *J. Phys. D*, **50**, 273001, 2017.
47. Yu, D., Goh, K., Wang, H., Wei, L., Jiang, W., Zhang, Q., Dai, L., Chen, Y., Scalable synthesis of hierarchically structured carbon nanotube-graphene fibres for capacitive energy storage. *Nat. Nanotechnol.*, **9**, 555, 2014.
48. Mao, N., Chen, W., Meng, J., Li, Y., Zhang, K., Qin, X., Zhang, H., Zhang, C., Qiu, Y., Wang, S., Enhanced electrochemical properties of hierarchically sheath-core aligned carbon nanofibers coated carbon fiber yarn electrode-based supercapacitor via polyaniline nanowire array modification. *J. Power Sources*, **399**, 406, 2018.
49. Li, S., Huang, D., Zhang, B., Xu, X., Wang, M., Yang, G., Shen, Y., Flexible supercapacitors based on bacterial cellulose paper electrodes. *Adv. Energy Mater.*, **4**, 1301655, 2014.
50. Chen, R., Yang, Y., Huang, Q., Ling, H., Li, X., Ren, J., Zhang, K., Sun, R., Wang, X., A multifunctional interface design on cellulose substrate enables high performance flexible all-solid-state supercapacitors. *Energy Storage Mater.*, **32**, 208, 2020.
51. Ren, M., Di, J., Chen, W., Recent progress and application challenges of wearable supercapacitors. *Batteries Supercaps*, **4**, 1279, 2021.
52. Manjakkal, L., Pullanchiyodan, A., Yogeswaran, N., Hosseini, E.S., Dahiya, R., A wearable supercapacitor based on conductive PEDOT:PSS-coated cloth and a sweat electrolyte. *Adv. Mater.*, **32**, 1907254, 2020.
53. Zhao, J., Li, H., Li, C., Zhang, Q., Sun, J., Wang, X., Guo, J., Xie, L., Xie, J., He, B., Zhou, Z., Lu, C., Lu, W., Zhu, G., Yao, Y., MOF for template-directed growth of well-oriented nanowire hybrid arrays on carbon nanotube fibers for wearable electronics integrated with triboelectric nanogenerators. *Nano Energy*, **45**, 420, 2018.

54. Wu, X., Xu, Y., Hu, Y., Wu, G., Cheng, H., Yu, Q., Zhang, K., Chen, W., Chen, S., Microfluidic-spinning construction of black-phosphorus-hybrid microfibres for non-woven fabrics toward a high energy density flexible supercapacitor. *Nat. Commun.*, **9**, 4573, 2018.
55. Yang, Z., Deng, J., Chen, X., Ren, J., Peng, H., A highly stretchable, fiber-shaped supercapacitor. *Angew. Chem.*, **125**, 13695, 2013.
56. Shao, G., Yu, R., Zhang, X., Chen, X., He, F., Zhao, X., Chen, N., Ye, M., Liu, X.Y., Making stretchable hybrid supercapacitors by knitting non-stretchable metal fibers. *Adv. Funct. Mater.*, **30**, 2003153, 2020.
57. Zang, J., Cao, C., Feng, Y., Liu, J., Zhao, X., Stretchable and high-performance supercapacitors with crumpled graphene papers. *Sci. Rep.*, **4**, 6492, 2014.
58. Moon, I.K., Ki, B., Oh, J., Three-dimensional porous stretchable supercapacitor with wavy structured PEDOT:PSS/graphene electrode. *Chem. Eng. J.*, **392**, 123794, 2020.
59. Chang, T.H., Zhang, T., Yang, H., Li, K., Tian, Y., Lee, J.Y., Chen, P.Y., Controlled crumpling of two-dimensional titanium carbide (MXene) for highly stretchable, bendable, efficient supercapacitors. *ACS Nano*, **12**, 8048, 2018.
60. White, S.R., Sottos, N.R., Geubelle, P.H., Moore, J.S., Kessler, M.R., Sriram, S.R., Brown, E.N., Viswanathan, S., Autonomic healing of polymer composites. *Nature*, **409**, 794, 2001.
61. Wool, R.P. and O'Connor, K.M., A theory of crack healing in polymers. *J. Appl. Phys.*, **52**, 5953, 1981.
62. Wool, R.P., Self-healing materials: A review. *Soft Matter*, **4**, 400, 2008.
63. Qin, H., Liu, P., Chen, C., Cong, H.P., Yu, S.H., A multi-responsive healable supercapacitor. *Nat. Commun.*, **12**, 4297, 2021.
64. Sun, H., You, X., Deng, J., Chen, X., Yang, Z., Chen, P., Fang, X., Peng, H., A twisted wire-shaped dual-function energy device for photoelectric conversion and electrochemical storage. *Angew. Chem.*, **126**, 6782, 2014.
65. Huang, Y., Zhong, M., Huang, Y., Zhu, M., Pei, Z., Wang, Z., Xue, Q., Xie, X., Zhi, C., A self-healable and highly stretchable supercapacitor based on a dual crosslinked polyelectrolyte. *Nat. Commun.*, **6**, 10310, 2015.
66. Wang, S., Liu, N., Su, J., Li, L., Long, F., Zou, Z., Jiang, X., Gao, Y., Highly stretchable and self-healable supercapacitor with reduced graphene oxide based fiber springs. *ACS Nano*, **11**, 2066, 2017.
67. Chen, C.R., Qin, H., Cong, H.P., Yu, S.H., A highly stretchable and real-time healable supercapacitor. *Adv. Mater.*, **31**, 1900573, 2019.
68. Kim, J., Kim, J.W., Kim, S., Keum, K., Park, J., Jeong, Y.R., Jin, S.W., Ha, J.S., Stretchable, self-healable, and photodegradable supercapacitor based on a polyelectrolyte crosslinked via dynamic host-guest interaction. *Chem. Eng. J.*, **422**, 130121, 2021.
69. Vu, V.P., Mai, V.D., Nguyen, D.C.T., Lee, S.H., Flexible and self-healable supercapacitor with high capacitance restoration. *ACS Appl. Energy Mater.*, **5**, 2211, 2022.
70. Huang, Y., Kershaw, S.V., Wang, Z., Pei, Z., Liu, J., Huang, Y., Li, H., Zhu, M., Rogach, A.L., Zhi, C., Highly integrated supercapacitor-sensor systems via material and geometry design. *Small*, **12**, 3393, 2016.
71. Pan, Z., Yang, J., Li, L., Gao, X., Kang, L., Zhang, Y., Zhang, Q., Kou, Z., Zhang, T., Wei, L., Yao, Y., Wang, J., All-in-one stretchable coaxial-fiber strain sensor integrated with high-performing supercapacitor. *Energy Storage Mater.*, **25**, 124, 2020.
72. Zhao, Y., Cao, J., Zhang, Y., Peng, H., Gradually crosslinking carbon nanotube array in mimicking the beak of giant squid for compression-sensing supercapacitor. *Adv. Funct. Mater.*, **30**, 1902971, 2020.

73. Park, H., Song, C., Jin, S.W., Lee, H., Keum, K., Lee, Y.H., Lee, G., Jeong, Y.R., Ha, J.S., High performance flexible micro-supercapacitor for powering a vertically integrated skin-attachable strain sensor on a bio-inspired adhesive. *Nano Energy*, **83**, 105837, 2021.
74. Mohanadas, D. and Sulaiman, Y., Recent advances in development of electroactive composite materials for electrochromic and supercapacitor applications. *J. Power Sources*, **523**, 231029, 2022.
75. Zhou, Y., Fang, J., Wang, H., Zhou, H., Yan, G., Zhao, Y., Dai, L., Lin, T., Multicolor electrochromic fibers with helix-patterned electrodes. *Adv. Electron. Mater.*, **4**, 1800104, 2018.
76. Nwanya, A.C., Jafta, C.J., Ejikeme, P.M., Ugwuoke, P.E., Reddy, M.V., Osuji, R.U., Ozoemena, K.I., Ezema, F.I., Electrochromic and electrochemical capacitive properties of tungsten oxide and its polyaniline nanocomposite films obtained by chemical bath deposition method. *Electrochim. Acta*, **128**, 218, 2014.
77. Guo, Q., Li, J., Zhang, B., Nie, G., Wang, D., High-performance asymmetric electrochromic-supercapacitor device based on poly(indole-6-carboxylicacid)/TiO_2 nanocomposites. *ACS Appl. Mater. Interfaces*, **11**, 6491, 2019.
78. Zhang, S., Chen, S., Cao, Y., Yang, F., Peng, H., Yan, B., Jiang, H., Gu, Y., Xiang, M., Polyaniline nanoparticle coated graphene oxide composite nanoflakes for bifunctional multicolor electrochromic and supercapacitor applications. *J. Mater. Sci. Mater*, **30**, 13497, 2019.
79. Jiao, X., Li, G., Yuan, Z., Zhang, C., High-performance flexible electrochromic supercapacitor with a capability of quantitative visualization of its energy storage status through electrochromic contrast. *ACS Appl. Energy Mater.*, **4**, 14155, 2021.
80. Liu, L., Shen, B., Jiang, D., Guo, R., Kong, L., Yan, X., Watchband-like supercapacitors with body temperature inducible shape memory ability. *Adv. Energy Mater.*, **6**, 1600763, 2016.
81. Deng, J., Zhang, Y., Zhao, Y., Chen, P., Cheng, X., Peng, H., A shape-memory supercapacitor fiber. *Angew. Chem. Int. Ed.*, **54**, 15419, 2015.
82. Du, P., Hu, X., Yi, C., Liu, H.C., Liu, P., Zhang, H.L., Gong, X., Self-powered electronics by integration of flexible solid-state graphene-based supercapacitors with high performance perovskite hybrid solar cells. *Adv. Funct. Mater.*, **25**, 2420, 2015.
83. Xu, X., Li, S., Zhang, H., Shen, Y., Zakeeruddin, S.M., Graetzel, M., Cheng, Y.B., Wang, M., A power pack based on organometallic perovskite solar cell and supercapacitor. *ACS Nano*, **9**, 1782, 2015.
84. Zhang, Z., Chen, X., Chen, P., Guan, G., Qiu, L., Lin, H., Yang, Z., Bai, W., Luo, Y., Peng, H., Integrated polymer solar cell and electrochemical supercapacitor in a flexible and stable fiber format. *Adv. Mater.*, **26**, 466, 2014.
85. Mondal, S., Zehra, N., Choudhury, A., Iyer, P.K., Wearable sensing devices for point of care diagnostics. *ACS Appl. Bio Mater.*, **4**, 47, 2021.
86. Bharathi Sankar Ammaiyappan, A. and Ramalingam, S., Self-powered supercapacitor for low power wearable device applications. *IOP Conf. Ser.: Earth Environ. Sci.*, **850**, 012016, 2021.
87. Liu, L., Yang, X., Zhao, L., Xu, W., Wang, J., Yang, Q., Tang, Q., Nanowrinkle-patterned flexible woven triboelectric nanogenerator toward self-powered wearable electronics. *Nano Energy*, **73**, 104797, 2020.
88. Wang, H., Ail, U., Gabrielsson, R., Berggren, M., Crispin, X., Ionic seebeck effect in conducting polymers. *Adv. Energy Mater.*, **5**, 1500044, 2015.
89. Khoso, N.A., Jiao, X., Guangyu, X., Tian, S., Wang, J., Enhanced thermoelectric performance of graphene based nanocomposite coated self-powered wearable e-textiles for energy harvesting from human body heat. *RSC Adv.*, **11**, 16675, 2021.

15

Electrospun Materials

Hina Sahar[1], Sania Naseer[2], Muhammad Rehan Hasan Shah Gilani[1], Syed Ali Raza Naqvi[3], Muhammad Aamir[2*] and Javeed Akhtar[2]

[1]*Institute of Chemical Sciences, Bahauddin Zakariya University, Multan, Pakistan*
[2]*Materials Laboratory, Department of Chemistry, Mirpur University of Science and Technology (MUST), Mirpur, Pakistan*
[3]*Department of Chemistry, GC University of Faisalabad, Mirpur (AJK), Faisalabad, Pakistan*

Abstract

Electrospinning, a widely-known technology for generating nanoscale fibers, has lately received widespread attention because of its several benefits, which includes large surface/volume ratio, easy of surface modification and variable porosity. Because of its versatility, utility, and simplicity, electrospinning is fast growing as a simple technique capable of creating different morphologies among a number of nanofiber preparation technologies. Herein, we described a brief history of electrospinning, the electrospinning method, and parameters and their impacts during the electrospinning process, like solution, processing, and ambient parameters. Furthermore, preparation methods for nanofibers based on electrospinning technology, pore formation mechanisms in electrospun polymer fibers, and various ways of modification of electrospun micro- and nanofibers are covered. We highlight the most recent and significant breakthroughs in electrospun nanofiber applications, with an emphasis on medicinal, electrochemical, environmental, and energy storage and conversion applications.

Keywords: Electrospinnig, electrospun materials, electrospun fibers, application of electrospun materials

15.1 Introduction

Electrospinning is a widely used technique for manufacturing nanofibers with diameters in the range of sub-nanometers to micrometers on a continuous basis. Because of their small diameters, those fibers have a large surface/volume ratio, making them ideal candidates for a huge number of applications in nearly every field, which include composites, tissue engineering, biomaterials, energy conversion and storage, packaging materials for food, drug delivery, catalysts, flexible electronics, sensors, reactors, and environmental protection [1, 2].

Corresponding author: aamirorg@gmail.com; amir.chem@must.edu.pk

Inamuddin, Tariq Altalhi and Sayed Mohammed Adnan (eds.) Sustainable Materials for Electrochemical Capacitors, (361–390) © 2023 Scrivener Publishing LLC

15.1.1 Brief History

Anton Formhals devised electrospinning as a revolutionary patented technology for creating thin nanofibers with a nano diameter in 1934, successfully spinning fibers of cellulose acetate from acetone solution, yet this idea was never commercialized because of the difficulties with solvent evaporation and fiber collecting [3]. Taylor also mathematically modeled the electrospinning process to better understand the mechanism of cone shape (Taylor cone) generation from the liquid droplet under the impact of an electric field. However, electrospinning technology acquired significant popularity after 1990, mostly due to advances in nanoscience research. Numerous studies on the topic have been conducted since then. Electrospinning technology research has shown a steep increasing trend over the last decade, as seen by an increase in scholarly papers and publications [4].

15.2 Electrospinning Process

Many teams and organizations have conducted theoretical studies on the electrospinning process. Sir William Gilbert described the nature of magnetic and electrostatic emission around the end of the 1500s. He discovered that by applying an electrostatic field to a water droplet, the water takes on the shape of a hopper or a cone, and a liquid droplet slowly releases through the hopper's head. This was the first electro-spraying operation. Electrospinning may be thought of as a form of electrospraying. Electrospinning's raw material, like electrospraying's, is connected to a power supply of high voltage to increase the electrostatic potential of the liquid. Because of their intermolecular interaction, greater molecular degree polymers have long been used as raw resources. Since the electrostatic potential and surface charge of a liquid are inextricably linked, increasing or decreasing one has the same impact on the other. Surface tension is commonly used to calculate the fluid's volume shape. By introducing the charge to the fluid, surface tension and charges act oppositely, allowing Taylor cone formation [5].

Electrospinning appears to be simple to use at first glance. Electrospinning is a simple procedure that requires just a few components: i) a syringe pump, ii) a power source of high voltage, iii) a capillary tube with a capillary tip (spinneret) or a needle, iv) and a fiber collector [6]. Electrospinning is the process of pulling a solution or melt jet of polymer to form nanofibers using electrostatic forces. The syringe pump pushes the polymer solution for electrospinning out of the syringe, because of gravity and surface tension, a droplet forms at the spinneret. At the tip of the capillary tube, the droplet of polymer solution deforms into a Gilbert-Taylor cone under the influence of electrostatic force [7]. In the 15th century, in an electric field, liquid droplet cone generation was initially explained by William Gilbert. In his book "De Magnete," he published his findings [8]. As electrostatic force suppresses the surface tension, a charged jet is released. The jet passes to a counter electrode, which is a ground plate [9–11]. There are two phases to electrospinning. The first phase involves the generation of a Taylor cone and a solution jet owing to polarization at high voltage. As the surface tension and electric force become equal, the liquid droplet takes on a cone shape, and the Taylor cone develops [12]. Nanofibers are formed on a collector consisting of a metal plate, screen, or drum [13]. As the electric force reaches a critical value (cone formation), a liquid jet is discharged, leading to suspended drop stretching and nanofiber formation [14].

A counter electrode (mash, wire, plate, conducting coagulation bath, or metal cylinder) is grounded and placed against the capillary. The collector is coated with a polymer fibers membrane. Due to its excellent surface/volume ratio and large surface area, electrospun nano-fiber has a lot of capacity for developing outstanding energy-restoring systems. They can provide an impressive number of active sites as well as a regulated porous shape to counteract the substantial volume changes that take place during the cycling of the battery and permeate the electrolyte [15]. The morphology of fiber can be affected by the polymer solution concentration, loading voltage, feeding rate, collecting distance, humidity, and ambient temperature during the electrospinning process [9, 16].

15.3 Advantages of Electrospinning Technique

Phase separation, template synthesis, self-assembly, drawing, and electrospinning are some of the current techniques for producing nanofibers. Electrospinning is a simple and effective technology for producing nanofibers on a large scale and in a continuous way. Low expenses, moderate experimental parameters, straightforward function and operation, and a diverse variety of building ingredients are just a few of the benefits. The spinning process can be controlled, and the conditions can be customized to meet the needs of various study areas. Electrospun nanofibers, for example, are made in specific shapes and orientations to study the link between molecular orientation and mechanical characteristics quantitatively [17]. Electrospinning method produced a material with an extraordinarily extended size, porosity, and a high surface area, which may considerably enhance its response time and sensibility, making it an ideal choice for sensor applications [18, 19].

Electrospun micro/nanofibers are currently being developed from a variety of materials, such as metals, metal oxides, ceramics, organic-inorganic composites, and polymers. These electrospun membranes, which are made up of organic fillers, blends, or copolymers, always have better mechanical properties, thermal stability, and permeability [20, 21].

15.4 Working Parameters of Electrospinning Process

Considering not just the mechanism of electrospinning but rather the electrospinning transformation of polymer solutions into nanofibers, requires working parameters. The three categories of parameters that may be identified are the solution, process, and ambient parameters (Figure 15.1) [22]. These parameters can have an impact on fiber quality, and by correctly managing these parameters, one can create electrospun fibers with desirable shapes, sizes, and diameters [23, 24].

15.4.1 Solution Parameters

Generally, the preparation of the polymer solution is the most significant part of making an electrospun nanofibrous membrane. The following are some of the solution properties that may have an impact on electrospinning [25].

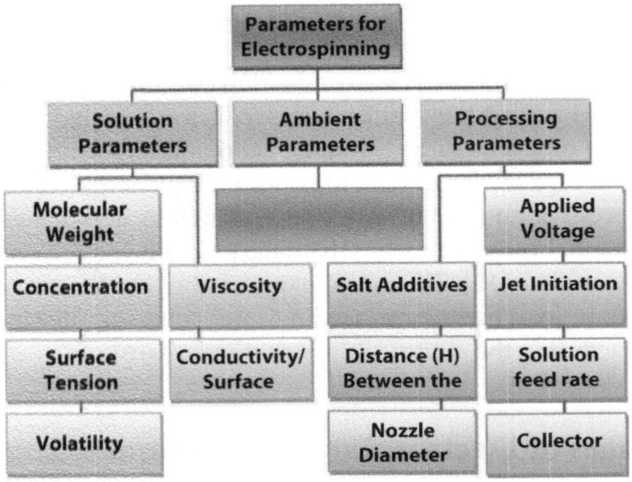

Figure 15.1 Parameters for electrospinning process.

- ***Molecular weight***
  Molecular weight has a considerable influence on electrical and rheological properties like dielectric strength, conductivity, viscosity, and surface tension of a polymer. High molecular weight polymer solutions are commonly utilized in electrospinning because they give the appropriate viscosity for fabrication. A low molecular weight solution forms beads rather than fibers, whereas solutions with a higher molecular weight produce fibers with a greater average diameter. The polymer's molecular weight indicates the number of entangled polymer chains in a solution and, consequently, the viscosity of the solution [26].

- ***Concentration***
  A small concentration of the solution is required in the electrospinning process for the fabrication of fiber.

  At low concentrations of the solution, a combination of fibers and beads is produced. However, at high concentrations, the bead topology changes from spheres to the spindles, and eventually, fibers of larger diameters and regular sizes are produced due to the higher viscosity barrier.

- ***Viscosity***
  During the spinning of polymeric fibers, the solution's viscosity plays an essential part in defining fiber shape and size. It has been observed that at extremely low viscosity, no continuous production of fiber occurs, and when the viscosity is excessively high, the polymer solution jet ejection is problematic, demanding the use of an optimum viscosity for electrospinning.

- ***Surface Tension***
  Surface tension is very important in the electrospinning process, and by lowering the nanofiber solution's surface tension, fibers may be generated without beads. This is most likely a function of the solution's solvent composition. Surface tensions may change depending on the solvent. The electrospinning process is generally affected by a solution's high surface tension due to sprayed droplet production and jet fluctuation [27].

- ***Conductivity/Surface Charge Density***
  Except for a few dielectric materials, polymers are typically conductive, and in the polymer solution, charged ions play a crucial part in the jet formation. Solution conductivity is affected by the kind of polymer employed, the solvent, and the availability of ionic salts. The electrospun nanofiber's diameter drops dramatically as the solution's electrical conductivity increases, but as the solution's electrical conductivity declines, under the electrical force, there is inadequate jet elongation to make a homogeneous fiber, so it's possible to see beads [28].

- ***Volatility***
  When it comes to solvent volatility, a polymeric solution made with low-volatility solvents might result in the collection of moist, fused, or even undesirable fibers. Due to the polymer solidification at the spinneret tip, a very volatile solution may cause irregular spinning [29].

15.4.2 Processing Parameters

- ***Applied Voltage***
  More liquid droplet stretching may be caused by a greater applied voltage. However, due to the increased potential difference, a greater voltage may also result in a quicker acceleration towards the substrate collector, allowing the jet to stretch more quickly before deposition. As a result, the fiber diameter increases [30].

- ***Jet initiation***
  The complete process of exposing the electrospinning solution to an electric applied field and inducing charges to it is known as jet initiation [31].

- ***Feed rate of solution***
  Fiber diameter dispersion will be smallest at the optimum solution feed rate, so any variation may result in higher fiber diameter spreading [32].

- ***Salt Additives***
  The addition of salt may have the following effects on the electrospinning procedure/technique:

(1) Increased Stretching of the spinning jet (because of greater conductivity), resulting in thinner fibers
(2) A more constant distribution of charges may result in higher fiber homogeneity
(3) Reduction of solution surface tension
(4) During electrospinning reduction of beads formation [33]

- ***Tip of the syringe and collector distance***
  Another method for controlling fiber diameter and shape has been to study the tip of the syringe and collector distance. It's been discovered that a minimal spacing is essential to provide the fibers with sufficient drying time before approaching the fiber collector; otherwise, the beads will be detected when the distance is too small or too far [27].

- ***Collector***
  Fiber collectors are critical in achieving aligned orientation. Many particularly designed collectors, like collector drums rotating at high speed [34], as collectors two strips of parallel conductors [35] tip (spinneret) collector [36], or metal plate or frame collectors [15] have been employed to get aligned fibers.

- ***Nozzle diameter***
  As the nozzle diameter increased, so did the fiber diameter, distribution, and productivity. This might be because there is more mass available for the electro-spinning process [37].

15.4.3 Ambient Parameters

- ***Temperature***
  A greater solution temperature is generally established to reduce the solution's viscosity. As a result, lower viscosity will almost certainly result in smaller fiber diameter [38].

- ***Humidity and Environment***
  The temperature and humidity of the surroundings have an impact on the physical features of the fibers and the whole electrospinning process because of their effect on the solvent and solution sensitivity to humidity and rate of vaporization [39]. The fiber with the largest diameter will come from higher humidity due to neutralization of charges on the jet, which leads to reduced stretching pressure. Low humidity, on the other hand, can speed up the evaporation process and completely dry the solvent. Casper *et al.* showed that the surface morphologies of electrospun PS fibers may be affected by a wide range of humidity [22].

15.5 Electrospinning-Based Preparation Methods for Nanofibers

One of two approaches can be utilized while electrospinning a polymer. The polymer can be electrospun directly from a melt or after being dissolved in a suitable solvent.

15.5.1 Melt Electrospinning

The necessity to prevent the immense amount of solvents involved in solution electrospinning and to raise the production rate are both critical topics, particularly in terms of technological applications. Melt electrospinning might be a solution [40, 41]. Unlike solution electrospinning, melt electrospinning has earned little interest from researchers in the past. The following are some of the key causes behind this:

(1) The high viscosity of the molten polymer and the difficulties in adjusting its viscosity in comparison to the polymer's solution
(2) The manufacture of larger fibers (in the micron range)
(3) The apparatus's complexity is mostly due to the interaction of the temperature and heating control systems with the power source of higher voltage, and
(4) Non-conductive characteristics of the molted polymer [42].

Zhou *et al.* prepared a literature-based list of melt electrospinning techniques. The majority of melt electrospinning research included polymers that at ambient temperature, are hard to electrospun. Polypropylene (PP) and polyethylene (PE) are two examples. According to the study, the bulk of these polymers were electrospun over 1500 degrees Celsius temperatures. High temperatures may necessitate the use of complex equipment and polymer decomposition during melt electrospinning. Melt electrospinning, on the other hand, may create very high fiber outputs and achieve solvent-free systems, lowering equipment costs by eliminating the need for ventilation equipment [43, 44].

Melt electrospinning might have evolved into melt-electrospinning writing (MEW), a form of 3D printing method that can accurately regulate the macrostructure. MEW is mainly focused on close-field electrospinning mixed with fused deposition modeling, which involves the development of a scaffold using layer-by-layer deposition and computer software. As a result, the MEW approach may be used to construct 3D structures and implants for blood vessels and bone tissue. The electrospinning instrument generates electrically charged molten jets that, by using a moveable collector, can write a 3D structure directly. The capacity to control pore size and interconnectivity balance is another feature of the MEW technique [45].

Polymer melts also eliminate the problem of insufficient evaporation of the solvent between the collector and capillary tip; however, in order to create cylindrical fibers, the polymer must be able to cool sufficiently over this distance.

15.5.2 Solution Electrospinning

It is now the most widely used electrospinning technology, where the remaining polymer fibers solidify and are collected after the solvent evaporates. Even though solution electrospinning is an adaptable and efficient technology for the production of micro to nanometers sized fibers from a range of polymer solutions, the main obstacles to solution electrospinning include possible environmental contamination from solvent buildup, solvent recovery, toxic or organic solvent residue in fibers, and microscopic holes in the fiber membrane. These disadvantages have a significant impact on mass manufacturing and biomedical uses of e-spun nanofibers. Furthermore, solution e-spinning cannot handle non-soluble polymers such as polyethylene (PE) and polypropylene (PP) [46].

15.6 Formation of Pore in Electrospun Polymer Fibers

We evaluate the study on porous electrospun polymer fibers within the context of the numerous methods for causing porosity in this section. The generation of hierarchically porous electrospun polymer fibers has been demonstrated using the following methods:

1) Breath Figures (BF),
2) Vapor induced phase separation (VIPS),
3) Non-solvent induced phase separation (NIPS),
4) Thermally induced phase separation (TIPS),
5) Selective removal of one specific component from blend of electrospun fibers.

Without the requirement for post-treatments, a liquid-rich phase generated by NIPS, VIPS, or TIPS can evaporate instantly to form multi-scale holes inside or on the surface of the fibers. They offer significant benefits in terms of efficiency, cost, and controllability [47].

15.6.1 Breath Figures (BF)

When using the BF mechanism to create porosity during electrospinning, two requirements must be satisfied: higher relative humidity and a large temperature decrease. Volatile

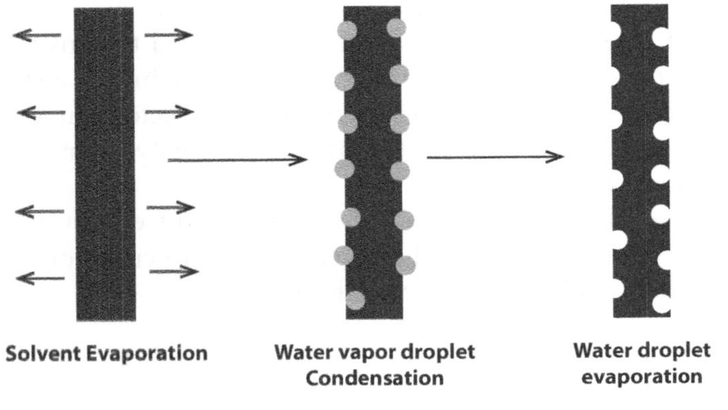

Figure 15.2 Pore formation by the BF mechanism. Reproduced with permission from Ref. [49].

solvents like, tetrahydrofuran (THF), CHL, and DCM [48] were utilized to cause this substantial temperature decrease. Figure 15.2 depicts a schematic representation of the development of breath figures. The volatile solvent evaporation causes a considerable temperature drop on the fiber surface, resulting in the water vapors condensation. Once the solvent and droplets evaporate, the condensed water vapors function as a template, taking up space on the moderately dried surface of the fiber and effectively leading to pore formation.

15.6.2 Vapor-Induced Phase Separation (VIPS)

In the VIPS process (Figure 15.3), water vapors are involved, similarly to how they are in the breath figures mechanism. As a result, this process necessitates a relatively higher RH (relative humidity) about 40 percent. The distinction is that a solvent is required for VIPS, with good water miscibility and low volatility, whereas BF requires a solvent with poor water miscibility and high volatility. Dimethylformamide (DMF) is a good solvent for inducing porosity via the VIPS mechanism [49, 50]. During the drying and electrospinning processes, a nonvolatile solvent slowly evaporates, enabling water vapors to interfere with the surface of the fiber and permeate inside. Vapor of water behaves as a non-solvent when electrospinning a hydrophobic polymer like PS or PLA [51].

15.6.3 Non-Solvent-Induced Phase Separation (NIPS)

The diffusion polymer solution and non-solvent and the original solvent replacement with the non-solvent cause phase inversion in NIPS. As a result, mass transfer events play a crucial part in phase inversion [52]. The NIPS technique (also known as the Loeb Sourirajan method or immersion precipitation) is perhaps the most adaptable and commonly used to create asymmetric structures in industry, hollow fiber membranes, and standard flat-sheet. The produced membranes are simple to process and perform well, that's what NIPS is seen

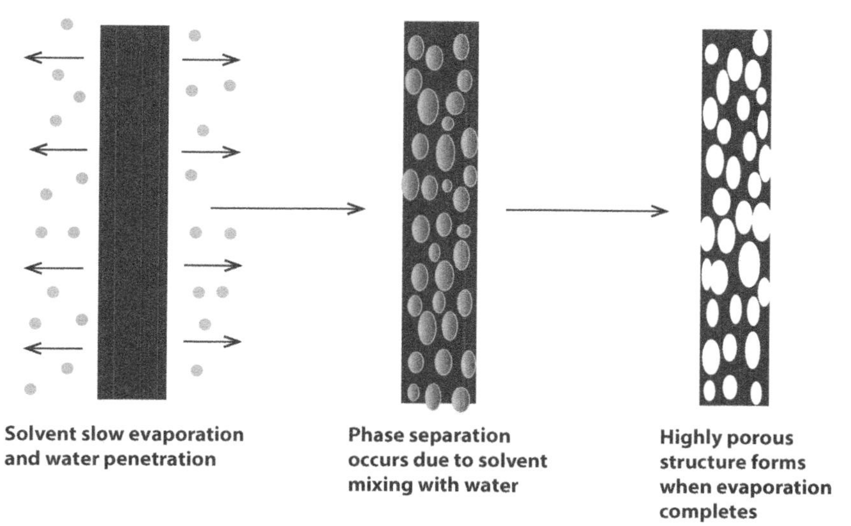

Figure 15.3 Porosity induced by the VIPS Process. Reproduced with permission from Ref. [49].

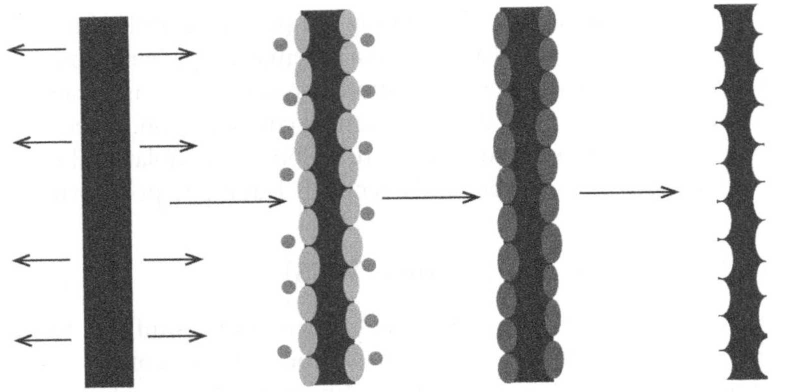

Figure 15.4 Non-solvent-induced phase separation (NIPS) process. Reproduced with permission from Ref. [49].

as useful and broadly applicable. At least three components are required for the NIPS technique: a solvent to dissolve the polymer, a polymer to assess membrane characteristics, and a nonsolvent to regulate phase inversion occurrences (Figure 15.4). A homogenized dope solution including a solvent, a polymer, and a tiny amount of nonsolvent is initially created in a conventional NIPS process. After that, the solution is molded into a fine layer or spun into a porous matrix, and then submerged in a non-solvent bath such as water. Polar aprotic solvents for example, dimethylacetamide, and dimethylformamide are commonly utilized as doping solutions due to their chemical affinity with polymers [7].

15.6.4 Thermally Induced Phase Separation (TIPS)

The idea behind TIPS is that the solvent's quality normally reduces when the temperature is reduced. Generally, after inducing separation, the solvent is evaporated, extracted, or freeze-dried. TIPS develops between the solvent-poor and solvent-rich areas of an electrospun fiber when it is immediately spun into an ultracold liquid due to a rapid temperature drop. Following the regulated solvent evaporation, pores develop throughout the fiber (not just on the surface). However, because of the quick-freezing, this process often produces larger pores and thicker (broken) fibers, which impairs fiber whipping and elongation [53].

15.6.5 Selective Removal

As discussed above, the phase separation method is mostly used to produce electrospun porous fibers, although removing a sacrificial component can be used as an alternate option. Selective removal is a two-stage procedure where the post-treatment stage is properly regulated to totally eliminate the sacrificing phase while maintaining the integrity of the fiber matrix. It is a powerful technique for creating porous structures from both polymers of hydrophilic and hydrophobic nature.

Silica, polymers, or salts that are easily removed by solvent or eliminated by various post-treatment procedures can be used as sacrificial components. Although most sacrificial components may be extracted and/or thermally treated out of the precursors, a tiny quantity may linger in the polymer matrix and impact the fiber characteristics [54].

15.7 Modification of Electrospun Micro- and Nanofibers

In several cases, fibers produced using the techniques outlined in the preceding sections lack all of the qualities and functions necessary for a specific application. Post-modification approaches for micro- and nanofibers are a promising option for promoting surface modifications (Figure 15.5) and providing required characteristics by i) modifying the micro- or nanofiber's surface with new chemical functionalities; ii) converting existing functionalities into more required functionalities; or iii) extracting specific functional groups. A number of materials and procedures can be used to execute such changes with various aims [55].

15.7.1 Chemical Modification

It is a widely used technique for modifying the micro/nanofibers surface because of its adaptability, effectiveness, and extensive application. Chemical modification, in particular, causes permanent or difficult-to-reverse alterations to the micro/nanofiber surface, generally by removing moieties or adding molecules or functional groups through covalent bonds. Furthermore, certain approaches can modify micro/nanofibers that are buried deep inside the fibers mat, leading to a far more uniform modification. There are several ways of performing such modification, including oxidation, wet chemistry, cross-linking, hydrolysis, grafting, and so on, which allow the alteration of a wide range of chemical groups and materials [56, 57].

Nonwovens are significantly more complex to modify chemically than they are to modify physically. Crosslinking of nonwoven soluble in water, stable surface modification, and

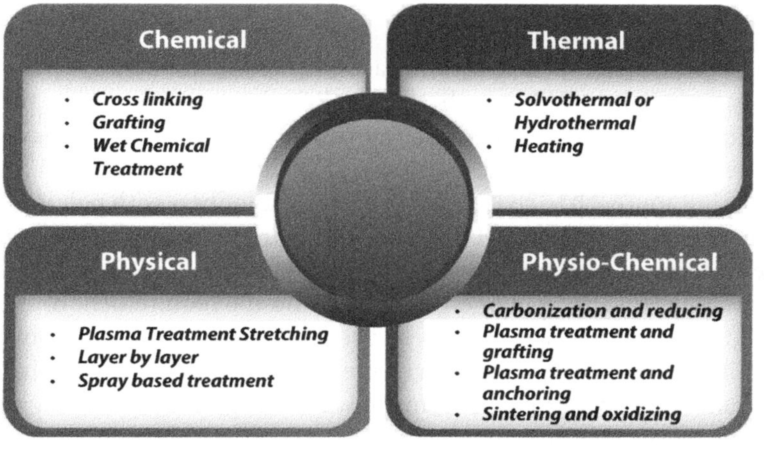

Figure 15.5 Modification techniques of electrospun micro and nanofibers.

mineral compound coating on the surface are the most common uses. The two most significant disadvantages are the requirement for medically nonpermitted chemicals and the acceleration of polymer deterioration.

Chiu and coworkers converted cyanide(–CN) to hydrophilic carboxyl groups (–COOH) using sodium hydroxide (NaOH) in electrospun PAN nanofibers [58]. They discovered an adsorption capability of lysozyme at approximately 105 mg/g (at pH 9), which is nearly twice as much as commercially accessible products. Schneiderman and coworkers [59] also used nitric acid at 90 degrees Celsius for 48 hours to carboxylate the carbon nanofiber's surface for the adsorption of proteins. The nanofiber mats' absorption capability was nearly ten times that of their equivalent microfiber. Li *et al.* [60] used lysine (LYS) functionalized polyacrylonitrile (PAN) nanofibers for the adsorption of selective proteins to develop pH-controllable electrospun nanofibers. By varying the pH of the nanofiber surface, they can produce negative and positive charges. isoelectric point for capturing lysozyme (IP = 10.8) and pepsin (IP = 1). Maximum adsorption capacities of 54.98 mg/g at pH 8 and 425.49 mg/g at pH 3 were found.

15.7.1.1 Cross-Linking

Cross-linking is another extensively used approach for chemically modifying the surface of micro/nanofibers via covalent bonding, allowing chemical functionalization. Cross-linkers that are often utilized include genipin, hexamethylene diisocyanate, glutaraldehyde, carbodiimide, and acyl azides [57, 61]. Dong and colleagues modified electrospun (PVA) poly (vinyl alcohol) nanofibers with glutaraldehyde (GA) before further functionalization with fluoroalkyl silane (FAS) [62]. Wang and colleagues created PVA/GA electrospun nanofiber mats which were cross-linked using acid treatment to produce water-insoluble PVA nanofibers [63]. The nanofibers were further modified with AuNPs for application in H_2O_2 sensing. Another study employed GA for activation of the surface of polyacrylonitrile (PAN)/ carbon nanotube (CNT) nanofibers and functionalized them with TiO_2 nanoparticles for photocatalytic degradation of pharmaceutical drugs [64].

15.7.1.2 Grafting

Grafting is commonly used to change the surface of micro/nanofibers by chemically linking polymer chains or initiators. This may be accomplished in two ways: i) on the surface of a nanofiber by attaching a premade polymer chain, or ii) by employing a radical initiator that permits polymer formation In order to achieve DNA adsorption. Demirci and colleagues employed a grafting approach to coat cellulose acetate nanofibers with poly[(ar-vinylbenzyl) trimethylammonium chloride] (poly(VBTAC)). A controlled/living radical polymerization is a reversible addition-fragmentation chain transfer (RAFT) (CLRP) that has benefits like the ability to build customized polymer topologies and the requirement of simple reaction conditions [65, 66]. Another approach is to use atom transfer radical polymerization to graft electrospun chitosan nanofibers with zwitterionic phosphorylcholine [67]. The ATRP enabled simultaneous polymer chain formation, resulting in a regulated grafting density on the nanofiber surface. Ahmadi and colleagues attached poly (methyl methacrylate) (PMMA) and poly(butyl acrylate) (PBA) onto cellulose nanofibers utilizing the free radical initiator technique [68].

15.7.1.3 Wet Chemical Treatment Technique

The electrospun mat surface reacts with a chemical component of the solution in wet-chemistry wettability modification procedures. Depending on the materials used, the alteration may increase hydrophilicity/hydrophobicity, and roughness. Aminolysis and hydrolysis of polyesters like PLGA to create hydrophilic polar groups at the surface are two common examples [6].

Polyester hydrolysis, in which the ester bonds break and create hydroxyl or carboxyl groups on the surface of degraded polymers, is the most common process employed to modify the wetting behavior of electrospun mats. A dilute alkali or acid solution is introduced to speed up the process, as hydrolysis with just water is very slow. PBGL (Poly (-benzyl-L-glutamate), PLGA (Poly(lactic-co-glycolic acid), PCL (Poly(-caprolactone), PCL (Poly(-caprolactone), electrospun mats have been treated with aqueous solutions containing NaOH to induce hydrolysis and form an RC mat [12].

15.7.2 Thermal Modifications

Nanofibrous membranes are typically constructed by a 3-dimensional system of nanofibers unevenly dispersed in space and have weak interconnections linking them, leading to low mechanical strength. Thermal modification is the simplest and most environmentally safe method for improving the nanofibrous membrane's mechanical characteristics, and it may also be used as a nanofiber surface post-modification treatment. Heat press, hydrothermal treatment, and pyrolysis/carbonization are common examples.

15.7.2.1 Hydrothermal/Solvothermal Modification

Hydrothermal/solvothermal treatment is mostly used as a post-modification strategy to impart new functional groups on micro/nanofibrous membranes, with applications including sensors, medical, water treatment, energy conversion, and solar cells [69–71]. The nanofibrous membranes are typically submerged in an aqueous phase that contains the required precursor or component in a sealable autoclave before being subjected to hydrothermal treatment under pressure and temperature. Solvothermal treatment is when a solvent other than water is employed. Then, the nucleation and precipitation processes are repeated, and nanoparticles with the required composition and phase are formed on the nanofiber surface [72, 73].

15.7.2.2 Heating

Temperature curing is one of the most basic physical ways of modifying the characteristics of electrospun nanofibers. Sullivan *et al.* [74] electrospun a poly(ethylene oxide) composition with lactoglobulin (BLG) or whey protein and employed heat to make the fibers water insoluble. The mat was inflated in water and kept its fibrous texture, suggesting its potential application in regeneration therapy. Rhodamine B (RhB) was put into the mat and then released as a model flavonoid.

Min *et al.* [75] used electrospinning to regenerate silk fibroin (SF) and steam treatment to denature the protein. The fibroin structure evolved from a random coil to a sheet that is insoluble in water and has higher mechanical strength. The adhesion of human fibroblasts and keratinocytes, as well as a dispersion study, revealed outstanding cellular compatibility.

The material might be used to make wound dressings or for tissue engineering scaffolds. Enayati *et al.* [76] investigated nonwoven poly(vinyl alcohol) (PVA) composites containing cellulose nanofibers (Cellulose NF) or hydroxyapatite nanoparticles. The composites were warmed-over 180°C to make them water-insoluble. The researchers investigated the effect of nano-particles on the crystallinity of polymers hardened under temperature.

15.7.3 Physical Modification

Physical surface alterations of electrospun nanofibers include absorption or leaching of water-soluble polymers, stretching to modify mechanical characteristics, ultraviolet and laser ablation to induce programmed changes in characteristics, ultrasonic treatment to increase porosity or fiber fragmentation, and plasma processing to promote hydrophilicity of the surface [56, 77].

15.7.3.1 Plasma Treatment

It is a technique for adding polar functionalities to a surface, increasing its hydrophilicity and stickiness. With the right plasma supply, bioactives including ECM proteins, cationized gelatin, and RGD peptides can be immobilized or linked to the material's surface, resulting in excellent biocompatibility or the ability to immobilize them. Plasma treatments with air, ammonia, or oxygen can form amine or carboxyl groups, which can increase cell proliferation and adherence [78–81].

Plasma treatment is an easy and efficient approach for increasing the hydrophilicity of a material's surface. It's commonly used in industry to change the properties of polymer sheets [82]. The gas type (e.g., O_2 or Ar), treatment duration, and power are the most important factors in plasma modification. Plasma-treated polymers needed to be processed further. The effects of plasma alteration on the surface are only transitory and fade away with time. The oxygen content of electrospun poly(caprolactam) (PA6) exposed to oxygen plasma for 1 minute was higher, the angle of contact was improved, and the surface roughness was lower than the sample treated for 5 minutes [83].

Radio-frequency (RF) plasma has been used to modify poly(caprolactone) nanofiber meshes to improve cell adhesion, proliferation, and differentiation. Treatment with oxygen plasma was the most effective for osteoblasts, chondrocytes, and fibroblasts. The angle of contact and material–cell interfaces were both significantly improved. According to scientists, surface roughness is also a critical element in the material–cell interface [84]. Methane or oxygen plasma has been used to modify silk fibroin nanofibers to create skin regeneration scaffolds. When compared to oxygen plasma, methane plasma modification had a minor effect on hydrophilicity. Human epidermal keratinocytes and fibroblasts' wettability and cellular activity increased dramatically as a result of the latter [85].

15.7.3.2 Stretching

Stretching can increase crystallinity and strength in spun nanofibers since they have a low degree of molecular orientation. A higher degree of crystallinity is associated with lower rates of bio-degradation and much lower elasticity. Stretching electrospun nanofibers is a really unusual form of modification. When compared to nonwoven mats, the modulus strength of a single fiber can increase by a factor of two [56].

15.7.3.3 Layer-by-Layer

This method includes gathering and creating layers with opposing charged surfaces on substrates. The steps are as follows: i) immersing a substrate in a polyelectrolyte solution; ii) successive rinsing to eliminate poorly adsorbed particles, and iii) immersing the substrate in an oppositely charged solution. This process is repeated several times, resulting in the overlap of many oppositely charged layers. At the conclusion of the procedure, an electrostatic attraction between layers of opposing charges forms a multilayered coated structure, which has the benefit of being independent of the topology and size of the substrate, simple, and in terms of thickness of the layer [86, 87]. The LbL technique has been utilized to alter micro and nanofibers for usage in a variety of applications [88].

15.7.3.4 Spray-Based Methods

Another possible approach for physical modification of nano/microfibers is air or air-brush spray, which involves releasing a solid particle suspension via a nozzle orifice using pressured gas, resulting in a tiny particle spray outside of the nozzle environment. Coating consistency during the air-brush procedure has been linked to the following experimental characteristics: i) distance between nozzle and substrate. This has an impact on the substrate's intended deposition area and evaporation of the solvent, permitting wet, dry, and wet-dry (intermediate) deposition; ii) allocated liquid phase, which increases phase evaporation along the spray trajectory; and iii) substrate and ambient conditions, which impact liquid evaporation. Chen and colleagues performed a descriptive analysis of characteristics that can have a big impact on the shape, thickness, homogeneity, and performance of perovskite solar cell films [89, 90]. Nanofibrous matrices exhibit pore permeability and an uneven surface (because of random alignment of micro and nanofibers), whereas certain other flat-surface substrates that may be modified with an airbrush, for example polymeric film, glass, etc., which allow spraying particles to penetrate and reach fibers located inside the membrane. Thus, depending on mesh and particle size, the linked open holes can allow sprayed particles to penetrate, resulting in a distribution gradient over the membrane thickness.

15.7.4 Physico-Chemical Modifications

The most common approach to nonwoven postprocessing is physicochemical alteration. It enables the development of nanomaterials made of totally new, non-spinnable materials (ceramic and graphite), the stability and attachment of biomolecules on nanofiber surfaces, and surface grafting with polymers.

15.8 Applications

15.8.1 Tissue Engineering

Multiple factors influence the utilization of electrospun meshes and fibers meshes in tissue engineering, such as material selection, surface modification, porosity, fiber orientation, and tissue applications (Figure 15.6). Synthetic and natural (non-degradable and

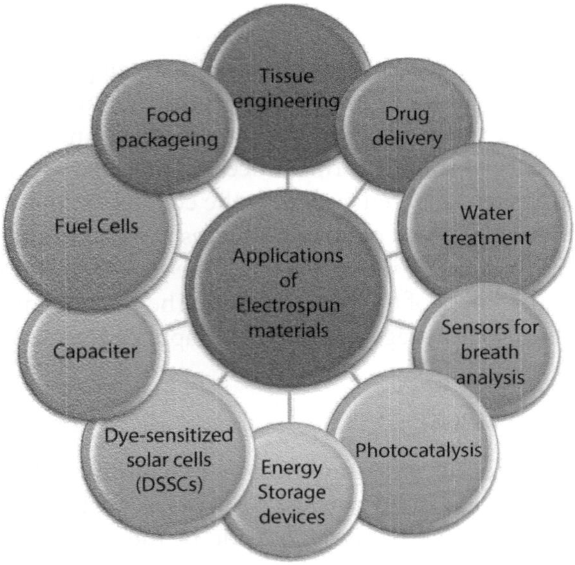

Figure 15.6 Applications of electrospun materials.

biodegradable) materials, as well as hybrid blends of the two, are available, allowing for a perfect balance of mechanical and biomimetic qualities. The fiber's pore size and porosity (cell infiltration) and fiber orientation (random vs. aligned) of the electrospun scaffold may be modified and optimized for every specific application by adjusting the already described and processing parameters. A high density of biologically active substances can be used to modify the surface of the scaffold, after fabrication, owing to its comparatively wide surface area. Due to their versatility in the selection of materials and capacity to control scaffold characteristics, electrospun scaffolds have been used in a wide range of tissue applications, such as neural, vascular, ligament/tendon, and bone repair [91].

Tissue engineering is the technique of producing three-dimensional scaffolds to facilitate cellular in-growth and expansion. The target of tissue engineering is to restore or heal wounded tissue by combining diverse biomaterials, bioactive agents, and cells [92]. A patient's isolated cells can be cultured *ex vivo* on a scaffold before being implanted at the wounded spot in the patient's body. On the one hand, a scaffold can be supplied directly to the patient's wounded region to stimulate tissue development *in situ* [93]. Aside from acting as an artificial and temporary extracellular matrix (ECM) for growing cells that decompose over time, scaffolds may also be employed as a reservoir to supply bioactive chemicals that stimulate tissue regeneration in wounded regions [94].

Electrospun fibers have a small pore size, which limits cellular penetration, prevents waste and nutrient exchange with the surroundings, and prevents vascularization [95]. Electrospun fibers' potential utility in tissue engineering is limited because developing cells must penetrate the porous structure and generate ECM to take over the structural role of fibers that would decay over time. Pore diameters of 300 lm have been proposed to support vascularization in bone and cell proliferation, while on the other hand, pore diameters of 6–20 lm in dermal fibroblasts are preferred [96].

While electrospun materials resemble natural tissues in terms of morphology, because synthetic polymers are generally only capable of morphological biomimesis rather than biochemical biomimesis, chemical similarities are no longer suitable. Attempts have been made to label polymer surfaces with particular chemical signaling components, also direct electrospinning of polymer and biomolecule [97].

Transanethole loaded on the surface of polycaprolactone/polyvinylpyrrolidone nanofiber to stimulate *in vitro* osteoblast development and for possible use in the regeneration of bone tissue using electrospinning [98]. For tissue engineering applications, electrospinning has lately been recognized as a key approach for creating biomimetic scaffolds from natural and synthesized polymers [99].

15.8.2 Wound Dressing

Electrospinning might create a platform for wound protection. The key goals of wound dressing are to promote the quickest pace of healing and the greatest aesthetic tissue regeneration. The experiments reveal that by applying an electric field, ultrafine nanofibers may be spun right onto the wounded area to produce a fibrous mat dressing. These nanofibers, with hole sizes ranging from 500 to 100 lm, are ideal for preventing bacterial growth in wounds [100]. Electrospun fibers are a great choice for wound dressing because they inhibit wound drying, soak wound exudates more effectively, enable gas permeability, protect the wound from bacterial infection, and have superior conformability [101]. Additionally, numerous polymeric materials may be employed for the manufacture of dressings, as well as the simple inclusion of bioactive molecules/agents to give additional functionalities such as tissue development and anti-inflammatory activity [102, 103]. These exceptional properties are assigned to the fibers' large surface-to-volume ratio, electrospinnability from hydrophilic polymers, porosity, and drug loading capacity.

15.8.3 Drug Delivery

The purpose of creating a drug delivery system is to allow the regulated release of medicine at a specified rate over a specified span of time. Electrospinning has also proven useful in drug delivery due to its capacity to generate nanofibers that can operate as drug carriers due to excellent functional attributes like higher surface area, which is linked with greater dissolving rate, easy drug absorption, and a specific amount of time for drug recrystallization due to quick solvent evaporation [104]. Aside from that, numerous biomolecules, for example, genes, proteins, and enzymes, have effectively been integrated into electrospun nanofibers for controlled delivery. In terms of processing, there are two techniques for integrating biomolecules into fibers: coaxial electrospinning and mixed electrospinning. In blend electrospinning, biomolecules and polymers are combined before electrospinning, but in coaxial electrospinning, both biomolecules and polymers are coaxially and concurrently electrospun to generate a core-shell fiber [105].

15.8.4 Water Treatment

Water is necessary for survival and ecological health. Electrospun porous fibers, with their high surface area and remarkable sorption capabilities, have the potential to cure water contaminated by oil leaks, heavy metals, or organic dyes.

15.8.4.1 Oil/Water Separation

The most important wastewater released in the oil and gas sector is produced water, and its administration has turned into a significant ecological issue. Because of its mass release, it has sparked the interest of researchers looking for the most valid answer. It has significant benefits and possibilities in wastewater treatment and recycling. Furthermore, electrospun nanomaterials may be readily functionalized by integrating multifunctional elements to provide various water treatment effects [106].

The advantages of employing an electrospun nanofiber membrane for produced water treatment have prompted researchers to develop and improve manufacturing procedures to upgrade the membrane's final performance. Electrospun nanofiber membrane (NFM) has a significant capacity for treating oily wastewater because of its favorable features like large surface area and porosity, adjustable wettability, low basis weight, and high permeability [91].

These are the important characteristics that make NFM an appealing material for the separation of oil/water emulsions. Furthermore, the particular properties of nanofiber webs, which have very tiny holes and high porosity, might increase the efficacy of standard technologies and materials currently utilized for filtration and particle separation [107].

Despite the tremendous potential of managing oil/water emulsions, one of the greatest drawbacks is the fouling of membranes. Membrane fouling is the deposition of microparticles, solute molecules, colloidal particles, or microorganisms on the membrane surface or into its pores. Membrane fouling reduces permeability and complicates the filtration process. As a result, surface modification procedures are used to prevent fouling and increase the nanofiber crosslinking in order to increase the mechanical strength of the membrane [108].

15.8.4.2 Organic Dyes Removal

Cuiru *et al.* effectively synthesized a PCL/PEO@PDA composite membrane by chemically modifying PDA, demonstrating multiple active sites for methyl orange adsorption (MO) and methylene blue (MB). Polydopamine (PDA) nanoparticles have a high adsorption capacity and can be utilized as an adsorbent. Nanosized adsorbents, on the other hand, are prone to aggregate and so have severe limitations in the field of adsorption. They effectively planned and fabricated a PCL/PEO@PDA composite membrane using *in-situ* oxidation self-polymerization of polydopamine on the surface of polycaprolactone (PCL)/polyethylene oxide (PEO) electrospun fiber following solvent vapor annealing (SVA). The SVA changed the microscopic shape of fine PCL/PEO electrospun fibers with a folded microstructure, providing a lot of active sites for PDA nanoparticles to stick to by increasing the specific surface area [109].

The adsorption of organic dyes from water has been investigated using PAN electrospun composite membranes loaded with active agents. The removal of dye using pre-surface modified PAN has been investigated by incorporating polyamidoamine into the spinning dope. For composite nanofibers containing 25% polyamidoamine, PAN's fiber average diameter increases from 240 to 355 nm. The proposed membrane has a maximum absorption for the anionic dye Direct Red 80 of 1666.66 mg/g and for the anionic dye Direct Red 23 of 2000 mg/g [110].

15.8.4.3 Heavy Metal Ions Removal

Heavy metal contaminants are progressively being dumped into the environment, either directly or indirectly, particularly in developing nations. Toxic heavy metals in polluted water, such as mercury (Hg), chromium (Cr), arsenic (As), plumbum (Pb), cadmium (Cd), zinc (Zn), and copper (Cu), pose serious risks to human health and agricultural production since they are non-degradable and prone to accumulating in the environment. Mining operations, metal plating, petroleum refining, battery production, tanneries, pigment and paint industries, pesticides, photography, printing, and so on are all causes of heavy metal pollution [111].

Sang and colleagues used a nanofiber membrane made from chloridized polyvinyl chloride and a high-voltage electrospinning technique to remove metal cations such as Pb2+, Cd2+, and Cu2+ from simulated groundwater. Several experimental approaches were described to achieve the optimum heavy metal removal, including direct filtration, static adsorption, micellar enhanced filtration (MEF), diatomic-addition filtration, and soil-addition filtration. The testing findings demonstrated that in simulated groundwater MEF rejections of Cu2+ may reach over 73 percent, Pb2+ rejections can reach over 82 percent, and Cd2+ rejections can reach over 91 percent. Membranes made of nanofibers have been shown to be efficient in treating groundwater containing Pb2+, Cd2+, and Cu2+ [109].

15.8.5 Sensors for Breath Analysis

Humidity measurement is significant for a number of reasons. Changing humidity levels are an issue for different industries like food, health, materials, metallurgy, and electronics [112]. The rate of survival of airborne infectious pathogens is also influenced by relative humidity levels [113]. As a result, they must be closely regulated and monitored. Relative humidity in exhaled breath, on the other hand, is a valid indicator of the quality of metabolic activities and health [114].

Sathvik Ajay Iyengar and his colleague in their study propose an efficient way to fabricate wearable devices for humidity sensing with a fast dynamic response and a reaction time of 1 s in order to record and detect relative humidity (% RH) in mouth and nasal exhaled gases. For this, they used low-cost successive ionic layer adsorption and reaction (SILAR) technique in which the sensor is made up of a flexible core of electrospun poly(vinylidene fluoride)/reduced graphene oxide (PVDF/rGO) nanofibers that have been preferentially made sensitive to humidity via polymerization of aniline surface. As an appealing characteristic for incorporation into wearable electronics, they claim a greater accuracy and a complete response ranging (0-95 % RH) assisted by a lower voltage supply and simple circuits [115].

Diagnostic sensing devices that use human exhaled breath offer significant benefits because they are non-invasive, easily portable, and easy to diagnosis procedure. A variety of volatile organic compounds (VOCs) can be found in exhaled human breath. Precise detection of certain VOCs in exhaled breath can give critical details for diagnosis. Acetone (CH_3COCH_3), NH_3, H_2S, NO, and toluene, for example, can be used to assess diabetes, renal dysfunction, halitosis, asthma, and lung cancer, respectively [116–118]. Exhaled human breath can be used to diagnose various illnesses. Choi *et al.* reported an extremely fast and ultrasensitive rough WO3 hemitube nanocomposite supported by oxygen plasma surface functionalization using graphene material for the sensing of hydrogen sulphide (H2S) and acetone (CH3COCH3), both of which are bioindicators for halitosis and diabetes diagnosis [119]. In our daily life, volatile amines are found as a consequence of industrial manufacturing, cultivation, decaying organic matter discharge, or exhaled breath during specific health conditions or disorders. Because these chemicals are frequently harmful, an effective, quick, and low-cost approach to identifying them is required [120]. Ulrich *et al.* developed ultra-sensitive sensors for volatile amine sensors using electrospun porous fiber grids with active furan adducts integrated within the core of polymeric fibers. The capacity of a sensor having multiple furan adducts to sense a variety of important ammonia and aliphatic amines were examined, and 3rd generation FAs demonstrated sensitivity below the ppb level [121].

15.8.6 Photocatalysis

Photocatalysis is a "green" technology that has widely been employed in the area of water treatment because photocatalytically active semiconductors may totally break down many types of contaminants when exposed to light [122]. Photocatalysis is a "green" technology that has widely been employed in the area of water treatment because photocatalytically active semiconductors may totally break down many types of contaminants when exposed to light [123]. To be effective, a catalyst must absorb light effectively, particularly in the near-UV or visible range of the electromagnetic spectrum. To avoid the electron-hole pair's recombination on exposure to light, a significant number of electron-vacant entities must be available. Another significant component of photo-catalysis is the possible use of sunlight, which is renewable energy, to avoid the need for fossil fuels. Research has recently focused on the photocatalytic activity of different semiconductor metal oxides, such as TiO_2 and ZnO-based nanofibers, in the breakdown of hazardous organic compounds contained in wastewater [123–125].

15.8.7 Energy Storage Devices

In the past few decades, the utilization of nanostructured materials for energy storage devices has become popular. A variety of energy storage devices like rechargeable high-performance lithium (Li) batteries (such as Li-sulfur, Li-air, and Li-ion batteries), magnesium (Mg), sodium (Na), zinc (Zn), and aluminum (Al), supercapacitors, and batteries have all been fabricated using nanostructured fibers made of polymers, ceramics, metals, and metal oxide composites [125–129]. Many characteristics of these nanofibers contribute to their fame, including high surface/volume ratio, adjustable fiber diameter, large pore volume and low density [130]. Compared to their nanowire, powder, thin film, crystal, and other equivalents, these characteristics make the nanofiber structure more favorable when employed in LIBs because they provide stable cycle efficiency, excellent electrochemical characteristics, better low-temperature efficiency, and increased capacity [131–135]. Nanofibers' excellent properties/performance have found widespread application in LIBs (as electrodes and separating materials), supercapacitors, and chemical sensors. A range of processes is used to make these nanofibers [136, 137].

Lithium-ion batteries (LIB) are often found in consumer electronics such as cellphones, laptops, and power tools. Yet, LIBs are gaining popularity in automotive industries such as electric car manufacturing [138]. Higher energy density, specific energy, and long cycle life are required for these applications, as well as considerations for safety, cost, and environmental issues. Lithium-ion batteries, which are extensively used in modern electronics, are distinguished by three main functional parts: anode, cathode, and electrolyte, which were developed to efficiently regulate the movement of ions from the anode to the cathode during discharge and again during recharging.

A series of studies have been conducted on nanofibers, namely core-shell nanofibers for use as anodes in LIBs. Anodes made from carbon–cobalt composite nanofibers with widths ranging from 100 to 300 nm were created by electrospinning and followed by heating [139]. These fibers have good electrochemical characteristics, with a reversible capacity of over 750 mA h g1. On reactions with Li, manganese oxide anode nanofibers with diameters ranging from 100–300 nm were observed to be formed of oxide granules with sizes of approximately 20–25 nm and to have a potential of above 450 mA h g1 [140]. When employed directly as anode materials, carbon-SnO_2 core sheet composite fibers formed by the combination of electrodeposition and electrospinning have been shown to have outstanding electrochemical activity [141].

15.8.8 Capacitors

The specific and volumetric capacitances of electrochemical double-layer capacitors (EDLCs), commonly called supercapacitors, are several orders of magnitude larger than ordinary capacitors. EDLCs have a higher power density than batteries, but they have a lower energy density. Due to nanometer-sized gaps between fibers, EDLC electrodes integrating an electrospinning processing step have a large surface area. This enables enormous charge storage regions and nanoscale thick double layers. The majority of studies on capacitor electrode electrospun materials concentrated on poly (amic acid), polybenzimidazole, and PAN fibers that were then carbonized [142, 143].

15.8.9 Dye-Sensitized Solar Cells (DSSCs)

Solar, wind, tide, and other renewable energy sources can be transformed into electrical energy. Solar cells, which employ the photovoltaic effect to directly convert sunlight into electrical energy, have long been viewed as one of the most potent energy sources due to the unlimited supply of sunlight. Silicon-based solar cells are first-generation photovoltaic devices with good performance but expensive material and production costs. The use of dye photosensitizer in dye-sensitized solar cells (DSSCs) leads to a 3rd-generation photovoltaic device as photovoltaic technology has progressed. O'Regan designed DSSCs in 1991, and they quickly gained popularity because of their lightweight, low cost, and ease of manufacture [15].

A DSSC is typically made up of a counter electrode, a photoanode, and an electrolyte that separates them. Transparent conducting oxide with a platinum thin film on top acts as a counter electrode, and the dye photosensitizer is layered in-between the two transparent conducting oxide glasses. The photoanode is made up of a TCO layer covered with a semiconductor film (often TiO_2) with a large bandgap. Photosensitizer (a charge-transfer dye) in a DSSC forms excited electron-hole pairs by capturing photons of light, which are subsequently dissociated to liberate the free electron-hole pairs. Free electrons are captured on the photoanode after entering the semiconductor conducting band. An external circuit is subsequently used to transport the photoelectrons to the counter electrode, where they combine to produce current. The greater the number of electrons gathered, the greater the efficiency. Up to this point, the maximum output of DSSCs has been around 11.9 percent, which falls well short of the semiconductor solar cells' 30% theoretical limit [144].

Additionally, studies indicate that grain boundaries can impair electron-conducting capacity, and 1-dimensional fibers are anticipated to have better charge transmission than sintered nanoparticles due to their reduced grain borders. Moreover, 1-dimensional fibers have a large specific surface area, which may boost dye sensitizer adsorption. As previously stated, the electrospinning technique is a potential way of obtaining 1-dimensional fiber; hence, the utilization of such materials in DSSCs has attracted considerable attention and has been extensively studied [145–147]. Similarly, electrospun materials may be used in all components of the DSSC, such as the photoanode, counter electrode, and electrolyte.

15.8.10 Fuel Cells

The anode, the cathode, and the electrolyte are the three main building components in a fuel cell. A fuel cell transforms chemical energy from a fuel into electrical energy via a chemical reaction with oxygen or other oxidizing agents. A catalyst must oxidize the fuel at the anode, converting it to a negatively charged electron and a positively charged ion. The most commonly used fuel is hydrogen, but other hydrocarbons like alcohols like methanol and natural gas are also utilized. The electrolyte must be built in such a way that ions can travel through it, but electrons must not. A proton-conducting polymer membrane commonly separates the anode and cathode sides. The liberated electrons must pass across a wire to generate an electric current. The ions must be rejoined with the electrons at the cathode, before reacting with a 3rd ion, generally oxygen, to form carbon dioxide or water. At full load, a common fuel cell generates about 0.6 to 0.7V of voltage. If we talk about electrospun nanofibers, they have been engineered to serve as anodes, cathodes, or proton conductive membranes [148].

Proton exchange fuel cells necessitate the use of specific membranes with higher proton transport and solvent absorption, along with mechanical qualities that are very responsive to membrane microstructure. Li and coworkers used electrospun fibers of sulfonated poly(ether ether ketone ketone) to produce 0.37 S/cm proton conductivity. The observed conductivity is greater than that of membranes manufactured using traditional casting processes [149, 150].

15.8.11 Food and Food Packaging

The surface qualities of nanofibers are significant in packaging materials. Although the electrospinning technique may produce fiber membranes with a unique internal structure, the surface qualities of fiber membranes, such as wettability and adsorption, are insufficient for use in food packaging. As a result, later modification is required to modify or change the electrospun nanofiber's surface to match the packing field characteristics and unique application needs [151].

15.9 Conclusion

Electrospinning has attracted extensive attention in recent years as a promising polymer processing technology for uses in medicine, tissue engineering, photocatalysis, sensors, water treatment, and many more. This newfound interest may be owing to its relative adaptability, ease of handling, and potential to generate nanoscale fibers. In this chapter, we have discussed the short history of electrospinning, the electrospinning methods, parameters, and their effects during the process. Moreover, preparation methods of nanofibers based on electrospinning technology, mechanism of pore formation, and modification of electrospun micro- and nanofibers converse. The most recent and noteworthy advancements in electrospun nanofiber applications are highlighted, with a focus on biomedical, environmental, electrochemical, and energy-related applications.

References

1. Zhao, L., Duan, G., Zhang, G., Yang, H., He, S., Jiang, S., Electrospun functional materials toward food packaging applications: A review. *J. Nanomater.*, 10, 1, 150, 2020.
2. Lu, P. and Ding, B., Applications of electrospun fibers. *Recent Pat. Nanotechnol.*, 2, 3, 169–182, 2008.
3. Anton, F., Process and apparatus for preparing artificial threads, Google Patents, 1934.
4. Kadam, V.V., Wang, L., Padhye, R., Electrospun nanofibre materials to filter air pollutants–A review. *J. Ind. Text.*, 47, 8, 2253–2280, 2018.
5. Mirjalili, M. and Zohoori, S., Review for application of electrospinning and electrospun nanofibers technology in textile industry. *J. Nanostructure Chem.*, 6, 3, 207–213, 2016.
6. Esfanjani, A.F. and Jafari, S.M., Biopolymer nano-particles and natural nano-carriers for nano-encapsulation of phenolic compounds. *Colloids Surf. B: Biointerfaces*, 146, 532–543, 2016.
7. Huang, C. and Thomas, N.L., Fabrication of porous fibers via electrospinning: Strategies and applications. *Polym. Rev.*, 60, 4, 595–647, 2020.
8. Gilbert, W., *De magnete*, Courier Corporation, Dover Publications Inc., New York, 1958.
9. Li, X., Chen, Y., Huang, H., Mai, Y.-W., Zhou, L., Electrospun carbon-based nanostructured electrodes for advanced energy storage–a review. *Energy Storage Mater.*, 5, 58–92, 2016.
10. Bürck, J., Aras, O., Bertinetti, L., Ilhan, C.A., Ermeydan, M.A., Schneider, R., Ulrich, A.S., Kazanci, M., Observation of triple helix motif on electrospun collagen nanofibers and its effect on the physical and structural properties. *J. Mol. Struct.*, 1151, 73–80, 2018.
11. Wang, X. and Hsiao, B.S., Electrospun nanofiber membranes. *Curr. Opin. Chem. Eng.*, 12, 62–81, 2016.
12. Fathi, M., Martín, Á., McClements, D.J., Nanoencapsulation of food ingredients using carbohydrate based delivery systems. *Trends Food Sci. Technol.*, 39, 1, 18–39, 2014.
13. Stijnman, A.C., Bodnar, I., Tromp, R.H., Electrospinning of food-grade polysaccharides. *Food Hydrocolloids*, 25, 5, 1393–1398, 2011.
14. Ghorani, B. and Tucker, N., Fundamentals of electrospinning as a novel delivery vehicle for bioactive compounds in food nanotechnology. *Food Hydrocolloids*, 51, 227–240, 2015.
15. Liu, Q., Zhu, J., Zhang, L., Qui, Y., Recent advances in energy materials by electrospinning. *Renewable Sustainable Energy Rev.*, 81, 1825–1858, 2018.
16. Haider, A., Haider, S., Kang, I.-K., A comprehensive review summarizing the effect of electrospinning parameters and potential applications of nanofibers in biomedical and biotechnology. *Arab. J. Chem.*, 11, 8, 1165–1188, 2018.
17. Yang, H., Jiang, S., Fang, X., Duan, G., Hou, H., Molecular orientation in aligned electrospun polyimide nanofibers by polarized FT-IR spectroscopy. *Spectrochim. Acta Part A: Mol. Biomol. Spectrosc.*, 200, 339–344, 2018.

18. Li, Y., Zhang, M., Zhang, X., Xie, G., Su, Z., Wei, G., Nanoporous carbon nanofibers decorated with platinum nanoparticles for non-enzymatic electrochemical sensing of H2O2. *J. Nanomater.*, 5, 4, 1891–1905, 2015.
19. Su, Z., Ding, J., Wei, G., Electrospinning: A facile technique for fabricating polymeric nanofibers doped with carbon nanotubes and metallic nanoparticles for sensor applications. *RSC Adv.*, 4, 94, 52598–52610, 2014.
20. Yu, Y., Kong, L., Li, L., Li., N., Yan, P., Antitumor activity of doxorubicin-loaded carbon nanotubes incorporated poly (lactic-co-glycolic acid) electrospun composite nanofibers. *Nanoscale Res. Lett.*, 10, 1, 1–9, 2015.
21. Gonçalves, R., Martins, P., Moya, X., Ghidini, M., Sencadas, V., Botelho, G., Mathur, N.D., Lanceros-Mendez, S., Magnetoelectric CoFe2O4/polyvinylidene fluoride electrospun nanofibres. *Nanoscale*, 7, 17, 8058–8061, 2015.
22. Casper, C.L., Stephens, J.S., Tassi, N.G., Chase, B.B., Rabolt, J.F., Controlling surface morphology of electrospun polystyrene fibers: Effect of humidity and molecular weight in the electrospinning process. *Macromolecules*, 37, 2, 573–578, 2004.
23. Ray, S.S., Chen, S.-S., Nguyen, N.C., Nguyen, H.T., Electrospinning: A versatile fabrication technique for nanofibrous membranes for use in desalination, in: *Nanoscale Materials in Water Purification*, pp. 247–273, Elsevier Sciences, Amsterdam, 2019.
24. Yadav, T.C., Srivastava, A.K., Mishra, P., Singh, D., Raghuwanshi, N., Singh, N.K., Singh, A.K., Tiwari, S.K., Parsad, R., Pruthi, V., Electrospinning: An efficient biopolymer-based micro- and nanofibers fabrication technique, in: *Next Generation Biomanufacturing Technologies*, pp. 209–241, ACS Publications, ACS Symposium Series, American Chemical Society, Washington, DC, 2019.
25. Hossain, M.F., Gong, R.H., Rigout, M., Optimization of the process variables for electrospinning of poly(ethylene oxide)-loaded hydroxypropyl-β-cyclodextrin nanofibres. *J. Text. Inst.*, 107, 1, 1–11, 2016.
26. Haghi, A. and Akbari, M., Trends in electrospinning of natural nanofibers. *Phys. Status Solidi (a)*, 204, 6, 1830–1834, 2007.
27. Bhardwaj, N. and Kundu, S.C., Electrospinning: A fascinating fiber fabrication technique. *Biotechnol. Adv.*, 28, 3, 325–347, 2010.
28. Zong, X., Kim, K., Fang, D., Ran, S., Hsiao, B.S., Chu, B., Structure and process relationship of electrospun bioabsorbable nanofiber membranes. *J. Polym.*, 43, 16, 4403–4412, 2002.
29. Mit-uppatham, C., Nithitanakul, M., Supaphol, P., Effects of solution concentration, emitting electrode polarity, solvent type, and salt addition on electrospun polyamide-6 fibers: A preliminary report. *Macromol. Symp.* Wiley Online Library, 216, 293–299, 2004.
30. Son, W.K., Youk, J.H., Lee, T.S., Park, W.H., The effects of solution properties and polyelectrolyte on electrospinning of ultrafine poly(ethylene oxide) fibers. *J. Polym.*, 45, 9, 2959–2966, 2004.
31. Tan, S.-H., Ini, R., Kotaki, M., Ramakrishna, S., Systematic parameter study for ultra-fine fiber fabrication via electrospinning process. *J. Polym.*, 46, 16, 6128–6134, 2005.
32. Nangrejo, M., Bragman, F., Ahmad, Z., Stride, E., Edirisinghe, M., Hot electrospinning of polyurethane fibres. *Mater. Lett.*, 68, 482–485, 2012.
33. Fashandi, H. and Karimi, M., Pore formation in polystyrene fiber by superimposing temperature and relative humidity of electrospinning atmosphere. *J. Polym.*, 53, 25, 5832–5849, 2012.
34. Li, D., Wang, Y., Xia, Y., Electrospinning of polymeric and ceramic nanofibers as uniaxially aligned arrays. *Nano Lett.*, 3, 8, 1167–1171, 2003.
35. Rafique, J., Yu, Yu, J., Fang, G., Wong, K.W., Zheng, Z., Ong, H.C., Lau, W.M., Electrospinning highly aligned long polymer nanofibers on large scale by using a tip collector. *Appl. Phys. Lett.*, 91, 6, 063126, 2007.

36. Dersch, R., liu, T., Schaper, A.K., Greiner, A., Wendorff, J.H., Electrospun nanofibers: Internal structure and intrinsic orientation. *J. Polym. Sci. Part A: Polym. Chem.*, 41, 4, 545–553, 2003.
37. Ding, B., Wang, M., Wang, X., Yu, J., Sun, G., Electrospun nanomaterials for ultrasensitive sensors. *J. Mater. Today*, 13, 11, 16–27, 2010.
38. Heikkilä, P. and Harlin, A., Parameter study of electrospinning of polyamide-6. *Eur. Polym. J.*, 44, 10, 3067–3079, 2008.
39. Li, Z. and Wang, C., *Springer briefs in materials one-dimensional nanostructures electrospinning technique and unique nanofibers*, Springer, Berlin/Heidelberg, Germany, 2013.
40. Dersch, R., Graeser, M., Greiner, A., Wendorff, J.H., Electrospinning of nanofibres: Towards new techniques, functions, and applications. *Aust. J. Chem.*, 60, 10, 719–728, 2007.
41. Brown, T.D., Dalton, P.D., Hutmacher, D.W., Melt electrospinning today: An opportune time for an emerging polymer process. *Prog. Polym. Sci.*, 56, 116–166, 2016.
42. Hutmacher, D.W. and Dalton, P.D., Melt electrospinning. *Chem.-Asian J.*, 6, 1, 44–56, 2011.
43. Angammana, C.J. and Jayaram, S.H., Fundamentals of electrospinning and processing technologies. *Part. Sci. Technol.*, 34, 1, 72–82, 2016.
44. Zhang, L.-H., Duan, X.-P., Yan, X., Yu, M., Ning, X., Zhao, Y., Long, Y.-Z., Recent advances in melt electrospinning. *RSC Adv.*, 6, 58, 53400–53414, 2016.
45. Abdal-hay, A., Abbasi, N., Gwiazda, M., Hamlet, S., Ivanoski, S., Novel polycaprolactone/hydroxyapatite nanocomposite fibrous scaffolds by direct melt-electrospinning writing. *Eur. Polym. J.*, 105, 257–264, 2018.
46. Huang, L., Manickam, S.S., McCutcheon, J.R., Increasing strength of electrospun nanofiber membranes for water filtration using solvent vapor. *J. Membr. Sci.*, 436, 213–220, 2013.
47. Demir, M.M., Horzum, N., Tasdemirci, A., Turan, K., Guden, M., Mechanical interlocking between porous electrospun polystyrene fibers and an epoxy matrix. *ACS Appl. Mater. Interfaces*, 6, 24, 21901–21905, 2014.
48. Ismail, N., Venault, A., Mikkola, J.-P., Bouyer, D., Drioli, E., Kiadeh, N.T.H., Investigating the potential of membranes formed by the vapor induced phase separation process. *J. Membr. Sci.*, 597, 117601, 2020.
49. Huang, C. and Thomas, N., Fabricating porous poly (lactic acid) fibres via electrospinning. *Eur. Polym. J.*, 99, 464–476, 2018.
50. Lu, P. and Xia, Y., Maneuvering the internal porosity and surface morphology of electrospun polystyrene yarns by controlling the solvent and relative humidity. *Langmuir*, 29, 23, 7070–7078, 2013.
51. Deitzel, J.M., Kleinmeyer, J., Harris, D., Tan, N., The effect of processing variables on the morphology of electrospun nanofibers and textiles. *Polymer*, 42, 1, 261–272, 2001.
52. Wang, H.H., Jung, J.T., Kim, J.F., Kim, S., Dtioli, E., Lee, Y.M., A novel green solvent alternative for polymeric membrane preparation via nonsolvent-induced phase separation (NIPS). *J. Membr. Sci.*, 574, 44–54, 2019.
53. Natarajan, L., New, J., Dasari, A., Yu, S., Manan, M.A., Surface morphology of electrospun PLA fibers: Mechanisms of pore formation. *RSC Adv.*, 4, 83, 44082–44088, 2014.
54. Zheng, J., Zhang, H., Zhao, Z., Han, C.C., Construction of hierarchical structures by electrospinning or electrospraying. *Polymer*, 53, 2, 546–554, 2012.
55. Sánchez, L.D., Brack, N., Postma, A., Pigram, P.J., Meagher, L., Surface modification of electrospun fibres for biomedical applications: A focus on radical polymerization methods. *J. Biomater.*, 106, 24–45, 2016.
56. Sagitha, P., Reshmi, C.R., Sundaran, S.P., Sujith, A., Recent advances in post-modification strategies of polymeric electrospun membranes. *Eur. Polym. J.*, 105, 227–249, 2018.
57. Udomluck, N., Koh, W.-G., Lim, D.-J., Park, H., Recent developments in nanofiber fabrication and modification for bone tissue engineering. *Int. J. Mol. Sci.*, 21, 1, 99, 2019.

58. Chiu, H.T., Lin, J.-M., Cheng, T.-H., Chou, S.-Y., Huang, C.-C., Direct purification of lysozyme from chicken egg white using weak acidic polyacrylonitrile nanofiber-based membranes. *J. Appl. Polym. Sci.*, 125, S2, E616–E621, 2012.
59. Schneiderman, S., Zhang, L., Fong, H., Menkhaus, T., Surface-functionalized electrospun carbon nanofiber mats as an innovative type of protein adsorption/purification medium with high capacity and high throughput. *J. Chromatogr. A*, 1218, 50, 8989–8995, 2011.
60. Li, G., Li, T., Li, Y., An, L., Li, W., Zhang, Z., Preparation of pH-controllable nanofibrous membrane functionalized with lysine for selective adsorption of protein. *Colloids Surf. A: Physicochem. Eng. Asp.*, 531, 173–181, 2017.
61. Han, N.K., Choi, Y.C., Park, D.U., Ryu, J.H., Jeong, Y.G., Core-shell type composites based on polyimide-derived carbon nanofibers and manganese dioxide for self-standing and binder-free supercapacitor electrode applications. *Compos. Sci. Technol.*, 196, 108212, 2020.
62. Dong, Z.-Q., Wang, B.-J., Ma, X.-H., Wei, Y.-M., Xu, Z.-L., FAS grafted electrospun poly (vinyl alcohol) nanofiber membranes with robust superhydrophobicity for membrane distillation. *ACS Appl. Mater. Interfaces*, 7, 40, 22652–22659, 2015.
63. Wang, J., Yao, H.-B., He, D., Zhang, C.-L., Yu, S.-H., Facile fabrication of gold nanoparticles-poly (vinyl alcohol) electrospun water-stable nanofibrous mats: Efficient substrate materials for biosensors. *ACS Appl. Mater. Interfaces*, 4, 4, 1963–1971, 2012.
64. Uheida, A., Mohamed, A., Belaqziz, M., Nasser, W.S., Photocatalytic degradation of Ibuprofen, Naproxen, and Cetirizine using PAN-MWCNT nanofibers crosslinked TiO2-NH2 nanoparticles under visible light irradiation. *Sep. Purif. Technol.*, 212, 110–118, 2019.
65. Demirci, S., Celebioglu, A., Uyar, T., Surface modification of electrospun cellulose acetate nanofibers via RAFT polymerization for DNA adsorption. *Carbohydr. Polym.*, 113, 200–207, 2014.
66. Yeole, N., Kutcherlapati, S.R., Jana, T., Polystyrene–graphene oxide (GO) nanocomposite synthesized by interfacial interactions between RAFT modified GO and core–shell polymeric nanoparticles. *J. Colloid Interface Sci.*, 443, 137–142, 2015.
67. Oktay, B., Kayaman-Apohan, N., Suleymanoglu, M., Erdem-Kurua, S., Zwitterionic phosphorylcholine grafted chitosan nanofiber: Preparation, characterization and *in-vitro* cell adhesion behavior. *Mater. Sci. Eng.: C*, 73, 569–578, 2017.
68. Ahmadi, M., Behzad, T., Bagheri, R., Ghiaci, M., Sain, M., Topochemistry of cellulose nanofibers resulting from molecular and polymer grafting. *Cellulose*, 24, 5, 2139–2152, 2017.
69. Zhang, M., Ma, W., Cui, J., Wu, S., Han, J., Zou, Y., Huang, C., Hydrothermal synthesized UV-resistance and transparent coating composited superoloephilic electrospun membrane for high efficiency oily wastewater treatment. *J. Hazard. Mater.*, 383, 121152, 2020.
70. Lou, L., Kendall, R.L., Smith, E., Ramkumar, S.S., Functional PVDF/rGO/TiO2 nanofiber webs for the removal of oil from water. *Polymer*, 186, 122028, 2020.
71. Li, L., Zhang, X., Fu, L., Wang, Ji, S., Wu, M., Wang, H., Zhang, W., Electrospun and hydrothermal techniques to synthesize the carbon-coated nickel sulfide microspheres/carbon nanofibers nanocomposite for high performance liquid-state solar cells. *Composites Part B: Eng.*, 173, 107026, 2019.
72. Abdal-hay, A., Hamdy, A.S., Mosri, Y., Khalil, K.A., Lim, J.K., Novel bone regeneration matrix for next-generation biomaterial using a vertical array of carbonated hydroxyapatite nanoplates coated onto electrospun nylon 6 nanofibers. *Mater. Lett.*, 137, 378–381, 2014.
73. Abdal-hay, A., Hamdy, A.S., Khalil, K.A., Fabrication of durable high performance hybrid nanofiber scaffolds for bone tissue regeneration using a novel, simple *in situ* deposition approach of polyvinyl alcohol on electrospun nylon 6 nanofibers. *Mater. Lett.*, 147, 25–28, 2015.
74. Sullivan, S.T., Tang, C., Kennedy, A., Talwar, S., Khan, S., Electrospinning and heat treatment of whey protein nanofibers. *Food Hydrocolloids*, 35, 36–50, 2014.

75. Min, B.M., Jeong, L., Lee, K.Y., Park, W.H., Regenerated silk fibroin nanofibers: Water vapor-induced structural changes and their effects on the behavior of normal human cells. *Macromol. Biosci.*, 6, 4, 285–292, 2006.
76. Enayati, M.S., Behzad, T., Sajkiewicz, P., Bagheri, R., Ghasemi-Mobarakeh, L., Lojkowski, W., Pahlevanneshan, Z., Ahmadi, M., Crystallinity study of electrospun poly (vinyl alcohol) nanofibers: Effect of electrospinning, filler incorporation, and heat treatment. *Iran. Polym. J.*, 25, 7, 647–659, 2016.
77. Kim, Y.-J., Ahn, C.H., Choi, M.O., Effect of thermal treatment on the characteristics of electrospun PVDF–silica composite nanofibrous membrane. *Eur. Polym. J.*, 46, 10, 1957–1965, 2010.
78. Ho, M.H., Hou, L.-T., Tu, C.-Y., Heieh, H.-J., Lai, J.-Y., Chen, W.-J., Wang, D.-M., Promotion of cell affinity of porous PLLA scaffolds by immobilization of RGD peptides via plasma treatment. *Macromol. Biosci.*, 6, 1, 90–98, 2006.
79. Sanders, J.E., Lamont, S.E., Karchin, A., Golledge, S.L., Ranterm, B.D., Fibro-porous meshes made from polyurethane micro-fibers: Effects of surface charge on tissue response. *Biomaterials*, 26, 7, 813–818, 2005.
80. Siow, K.S., Britcher, L., Kumar, S., Griesser, H.J., Plasma methods for the generation of chemically reactive surfaces for biomolecule immobilization and cell colonization-a review. *Plasma Processes Polym.*, 3, 6-7, 392–418, 2006.
81. Chen, J.-P. and Su, C.-H., Surface modification of electrospun PLLA nanofibers by plasma treatment and cationized gelatin immobilization for cartilage tissue engineering. *Acta Biomater.*, 7, 1, 234–243, 2011.
82. Kogelschatz, U., Dielectric-barrier discharges: Their history, discharge physics, and industrial applications. *Plasma Chem. Plasma Process.*, 23, 1, 1–46, 2003.
83. Wei, Q., Gao, W.D., Hou, D.Y., Wang, X.Q., Surface modification of polymer nanofibres by plasma treatment. *Appl. Surf. Sci.*, 245, 1-4, 16–20, 2005.
84. Martins, A., Pinho, E.D., Fari, S., Pashkuleva, I., Marques, A.P., Reis, R.L., Neves, N.M., Surface modification of electrospun polycaprolactone nanofiber meshes by plasma treatment to enhance biological performance. *Small*, 5, 10, 1195–1206, 2009.
85. Jeong, L., Yeo, I.-S., Kim, H.N., Yoon, Y., Jung, S.Y., Min, B.-M., Park, W.-H., Plasma-treated silk fibroin nanofibers for skin regeneration. *Int. J. Biol. Macromol.*, 44, 3, 222–228, 2009.
86. Becker, A.L., Johnston, A.P., Caruso, F., Peptide nucleic acid films and capsules: Assembly and enzymatic degradation. *Macromol. Biosci.*, 10, 5, 488–495, 2010.
87. Karuppuswamy, P. *et al.*, Functionalized hybrid nanofibers to mimic native ECM for tissue engineering applications. *Appl. Surf. Sci.*, 322, 162–168, 2014.
88. Luo, Y., Wang, S., Shem, M., Qi, R., Fang, Y., Guo, R., Cai, H., Cao, X., Tomas, H., Zhu, M., Carbon nanotube-incorporated multilayered cellulose acetate nanofibers for tissue engineering applications. *Carbohydr. Polym.*, 91, 1, 419–427, 2013.
89. Aziz, F. and Ismail, A.F., Spray coating methods for polymer solar cells fabrication: A review. *Mater. Sci. Semicond. Process.*, 39, 416–425, 2015.
90. Chen, H., Ding, X., Pan, X., Hayat, T., Alsaedi, A., Ding, D., Dai, S., Comprehensive studies of air-brush spray deposition used in fabricating high-efficiency CH3NH3PbI3 perovskite solar cells: Combining theories with practices. *J. Power Sources*, 402, 82–90, 2018.
91. Abd Halim, N.S., Wirzal, M.D.H., Hizam, S.M., Bilad, M.R., Nordin, N.A.H.M., Sambudi, N.S., Putra, Z.A., Yusoff, A.R.M., Recent development on electrospun nanofiber membrane for produced water treatment: A review. *J. Environ. Chem. Eng.*, 9, 1, 104613, 2021.
92. Fong, E.L., Building bridges: Leveraging interdisciplinary collaborations in the development of biomaterials to meet clinical needs. *Adv. Mater.*, 24, 36, 4995–5013, 2012.

93. Rezwan, K., Chen, Q.Z., Blaker, J.J., Boccaccini, A.R., Biodegradable and bioactive porous polymer/inorganic composite scaffolds for bone tissue engineering. *Biomaterials*, 27, 18, 3413–3431, 2006.
94. Nair, L.S. and Laurencin, C.T., Biodegradable polymers as biomaterials. *Prog. Polym. Sci.*, 32, 8-9, 762–798, 2007.
95. Lowery, J.L., Datta, N., Rutledge, G.C., Effect of fiber diameter, pore size and seeding method on growth of human dermal fibroblasts in electrospun poly (ε-caprolactone) fibrous mats. *Biomaterials*, 31, 3, 491–504, 2010.
96. Sill, T.J. and Von Recum, H.A., Electrospinning: Applications in drug delivery and tissue engineering. *Biomaterials*, 29, 13, 1989–2006, 2008.
97. Lannutti, J., Reneker, D., Ma, T., Tomasko, D., Farson, D., Electrospinning for tissue engineering scaffolds. *J. Mater. Sci. Eng. C*, 27, 3, 504–509, 2007.
98. Shadamarshan, R.P., Balaji, H., Rao, H.S., Balagangadharan, K., Chandran, S.V., Selvamurugan, N., Fabrication of PCL/PVP electrospun fibers loaded with trans-anethole for bone regeneration *in vitro*. *J. Colloids Surf. B: Biointerfaces*, 2171, 698–706, 018.
99. Li, M., Mondrinos, M.J., Gandhi, M.R., Ko, F.K., Weiss, A.S., Lelkes, P., Electrospun protein fibers as matrices for tissue engineering. *Biomaterials*, 26, 30, 5999–6008, 2005.
100. Jin, H.-J., Frideikh, S.V., Rutledge, G.C., Kaplan, D.L., Electrospinning Bombyx mori silk with poly (ethylene oxide). *Biomacromolecules*, 3, 6, 1233–1239, 2002.
101. Zahedi, P., Rezaeian, I., Ranaei-Siadat, S.-O., Jafari, S.-H., Supaphol, P., A review on wound dressings with an emphasis on electrospun nanofibrous polymeric bandages. *Polym. Adv. Technol.*, 21, 2, 77–95, 2010.
102. Liu, S.-J., Kau, Y.-C., Chou, C.-Y., Chen, J.-K., Wu, R.-C., Yeh, W.-L., Electrospun PLGA/collagen nanofibrous membrane as early-stage wound dressing. *J. Membr. Sci.*, 355, 1-2, 53–59, 2010.
103. Vargas, E.T., Do Vale Baracho, N.C., De Brito, J., De Queiroz, A.A.A., Hyperbranched polyglycerol electrospun nanofibers for wound dressing applications. *Acta Biomater.*, 6, 3, 1069–1078, 2010.
104. Verreck, G., Chun, I., Peeters, J., Rosenblatt, J., Brewster, M., Preparation and characterization of nanofibers containing amorphous drug dispersions generated by electrostatic spinning. *Pharm. Res.*, 20, 5, 810–817, 2003.
105. Soliman, S., Sant, S., Nichol, J.W., Khabiry, M., Traversa, E., Khademhosseini, A., Controlling the porosity of fibrous scaffolds by modulating the fiber diameter and packing density. *J. Biomed. Mater. Res. Part A*, 96, 3, 566–574, 2011.
106. Chen, H., Huang, M., Liu, Y., Meng, L., Ma, M., Functionalized electrospun nanofiber membranes for water treatment: A review. *Sci. Total Environ.*, 739, 139944, 2020.
107. Bae, J., Baek, I., Choi, H., Mechanically enhanced PES electrospun nanofiber membranes (ENMs) for microfiltration: The effects of ENM properties on membrane performance. *Water Res.*, 105, 406–412, 2016.
108. Abdelrasoul, A., Doan, H., Lohi, A., Fouling in membrane filtration and remediation methods, in: *Mass Transfer-Advances in Sustainable Energy and Environment Oriented Numerical Modeling*, p. 195, 2013.
109. Wang, C., Yin, J., Wang, R., Jiao, T., Hiang, H., Zhou, J., Zhang, L., Peng, Q., Facile preparation of self-assembled polydopamine-modified electrospun fibers for highly effective removal of organic dyes. *Nanomaterials*, 9, 1, 116, 2019.
110. Almasian, A., Olya, M.E., Mahmoodi, N.M., Synthesis of polyacrylonitrile/polyamidoamine composite nanofibers using electrospinning technique and their dye removal capacity. *J. Taiwan Inst. Chem. Eng.*, 49, 119–128, 2015.
111. Huang, Y., Miao, Y.E., Liu, T., Electrospun fibrous membranes for efficient heavy metal removal. *J. Appl. Polym. Sci.*, 131, 19, 40864, 2014.

112. Vanhoestenberghe, A. and Donaldson, N., Corrosion of silicon integrated circuits and lifetime predictions in implantable electronic devices. *J. Neural Eng.*, 10, 3, 031002, 2013.
113. Tang, J.W., The effect of environmental parameters on the survival of airborne infectious agents. *J. R. Soc. Interface*, 6, suppl_6, S737–S746, 2009.
114. Oki, Y., Sakai, H., Misu, S., Mitani, Y., Ishikawa, A., Usefulness of the 6-minute walk test as a screening test for pulmonary arterial enlargement in COPD. *Int. J. Chron. Obstruct. Pulmon. Dis.*, 11, 2869, 2016.
115. Iyengar, S.A., Jana, S.K., Islam, M.R., Ahuja, T., Pradeep, T., Surface-treated nanofibers as high current yielding breath humidity sensors for wearable electronics. *ACS Appl. Electron. Mater.*, 1, 6, 951–960, 2019.
116. Shin, J., Choi, S.-J., Youn, D.Y., Kim, D., Exhaled VOCs sensing properties of WO3 nanofibers functionalized by Pt and IrO2 nanoparticles for diagnosis of diabetes and halitosis. *J. Electroceram.*, 29, 2, 106–116, 2012.
117. Awano, S., Ansai, T., takata, Y., Soh, I., Sonoki, K., Takehara, T., Relationship between volatile sulfur compounds in mouth air and systemic disease. *J. Breath Res.*, 2, 1, 017012, 2008.
118. Peng, G., Tisch, U., Adams, O., Broza, Y.Y., Kuten, A., Haick, H., Diagnosing lung cancer in exhaled breath using gold nanoparticles. *Nat. Nanotechnol.*, 4, 10, 669–673, 2009.
119. Choi, S.-J., Fuchs, F., Janf, B.-H., Lee, S.-J., Lee, J.-H., Tuller, H.L., Kim, D., Fast responding exhaled-breath sensors using WO3 hemitubes functionalized by graphene-based electronic sensitizers for diagnosis of diseases. *ACS Appl. Mater. Interfaces*, 6, 12, 9061–9070, 2014.
120. Gong, W.L., Sears, K.J., Alleman, J.E., Balthcley, E.R., Toxicity of model aliphatic amines and their chlorinated forms. *Environ. Toxicol. Chem.: An Int. J.*, 23, 2, 239–244, 2004.
121. Ulrich, S., Moura, S.O., Diaz, Y., Clerc, M., Guex, A.G., Neves, N.M., Rossi, R.M., Electrospun colourimetric sensors for detecting volatile amines. *Sens. Actuators B: Chem.*, 322, 128570, 2020.
122. Shavisi, Y., Sharifinia, S., Zendehzaban, M., Kakehazar, S., Application of solar light for degradation of ammonia in petrochemical wastewater by a floating TiO2/LECA photocatalyst. *J. Ind. Eng. Chem.*, 20, 5, 2806–2813, 2014.
123. Marwani, H.M., Bakhsh, E.M., Khan, S.B., Danish, E.Y., Asiri, A.M., Cerium oxide-cadmium oxide nanomaterial as efficient extractant for yttrium ions. *J. Mol. Liq.*, 269, 252–259, 2018.
124. Rahman, M.M., Asiri, A.M., Youssef, T.E., Marwani, H.M., Photocatalytic degradation of remazol brilliant orange 3R using wet-chemically prepared CdO-ZnO nanofibers for environmental remediation. *Mater. Express*, 6, 2, 137–148, 2016.
125. Fang, S., Shen, L., Tong, Z., Zheng, H., Zhang, F., Zhang, X., Si nanoparticles encapsulated in elastic hollow carbon fibres for Li-ion battery anodes with high structural stability. *Nanoscale*, 7, 16, 7409–7414, 2015.
126. Lee, H.-G., Gopalan, A.-I., Sai-Anand, G., Lee, B.-C., Kang, S.-W., Lee, K., Facile synthesis of functionalized graphene-palladium nanoparticle incorporated multicomponent TiO2 composite nanofibers. *Mater. Chem. Phys.*, 154, 125–136, 2015.
127. Kim, Y.J., Lee, H., Lee, D.J., Park, J.-K., Kim, H.-J., Reduction of charge and discharge polarization by cobalt nanoparticles-embedded carbon nanofibers for Li–O2 batteries. *ChemSusChem*, 8, 15, 2496–2502, 2015.
128. Jiang, H., Ge, Y., Fu, Y., Fu, K., Lu, Y., Chen, C., Zhang, X., Centrifugally-spun tin-containing carbon nanofibers as anode material for lithium-ion batteries. *J. Mater. Sci.*, 50, 3, 1094–1102, 2015.
129. Feng, J., Zhang, Z., Ci, L., Zhai, W., Ai, W., Ai, Q., Xiong, S., Chemical dealloying synthesis of porous silicon anchored by *in situ* generated graphene sheets as anode material for lithium-ion batteries. *J. Power Sources*, 287, 177–183, 2015.
130. Kim, H., Huang, X., Wen, Z., Cui, S., Guo, X., Chen, J., Novel hybrid Si film/carbon nanofibers as anode materials in lithium-ion batteries. *J. Mater. Chem. A*, 3, 5, 1947–1952, 2015.

131. McCormac, K., Byrad, I., Brannen, R., Seymour, R., Seymour, B., Li, J., Wu, J., Preparation of porous Si and TiO2 nanofibres using a sulphur-templating method for lithium storage. *Phys. Status Solidi (a)*, 212, 4, 877–881, 2015.
132. Peng, Y.-T. and Lo, C.-T., Effect of microstructure and morphology of electrospun ultra-small carbon nanofibers on anode performances for lithium ion batteries. *J. Electrochem. Soc.*, 162, 6, A1085, 2015.
133. Tang, W.S., Chotard, J.-N., Janot, R., Synthesis of single-phase LiSi by ball-milling: Electrochemical behavior and hydrogenation properties. *J. Electrochem. Soc.*, 160, 8, A1232, 2013.
134. Li, B., Yang, S., Li, S., Wang, B., Liu, J., From commercial sponge toward 3D graphene–silicon networks for superior lithium storage. *Adv. Energy Mater.*, 5, 15, 1500289, 2015.
135. Jing, S., Jiang, H., Hu, Y., Shen, J., Li, C., Face-to-face contact and open-void coinvolved Si/C nanohybrids lithium-ion battery anodes with extremely long cycle life. *Adv. Funct. Mater.*, 25, 33, 5395–5401, 2015.
136. Lee, J.-H., Katoch, A., Choi, S.-W., Kin, S.S., Extraordinary improvement of gas-sensing performances in SnO2 nanofibers due to creation of local p–n heterojunctions by loading reduced graphene oxide nanosheets. *ACS Appl. Mater. Interfaces*, 7, 5, 3101–3109, 2015.
137. Xu, S., Kan, K., Yang, Y., Jiang, C., Gao, J., Li, L., Shi, K., Enhanced NH3 gas sensing performance based on electrospun alkaline-earth metals composited SnO2 nanofibers. *J. Alloys Compd.*, 618, 240–247, 2015.
138. Armand, M. and Tarascon, J.-M., Building better batteries. *Nature*, 451, 7179, 652–657, 2008.
139. Wang, L., Yu, Y., Chen, P.-C., Chen, C.-H., Electrospun carbon–cobalt composite nanofiber as an anode material for lithium ion batteries. *Scr. Mater.*, 58, 5, 405–408, 2008.
140. Fan, Q. and Whittingham, M.S., Electrospun manganese oxide nanofibers as anodes for lithium-ion batteries. *Electrochem. Solid-State Lett.*, 10, 3, A48, 2006.
141. Ji, L., Lin, Z., Guo, B., Medford, A.J., Zhang, X., Assembly of carbon–SnO2 core–sheath composite nanofibers for superior lithium storage. *Chem.–Eur. J.*, 16, 38, 11543–11548, 2010.
142. Larcher, D., Beattie, S., Morcrette, M., Jumas, J.-C., Tarascon, J.-M., Recent findings and prospects in the field of pure metals as negative electrodes for Li-ion batteries. *J. Mater. Chem.*, 17, 36, 3759–3772, 2007.
143. Winter, M. and Brodd, R.J., What are batteries, fuel cells, and supercapacitors? *Chem. Rev.*, 104, 10, 4245–4270, 2004.
144. Green, M., Dunlop, E., Hohl-E., J., Yoshita, M., Hao, X., Solar cell efficiency tables (version 57). *Prog. Photovoltaics: Res. Appl.*, 29, 1, 3–15, 2021.
145. Kumar, E.N., Jose, R., Archana, P.S., Vijila, C., RAmakrishna, S., High performance dye-sensitized solar cells with record open circuit voltage using tin oxide nanoflowers developed by electrospinning. *Energy Environ. Sci.*, 5, 1, 5401–5407, 2012.
146. Krysova, H., Zukal, A., Chandiran, A.K., Gratzel, M., Kavan, L., The application of electrospun titania nanofibers in dye-sensitized solar cells. *CHIMIA Int. J. Chem.*, 67, 3, 149–154, 2013.
147. Wang, G., Xiao, W., Yu, J., High-efficiency dye-sensitized solar cells based on electrospun TiO2 multi-layered composite film photoanodes. *Energy*, 86, 196–203, 2015.
148. Agarwal, G., Greiner, A., Wendorff, *Electrospinning: Materials, processing and applications*, Wiley-VCH, Weinheim, Germany, 2013.
149. Zhong, S., Liu, C., Dou, Z., Li, X., Zhao, C., Fu, T., Na, H., Synthesis and properties of sulfonated poly (ether ether ketone ketone) containing tert-butyl groups as proton exchange membrane materials. *J. Membr. Sci.*, 285, 1–2, 404–411, 2006.
150. Li, X., Ao, X., Xu, D., Zhang, G., Zhong, S., Na, H., Wang, D., Fabrication of sulfonated poly (ether ether ketone ketone) membranes with high proton conductivity. *J. Membr. Sci.*, 281, 1–2, 1–6, 2006.
151. Wang, H.-S., Fu, G.-D., Li, X.-S., Functional polymeric nanofibers from electrospinning. *Recent Pat. Nanotechnol.*, 3, 1, 21–31, 2009.

16

Polysaccharide Biomaterials for Electrochemical Applications

Neelam Srivastava* **and Dipti Yadav**

Department of Physics (MMV Section) Banaras Hindu University, Varanasi, India

Abstract

Today's lifestyle and energy devices are synonyms to each other hence for sustainable developments optimum use of renewable energy sources is the only way out. For better utilization of renewable energy resources, efficient energy storage systems are obligatory. Unfortunately, excessive and inefficient use of energy storage system leads to environmental threat because they add up to global chemical garbage. Hence scientific community is optimistic towards renewable chemicals to replace the synthetic chemical components in energy devices. When such alternate options are sought to reach the commercial level then the cost effectiveness and abundance become an inevitable component. Polysaccharides, being abundant, cost effective and being available in variety, have allured electrochemists for its application as electrolyte, electrode, binder, separator etc. Variety of polysaccharides such as cellulose and its derivatives, chitosan/chitin, starch, alginate, gum, pectin, etc., have been successfully explored. Polymer-in-salt-electrolytes (PISEs) have also been successfully developed using polysaccharides. PISEs are the polymer electrolytes where amount of salt is beyond certain threshold so that ion motion is decoupled from polymer segmental motion and hence in such systems possibility of achieving single and fast ion transport is very high. Literature indicates that synthetic polymers have substantially failed to hold large amount of salt in dissociated form along with flexible morphology. Different methods of electrode preparation used with synthetic chemical as precursor are applicable to polysaccharides also. This chapter reviews the use of polysaccharides for electrochemical applications.

***Keywords*:** Cellulose, pectin, chitosan/chitin, starch, gum, alginate, electrolyte, activated-carbon

16.1 Introduction

Multifunctional and high-performance energy storage devices are needed for the ever-increasing demand of energy. Transformation of electronic devices from solid, single shape physique to wearable, versatile devices have further sprung up the need to explore newer energy devices fulfilling the needs of these advanced electronic devices. To satisfy the insatiable need of energy demand we have lagged in thinking about the harms done to the environment. Owing to increased environmental exploitation and ruthless use of

**Corresponding author*: neel@bhu.ac.in

Inamuddin, Tariq Altalhi and Sayed Mohammed Adnan (eds.) Sustainable Materials for Electrochemical Capacitors, (391–414) © 2023 Scrivener Publishing LLC

natural energy resources, like fossil fuels, without considering the side effects of irresponsible decompositions back to nature, especially, back to soil, has led to major disasters. For sustainable survival of human being, maximizing the use of renewable energy sources is the only way out, but the renewable energy sources depend upon weather, day-night conditions, geographical status, etc. To optimize the utilization of renewable energy in a cost effective and environment benign way we require energy storage devices. Among variety of energy storage devices, electrochemical capacitors (ECs) have proved to be credible in satisfying the demand of storing and delivering of electrical energy. The first commercial farad sized electrochemical capacitors for computer memory backup applications came into existence in 1978 and after that it has seen tremendous advancements.

For sustainable development all the components of energy devices should be made up of renewable materials because when the substrates of the electrochemical devices are non-degradable, they enhance global chemical waste. As per an estimation in 2020, the e-waste production has reached to 53.6 million tons in which an average contribution by each of us is 7.3 kg [1]. It has been predicted that chemical waste just doubles in a short time span of 16 years. The recycling rate of e-waste is very poor (~17.4% as per the data available for 2019) due to imperfect policies, complicated and tedious process involved. Being responsible part of scientific community, it is our sole responsibility to do everything that should be done to manage natural resources to the best extent else it is inevitable to receive reactions from nature itself.

Hence, presently a variety of environment-benign, renewable, and economical electrochemical products are being explored, which are made of materials of biological origin, including cellulose, lignin, Chitin/chitosan, and starch, etc. Being natural materials, they are abundant and easily degradable. They are also acceptable for healthcare and medical applications such as implantable devices. Electrodes and electrolytes are two basic components of any energy device. Hence, before discussing the use of polysaccharides for their synthesis, it is important to have a look at state of art of these components.

Liquid electrolytes are the best ion conducting material but due to many disadvantages associated with them, such as leakage, corrosion, design problems, low electrochemical-stability-window (ESW) etc., scientific community started exploring the solid electrolytes. Ions, being bigger in size, needs special lattice for comfortable mobility. Being highly amorphous in nature polymer-electrolytes came as the compromised version of solid and liquid electrolytes. The technological importance of polymer electrolytes was recognized in 1970s by Armand and coworker [2, 3] and after that polymer electrolytes have seen much advancement and presently, they are dominating the market of energy devices. Still, they are having many issues to be addressed, such as poor targeted ion transport, slow ion motion, need of liquid component (as plasticizer) to achieve desired electrochemical properties etc. Such polymer electrolytes, where the salt concentration is lower than that of polymer host, are called salt-in-polymer electrolytes (SIPEs). In these polymer electrolytes the ion conduction is coupled with polymer segmental motion, the polymer segments being bulky moves slowly and hence effectively reduces ion mobility. To achieve better performing electrolytes tremendous research is going on to get suitable polymer host and salt combination and/or to recognize required additives (filler and/or plasticized). SIPE related problems have been recognized in 1990s, and Vincent and his group [4] proposed that single and fast ion transport can be achieved if the salt concentration is increased beyond the threshold value after which an ion-cluster will be formed, and then ion movement will be decoupled from polymer-segmental motion and ion will diffuse through the ion-cluster resulting in much faster ion transport [5]. These materials are called polymer-in-salt-electrolytes (PISE). Desired success (a minimum conductivity of the order of 10^{-4} S/cm with morphology suitable for device fabrication) is still to be achieved in PISEs. The main hindrance is unavailability of a suitable host which can hold the large amount of salt (above the required threshold) in dissociated form. It has been generally observed that due to re-crystallization of salt it results in poor conductivity and/or morphology becomes brittle, and hence these PISEs could not be used for device fabrication. So, at present the pure solid-state-polymer-electrolyte is still to be achieved and research is being carried out in this direction.

The second component of the energy devices is electrode. In supercapacitors the charge exchange mechanism depends upon the electrode/electrolyte combination and classification of supercapacitors based on the charge storage mechanism supercapacitors are i) Electrochemical Double layer capacitor, ii) Faradic or pseudo-capacitors and iii) Hybrid capacitors. Supercapacitor charge storage is quite high (100–1000F) in comparison to conventional capacitors (micro-farad to milli-farad). To achieve the desired capacitance and proper placing in Ragone plot, electrode materials must simultaneously fulfill many criteria such as high specific surface area (SSA), corrosion resistance, chemical stability during charging/discharging cycles, high conductivity to reduce the series equivalent resistance (ESR), surface wettability etc. In achieving the high value of capacitance SSA plays an important role but there are many other parameters such as shape & size of pore, pore size distribution, electrode/electrolyte compatibility etc. For example, smaller pores result in high ESR resulting in low power density whereas with decreasing pore size energy density increasing i.e., the appropriate distribution of pore size (micro and mesopores) is important to optimize the parameters.

With the aim to improve the long lasting and high performing characteristics, new electrodes are continuously being explored, such as carbon-based electrodes [6, 7], metal oxide/hydroxide-based electrodes [8, 9]/conducting polymer-based electrodes [10, 11]/MXene 2D electrodes [12, 13]/graphene based [14, 15] etc. The nanocomposites [16, 17] are also being studied as these materials provide better active surface area and reduced electron/ion diffusion path. The use of redox-active [18, 19] and bio-derived functional group materials [20, 21] has also been found to improve the electrochemical properties of supercapacitors. For commercial application of any material, its cost effectiveness, and environment benign nature is of immense importance. Hence along with many chemically synthesized electrodes the renewable polymers like polysaccharides-based electrodes are also being explored.

16.2 Polysaccharides in Energy Devices

Polysaccharides are naturally abundant and low-cost materials, which can be extracted from botanical sources and from living organisms. They are used in energy storage devices in various forms [22, 23] such as i) binders for electrodes, ii) separators, iii) electrolytes, iv) electrodes etc. In present chapter only the last two are covered. Literature survey indicates that lots of polysaccharide-based electrode and electrolyte materials have been studied for different kinds of applications in energy devices. For supercapacitive applications electrode-based reports dominate in literature whereas comparatively few polysaccharide-based electrolytes have been tested for their supercapacitive applications. Still authors find it important to summarize few recent polysaccharide-based electrolytes for completeness of the information and making it available to the reader interested in exploring it for supercapacitive applications. The chapter has been divided into two parts i) polysaccharide-based electrolytes and ii) polysaccharide-based electrodes.

16.2.1 Polysaccharide-Based Electrolytes

Since when the polymer electrolytes have been explored, efforts are being made to improve the electrochemical properties along with 3E properties (economical, easy to handle and environment-benign). In this effort polysaccharides like chitosan, algal, gum arabic, agar-agar, starch etc. are being explored for their electrolyte applications [24–28]. The abundance of the materials is an important criterion for its long-term use and according to this criterion cellulose comes at the top followed by chitosan and starch respectively. Cellulose and its derivatives [29–31] (where the OH are replaced by another ion) have been exclusively studied for electrolyte synthesis. The most popular derivative is cellulose acetate where one OH is replaced by acetate group. Such variations bring great changes in the basic properties of host for example cellulose is hydrophilic whereas cellulose acetate is more hydrophobic and many other properties like solubility in particular solvent, their mechanical nature etc. are also changed. Few of cellulose and its derivative based electrolytes and their conductivities are given in Table 16.1. Quite high conductivity (10^{-3} S/cm) with good mechanical properties is reported with cellulose and its derivatives [32, 33]. In few cases, when both

Table 16.1 Few polysaccharide-based electrolytes along with their conductivities.

Polysaccharides used	Additives	Conductivity (S/cm)	References
Pectin	Methylcellulose + Potassium phosphate	7.5×10^{-4}	[34]
Pectin	Ammonium thiocyanate (NH_4SCN) salt	1.5×10^{-3}	[35]
Chitosan	Sodium triflate	2.41×10^{-4}	[36]
Chitosan	Magnesium chloride	1.03×10^{-3}	[37]
Chitosan	Magnesium acetate	1.09×10^{-5}	[38]
Chitosan	$LiCF_3SO_3$	2.75×10^{-5}	[39]
Phthaloyl chitosan	Ammonium thiocyanate (NH_4SCN)	2.42×10^{-5}	[40]
Chitosan	$LiCO_2CH_3$	5.19×10^{-4}	[41]
Chitosan	$LiClO_4$	1.20×10^{-3}	[42]
Chitosan	CS-coated CNTs	3.46×10^{-2}	[43]
Sulfonated chitosan	PEO with s-SiO_2	10^{-2}	[44]

(Continued)

Table 16.1 Few polysaccharide-based electrolytes along with their conductivities. (*Continued*)

Polysaccharides used	Additives	Conductivity (S/cm)	References
Cellulose acetate	NH_4NO_3	1.02×10^{-3}	[45]
Cellulose	Polyethylene glycol	3.31×10^{-3}	[46]
Carboxymethyl cellulose	Ammonium bromide	3.31×10^{-3}	[47]
Cellulose	Butylmethylimidazolium chloride	0.325	[48]
Hydroxyethyl cellulose	Soybean protein isolate and Li_2SO_4	8.40×10^{-3}	[49]
Methyl cellulose	Lithium trifluoromethanesulfonate	10^{-3}	[50]
Xanthan gum	Sodium sulfate	1.12	[51]
Alginate	$LiNO_3$	1.14×10^{-4}	[52]
Alginate	NH_4NO_3	5.56×10^{-5}	[53]
Sodium alginate	Dopamine functionalized polypyrrole (DAPPy)	1.33	[54]
Sodium alginate	PVA	7.82	[55]
Corn starch	Citric acid	2.30×10^{-3}	[56]
Corn starch	Lithium acetate	8.37×10^{-4}	[57]
Corn starch	$NaHSO_3$	2.22×10^{-4}	[58]
Potato starch	Lithium triflate	4.29×10^{-4}	[59]
Corn starch	Sodium iodate	1.08×10^{-4}	[60]
Potato starch	Magnesium acetate	6.28×10^{-4}	[61]
Potato starch	Sodium iodide	9.59×10^{-3}	[62]
Potato starch	Ammonium iodide	2.4×10^{-4}	[63]

electrode and electrolytes are cellulose based, performance of fabricated supercapacitor has been found to be superior to similar studies [33]. Cellulose nanoparticles are also added to typical polymer electrolytes to increase the conductivity of electrolyte materials [32].

Chitosan and its derivatives have also attracted many electrochemists for their use as different components in different energy devices. For supercapacitors also, chitosan has been explored as electrode and electrolyte both. The use of chitosan as polymer host has not been very successful in its pure form but when blended with other polymers and/or plasticizers are added to them, the observed electrochemical parameters were quite favorable for device fabrication. Few recent chitosan-based electrolytes are reported in Table 16.1.

Agar is a Polysaccharide extracted from algae, has agarose and agaropectin as its main components. Agarose is the gelling component, which is a linear Polysaccharide having 1, 3-linked β-d-galactose and 1, 4-linked 3–6-anhydroL-galactoseas its basic units. It has been studied with different salts especially with ammonium salts such as NH_4SCN [64], NH_4I [65], NH_4Br [66] etc., with target to get proton conducting electrolytes for fuel cells. These combinations resulted in high conducting electrolytes. Gel electrolytes based on agar/agarose are successfully tested for photovoltaic applications, dye sensitized solar cell. Few agar and sodium salts based gel electrolytes have also been successfully studied for supercapacitor applications [67, 68].

Alginate is another polysaccharide which is extracted from Algae. Agar/agarose (discussed above) is extracted from red algae (Rhodophyceae class) whereas alginate comes from cellwall of brown algae of phaeophyceae class. Alginate is basically made up of (1→4')-linked β-d-mannuronic acid and α-l-guluronic acid units. Because of their low cost and easy biodegradability, Alginates are studied with variety of lithium and sodium salts [69–71]. Good electrochemical properties have been demonstrated by the self-healing hydrogel, synthesized by crosslinking procedures [72, 73]. Many of the synthesized materials having sufficiently high conductivity (mS/cm) are being tested for different devices. Some electrolyte combinations are given in Table 16.1. Few of its hydrogel-based electrolytes [74] have demonstrated good supercapacitive behavior.

As mentioned in introduction section, depending upon the salt concentration, polymer electrolytes can be classified as i) salt-in-polymer-electrolytes (SIPEs) and polymer-in-salt-electrolytes (PISEs). All the above discussed polysaccharide-based electrolytes belong to the SIPE group. Recent starch based SIPEs are also summarized in Table 16.1. In literature, polysaccharide based PISE are mostly reported for starch as host. Our group has developed a synthesis protocol for PISE by crosslinking the starch with glutaraldehyde. The synthesis protocol is applicable to variety of sodium and magnesium salts. Cross-linked starch-based electrolytes have small amount of water (<25%). This water, in presence of large amount of salt (starch to salt ratio 1:3), is playing an important role in hydrogen bonding which seems to keep the salt in dissociated state and simultaneously keeping the matrix flexible. Table 16.2 summarizes few latest reports of PISE materials synthesized using the glutaraldehyde cross-linked starch as host material. The latest trend in polymer electrolytes is water-in-salt-polymer-electrolytes [75]. Here it is important to understand that aqueous electrolytes have high conductivity and are supposed to be very good but electrochemically they have the disadvantage of small ESW (generally less than <1.5V) resulting in low energy/power density. Hence now efforts are being made to stabilize the water molecules in solid matrix having large amount of salt so that a wider ESW can be achieved along with high conductivity. Since the PISE developed by our group has some absorbed water component in it and FTIR studies have indicated the presence of variety of hydrogen bonding, which confirms that water through its hydrogen bonding capability is playing important role in acceptance of salt in matrix and simultaneously keeping the matrix flexible. The initial studies of its use in supercapacitor [76] indicated that these materials are potential candidate for energy storage devices. All starch based PISEs have high conductivity (>0.01 S/cm) and flexible morphology. The ion relaxation time of these electrolytes is in microsecond range which ensures the fast

Table 16.2 Few starch-based PISE electrolytes with their conductivities and relaxation time.

Electrolyte (Starch: Salt)	Conductivity (S/cm)	Relaxation time (μs)	Reference
PS + NaI (1:3)	3.41×10^{-2}	15	[76]
AS: NaI (1:1)	10^{-2}	14	[77]
CS + $NaClO_4$ (1:1)	1.0×10^{-2}	65	[78]
WS + NaI (1:3)	3.61×10^{-2}	13	[79]
RS + $NaClO_4$ (1:2)	1.59×10^{-2}	62	[80]

device performance. These electrolytes are synthesized using a facile protocol and hence are cost effective also.

16.2.2 Polysaccharide-Based Electrodes

For supercapacitors, carbon-based electrode materials are very popular because of their availability in various allotropes (Diamond, graphite and fullerenes, fullerene can be subdivided into nanotubes, buckyballs, etc.) degree of graphitization and structures. They have wide variety of micro textures and dimensionality (0 to 3D). Because of its amphoteric nature along with wide temperature range applicability, and stability in both acidic and basic solutions, it is very popular for electrochemical applications. The activated carbon in form of nano structure has excellent surface functionality resulting in large capacity. Hence, nanoporous carbons are very popular. Generally, template methods are used to control the pore size. These synthesized carbons are further activated through different processes to increase the surface area. Either of physical or chemical activation process can be employed. The synthesized activated carbon nanospheres pore size varies from microspheres (<2 nm) mesopores (2–5 nm) and macropores (>50nm). Hierarchal porous carbon with combination of micropores, mesopores and/or macropores is becoming popular now-a-days, because of their positive synergistic effect on device performance. Considering the economic and environmental constraints use of biomaterials for carbon synthesis has received great attention of scientific community. Polysaccharides, being easily available in various forms and cost effective, have been extensively explored for carbon synthesis for supercapacitor electrodes [81–90]. Polysaccharide based electrodes have been extensively studied for supercapacitive applications. Literature indicates that per gram capacitance of supercapacitors fabricated using polysaccharide-based electrodes is quite high, ranging in hundreds to thousands of farads. Few important classes of polysaccharides are discussed below, which are being explored using different synthesis process to increase the specific surface area and proper ratio of micro, meso and macropores distribution so that electrochemical properties of fabricated device can be optimized.

16.2.2.1 Cellulose-Based Electrode Materials

Cellulose being abundant, inexpensive, versatile, and sustainable material has become an important class of polysaccharides for synthesis of various part of energy storage devices such as separator, binder, substrate, current collector etc. Cellulose themselves are insulating material having negligible electrochemical activity but using different paper making techniques they can be molded as per requirement. The pore size being very important parameter for energy storage devices and hence cellulose becomes important as it felicitates the possibility of variety in manufacturing new separators with adjustable porosities. Cellulose's structural configuration along with its natural abundance makes it a promising candidate for its use as separators and substrates in supercapacitors. Because of their porosity, stability, and mechanical properties the cellulose microfibers (CMFs), nanocellulose (NCFs), and cellulose nanocrystals (CNCs) are used in sustainable supercapacitors. The most common substrate are CMFs which are the main constituents' part of commercially available papers. In paper form CMFs are economical, recyclable and have mechanical flexibility and they have rough texture, large number of surface functional groups and provide a mesoporous platform. The large number of functional groups available at surface are helpful in binding of electroactive materials and providing good contact whereas the porous structure helps in ion transport. CMFs in CMF-based composite electrodes provide effective pathways to ion transport and act as an interior electrolyte reservoir [91, 92].

Variety of combinations have been tested to achieve the best of capacitance, some of them are listed in Table 16.3. $Ni(OH)_2$-CFG composite, where CFG stands for cellulose-fiber covered with graphene on which nano structure of $Ni(OH)_2$ has been grown, resulted

Table 16.3 Cellulose based electrodes with respective value of capacitance achieved.

Electrode composition	Electrolyte	Specific capacitance (F/g)	References
Cross-linked Ni(OH)2-CFs	2 M NaOH	2276	[93]
CNFs/MWCNTs/PANI aerogel	1 M H_2SO_4	791	[94]
NiS/N-CFA	2 M KOH	1612	[95]
Cellulose/graphite/Ni/MnO2		680	[96]
Paper/CNT/MnO2	0.1 M Na_2SO_4	540	[97]
Cotton/MWCNT/PPy	2 M NaCl	535	[98]
3D network of BC/GO/PPy	1 M $EMIMBF_4$	556	[99]
Paper/Au/PANI composite	1 M H_2SO_4	560	[100]
SWCNTs and PANI nanoribbons wrapped around interconnected cellulose fiber	1 M H_2SO_4	533	[101]
Nitrogen-doped BC nanofibers	6 M KOH	1950	[102]

(*Continued*)

Table 16.3 Cellulose based electrodes with respective value of capacitance achieved. (*Continued*)

Electrode composition	Electrolyte	Specific capacitance (F/g)	References
CNFs/graphene oxide (GO) aerogel	H_2SO_4/PVA	207	[103]
CNFs/reduced GO (RGO)/ carbon nanotubes (CNTs) aerogel	H_2SO_4/PVA	252	[104]
Ag-fibers/RGO-fibers	CMC/Na_2SO_4	222	[105]
poly(arylene ether sulfone)/ poly(vinylpyrrolidone) (PAES-Q/PVP)	hydroxypropyl cellulose/poly(vinyl alcohol)/$LiClO_4$ water/ glycerol	194	[106]
CNFs/PANI aerogel	H_2SO_4	291	[107]
CNFs/PPy aerogel	1 M NaCl	392	[108]
PPy/CNFs aerogel	LiCl/PVA	215	[109]
carbon fibers (CFs)-CNFs/ MWCNTs/MnO2/CFsCNFs aerogel/MWCNTs/ active carbon (AC) aerogel C	LiCl/PVA	357	[110]
B,N-doped methyl cellulose CA/MnO2	1 M Na_2SO_4	338	[111]
TOCN/RGO CA	1 M H_2SO_4	398	[112]

in a binder-free electrode which has good cycling stability and high rate capability. This electrode resulted in high capacitance (2276 F g^{-1}, at 1 A g^{-1}), [93] which, for a high current density of 5 A g^{-1}, shows no decay even after 1000 cycles. The beauty of this electrode was its extremely high surface area which has been achieved via vertical and cross-linked structure. A large electrode/electrolyte interface area is achieved due to this special cross-linked structure which helped in formation of the electric double layer. The nano structure has also provided a unique hierarchical pore texture which leads to fast electrolyte ion diffusion in/ out of its pore channels. Few electrodes composite materials based on cellulose are summarized in Table 16.3 along with the capacitance value obtained.

16.2.2.2 Chitosan/Chitin-Based Electrode Materials

Another highly abundant biopolymer is chitin. It is basically a linear polysaccharide biopolymer, 2-amino-2-deoxy-D-glucopyranose having 1, 4-glycosidic bonds [113], are its building blocks. It is found in fungal cell walls, of mollusk radula, cephalopod beaks insect and crustacean exoskeletons. Seafood processing is the main source of man-made chitin waste, which after passing through rigorous chemical process is supplied as commercial chitin. The alternate routes to obtain the chitin are biowaste from other sources, such as mushroom, silk, and honey-harvesting industries [114]. Table 16.4 summarizes few supercapacitor fabricated using chitosan/chitin based electrodes.

With some optimized quantity of constituents, the nanocomposite of Zinc doped Iron oxide/Graphene Oxide/Chitosan can be easily produced by solution-mixture evaporation

Table 16.4 Chitosan/Chitin based electrodes with respective value of capacitance achieved.

Electrode composition	Electrolyte	Specific capacitance (F/g)	References
CH/Spinal (nickel ferrite) + N-doped C matrix	6 M KOH	958	[116]
CH-G polyaniline	1 M H_2SO_4	1519	[117]
Chitin/graphene oxide/zinc oxide/ polyaniline	0.2 M H_2SO_4	543	[118]
Polyionic liquid/carboxymethyl chitosan (nitrogen- and sulfur-codoped porous carbon)	1.0 M H_2SO_4	633	[119]
CuO decorated graphene oxide/ chitosan	0.1 M KCl	772	[120]
CH-ZnO/polyaniline	1 M H_2SO_4	587	[121]
CNT/CH-ferrocene	1 M H_2SO_4	695	[122]
N-doped nanofibrous chitin microspheres	EMIM TFSI	113	[123]
Chitosan-derived carbonaceous materials	0.5 M K_2SO_4	231	[124]
Chitosan-derived carbonaceous materials	6 M KOH	305	[124]
Chitosan-derived carbonaceous materials	0.5 M K_2SO_4	154	[124]
Nitrogen and phosphorus dual-doped hierarchical porous carbon	6 M KOH	312	[125]
Chitosan-based layered carbon materials	1.0 M H_2SO_4	355	[126]
Chitosan-based layered carbon materials	6 M KOH	275	[126]
Cellulose-derived porous carbons/PANI	1.0 M H_2SO_4	765	[127]

method [115]. This electrode resulted in 830 Fg^{-1} specific capacitance when tested with 1M H_2SO_4 as electrolyte.

A complex method was used to N_2 doped electrode from chitosan biowaste. To get nitrogen and carbon sources in one go chitosan-urea-formaldehyde polymer ligand were prepared which were converted into bimetallic polymer metal complex using $NiCl_2.6H_2O$ and Fe $(Cl_3)_3.6H_2O$. This complex on annealing at 800°C resulted in carbon which was used as electrode and its other version with polyaniline was also synthesized. The latter version of electrode resulted in supercapacitor having ~958 Fg^{-1} capacitance with high cycling stability [116]. Anand Torvi and group synthesized series of binary and tertiary composites, Chitosan-polyaniline and Chitosan-Graphene-Polyaniline. For this synthesis an *in-situ* polymerization of aniline [117] in presence of other additives has been followed. Both have high conductivity and especially ternary composite's conductivity is ~ 3 S/cm which resulted in 1519 F g^{-1} capacitance with 1M H_2SO_4 electrolyte, at 5 mV/s scan rate.

16.2.2.3 Starch-Based Electrode Materials

Starch a natural-derived polysaccharide having glucose monomers as basic units, is abundant, environment-benign renewable polymer, and this is the reason why it is considered to have great potential. Being economical and having good processability it is being explored for many applications including supercapacitor. It can be obtained from different botanical origins and also from different parts of tree/fruits. Abundantly it is available in roots, seeds, and stems of the rice, sweet potato, corns, wheat, arrowroot, potatoes, cassavas, and other crops. Its exceptional environmental degradability and biocompatibility makes it a potential candidate as starting material for fabricating green energy storage devices.

A high specific capacitance (1120 F g^{-1} at 2 A g^{-1}) with good rate capability along with a retention ratio of 81.3% is achieved by [128] Chen and Liu. Authors have used solution of starch and de-ionized water along with urea and Ni $(NO_3)_2{\cdot}6H_2O$. After heating at 120°C for 6 hours in autoclave, the collected green precursor was washed, followed by drying at 60°C. In nitrogen atmosphere, precursor was annealed at 500°C for 2 hours and to obtain black C-Ni-QDs hybrids. Table 16.5 summarizes few starch-based electrodes and component details of fabricated devices.

16.2.2.4 Gum-Based Electrode Materials

Gum, an extensively diverse and multifaceted polysaccharide, is available in different forms such as Xanthan, Peach, Acacia (which is also known as Senegal gum, Gum Arabia, Indian gum), Arabic etc. [139]. Depending upon the source of extraction they can be either hydrophilic and/or hydrophobic and generally have colloidal properties and high molar mass [140]. They can be extracted from different parts of the plants such as i) from seed endosperm-guar gum ii) from shrubs and trees-arabic, karaya gum, cashew gum [145–147], iii) from bacteria-Xanthan Gum and others. They are comprised of negatively charged carboxylate and hydroxyl groups. For the facilitation of compatibility between electrode and electrolytes, inherited properties of the polysaccharides play an important role in device performance [141–147]. Few gum-based electrodes and the capacity of the fabricated supercapacitors is summarized in Table 16.6. Gum based carbon electrodes have been

Table 16.5 Starch based electrodes with respective value of capacitance achieved.

Electrode composition	Electrolyte	Specific capacitance (F/g)	Reference
C-Ni-QDs hybrids	-	1120	[128]
Dialdehyde starch/GO/ polyaniline (DAS-RGO/PANI)	1.0 M H_2SO_4	499	[129]
Starch magnesium nitrate derived HPCs	6 M KOH	229	[130]
Starch-derived 3D-reticular porous carbon	2M KOH	372	[131]
Starch-derived porous carbon	1 M Et_4NBF_4	272	[132]
Starch-derived HPCs	$EmimBF_4$	385	[133]
Composite of nanostructured Co_3O_4 and corn starch	PVA-KOH	473	[134]
N-doped wheat flour-derived porous carbon nanospheres/ PANI composite	1.0 M H_2SO_4	783	[135]
Artocarpus heterophyllus starch	*Manihot esculenta* starch film	240	[136]
Po Monodisperse C microspheres CMSs/RGO	1M KOH	245	[137]
Porous carbon derived from Corn starch activated by H_3PO_4	6 M KOH	162	[138]

synthesized using different processes or following combinations of processes simultaneously such as Nitrogen doping, chemically blowing of urea, and chemically activating with KOH. Swelling and chemical blowing processes [143], composite synthesis with synthetic polymerization [145], pyrolyzing and/or hydrothermal treatment has also been successfully employed [146]. The morphology of electrodes is vital for device fabrication. For thin double walled carbon nanotubes dispersed in gum derived activated carbon resulted [144] in good free standing hybridized electrode. A very high specific capacitance, 2150 F/g with 2M KOH liquid electrolyte, is achieved by using Arabic gum as polysaccharide binder in fabrication of electrode for supercapacitor.

16.2.2.5 Alginates-Based Electrode Materials

Alginates are also one of the attractive prospects for the energy storage applications. Alginates are biopolymer composed of series of block structures of monomers. It comprises

Table 16.6 Gum based electrodes with respective value of capacitance achieved.

Electrode composition	Electrolyte	Specific capacitance (F/g)	References
Peach Gum-Derived Heteroatom-Doped Porous Carbon	1 M H_2SO_4	379	[141]
Acacia gum	1 M HCl	272	[142]
Nitrogen doped porous carbon nanosheet material using swelled peach gum	6 M KOH	350	[143]
Xanthan gum	1 M H_2SO_4	142	[144]
PPy/Gum Arabic	1 M H_2SO_4	368	[145]
Peach Gum	6 M KOH	406	[146]
Mo/C/Ni/XG	2.0 M KOH	1735	[147]
Mo/C/Ni/AG	2.0 M KOH	2150	[147]

of unbranched (1-4)-linked glycuronans consisting of residues of β-D-mannosyluronic acid and the C5 epimer α-L-gulosyluronic acid. In the presence of specific divalent (or multivalent) cations, alginates exhibit gel forming properties [148]. Alginates have been successfully used as precursor for synthesis of 3D hierarchal structure of well interconnected micro-mesopores. Such structures are beneficial for specific capacitance and rate performance [149]. Porous carbons are also prepared by microwave heating of alginates-based hydrogel which are cross-linked via. bivalent ions, such as Cu^{2+}, Zn^{2+}, Ca^{2+}, etc. [150]. Oxygen doped micropores electrodes synthesized from alginates resulted in supercapacitor fabrication which retains more than 90% of capacity even after 20000 cycles of charging and discharging [151]. These electrodes also have hydrogen storage capability. Sometimes, nitrogen and oxygen co-doped porous carbon are synthesized using alginates cross-linked by polyvalent atoms [152]. It has been reported that sodium alginates-based oxygen doped porous carbon electrodes does not require binder and hence, result in fabrication ease and reduced cost [153].

Table 16.7 summarizes few of recent reports on alginate-based electrodes. Supercapacitor fabricated using aqueous electrolyte 1M H_2SO_4 and alginate-based electrodes demonstrated high specific capacitance 2093 F g^{-1} [155]. Electrode consisted of composition of polyaniline and sodium alginate where solution method has been utilized to synthesize nanostructured polyaniline and sodium alginate composite.

16.2.2.6 *Pectin-Based Electrode Materials*

Pectin is mixture of polysaccharides that comes from plants having pectinic acids as major components. Pectin is water soluble and can form gels under favorable conditions. The physical property of pectin, which makes it important, is its ability to form spreadable gels [158].

Table 16.7 Alginate based electrodes with respective value of capacitance achieved.

Electrode composition	Electrolyte	Specific capacitance (F/g)	Reference
3D activated hierarchical macro-meso-microporous carbon aerogels	6.0 M KOH	188	[149]
alginate-based hydrogel beads cross-linked by bivalent ions (Cu2+, Zn2+, and Ca2+) are converted into porous carbons (SPCs)	6 M KOH	804	[150]
Sodium alginate was applied to assisted preparation of oxygen-doped microporous carbon	6 M KOH	385	[151]
Sodium alginate (SA) beads cross-linked by polyvalent metal ions	6 M KOH	269	[152]
Oxygen-doped porous carbon sodium alginate	6 M KOH	426	[153]
CuS/ZnS/sodium alginate/reduced graphene oxide nanocomposites	2 M KOH	992	[154]
PANI/SA	1 M H_2SO_4	2093	[155]
Cu-doped carbon aerogels sodium alginate (SA) as the carbon precursor	6 M KOH	414	[156]
Defect-rich N doped porous carbon (DNPC) derived from alginate	1 M Na_2SO_4	296	[157]

For implantable electronic medical devices, it is of immense important to check the operation of devices in physiological fluids [159]. PANI, (a conducting polymer) and PECTIN (a biopolymer) has been used to achieve an electrode having excellent combination of antibacterial property, electroactivity, biocompatibility, processability along with the ability to operate in physiological fluids. At initial stage PANI was modified by pectin were to obtain water dispersible PANI-PEC, followed by treatment with $AgNO_3$ solution to achieve. Ag nanoparticle decorated PANI-PEC (Ag-PANI-PEC). Amarnath and group got success in testing the Ag-PANI-PEC physiological fluids like serum, blood, PBS, and urine. They obtained capacitance of the order of few hundreds of farads per gram. Porous carbon sphere of high specific surface area is easily prepared using pectin as biomass precursor and following hydrothermal carbonization and chemical activation route [160]. Carbon electrodes developed by the process are reported to have excellent micro/meso porosity and the pore size distribution lies into a very narrow band. Pectin biomass precursor is also used to synthesize N/O doped carbon electrodes having hierarchal pores, by the process of *in-situ* template application. Yuhao *et al.* synthesized an electrode using $CaCO_3$ as *in-situ* template and urea as dopant, where it is reported that activation temperature and additive urea have their signature on performance, pore size morphology, and ratio of micropores/mesopores. Specific surface area of such

Table 16.8 Pectin based electrodes with respective value of capacitance achieved.

Electrode composition	Electrolyte	Specific capacitance (F/g)	References
Ag-PANI-PECTIN	phosphate buffer saline	140	[159]
Ag-PANI-PECTIN	Blood	290	[159]
Ag-PANI-PECTIN	Urine	144	[159]
Ag-PANI-PECTIN	Serum	121	[159]
Porous carbon spheres derived from pectin at 700° C	6 M KOH	120	[160]
Porous carbon spheres derived from pectin at 800° C	6 M KOH	173	[160]
Porous carbon spheres pectin at 900° C	6 M KOH	231	[160]
N/O co-doped 3D carbon architecture	1 M Na_2SO_4	338	[161]
Mg^{2+}/Ca^{2+}-Pectin-Graphene-Hydrogel	1.0 M H_2SO_4	839.2	[162]

electrodes is very high (~ 2928 m^2/gm) [161]. Pectin is also used to develop self-assembled graphene hydrogel electrodes. For this firstly pectin was chelated with cations (Mg^{2+}/Ca^{2+}) to get an interwoven framework. Such type of electrode resulted in supercapacitor retaining 98.5% of its capacity even after 2000 cycles. The flexibility of the device is also reported to be encouraging (retaining 96.4% capacity even after 500 folding/unfolding) [162]. Table 16.8 summarizes capacity of some supercapacitor using electrodes synthesized from pectin.

16.2.3 Conclusion

In this chapter the uses of polysaccharides for synthesis of electrolytes and electrodes have been reviewed. It has been found that polysaccharides are suitable precursor to develop electrode materials using all the possible available techniques. Supercapacitors' capacitance ranges from few hundred farads to thousands of farads, and similar type of results have been obtained with polysaccharide-based electrodes. Hierarchical electrodes with suitable distribution of micro and mesopores have been successfully obtained using external template and/or self-template. Synthesis of nitrogen and/or oxygen doped electrodes is also feasible with polysaccharide. They have resulted in high conducting (> 0.01 S/cm) flexible PISE having wide electrochemical stability window (> 2.5 V) when cross-linked with glutaraldehyde. It is relevant to mention that a suitable synthetic polymer host, which can hold the minimum required salt for ion-cluster formation in PISE, is still to be identified. Availability of polysaccharide in variety has offered the possibility to synthesize the economical electrode materials as per requirement just by changing the precursor biowaste.

Acknowledgement: Author NS is thankful to BHU for providing incentive grant under "Research Grant for Faculty" (IoE Scheme), Development Scheme No 6031. Author (DY) is thankful to BHU for BHU-UGC research fellowship.

References

1. Vanessa, F., Balde, C.P., Ruediger, K., Garam, B., *The global e-waste monitor 2020*, United Nations University (UNU)/United Nations Institute for Training and Research (UNITAR)—Co-hosted SCYCLE Programme, International Telecommunication Union (ITU) & International Solid Waste Association (ISWA), Bonn/Geneva/Rotterdam, 2020.
2. Armand, M., Polymer solid electrolytes-an overview. *Solid State Ionics*, 9, 745–754, 1983.
3. Armand, M.B., Polymer electrolytes. *Annu. Rev. Mater. Sci.*, 16, 1, 245–261, 1986.
4. Vincent, C.A., Polymer electrolytes. *Prog. Solid State Chem.*, 17, 3, 145–261, 1987.
5. Chen, F., Wang, X., Armand, M., Forsyth, M., Cationic polymer-in-salt electrolytes for fast metal ion conduction and solid-state battery applications. *Nat. Mater.*, 21, 1175–1182, 2022, 10.21203/rs.3.rs-532893/v1.
6. Bo, Z., Wen, Z., Kim, H., Lu, G., Yu, K., Chen, J., One-step fabrication and capacitive behavior of electrochemical double layer capacitor electrodes using vertically-oriented graphene directly grown on metal. *Carbon*, 50, 12, 4379–4387, 2012.
7. Karthika, P., Rajalakshmi, N., Dhathathreyan, K.S., Functionalized exfoliated graphene oxide as supercapacitor electrodes. *Soft Nanosci. Letters*, 2, 59–66, 2012, 2012.
8. Wu, H.Y. and Wang, H.W., Electrochemical synthesis of nickel oxide nanoparticulate films on nickel foils for high-performance electrode materials of supercapacitors. *Int. J. Electrochem. Sci.*, 7, 4405–4417, 2012.
9. Namisnyk, A. and Zhu, J., A survey of electrochemical super-capacitor technology, in: *Australian Universities Power Engineering Conference*, University of Canterbury, New Zealand, 2003.
10. Wang, D.-W. *et al.*, Fabrication of graphene/polyaniline composite paper via *in situ* anodic electropolymerization for high-performance flexible electrode. *ACS Nano*, 3, 7, 1745–1752, 2009.
11. Mastragostino, M., Arbizzani, C., Soavi, F., Conducting polymers as electrode materials in supercapacitors. *Solid State Ionics*, 148, 3-4, 493–498, 2002.
12. Lyu, B., Kim, M., Jing, H., Kang, J., Qian, C., Lee, S., Cho, J.H., Large-area MXene electrode array for flexible electronics. *ACS Nano*, 13, 10, 11392–11400, 2019.
13. Chen, J., Li, Z., Ni, F., Ouyang, W., Fang, X., Bio-inspired transparent MXene electrodes for flexible UV photodetectors. *Mater. Horiz.*, 7, 7, 1828–1833, 2020.
14. Kim, T.Y., Lee, H.W., Stoller, M., Dreyer, D.R., Bielawski, C.W., Ruoff, R.S., Suh, K.S., High-performance supercapacitors based on poly (ionic liquid)-modified graphene electrodes. *ACS Nano*, 5, 1, 436–442, 2011.
15. Cai, Z. *et al.*, Engineering of electrolyte ion channels in MXene/holey graphene electrodes for superior supercapacitive performances. *Rare Met.*, 41, 6, 2084–2093, 2022.
16. Lu, W., Hartman, R., Qu, L., Dai, L., Nanocomposite electrodes for high-performance supercapacitors. *J. Phys. Chem. Lett.*, 2, 6, 655–660, 2011.
17. Zhu, W.C., He, P.Q., Tien, H.C., Liu, H.L., Chen, W.C., Lv, W., Lee, W.Y., Solvent-enhanced transparent stretchable polymer nanocomposite electrode for supercapacitors. *ACS Appl. Energy Mater.*, 4, 3, 2266–2274, 2021.
18. Zhang, X., Xiao, Z., Liu, X., Mei, P., Yang, Y., Redox-active polymers as organic electrode materials for sustainable supercapacitors. *Renewable Sustainable Energy Rev.*, 147, 111247, 2021.
19. Li, M., Liu, J., Zhang, T., Song, X., Chen, W., Chen, L., 2D redox-active covalent organic frameworks for supercapacitors: Design, synthesis, and challenges. *Small*, 17, 22, 2005073, 2021.

20. Wang, L., Su, B., Zhao, X., Li, X., Han, S., Jiang, J., Appropriate design of electrode material by bio-derived nanopores structure for symmetric supercapacitors. *Energy Fuels*, 34, 7, 8956–8965, 2020.
21. Atchudan, R., Edison, T.N.J.I., Perumal, S., Parveen, A.S., Lee, Y.R., Electrocatalytic and energy storage performance of bio-derived sulphur-nitrogen-doped carbon. *J. Electroanal. Chem.*, 833, 357–369, 2019.
22. Schlemmer, W., Selinger, J., Hobisch, M.A., Spirk, S., Polysaccharides for sustainable energy storage–A review. *Carbohydr. Polym.*, 265, 118063, 2021.
23. Wang, Z., Lee, Y.H., Kim, S.W., Seo, J.Y., Lee, S.Y., Nyholm, L., Why cellulose-based electrochemical energy storage devices? *Adv. Mater.*, 33, 28, 2000892, 2021.
24. Anthony, L.S., Vasudevan, M., Perumal, V., Ovinis, M., Raja, P.B., Edison, T.N.J.I., Bioresource-derived polymer composites for energy storage applications: Brief review. *J. Environ. Chem. Eng.*, 9, 5, 105832, 2021.
25. Schlemmer, W., Selinger, J., Hobisch, M.A., Spirk, S., Polysaccharides for sustainable energy storage–A review. *Carbohydr. Polym.*, 265, 118063, 2021.
26. Ngai, K.S., Ramesh, S., Ramesh, K., Juan, J.C., A review of polymer electrolytes: Fundamental, approaches and applications. *Ionics*, 22, 8, 1259–1279, 2016.
27. Torres, F.G. and De-la-Torre, G.E., Algal-based polysaccharides as polymer electrolytes in modern electrochemical energy conversion and storage systems: A review. *Carbohydr. Polym. Technol. Appl.*, 2, 100023, 2021.
28. Roy, B.K., Tahmid, I., Rashid, T.U., Chitosan-based materials for supercapacitor application-a review. *J. Mater. Chem. A*, 9, 33, 17592–17642, 2021.
29. Mahalakshmi, M., Selvanayagam, S., Selvasekarapandian, S., Moniha, V., Manjuladevi, R., Sangeetha, P., Characterization of biopolymer electrolytes based on cellulose acetate with magnesium perchlorate (Mg (ClO4) 2) for energy storage devices. *J. Sci.: Adv. Mater. Devices*, 4, 2, 276–284, 2019.
30. Rani, M.S.A., Rudhziah, S., Ahmad, A., Mohamed, N.S., Biopolymer electrolyte based on derivatives of cellulose from kenaf bast fiber. *Polymers*, 6, 9, 2371–2385, 2014.
31. Aziz, S.B. *et al.*, Compatible solid polymer electrolyte based on methyl cellulose for energy storage application: Structural, electrical, and electrochemical properties. *Polymers*, 12, 10, 2257, 2020.
32. Zhu, M., Yu, L., He, S., Hong, H., Liu, J., Gan, L., Long, M., Highly efficient and stable cellulose-based ion gel polymer electrolyte for solid-state supercapacitors. *ACS Appl. Energy Mater.*, 2, 8, 5992–6001, 2019.
33. Li, K. *et al.*, All-Cellulose-based quasi-solid-state supercapacitor with nitrogen and boron dual-doped carbon electrodes exhibiting high energy density and excellent cyclic stability. *Green Energy Environ.*, 2022.
34. Adam, A.A. *et al.*, Physicochemical and electrochemical performance of glycerol plasticized solid biopolymer electrolytes based on methylcellulose/pectin blend doped with potassium salt, SETA-D-22-00436, http://dx.doi.org/10.2139/ssrn.4032235.
35. Muthukrishnan, M., Shanthi, C., Selvasekarapandian, S. *et al.*, Synthesis and characterization of pectin-based biopolymer electrolyte for electrochemical applications. *Ionics*, 25, 203–214, 2019.
36. Aziz, S.B., Abdullah, O.G., Rasheed, M.A., Ahmed, H.M., Effect of high salt concentration (HSC) on structural, morphological, and electrical characteristics of chitosan based solid polymer electrolytes. *Polymers*, 9, 187, 2017.
37. Hamsan, M.H. *et al.*, Characteristics of EDLC device fabricated from plasticized chitosan: MgCl2 based polymer electrolyte. *J. Mater. Res. Technol.*, 9, 5, 10635–10646, 2020.

38. Aziz, S.B., Dannoun, E.M.A., Hamsan, M.H., Abdulwahid, R.T., Mishra, K., Nofal, M.M., Kadir, M.F.Z., Improving EDLC device performance constructed from plasticized magnesium ion conducting chitosan based polymer electrolytes via metal complex dispersion. *Membranes*, 11, 289, 2021.
39. Winie, T., Ramesh, S., Arof, A.K., Studies on the structure and transport properties of hexanoyl chitosan-based polymer electrolytes. *Physica B*, 404, 4308–4311, 2009.
40. Aziz, N.A., Majid, S.R., Arof, A.K., Synthesis and characterizations of phthaloyl chitosan-based polymer electrolytes. *J. Non-Cryst. Solids*, 358, 1581–1590, 2012.
41. Asnawi, A.S.F.M. *et al.*, Glycerolized Li+ Ion conducting chitosan-based polymer electrolyte for energy storage EDLC device applications with relatively high energy density. *Polymers*, 12, 1433, 2020.
42. Aziz, S.B. *et al.*, Impedance, circuit simulation, transport properties and energy storage behavior of plasticized lithium ion conducting chitosan based polymer electrolytes. *Polym. Test.*, 101, 107286, 2021.
43. Ou, Y. *et al.*, Chitosan-based composite membranes containing chitosan-coated carbon nanotubes for polymer electrolyte membranes. *Polym. Adv. Technol.*, 29, 1, 612–622, 2018.
44. Kalaiselvimary, J., Sundararajan, M., Prabhu, M.R., Preparation and characterization of chitosan-based nanocomposite hybrid polymer electrolyte membranes for fuel cell application. *Ionics*, 24, 3555–3571, 2018.
45. Monisha, S. *et al.*, Investigation of bio polymer electrolyte based on cellulose acetate-ammonium nitrate for potential use in electrochemical devices. *Carbohydr. Polym.*, 157, 38–47, 2017.
46. Zhao, L. *et al.*, High-strength and flexible cellulose/PEG based gel polymer electrolyte with high performance for lithium ion batteries. *J. Membr. Sci.*, 593, 117428, 2020.
47. Samsudin, A.S. and Saadiah, M.A., Ionic conduction study of enhanced amorphous solid bio-polymer electrolytes based carboxymethyl cellulose doped NH4Br. *J. Non-Cryst. Solids*, 497, 19–29, 2018.
48. Zhao, D. *et al.*, High performance, flexible, solid-state supercapacitors based on a renewable and biodegradable mesoporous cellulose membrane. *Adv. Energy Mater.*, 7, 18, 1700739, 2017.
49. Xun, Z. *et al.*, Construction of polymer electrolyte based on soybean protein isolate and hydroxyethyl cellulose for a flexible solid-state supercapacitor. *Polymers*, 11, 1895, 2019.
50. Kim, S.K. *et al.*, Recyclable high-performance polymer electrolyte based on modified methyl cellulose-lithium trifluoromethanesulfonate salt composite for sustainable energy systems. *ChemSusChem*, 10, 2756, 2019.
51. Yu, N. *et al.*, All-solid-state supercapacitors using a highly conductive neutral gum electrolyte. *RSC Adv.*, 9, 8169, 2019.
52. Fuzlin, A.F. *et al.*, Study on the effect of lithium nitrate in ionic conduction properties based alginate biopolymer electrolytes. *Mater. Res. Express*, 7, 015902, 2020.
53. Rasali, N.M.J., Nagao, Y., Samsudin, A.S., Enhancement on amorphous phase in solid biopolymer electrolyte based alginate doped NH4NO3. *Ionics*, 25, 641–654, 2019.
54. Li, Y. *et al.*, Facile preparation of stretchable and self-healable conductive hydrogels based on sodium alginate/polypyrrole nanofibers for use in flexible supercapacitor and strain sensors. *Int. J. Biol. Macromol.*, 172, 41–54, 2021.
55. Hu, O. *et al.*, A facile preparation method for anti-freezing, tough, transparent, conductive and thermoplastic poly(vinyl alcohol)/sodium alginate/glycerol organohydrogel electrolyte. *Int. J. Biol. Macromol.*, 164, 2512–2523, 2020.
56. Willfahrt, A. *et al.*, Printable acid-modifed corn starch as non-toxic, disposable hydrogel-polymer electrolyte in supercapacitors. *Appl. Phys. A*, 125, 474, 2019.
57. Ong, A.C.W. *et al.*, Nanocomposite polymer electrolytes comprising starch-lithium acetate and titania for all-solid-state supercapacitor. *Ionics*, 27, 853–865, 2021.

58. Awang, F.F., Hassan, M.F., Chan, K.S., Kamarudin, K.H., The ion transport study in corn starch-$NaHSO_3$ based polymer electrolytes. *Sci. Res. J.*, 18, 2, 17–36, 2021.
59. Selvanathan, V., Ruslan, M.H., Aminuzzaman, M., Muhammad, G., Amin, N., Sopian, K., Akhtaruzzaman, M., Resorcinol-formaldehyde (RF) as a novel plasticizer for starch-based solid biopolymer electrolyte. *Polymers*, 12, 9, 2170, 2020.
60. Awang, F.F., Kamarudin, K.H., Hassan, M.F., Employing an electrochemical impedance spectroscopy technique to estimate the ion transport parameters in corn starch based solid polymer electrolyte. *Int. J. Adv. Res. Eng. Innovation*, 2, 3, 78–88, 2020.
61. Sravanthi, K., Sundari, G.S., Erothu, H., Development of bio-degradable based polymer electrolytes for EDLC application. *Optik*, 241, 166229, 2021.
62. Tiwari, T., Kumar, M., Srivastava, N., Srivastava, P.C., Electrical transport study of potato starch-based electrolyte system-II. *Mater. Sci. Eng.: B*, 182, 6–13, 2014.
63. Kumar, M., Tiwari, T., Srivastava, N., Electrical transport behaviour of bio-polymer electrolyte system: Potato starch+ ammonium iodide. *Carbohydr. Polym.*, 88, 1, 54–60, 2012.
64. Selvalakshmi, S., Vijaya, N., Selvasekarapandian, S., Premalatha, M., Biopolymer agar-agar doped with NH_4SCN as solid polymer electrolyte for electrochemical cell application. *J. Appl. Polym. Sci.*, 15, 134, 2017.
65. Selvalakshmi, S., Mathavan, T., Selvasekarapandian, S., Premalatha, M., Study on NH4I composition effect in agar–agar-based biopolymer electrolyte. *Ionics*, 23, 10, 2791–2797, 2017.
66. Selvalakshmi, S., Mathavan, T., Selvasekarapandian, S., Premalatha, M., Effect of ethylene carbonate plasticizer on agar-agar: NH4Br-based solid polymer electrolytes. *Ionics*, 24, 8, 2209–2217, 2018.
67. Cano, E., Crespo, A., Lafuente, D., Ramirez Barat, B., A novel gel polymer electrolyte cell for *in-situ* application of corrosion electrochemical techniques. *Electrochem. Commun.*, 41, 16–19, 2014.
68. Ramírez Barat, B. and Cano, E., The use of agar gelled electrolyte for *in situ* electrochemical measurements on metallic cultural heritage. *Electrochim. Acta*, 182, 751–762, 2015.
69. Fuzlin, A.F., Nagao, Y., Misnon, I.I., Samsudin, A.S., Studies on structural and ionic transport in biopolymer electrolytes based on alginate-LiBr. *Ionics*, 26, 4, 1923–1938, 2020.
70. Jiang, X., Xiang, N., Zhang, H., Sun, Y., Lin, Z., Hou, L., Preparation and characterization of poly (vinyl alcohol)/sodium alginate hydrogel with high toughness and electric conductivity. *Carbohydr. Polym.*, 186, 377–383, 2018.
71. Lu, Y., Zhu, T., Xu, N., Huang, K., A semisolid electrolyte for flexible Zn-ion batteries. *ACS Appl. Energy Mater.*, 2, 9, 6904–6910, 2019.
72. Huang, J., Chi, X., Yang, J., Liu, Y., An ultrastable Na–Zn solid-state hybrid battery enabled by a robust dual-cross-linked polymer electrolyte. *ACS Appl. Mater. Interfaces*, 12, 15, 17583–17591, 2020.
73. Tao, F., Qin, L., Wang, Z., Pan, Q., Self-healable and cold-resistant supercapacitor based on a multifunctional hydrogel electrolyte. *ACS Appl. Mater. Interfaces*, 9, 18, 15541–15548, 2017.
74. Cheng, T. *et al.*, Conductive hydrogel-based electrodes and electrolytes for stretchable and self-healable supercapacitors. *Adv. Funct. Mater.*, 31, 24, 2101303, 2021.
75. Zhang, J. *et al.*, "Water-in-salt" polymer electrolyte for Li-ion batteries. *Energy Environ. Sci.*, 13, 9, 2878–2887, 2020.
76. Yadav, M., Kumar, M., Srivastava, N., Supercapacitive performance analysis of low cost and environment friendly potato starch based electrolyte system with anodized aluminium and teflon coated carbon cloth as electrode. *Electrochim. Acta*, 283, 1551–1559, 2018.
77. Tiwari, T., Chauhan, J.K., Yadav, M. *et al.*, Arrowroot + NaI: A low-cost, fast ion conducting eco-friendly polymer electrolyte system. *Ionics*, 23, 2809–2815, 2017.
78. Chauhan, J.K., Kumar, M., Yadav, M. *et al.*, Effect of $NaClO_4$ concentration on electrolytic behaviour of corn starch film for supercapacitor application. *Ionics*, 23, 2943–2949, 2017.

79. Yadav, M., Kumar, M., Tiwari, T. *et al.*, Wheat starch + NaI: A high conducting environment friendly electrolyte system for energy devices. *Ionics*, 23, 2871–2880, 2017.
80. Yadav, M., Nautiyal, G., Verma, A. *et al.*, Electrochemical characterization of NaClO4–mixed rice starch as a cost-effective and environment-friendly electrolyte. *Ionics*, 25, 2693–2700, 2019.
81. T.J. Calver, S.E. Male, P.J. Mitchell, I. Whyte, Carbon based electrodes. U.S. Patent 6,511,767, issued January 28, 2003.
82. Lee, S.W., Gallant, B.M., Byon, H.R., Hammond, P.T., Shao-Horn, Y., Nanostructured carbon-based electrodes: Bridging the gap between thin-film lithium-ion batteries and electrochemical capacitors. *Energy Environ. Sci.*, 4, 6, 1972–1985, 2011.
83. Lota, G., Fic, K., Frackowiak, E., Carbon nanotubes and their composites in electrochemical applications. *Energy Environ. Sci.*, 4, 5, 1592–1605, 2011.
84. Zhang, J. and Zhao, X.S., On the configuration of supercapacitors for maximizing electrochemical performance. *ChemSusChem*, 5, 5, 818–841, 2012.
85. Rai, P., Pandey, S., Arabale, G., Nikolaev, P., Arepalli, S., Nanotechnology (IEEENANO), pp. 76–79, 2011.
86. Niu, C., Sichel, E.K., Hoch, R., Moy, D., Tennent, H., High power electrochemical capacitors based on carbon nanotube electrodes. *Appl. Phys. Lett.*, 70, 11, 1480–1482, 1997.
87. Hoffman, P. and Ferrante, J., Energy storage capacitors of high energy density. *IEEE Trans. Nucl. Sci.*, 18, 4, 235–239, 1971.
88. Che, G., Lakshmi, B.B., Martin, C.R., Fisher, E.R., Ruoff, R.S., Chemical vapor deposition based synthesis of carbon nanotubes and nanofibers using a template method. *Chem. Mater.*, 10, 1, 260–267, 1998.
89. Lodewyckx, P., Rufford, T.E., Zhu, J., Hulicova-Jurcakova, D., Adsorption of volatile organic vapours, in: *Green Carbon Materials: Advances and Applications*, p. 235, 2014.
90. Hoefer, M. and Bandaru, P.R., Determination and enhancement of the capacitance contributions in carbon nanotube based electrode systems. *Appl. Phys. Lett.*, 95, 18, 183108, 2009.
91. Gui, Z., Zhu, H., Gillette, E., Han, X., Rubloff, G.W., Hu, L., Lee, S.B., Natural cellulose fiber as substrate for supercapacitor. *ACS Nano*, 7, 7, 6037–6046, 2013.
92. Olsson, H., Carlsson, D.O., Nyström, G., Sjödin, M., Nyholm, L., Strømme, M., Influence of the cellulose substrate on the electrochemical properties of paper-based polypyrrole electrode materials. *J. Mater. Sci.*, 47, 13, 5317–5325, 2012.
93. Zhang, L.L., Li, H.H., Fan, C.Y., Wang, K., Wu, X.L., Sun, H.Z., Zhang, J.P., A vertical and cross-linked Ni(OH)2 network on cellulose-fiber covered with graphene as a binder-free electrode for advanced asymmetric supercapacitors. *J. Mater. Chem. A*, 3, 37, 19077–19084, 2015.
94. Yang, C. *et al.*, Flexible highly specific capacitance aerogel electrodes based on cellulose nanofibers, carbon nanotubes and polyaniline. *Electrochim. Acta*, 182, 264–271, 2015.
95. Zhang, Y., Zuo, L., Zhang, L., Yan, J., Lu, H., Fan, W., Liu, T., Immobilization of NiS nanoparticles on N-doped carbon fiber aerogels as advanced electrode materials for supercapacitors. *Nano Res.*, 9, 9, 2747–2759, 2016.
96. Feng, J.X., Li, Q., Lu, X.F., Tong, Y.X., Li, G.R., Flexible symmetrical planar supercapacitors based on multi-layered MnO2/Ni/graphite/paper electrodes with high-efficient electrochemical energy storage. *J. Mater. Chem. A*, 2, 9, 2985–2992, 2014.
97. Kang, Y.J., Kim, B., Chung, H., Kim, W., Fabrication and characterization of flexible and high capacitance supercapacitors based on MnO2/CNT/papers. *Synth. Met.*, 160, 23-24, 2510–2514, 2010.
98. Liu, C., Cai, Z., Zhao, Y., Zhao, H., Ge, F., Potentiostatically synthesized flexible polypyrrole/multi-wall carbon nanotube/cotton fabric electrodes for supercapacitors. *Cellulose*, 23, 1, 637–648, 2016.

99. Wang, F., Kim, H.J., Park, S., Kee, C.D., Kim, S.J., Oh, I.K., Bendable and flexible supercapacitor based on polypyrrole-coated bacterial cellulose core-shell composite network. *Compos. Sci. Technol.*, 128, 33–40, 2016.
100. Yuan, L., Xiao, X., Ding, T., Zhong, J., Zhang, X., Shen, Y., Hu, B., Huang, Y., Zhou, J., Wang, Z.L., Paper-based supercapacitors for self-powered nanosystems. *Angew. Chem. Int. Ed.*, 51, 20, 4934–4938, 2012.
101. Ge, D., Yang, L., Fan, L., Zhang, C., Xiao, X., Gogotsi, Y., Yang, S., Foldable supercapacitors from triple networks of macroporous cellulose fibers, single-walled carbon nanotubes and polyaniline nanoribbons. *Nano Energy*, 11, 568–578, 2015.
102. Lai, F., Miao, Y.E., Zuo, L., Lu, H., Huang, Y., Liu, T., Biomass-derived nitrogen-doped carbon nanofiber network: A facile template for decoration of ultrathin nickel-cobalt layered double hydroxide nanosheets as high-performance asymmetric supercapacitor electrode. *Small*, 12, 24, 3235–3244, 2016.
103. Gao, K., Shao, Z., Li, J., Wang, X., Peng, X., Wang, W., Wang, F., Cellulose nanofiber–graphene all solid-state flexible supercapacitors. *J. Mater. Chem. A*, 1, 1, 63–67, 2013.
104. Zheng, Q., Cai, Z., Ma, Z., Gong, S., Cellulose nanofibril/reduced graphene oxide/carbon nanotube hybrid aerogels for highly flexible and all-solid-state supercapacitors. *ACS Appl. Mater. Interfaces*, 7, 5, 3263–3271, 2015.
105. Huang, Q., Yang, Y., Chen, R., Wang, X., High performance fully paper-based all-solid-state supercapacitor fabricated by a papermaking process with silver nanoparticles and reduced graphene oxide-modified pulp fibers. *EcoMat*, 3, 1, 12076, 2021.
106. Lu, N. *et al.*, Rational design of antifreezing organo hydrogel electrolytes for flexible supercapacitors. *ACS Appl. Energy Mater.*, 3, 2, 1944–1951, 2020.
107. Wang, D.-C. *et al.*, Supramolecular self-assembly of 3D conductive cellulose nanofiber aerogels for flexible supercapacitors and ultrasensitive sensors. *ACS Appl. Mater. Interfaces*, 11, 27, 24435–24446, 2019.
108. Liu, S., He, K., Wu, X., Luo, X., Li, B., Surface modification of cellulose scaffold with polypyrrole for the fabrication of flexible supercapacitor electrode with enhanced capacitance. *RSC Adv.*, 5, 106, 87266–87276, 2015.
109. Niu, Q., Guo, Y., Gao, K., Shao, Z., Polypyrrole/cellulose nanofiber aerogel as a supercapacitor electrode material. *RSC Adv.*, 6, 110, 109143–109149, 2016.
110. Xia, L., Li, X., Wu, Y., Hu, S., Liao, Y., Huang, L., Qing, Y., Lu, X., Electrodes derived from carbon fiber-reinforced cellulose nanofiber/multiwalled carbon nanotube hybrid aerogels for high-energy flexible asymmetric supercapacitors. *Chem. Eng. J.*, 379, 122325, 2020.
111. Jia, L., Shi, Y., Zhang, Q., Xu, X., Green synthesis of ultrafine methyl-cellulose-derived porous carbon/MnO2 nanowires for asymmetric supercapacitors and flexible pattern stamping. *Appl. Surf. Sci.*, 462, 923–931, 2018.
112. Yang, Q., Yang, J., Gao, Z., Li, B., Xiong, C., Carbonized cellulose nanofibril/graphene oxide composite aerogels for high-performance supercapacitors. *ACS Appl. Energy Mater.*, 3, 1, 1145–1151, 2019.
113. Antony, R., Arun, T., Manickam, S.T.D., A review on applications of chitosan-based Schiff bases. *Int. J. Biol. Macromol.*, 129, 615–633, 2019.
114. Suginta, W., Khunkaewla, P., Schulte, A., Electrochemical biosensor applications of polysaccharides chitin and chitosan. *Chem. Rev.*, 113, 7, 5458–5479, 2013.
115. Suneetha, R.B., Selvi, P., Vedhi, C., Synthesis, structural and electrochemical characterization of Zn doped iron oxide/grapheneoxide/chitosan nanocomposite for supercapacitor application. *Vacuum*, 164, 396–404, 2019.
116. Al-Farraj, E.S. *et al.*, Fabrication of hybrid nanocomposite derived from chitosan as efficient electrode materials for supercapacitor. *Int. J. Biol. Macromol.*, 120, 2271–2278, 2018.

117. Torvi, A., Naik, S., Kariduraganavar, M., Development of supercapacitor systems based on binary and ternary nanocomposites using chitosan, graphene and polyaniline. *Chem. Data Collect.*, 17, 459–471, 2018.
118. Anandhavelu, S., Dhanasekaran, V., Sethuraman, V., Park, H.J., Chitin and chitosan based hybrid nanocomposites for super capacitor applications. *J. Nanosci. Nanotechnol.*, 17, 2, 1321–1328, 2017.
119. Qian, L., Fan, Y., Song, H., Zhou, X., Xiong, Y., Poly (ionic liquid)/carboxymethyl chitosan complex-derived nitrogen and sulfur codoped porous carbon for high-performance supercapacitors. *Ionics*, 25, 10, 4915–4924, 2019.
120. Gopalakrishnan, A., Vishnu, N., Badhulika, S., Cuprous oxide nanocubes decorated reduced graphene oxide nanosheets embedded in chitosan matrix: A versatile electrode material for stable supercapacitor and sensing applications. *J. Electroanal. Chem.*, 834, 187–195, 2019.
121. Pandiselvi, K. and Thambidurai, S., Chitosan-ZnO/polyaniline ternary nanocomposite for high-performance supercapacitor. *Ionics*, 20, 4, 551–561, 2014.
122. Teimuri-Mofrad, R., Hadi, R., Abbasi, H., Fadakar Bajeh Baj, R., Synthesis, characterization and electrochemical study of carbon nanotube/chitosan-ferrocene nanocomposite electrode as supercapacitor material. *J. Electron. Mater.*, 48, 7, 4573–4581, 2019.
123. Duan, B., Gao, X., Yao, X., Fang, Y., Huang, L., Zhou, J., Zhang, L., Unique elastic N-doped carbon nanofibrous microspheres with hierarchical porosity derived from renewable chitin for high rate supercapacitors. *Nano Energy*, 27, 482–491, 2016.
124. Zhu, L., Shen, F., Smith Jr., R.L., Qi, X., High-performance supercapacitor electrode materials from chitosan via hydrothermal carbonization and potassium hydroxide activation. *Energy Technol.*, 5, 3, 452–460, 2017.
125. Zhao, X., Wang, S., Wu, Q., Nitrogen and phosphorus dual-doped hierarchical porous carbon with excellent supercapacitance performance. *Electrochim. Acta*, 247, 1140–1146, 2017.
126. Wu, Q., Gao, M., Cao, S., Hu, J., Huang, L., Yu, S., Ragauskas, A.J., Chitosan-based layered carbon materials prepared via ionic-liquid-assisted hydrothermal carbonization and their performance study. *J. Taiwan Inst. Chem. Eng.*, 101, 231–243, 2019.
127. Zhang, Y. *et al.*, Synergistically reinforced capacitive performance from a hierarchically structured composite of polyaniline and cellulose-derived highly porous carbons. *Mater. Lett.*, 244, 62–65, 2019.
128. Chen, T. and Liu, Z., Starch-assistant synthesis of Ni quantum dots/ultrathin carbon nanosheet hybrids for high performance supercapacitor. *Mater. Lett.*, 236, 248–251, 2019.
129. Wu, W., Li, Y., Yang, L., Ma, Y., Pan, D., Li, Y., A facile one-pot preparation of dialdehyde starch reduced graphene oxide/polyaniline composite for supercapacitors. *Electrochim. Acta*, 139, 117–126, 2014.
130. Cao, J., Zhu, C., Aoki, Y., Habazaki, H., Starch-derived hierarchical porous carbon with controlled porosity for high performance supercapacitors. *ACS Sustainable Chem. Eng.*, 6, 6, 7292–7303, 2018.
131. Liu, M.C. *et al.*, Three-dimensional interconnected reticular porous carbon from corn starch by a sample sol–gel method toward high-performance supercapacitors with aqueous and ionic liquid electrolytes. *ACS Sustainable Chem. Eng.*, 7, 22, 18690–18699, 2019.
132. Zhong, Y., Shi, T., Huang, Y., Cheng, S., Liao, G., Tang, Z., One-step synthesis of porous carbon derived from starch for all-carbon binder-free high-rate supercapacitor. *Electrochim. Acta*, 269, 676–685, 2018.
133. Guo, J., Guo, H., Zhang, L., Yang, B., Cui, J., Hierarchically porous carbon as a high-rate and long-life electrode material for high-performance supercapacitors. *ChemElectroChem*, 5, 5, 770–777, 2018.

134. Samdani, K.J., Kim, S.H., Park, J.H., Hong, S.H., Lee, K.T., Morphology-controlled synthesis of Co3O4 composites with bio-inspired carbons as high-performance supercapacitor electrode materials. *J. Ind. Eng. Chem.*, 74, 96–102, 2019.
135. Yu, P., Wang, Q., Zheng, L., Jiang, Y., Construction of ultrathin nitrogen-doped porous carbon nanospheres coated with polyaniline nanorods for asymmetric supercapacitors. *Front. Chem.*, 7, 455, 2019.
136. Kasturi, P.R., Ramasamy, H., Meyrick, D., Lee, Y.S., Selvan, R.K., Preparation of starch-based porous carbon electrode and biopolymer electrolyte for all solid-state electric double layer capacitor. *J. Colloid Interface Sci.*, 554, 142–156, 2019.
137. Ruibin, Q. *et al.*, Monodisperse carbon microspheres derived from potato starch for asymmetric supercapacitors. *Electrochim. Acta*, 167, 303–310, 2015.
138. Pang, L., Zou, B., Han, X., Cao, L., Wang, W., Guo, Y., One-step synthesis of high-performance porous carbon from corn starch for supercapacitor. *Mater. Lett.*, 184, 88–91, 2016.
139. Kurt, A., Cengiz, A., Kahyaoglu, T., The effect of gum tragacanth on the rheological properties of salep based ice cream mix. *Carbohydr. Polym.*, 143, 116–123, 2016.
140. Di Palma, T.M., Migliardini, F., Caputo, D., Corbo, P., Xanthan and κ-carrageenan based alkaline hydrogels as electrolytes for Al/air batteries. *Carbohydr. Polym.*, 157, 122–127, 2017.
141. Liu, X., Yu, C., Chen, Z., Xu, F., Liao, W., Zhong, W., Biomass peach gum-derived heteroatom-doped porous carbon via *in situ* molten salt activation for high-performance supercapacitors. *Energy Fuels*, 35, 23, 19801–19810, 2021.
142. Fan, Y., Liu, P., Zhu, B., Chen, S., Yao, K., Han, R., Microporous carbon derived from acacia gum with tuned porosity for high-performance electrochemical capacitors. *Int. J. Hydrogen Energy*, 40, 18, 6188–6196, 2015.
143. Ma, Y., Chen, M., Zheng, X., Yu, D., Dong, X., Synergetic effect of swelling and chemical blowing to develop peach gum derived nitrogen-doped porous carbon nanosheets for symmetric supercapacitors. *J. Taiwan Inst. Chem. Eng.*, 101, 24–30, 2019.
144. Kang, C.S. *et al.*, Hybridized double-walled carbon nanotubes and activated carbon as free-standing electrode for flexible supercapacitor applications. *Carbon Lett.*, 30, 5, 527–534, 2020.
145. Ullah, R., Khan, N., Khattak, R., Khan, M., Khan, M.S., Ali, O.M., Preparation of electrochemical supercapacitor based on polypyrrole/gum arabic composites. *Polymers*, 14, 2, 242, 2022.
146. Lin, Y., Chen, Z., Yu, C., Zhong, W., Heteroatom-doped sheet-like and hierarchical porous carbon based on natural biomass small molecule peach gum for high-performance supercapacitors. *ACS Sustainable Chem. Eng.*, 7, 3, 3389–3403, 2019.
147. Bayatpour, S., Afsharpour, M., Dini, Z., Naderi, H.R., High performance supercapacitor electrodes using functionalized CNTs/MoO3 with natural polysaccharide binders. *J. Mater. Sci.: Mater. Electron.*, 31, 8, 6150–6159, 2020.
148. Gacesa, P., Alginates. *Carbohydr. Polym.*, 8, 3, 161–182, 1988.
149. Wang, B. *et al.*, Alginate-based hierarchical porous carbon aerogel for high-performance supercapacitors. *J. Alloys Compd.*, 749, 517–522, 2018.
150. Chen, W., Luo, M., Yang, K., Zhou, X., Simple pyrolysis of alginate-based hydrogel cross-linked by bivalent ions into highly porous carbons for energy storage. *Int. J. Biol. Macromol.*, 158, 265–274, 2020.
151. Hu, W. *et al.*, Sodium alginate assisted preparation of oxygen doped microporous carbons with enhanced electrochemical energy storage and hydrogen uptake. *Int. J. Hydrogen Energy*, 46, 896–905, 2021.
152. Zhao, Y., Wei, M., Zhu, Z., Zhang, J., Xiao, L., Hou, L., Facile preparation of NO codoped hierarchically porous carbon from alginate particles for high performance supercapacitor. *J. Colloid Interface Sci.*, 563, 414–425, 2020.

153. Xia, L. *et al.*, Hierarchical macro-/meso-/microporous oxygen-doped carbon derived from sodium alginate: A cost-effective biomass material for binder-free supercapacitors. *Mater. Des.*, 182, 108048, 2019.
154. You, Y. *et al.*, Binder-free CuS/ZnS/sodium alginate/rGO nanocomposite hydrogel electrodes for enhanced performance supercapacitors. *Int. J. Biol. Macromol.*, 162, 310–319, 2020.
155. Li, Y., Zhao, X., Xu, Q., Zhang, Q., Chen, D., Facile preparation and enhanced capacitance of the polyaniline/sodium alginate nanofiber network for supercapacitors. *Langmuir*, 27, 10, 6458–6463, 2011.
156. Zhai, Z., Ren, B., Xu, Y., Wang, S., Zhang, L., Liu, Z., Green and facile fabrication of Cu-doped carbon aerogels from sodium alginate for supercapacitors. *Org. Electron.*, 70, 246–251, 2019.
157. Qin, L. *et al.*, Defect-rich N-doped porous carbon derived from alginate by HNO3 etching combined with a hard template method for high-performance supercapacitors. *Mater. Chem. Phys.*, 260, 124121, 2021.
158. BeMiller, J.N., An introduction to pectins: Structure and properties, 1986.
159. Amarnath, C.A., Venkatesan, N., Doble, M., Sawant, S.N., Water dispersible Ag@ polyaniline-pectin as supercapacitor electrode for physiological environment. *J. Mater. Chem. B*, 2, 31, 5012–5019, 2014.
160. Fan, Y. *et al.*, Bi-functional porous carbon spheres derived from pectin as electrode material for supercapacitors and support material for Pt nanowires towards electrocatalytic methanol and ethanol oxidation. *Electrochim. Acta*, 163, 140–148, 2015.
161. Zhou, Y., Ren, X., Du, Y., Jiang, Y., Wan, J., Ma, F., *In-situ* template cooperated with urea to construct pectin-derived hierarchical porous carbon with optimized pore structure for supercapacitor. *Electrochim. Acta*, 355, 136801, 2020.
162. Xu, L., Cui, L., Jia, M., Li, Y., Gao, J., Jin, X., Self-assembly of flexible graphene hydrogel electrode based on crosslinked pectin-cations. *Carbohydr. Polym.*, 195, 593–600, 2018.

17

Polymer Inks for Printable Supercapacitors

Yurui Liu[1], Yijie Zhou[1] and Yanfei Xu[1,2]*

[1]Mechanical & Industrial Engineering Department, University of Massachusetts, Amherst, MA, USA

[2]Chemical Engineering Department, University of Massachusetts, Amherst, MA, USA

Abstract

Printable polymer-based supercapacitors have raised attention, thanks to their unique characteristics such as lightweight, easy processing integration, high scalability, high discharge and charge rate, high energy, and high power density, to name a few. In this chapter, we review recent progress in the developments of polymer-based inks for printable supercapacitor electrodes. We highlight the design principles of polymer ink formulations. We show several selected examples for printed supercapacitors with high performance. We believe that functional polymer inks will play key roles in achieving next-generation printed supercapacitor with outstanding performance.

Keywords: Printable supercapacitors, polymer inks, electrical double-layer capacitors, pseudocapacitors, printing technology

17.1 Introduction

The contemporary development trends towards printable modern devices have raised the need for printable and rechargeable power sources [1]. The next-generation of supercapacitors with thin, lightweight, cheap, and flexible characteristics are highly desired for wearable electronics [2, 3]. Moreover, the "printed electronics", such as printable photovoltaic devices, [4] transistors, [5] and organic light-emitting diodes [6] are expected to revolutionize the way of manufacturing wearable electronics in the future. These applications require compatible energy sources. Fully printable charge storage devices are needed to be fully integrated into printed electronics in the process of device manufacturing [7–11].

Supercapacitors possess properties such as high power density, high energy density, long cycle life, high discharge and charge rate, low cost, and easy maintenance will provide new possibilities for energy storage devices [7, 12, 13]. In comparison to other rechargeable batteries, the life cycle and energy per unit volume of supercapacitors are increased by a factor of 500,000 and 100, respectively [14]. Supercapacitors also show variable capacitance performance with lower equivalent series resistance [14, 15]. In this book chapter, we focus on functional polymer inks for printable supercapacitors with high performance. We first discuss the current understanding of the charge storage mechanisms of supercapacitors. Then, we focus on polymer inks for printable supercapacitors. We use printable

**Corresponding author*: yanfeixu@umass.edu

Inamuddin, Tariq Altalhi and Sayed Mohammed Adnan (eds.) Sustainable Materials for Electrochemical Capacitors, (415–426) © 2023 Scrivener Publishing LLC

polypyrrole-based supercapacitors as illustrative examples of innovative electrochemical energy storage devices [16]. We discuss future research directions related to polymer inks for supercapacitors. We conclude future key applications of printable supercapacitors.

The basic principles of supercapacitor energy storage mechanisms are the double layer capacitance formed by charge separation and/or reversible Faradaic redox reactions at or near interfaces between electrodes and electrolytes. According to their charge storage mechanisms, supercapacitors can be classified into three types: pseudocapacitors, electrical double-layer capacitors, and hybrid of both (Figure 17.1) [17–20].

Electrical double-layer capacitors utilize charges accumulated on the interfacial electrolyte/electrode surfaces that largely include carbon-based materials (e.g., carbon nanotubes, templated carbons, and graphene) with large specific surface areas [21]. Pseudocapacitors utilize metal oxides or conducting polymers as electrode materials to store charges based on Faradaic redox reactions. The pseudocapacitive materials can store charges within a few minutes which is much faster than battery materials [22]. Numerous polymers with good electrical conductivities such as polyaniline, poly(3,4-ethylenedioxythiophene), and polypyrrole [16, 23, 24] have been investigated to show higher capacitances than those of electrode materials used for electrical double-layer capacitors. But most of pseudocapacitors based on pure polymers electrode materials do not possess satisfied cyclic stability (Figure 17.2) [21].

The charge accumulation at interfaces between electrode and electrolyte in electrical double-layer capacitors can be calculated by equation 17.1 [21, 25].

$$C = \frac{\varepsilon}{4\pi\delta} \int dS \qquad (17.1)$$

where C is the capacitance, δ represents the distance between ion center and electrolyte interface, ε represents the electrolyte dielectric constant, and S is surface area for electrode interface. Accordingly, equation 17.2 expresses the calculation of specific capacitance.

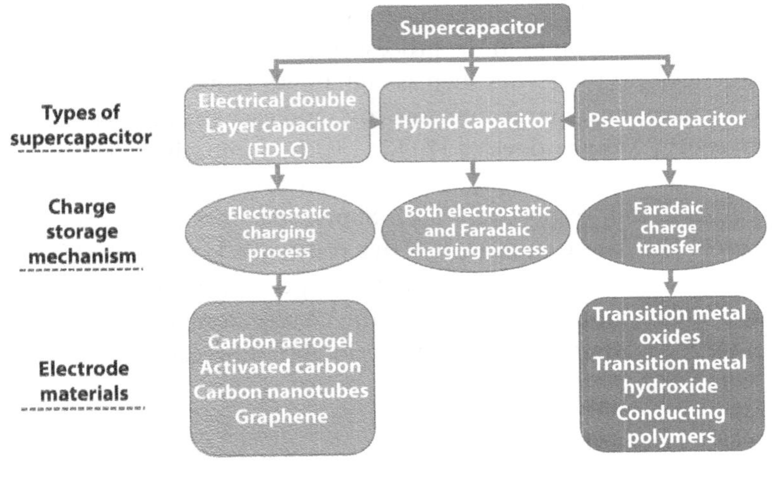

Figure 17.1 Classification of supercapacitors.

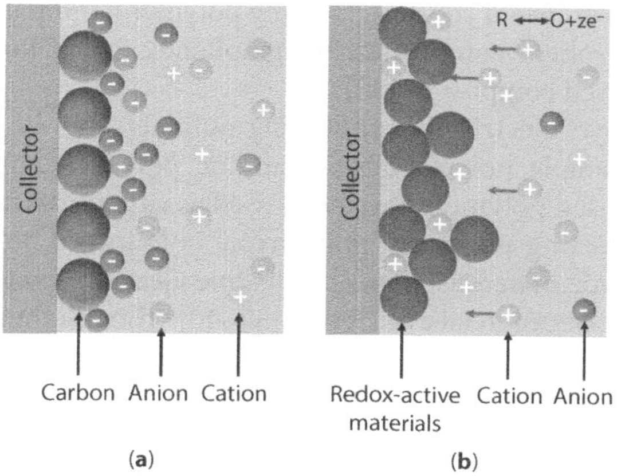

Figure 17.2 Illustration of the charge-storage mechanisms in (a) electrical double-layer capacitors (take carbon-based electrode as an example) and (b) pseudocapacitors [22].

$$C_m = k\frac{\int i}{ms} \tag{17.2}$$

where C_m represents the specific capacitance, m is electrode mass, k is a constant, S is the scan rate in the cyclic voltammetry experiment, and $\int i$ is the integrated area of the cyclic voltammetry curve [21, 26–28].

Besides, the calculation of power density and energy density can be based on the cyclic voltammetry curve and charge-discharge curve using equations 17.3 and 17.4 [21].

$$E_{cell} = \frac{1}{2}C_{CV/C-DC}V^2 \tag{17.3}$$

where $C_{CV/C\text{-}DC}$ represents the specific capacitance of an electrochemical cell based on charge/discharge measurements and the profile of cyclic voltammetry, V represents an operation potential window.

$$P_{cell} = E_{cell}/\Delta t \tag{17.4}$$

where Δt represents the discharge time.

Electrically conducting polymers are one of the most promising candidates for supercapacitor electrode materials, thanks to their excellent pseudocapacitive behaviors. Some pseudocapacitors have shown disadvantages of low power densities, poor cyclic stability, and low power densities. These disadvantages may be caused by the poor mechanical strength of polymers (weak or brittle) [8]. Furthermore, relatively poor electrical conductivity of polymers becomes another disadvantage that hinders the development of polymer-based pseudocapacitive electrodes. To overcome these disadvantages of pure polymers, graphene, an excellent substrate hosting polymer nanomaterial, [29] can be used to fabricate polymer/graphene composites. The graphene can also form a stable conductive network in polymer-based composites [30].

Like mentioned above, electrically conducting polymers can be used as electrodes in supercapacitors. For example, polypyrrole, polyaniline, poly (3,4-ethylenedioxythiophene), and other thiophene-based polymers have been used as supercapacitor electrodes [30–36]. Among different conducting polymers, polypyrrole is found to be a good candidate for supercapacitors' electrode materials, thanks to its fine electrical conductivity, easy synthesis, good flexibility, and commercial availability [37–40]. Though the specific capacitance of polypyrrole is 620 $F{\cdot}g^{-1}$ theoretically, when the doping level reaches 0.33 [41, 42]. The measured specific capacitance of polypyrrole normally ranges from 100 $F{\cdot}g^{-1}$ to 400 $F{\cdot}g^{-1}$ [41, 42] Such discrepancy of specific capacitance between theoretical value and experimental result is influenced by the electrical conductivity, density, porosity, morphology, and synthesis method of polypyrrole [43]. Increased capacitances of polypyrrole by applying different techniques have been reported. Zhao *et al.* synthesized polypyrrole nanowires using $FeCl_3$ as the oxidant, polypyrrole nanowires-based electrode exhibited a large specific capacitance of 420 $F{\cdot}g^{-1}$ [42]. Yang *et al.* fabricated self-assembled polypyrrole films with the aid of surfactant [44]. The specific capacitance of the polypyrrole reached 368.18 $F{\cdot}g^{-1}$ [44]. Xu *et al.* used *in-situ* polymerization to deposit polypyrrole on cotton fabrics [45]. The fabrics were applied as electrode materials in supercapacitors, which showed a measured energy density of 24.7 $W{\cdot}h{\cdot}kg^{-1}$ and high specific capacitance of 325 $F{\cdot}g^{-1}$ [45]. However, the cyclic stability in supercapacitor electrodes is reduced to 63% of the original value that needs efforts to address the stability issues [45].

To address the poor cyclic stability of pure conducting polymers during practical usage [32, 46], polymer composites have been fabricated by introducing carbon materials, metal oxides, and metal sulfides into electrical conducting polymers. Such composites have shown improved electrochemical properties (e.g., specific capacitance and cyclic stability) and increased mechanical strength of polypyrrole [32, 47–50]. Moreover, the synergistic effects of the pseudocapacitance of polypyrrole and the electric double layer capacitance of graphene-based materials could result in an enhancement of electrode capacitance [47, 51–53]. Regarding the synergistic effects and improvement of electrochemical properties brought by polypyrrole composites, printable polypyrrole-based supercapacitors are developed by different printing techniques.

Printing supercapacitors using the above polymer-based materials could offer simple and economical ways to make energy storage and conversion systems. Innovative polymers for high-performance electrode materials and polymeric ink formulations play key roles in printing high-performance supercapacitors at a large scale and low cost. Different printing platforms require tailored polymer inks with specific properties such as rheology and viscosity [54–56]. For instance, mole ratios of monomers, photo-initiators, and cross-linkers need to be tuned to enable effective photopolymerization during stereolithography processes [57]. Inhibitor concentrations are required to be optimized, in order to limit over-polymerization [57–59]. Glass transition temperatures of thermoplastic filaments are needed to be adjusted in fused deposition modeling-based 3D printing processes [60]. Direct ink writing systems generally generate stress fluids, inks are required to have optimized viscoelastic properties [61, 62]. The screen printing systems require ink stabilities and ink viscosities to be optimized. The viscosity of the inks influenced by the applied shear rate is needed to be tuned, to have scalable, reliable, and high-quality screen-printed polymer thin films [7, 63]. Table 17.1 shows selected polymer inks for printed supercapacitors accompanied by their optimized power density, energy density, and specific capacitance performance.

17.2 Screen Printing

As a well-known scalable fabrication approach, screen printing technique is suited to printing low-priced supercapacitors on various substrates such as flexible plastic, paper, and textiles, etc. Keskinen *et al.* developed asymmetric supercapacitors combining polypyrrole and activated carbon electrodes based on screen printing method. The yielding devices were proved to facilitate a longer lifetime with efficiency up to 91% during 14300 cycles [64]. Liu *et al.* developed silver/polypyrrole nanocomposites as electrode materials. These nanocomposite electrode materials have been integrated into solid-state micro-supercapacitors by a self-templated approach [65]. The supercapacitors integrated based on screen printing technique exhibits excellent capacitive feature, superior energy density, and outstanding recyclability [65]. Liang *et al.* reported a facile single-step approach to developing silver/polypyrrole/graphene composites as screen-printing ink for supercapacitors electrodes, which shows high areal capacitance (294.2 mF·cm^{-2}) and good capacitance retention (80.1% after 5000 cycles) [66]. Liu *et al.* fabricated fully printed asymmetric in-plane supercapacitors using silver/polypyrrole/MnO_2 composites as electrode materials [67]. This device exhibited ultra-high energy density (33.7 μWh·cm^{-2}) and power density (0.38 mW·cm^{-2}) with extraordinary stretchability and mechanical stability. The capacitance retention of yielding devices maintains 90.8% over 5000 cycles and 86.2% under 40% stretching strain [67]. Recently, Wei *et al.* demonstrated multifunctional transient supercapacitors based on Zn/polypyrrole composites fabricated by screen printing. The implantable supercapacitors exhibit an energy density as high as 0.394 mW·h·cm^{-2} [49].

17.3 Inkjet Printing

Inkjet printing offers unique benefits for large-scale supercapacitors fabrication, providing the ability to develop films with desirable patterns on a variety of different substrates with highly controlled geometries and specific locations [63]. Stempien *et al.* presented a promising approach for the development of polypyrrole films on a substrate based on the inkjet printing method [68]. The supercapacitor with polypyrrole-gold-based electrodes exhibited high specific capacitances of 72.3 F·g^{-1} over 2000 cycles. More details related to power density and stability can be found in the reference [68].

17.4 3D Printing

3D printing (as a typical additive manufacturing process), has the ability to convert a virtual three-dimensional model into a physical object through computer-aided design. 3D printing allows objects to be built layer by layer without the need for extra post-processing, such as molding or machining [69]. Additive manufacturing techniques include powder bed fusion, vat photopolymerization (stereolithography), material and binder jetting, extrusion (fused deposition modeling, 3D fiber deposition, 3D dispensing, and 3D plotting) [69].

Table 17.1 Selected polypyrrole-based inks for printable supercapacitor and the yielding devices performances, include cladophora sp. Algae cellulose/polypyrrole (CAC/polypyrrole), [64] silver/polypyrrole (Ag/polypyrrole), [65] graphene/polypyrrole/silver (GE/polypyrrole/Ag), [66] silver/polypyrrole/manganese dioxide (Ag/polypyrrole/MnO_2), [67] and polypyrrole/zinc (Polypyrrole/Zn) [49] inks for screen printing; Polypropylene/polypyrrole/gold (Polypropylene/polypyrrole/Au) [68] ink for inkjet printing; Polypyrrole/reduced graphene oxide (Polypyrrole/rGO), [71] polypyrrole nanotubes, [72] polypyrrole/titanium (Polypyrrole/Ti), [78] polypyrrole-coated graphene aerogel (Polypyrrole/GA), [75] and polypyrrole/black phosphorous (Polypyrrole/BPNS) [74] inks for 3D printing.

Printing methods	Electrode materials	Electrolytes	Specific capacitance	Energy density	Power density	Retention (cycles)	Ref.
Screen printing	CAC/polypyrrole	$NaNO_3$-H_3PO_4	211 F·g	-	-	91% (14300)	[64]
Screen printing	Ag/polypyrrole	PVA-H_2SO_4 gel	47.5 mF·cm^2	4.33 μWh·cm^2	0.6 mW·cm^2	82.6% (10000)	[65]
Screen printing	GE/polypyrrole/Ag	PVA-H_2SO_4 gel	294.2 mF·cm^2	2.81 μWh·cm^2 (1.40 Wh·kg)	1.68 mW·cm^2 835.98 W·kg	80.1% (5000)	[66]
Screen printing	Ag/polypyrrole/ MnO_2	PVA-Na_2SO_4	426.3 mF·cm^2 (at 0.5 mA·cm^2)	33.7 μWh·cm^2	0.38 mW·cm^2	90.8% (5000)	[67]
Screen printing	Polypyrrole/Zn	NaCl/agarose	-	394 μWh·cm^2	3.4 μW·cm^2	-	[49]
Inkjet Printing	Polypropylene/ polypyrrole/Au	PVA-H_2SO_4 gel	72.3 F·g (at 0.6 A·g)	6.12 Wh·kg	139 W·kg	70.8% (500)	[68]
3D printing fused deposition modeling	Polypyrrole/rGO	PVA-KOH gel	98.37 F·g	-	-	80% (1000)	[71]
3D printing	Polypyrrole nanotubes	PVA-H_2SO_4 gel	297 mF·cm^2 (at 1 mA·cm^2)	42 μWh·cm^2	0.5 mW·cm^2	93% (120° bending)	[72]

(Continued)

Table 17.1 Selected polypyrrole-based inks for printable supercapacitor and the yielding devices performances, include cladophora sp. Algae cellulose/polypyrrole (CAC/polypyrrole), [64] silver/polypyrrole (Ag/polypyrrole), [65] graphene/polypyrrole/silver (GE/polypyrrole/Ag), [66] silver/polypyrrole/manganese dioxide (Ag/polypyrrole/MnO_2), [67] and polypyrrole/zinc (Polypyrrole/Zn) [49] inks for screen printing; Polypropylene/polypyrrole/gold (Polypropylene/polypyrrole/Au) [68] ink for inkjet printing; Polypyrrole/reduced graphene oxide (Polypyrrole/rGO), [71] polypyrrole nanotubes, [72] polypyrrole/titanium (Polypyrrole/Ti), [78] polypyrrole-coated graphene aerogel (Polypyrrole/GA), [75] and polypyrrole/black phosphorous (Polypyrrole/BPNS) [74] inks for 3D printing. (*Continued*)

Printing methods	Electrode materials	Electrolytes	Specific capacitance	Energy density	Power density	Retention (cycles)	Ref.
3D printing selective laser melting	Polypyrrole/Ti	PVA-H_3PO_4 gel	162.07 ± 12.4 mF·cm^2	2.98 μWh·cm^2	15.0 kW·m^3	78% (1000)	[75]
3D printing	Polypyrrole/GA	KOH	2 F·cm^2 (at 6 A·g)	110 μWh·cm^2	0.4 mW·cm^2	90% (5000)	[48]
3D printing	Polypyrrole/BPNS	Ionic liquid (EMI-TFSI)	157 F·g (at 0.20 A·g)	6.5 Wh·kg	0.0374 W·kg	87% (10000)	[74]

3D printing is an emerging advanced layer-by-layer additive manufacturing technique for making supercapacitors, [11, 48, 70–75] which facilitates the production of supercapacitors in a scalable and low-cost way [70, 73, 76, 77], It has drawn great interest for supercapacitors applications due to its inherent advantages including controllable 3D structural prototyping ability [71, 75, 77]. Great endeavors have been devoted to the investigation of supercapacitors with superior performance and recyclability. However, current resulting supercapacitors are still suffering from the low active mass loading problem and thus result in low energy capacities.

Foo *et al.* developed 3D printed sputtered Au/polypyrrole-rGO composites as supercapacitors' electrodes [71]. The supercapacitors exhibited excellent capacitive features (98.37 $F{\cdot}g^{-1}$) [71]. The gold thin layer on polypyrrole-rGO electrodes' surfaces can reduce the surface resistivity, thus resulting in the decreased resistance at interfaces between electrolytes and electrolytes. The 3D printing methods are important in manufacturing supercapacitor devices. Gu *et al.* have fabricated polypyrrole-based supercapacitors by 3D printing technology and measured the mass loading as high as 337 $mg{\cdot}cm^{-2}$ [72]. Zhao *et al.* developed 3D interdigitated supercapacitors using a polypyrrole-based electrode [74]. The supercapacitors showed a high specific capacitance around 395 $F{\cdot}g^{-1}$. Xing *et al.* reported 3D printed supercapacitors with polypyrrole/black phosphorous nanocomposites, which show both superior specific capacitance (~ 417 $F{\cdot}g^{-1}$) and better mechanical properties [74].

17.5 Conclusion and Outlook

Developing printable high-performance supercapacitors will accelerate evolving of energy storage and conversion devices. From polymer inks design perspective, ink formulations of advanced materials are key components in the printing process. This is because ink formulation is essential for supercapacitors' performance and the possible printable resolution and geometries. From printing technology perspective, unique printing techniques are regarded as modern fabrication technologies in both academia and industry. Printing supercapacitor devices have shown outstanding charge storage functionalities and excellent mechanical properties. The convergence of advanced polymer inks design, advanced digital simulation, and developments of ink formulations for printing platforms will provide new opportunities for the fabrication of smart devices beyond our current reach.

References

1. Su, M. and Song, Y., Printable smart materials and devices: Strategies and applications. *Chem. Rev.*, 122, 5, 5144, 2022.
2. Rim, Y.S., Bae, S.H., Chen, H., De Marco, N., Yang, Y., Recent progress in materials and devices toward printable and flexible sensors. *Adv. Mater.*, 28, 22, 4415, 2016.
3. Zandi, S., Baharfar, M., Moazzam, P., Razmjou, A., Beheshti, A., Asadnia, M., Advances in printable devices for biomedical applications, in: *Printed and Flexible Sensor Technology*, pp. 4–1 to 4–28, IOP Publishing, Bristol, UK, 2021.
4. Scharber, M.C. and Sariciftci, N.S., Efficiency of bulk-heterojunction organic solar cells. *Prog. Polym. Sci.*, 38, 12, 1929, 2013.

5. Franklin, A.D., Nanomaterials in transistors: From high-performance to thin-film applications. *Science*, 349, 6249, aab2750, 2015.
6. Sekitani, T., Nakajima, H., Maeda, H., Fukushima, T., Aida, T., Hata, K., Someya, T., Stretchable active-matrix organic light-emitting diode display using printable elastic conductors. *Nat. Mater.*, 8, 6, 494, 2009.
7. Xu, Y., Schwab, M.G., Strudwick, A.J., Hennig, I., Feng, X., Wu, Z., Müllen, K., Screen-printable thin film supercapacitor device utilizing graphene/polyaniline inks. *Adv. Energy Mater.*, 3, 8, 1035, 2013.
8. Xu, Y., Hennig, I., Freyberg, D., Strudwick, A.J., Schwab, M.G., Weitz, T., Cha, K.C.P., Inkjet-printed energy storage device using graphene/polyaniline inks. *J. Power Sources*, 248, 483, 2014.
9. Sajedi-Moghaddam, A., Rahmanian, E., Naseri, N., Inkjet-printing technology for supercapacitor application: Current state and perspectives. *Appl. Mater. Interfaces*, 12, 31, 34487, 2020.
10. Choi, K.-H., Ahn, D.B., Lee, S.-Y., Current status and challenges in printed batteries: Toward form factor-free, monolithic integrated power sources. *ACS Energy Lett.*, 3, 1, 220, 2018.
11. Browne, M.P., Redondo, E., Pumera, M., 3D printing for electrochemical energy applications. *Chem. Rev.*, 120, 5, 2783, 2020.
12. Kaempgen, M., Chan, C.K., Ma, J., Cui, Y., Gruner, G., Printable thin film supercapacitors using single-walled carbon nanotubes. *Nano Lett.*, 9, 5, 1872, 2009.
13. Gogotsi, Y. and Simon, P., True performance metrics in electrochemical energy storage. *Science*, 334, 6058, 917, 2011.
14. Tomy, M., Ambika Rajappan, A., VM, V., Thankappan Suryabai, X., Emergence of novel 2D materials for high-performance supercapacitor electrode applications: A brief review. *Energy Fuels*, 35, 24, 19881, 2021.
15. Revankar, S.T., Chemical energy storage, in: *Storage and Hybridization of Nuclear Energy*, pp. 177–227, Academic Press, Cambridge, Massachusetts, 2019.
16. Shimoga, G., Palem, R.R., Choi, D.S., Shin, E.J., Ganesh, P.S., Saratale, G.D., Saratale, R.G., Lee, S.H., Kim, S.Y., Polypyrrole-based metal nanocomposite electrode materials for high-performance supercapacitors. *Metals*, 11, 6, 905, 2021.
17. Najib, S. and Erdem, E., Current progress achieved in novel materials for supercapacitor electrodes: Mini review. *Nanoscale Adv.*, 1, 8, 2817, 2019.
18. Xiong, G., Kundu, A., Fisher, T.S., Thermal management in electrochemical energy storage systems, in: *Thermal Effects in Supercapacitors*, pp. 1–10, Springer, Cham, 2015.
19. Raghavendra, K.V.G., Vinoth, R., Zeb, K., Gopi, C.V.M., Sambasivam, S., Kummara, M.R., Obaidat, I.M., Kim, H.J., An intuitive review of supercapacitors with recent progress and novel device applications. *J. Energy Storage*, 31, 101652, 2020.
20. Prasad, G.G., Shetty, N., Thakur, S., Bommegowda, K.B., Supercapacitor technology and its applications: A review. *IOP Conf. Ser.: Mater. Sci. Eng.*, 561, 1, 012105, 2019.
21. Chee, W.K., Lim, H.N., Zainal, Z., Huang, N.M., Harrison, I., Andou, Y., Flexible graphene-based supercapacitors: A review. *J. Phys. Chem. C*, 120, 8, 4153, 2016.
22. Choi, C., Ashby, D.S., Butts, D.M., DeBlock, R.H., Wei, Q., Lau, J., Dunn, B., Achieving high energy density and high power density with pseudocapacitive materials. *Nat. Rev. Mater.*, 5, 1, 5, 2020.
23. Bonnassieux, Y., Brabec, C.J., Cao, Y., Carmichael, T.B., Chabinyc, M.L., Cheng, K.T., Cho, G., Chung, A., Cobb, C.L., Distler, A., Egelhaaf, H.J., The 2021 flexible and printed electronics roadmap. *Flexible Printed Electron.*, 6, 2, 023001, 2021.
24. Snook, G.A., Kao, P., Best, A.S., Conducting-polymer-based supercapacitor devices and electrodes. *J. Power Sources*, 196, 1, 1, 2011.

25. Endo, M., Takeda, T., Kim, Y.J., Koshiba, K., Ishii, K., High power electric double layer capacitor (EDLC's); From operating principle to pore size control in advanced activated carbons. *Carbon Lett.*, 1, 3-4, 117, 2001.
26. Lim, Y.S., Tan, Y.P., Lim, H.N., Huang, N.M., Tan, W.T., Yarmo, M.A., Yin, C.Y., Potentiostatically deposited polypyrrole/graphene decorated nano-manganese oxide ternary film for supercapacitors. *Ceram. Int.*, 40, 3, 3855, 2014.
27. Vidhyadharan, B., Zain, N.K.M., Misnon, I.I., Abd Aziz, R., Ismail, J., Yusoff, M.M., Jose, R., High performance supercapacitor electrodes from electrospun nickel oxide nanowires. *J. Alloys Compd.*, 610, 143, 2014.
28. Stoller, M.D. and Ruoff, R.S., Best practice methods for determining an electrode material's performance for ultracapacitors. *Energy Environ. Sci.*, 3, 9, 1294, 2010.
29. Das, S., Wajid, A.S., Shelburne, J.L., Liao, Y.C., Green, M.J., Localized *in situ* polymerization on graphene surfaces for stabilized graphene dispersions. *Appl. Mater. Interfaces*, 3, 6, 1844, 2011.
30. Zhang, K., Zhang, L.L., Zhao, X.S., Wu, J., Graphene/polyaniline nanofiber composites as supercapacitor electrodes. *Chem. Mater.*, 22, 4, 1392, 2010.
31. Eftekhari, A., Li, L., Yang, Y., Polyaniline supercapacitors. *J. Power Sources*, 347, 86, 2017.
32. Huang, Y., Li, H., Wang, Z., Zhu, M., Pei, Z., Xue, Q., Huang, Y., Zhi, C., Nanostructured polypyrrole as a flexible electrode material of supercapacitor. *Nano Energy*, 22, 422, 2016.
33. Fan, L.-Z. and Maier, J., High-performance polypyrrole electrode materials for redox supercapacitors. *Electrochem. Commun.*, 8, 6, 937, 2006.
34. Laforgue, A., Simon, P., Sarrazin, C., Fauvarque, J.F., Polythiophene-based supercapacitors. *J. Power Sources*, 80, 1-2, 142, 1999.
35. Ryu, K.S., Lee, Y.G., Hong, Y.S., Park, Y.J., Wu, X., Kim, K.M., Kang, M.G., Park, N.G., Chang, S.H., Poly (ethylenedioxythiophene)(PEDOT) as polymer electrode in redox supercapacitor. *Electrochim. Acta*, 50, 2-3, 843, 2004.
36. Anothumakkool, B., Soni, R., Bhange, S.N., Kurungot, S., Novel scalable synthesis of highly conducting and robust PEDOT paper for a high performance flexible solid supercapacitor. *Energy Environ. Sci.*, 8, 4, 1339, 2015.
37. Diaz, A.F., Castillo, J.I., Logan, J.A., Lee, W.Y., Electrochemistry of conducting polypyrrole films. *J. Electroanal. Chem. Interfacial Electrochem.*, 129, 1-2, 115, 1981.
38. Ramanavičius, A., Ramanavičienė, A., Malinauskas, A., Electrochemical sensors based on conducting polymer—Polypyrrole. *Electrochim. Acta*, 51, 27, 6025, 2006.
39. Wang, X., Liu, Q., Wu, S., Xu, B., Xu, H., Multilayer polypyrrole nanosheets with self-organized surface structures for flexible and efficient solar–thermal energy conversion. *Adv. Mater.*, 31, 19, 1807716, 2019.
40. Yang, P. and Mai, W., Flexible solid-state electrochemical supercapacitors. *Nano Energy*, 8, 274, 2014.
41. Lota, K., Khomenko, V., Frackowiak, E., Capacitance properties of poly (3, 4-ethylenedioxythiophene)/carbon nanotubes composites. *J. Phys. Chem. Solids*, 65, 2-3, 295, 2004.
42. Zhao, J., Wu, J., Li, B., Du, W., Huang, Q., Zheng, M., Xue, H., Pang, H., Facile synthesis of polypyrrole nanowires for high-performance supercapacitor electrode materials. *Prog. Nat. Sci.: Mater. Int.*, 26, 3, 237, 2016.
43. Choudhary, R.B., Ansari, S., Purty, B., Robust electrochemical performance of polypyrrole (Polypyrrole) and polyindole (PIn) based hybrid electrode materials for supercapacitor application: A review. *J. Energy Storage*, 29, 101302, 2020.
44. Yang, Q., Tu, Z., Zhao, N., Research on interfacial polymerization of pyrrole assist with Span80 system. *IOP Conf. Ser.: Mater. Sci. Eng.*, 137, 012070, 2016, IOP Publishing.

45. Xu, J., Wang, D., Fan, L., Yuan, Y., Wei, W., Liu, R., Gu, S., Xu, W., Fabric electrodes coated with polypyrrole nanorods for flexible supercapacitor application prepared via a reactive self-degraded template. *Org. Electron.*, 26, 292, 2015.
46. Wang, K., Wu, H., Meng, Y., Wei, Z., Conducting polymer nanowire arrays for high performance supercapacitors. *Small*, 10, 1, 14, 2014.
47. Liu, Y., Zhang, Y., Ma, G., Wang, Z., Liu, K., Liu, H., Ethylene glycol reduced graphene oxide/polypyrrole composite for supercapacitor. *Electrochim. Acta*, 88, 519, 2013.
48. Liu, L., Lu, Q., Yang, S., Guo, J., Tian, Q., Yao, W., Guo, Z., Roy, V.A., Wu, W., All-printed solid-state microsupercapacitors derived from self-template synthesis of Ag@polypyrrole nanocomposites. *Adv. Mater. Technol.*, 3, 1, 1700206, 2018.
49. Tian, W., Li, Y., Zhou, J., Wang, T., Zhang, R., Cao, J., Luo, M., Li, N., Zhang, N., Gong, H., Zhang, J., Implantable and biodegradable micro-supercapacitor based on a superassembled three-dimensional network Zn@polypyrrole hybrid electrode. *ACS Appl. Mater. Interfaces*, 13, 7, 8285, 2021.
50. Meng, Q., Cai, K., Chen, Y., Chen, L., Research progress on conducting polymer based supercapacitor electrode materials. *Nano Energy*, 36, 268, 2017.
51. Zhao, Y., Liu, J., Hu, Y., Cheng, H., Hu, C., Jiang, C., Jiang, L., Cao, A., Qu, L., Highly compression-tolerant supercapacitor based on polypyrrole-mediated graphene foam electrodes. *Adv. Mater.*, 25, 4, 591, 2013.
52. Wang, W., Hao, Q., Lei, W., Xia, X., Wang, X., Graphene/SnO 2/polypyrrole ternary nanocomposites as supercapacitor electrode materials. *RSC Adv.*, 2, 27, 10268, 2012.
53. Zhang, F., Xiao, F., Dong, Z.H., Shi, W., Synthesis of polypyrrole wrapped graphene hydrogels composites as supercapacitor electrodes. *Electrochim. Acta*, 114, 125, 2013.
54. Zhan, Z., An, J., Wei, Y., Du, H., Inkjet-printed optoelectronics. *Nanoscale*, 9, 3, 965, 2017.
55. Glasser, A., Cloutet, E., Hadziioannou, G., Kellay, H., Tuning the rheology of conducting polymer inks for various deposition processes. *Chem. Mater.*, 31, 17, 6936, 2019.
56. Schlisske, S., Rosenauer, C., Rödlmeier, T., Giringer, K., Michels, J.J., Kremer, K., Lemmer, U., Morsbach, S., Daoulas, K.C., Hernandez-Sosa, G., Ink formulation for printed organic electronics: Investigating effects of aggregation on structure and rheology of functional inks based on conjugated polymers in mixed solvents. *Adv. Mater. Technol.*, 6, 2, 2000335, 2021.
57. Bagheri, A. and Jin, J., Photopolymerization in 3D printing. *ACS Appl. Polym. Mater.*, 1, 4, 593, 2019.
58. Li, L., Lin, Q., Tang, M., Duncan, A.J., Ke, C., Advanced polymer designs for direct-ink-write 3D printing. *Chem. Eur. J.*, 25, 46, 10768, 2019.
59. Tumbleston, J.R., Shirvanyants, D., Ermoshkin, N., Janusziewicz, R., Johnson, A.R., Kelly, D., Chen, K., Pinschmidt, R., Rolland, J.P., Ermoshkin, A., Samulski, E.T., Continuous liquid interface production of 3D objects. *Science*, 347, 6228, 1349, 2015.
60. Tully, J.J. and Meloni, G.N., A scientist's guide to buying a 3D printer: How to choose the right printer for your laboratory. *Anal. Chem.*, 92, 22, 14853, 2020.
61. Tagliaferri, S., Panagiotopoulos, A., Mattevi, C., Direct ink writing of energy materials. *Adv. Mater.*, 2, 2, 540, 2021.
62. Malda, J., Visser, J., Melchels, F.P., Jüngst, T., Hennink, W.E., Dhert, W.J., Groll, J., Hutmacher, D.W., 25th anniversary article: Engineering hydrogels for biofabrication. *Adv. Mater.*, 25, 36, 5011, 2013.
63. Liang, J., Jiang, C., Wu, W., Printed flexible supercapacitor: Ink formulation, printable electrode materials and applications. *Appl. Phys. Rev.*, 8, 2, 021319, 2021.
64. Keskinen, J., Tuurala, S., Sjödin, M., Kiri, K., Nyholm, L., Flyktman, T., Strømme, M., Smolander, M., Asymmetric and symmetric supercapacitors based on polypyrrole and activated carbon electrodes. *Synth. Met.*, 203, 192, 2015.

65. Liu, L., Lu, Q., Yang, S., Guo, J., Tian, Q., Yao, W., Guo, Z., Roy, V.A., Wu, W., All-printed solid-state microsupercapacitors derived from self-template synthesis of Ag@polypyrrole nanocomposites. *Adv. Mater. Technol.*, 3, 1, 1700206, 2018.
66. Liang, C., Yang, C., Zang, L., Liu, Q., Qiu, J., Li, Y., Zuo, W., Liu, X., UV-assisted one-step synthesis of ternary graphene/polypyrrole/silver nanocomposites for supercapacitors. *Energy Technol.*, 9, 3, 2000966, 2021.
67. Liu, L., Tian, Q., Yao, W., Li, M., Li, Y., Wu, W., All-printed ultraflexible and stretchable asymmetric in-plane solid-state supercapacitors (ASCs) for wearable electronics. *J. Power Sources*, 397, 59, 2018.
68. Stempien, Z., Khalid, M., Kozanecki, M., Filipczak, P., Wrzesińska, A., Korzeniewska, E., Sąsiadek, E., Inkjet printing of polypyrrole electroconductive layers based on direct inks freezing and their use in textile solid-state supercapacitors. *Mater. (Basel)*, 14, 13, 3577, 2021.
69. Ligon, S.C., Liska, R., Stampfl, J., Gurr, M., Mülhaupt, R., Polymers for 3D printing and customized additive manufacturing. *Chem. Rev.*, 117, 15, 10212, 2017.
70. Zhang, F., Wei, M., Viswanathan, V.V., Swart, B., Shao, Y., Wu, G., Zhou, C., 3D printing technologies for electrochemical energy storage. *Nano Energy*, 40, 418, 2017.
71. Foo, C.Y., Lim, H.N., Mahdi, M.A., Wahid, M.H., Huang, N.M., Three-dimensional printed electrode and its novel applications in electronic devices. *Sci. Rep.*, 8, 1, 7399, 2018.
72. Gu, Y., Zhang, Y., Shi, Y., Zhang, L., Xu, X., 3D all printing of polypyrrole nanotubes for high mass loading flexible supercapacitor. *ChemistrySelect*, 4, 36, 10902, 2019.
73. Li, H. and Liang, J., Recent development of printed micro-supercapacitors: Printable materials, printing technologies, and perspectives. *Adv. Mater.*, 32, 3, e1805864, 2020.
74. Xing, R., Xia, Y., Huang, R., Qi, W., Su, R., He, Z., Three-dimensional printing of black phosphorous/polypyrrole electrode for energy storage using thermoresponsive ink. *Chem. Commun. (Camb)*, 56, 21, 3115, 2020.
75. Zhao, C., Wang, C., Gorkin Iii, R., Beirne, S., Shu, K., Wallace, G.G., Three dimensional (3D) printed electrodes for interdigitated supercapacitors. *Electrochem. Commun.*, 41, 20, 2014.
76. Areir, M., Xu, Y., Harrison, D., Fyson, J., 3D printing of highly flexible supercapacitor designed for wearable energy storage. *Mater. Sci. Eng. B: Solid-State Mater. Adv. Technol.*, 226, 29, 2017.
77. Park, S.H., Goodall, G., Kim, W.S., Perspective on 3D-designed micro-supercapacitors. *Mater. Des.*, 193, 108797, 2020.
78. Zhang, H., Qiao, Y., Lu, Z., Fully printed ultraflexible supercapacitor supported by a single-textile substrate. *ACS Appl. Mater. Interfaces*, 8, 47, 32317, 2016.

18

Biomass-Derived Carbon for Supercapacitors

Priyadharshini M., Pazhanivel T.* and Hariprasath K. R.

Smart Materials Interface Laboratory, Department of Physics, Periyar University, Salem, Tamil Nadu, India

Abstract

This chapter describes the fundamental sources and constituents of different types of biomass and signifies the importance of biomass-derived activated carbon for supercapacitor application. Moreover, a detailed description of synthesis and application of a few major sources like plant animal and microorganism-derived carbon as electrodes reported in the literature were summarized. The advantages, limitations and its ability to meet the requirement of energy crisis were focused on.

The preparation and utilization of carbon obtained from biomass for energy storage have drawn research attention for ease of production, economy, and sustainability of porous carbon from different biological precursors. In comparison with artificial nanostructural carbon, like 1D carbon nanotubes and 0D buck balls, activated carbon derived from biomass also would have a more specific capacity and better electrochemical properties suitable for supercapacitor applications. Nevertheless, problems persist with synthesis methods to acquire carbon obtained from biomass with high yield, efficient electrochemical properties, and controlled sp^2 microstructures. The above requirement may need an interpretation of the chemical, elemental composition and pure structural properties of precursors. Here, a complete analysis of the effects of chemical composition, morphology of biomass over the carbon yield and electrochemical behavior were elaborately explained. In addition, applications of activated carbon derived from various biomasses for supercapacitor applications reported in recent literature were also discussed and compared in detail. Finally, useful findings are stated for improving biomass-derived carbon efficiency for supercapacitor applications.

***Keywords*:** Biomass-derived carbon, biological precursor, biomass compositions, activation, porosity, supercapacitor

**Corresponding author*: pazhanit@gmail.com

Inamuddin, Tariq Altalhi and Sayed Mohammed Adnan (eds.) Sustainable Materials for Electrochemical Capacitors, (427–440) © 2023 Scrivener Publishing LLC

18.1 Introduction

The increasing population and industrialization exert pressure on energy sources. Since fossil fuels were the source of energy for the last several decades it gets depleted in addition to depletion [1, 2]. It also causes several environmental impacts which directly affect the health of humankind. So in search of alternate energy sources, renewable energy sources, renewable energy has acquired a great interest because of eco–friendly nature, cost-effectiveness and easy availability [3–5].

Although several renewable energy sources like the sun, wind and so generate energy its geometric contain forbids its wide application. Hence an efficient energy storage device is needed to access the full potential of renewable energy sources. Among various energy storage systems, electrochemical capacitors also called supercapacitors (sc_s) were considered a potential candidate with desirable features like high power density, wide operating temperature and soon [6, 7]. Supercapacitors act as a bridge between modern rechargeable Li-ion batteries to conventional capacitors in terms of both power and energy density. Due to its favorable properties, it has a wide range of applications in various fields including military defense to portable electronic devices [8, 9].

Briefly, based on the mechanism supercapacitors are classified into two classes, namely (i) Electric double-layer capacitors (EDLC) and (ii) Psuedocapacitor. In EDLC, the charge is stored via electrostatic attraction whereas in supercapacitors stores charge via redox reaction [10]. The commonly used electrode material for EDLC are mostly carbon-based materials like reduced graphene oxide (r-GO), graphene, activated carbon, carbon fiber etc. while pseudocapacitors have materials like metal oxide, sulphides, and conducting polymers [11, 12]. Although pseudocapacitive material possesses high energy density, carbon-based material yields high power density and superior cycling ability due to surface adsorption of charged ions. Therefore recently carbon-based material was considered a state of art for EDLC-based electrode materials [13].

Biomass was considered a promising source of carbon due to its eco-friendly properties, renewable nature, and easy availability. The major source of biomass is animal waste, plants, and so on [14]. The major advantage of converting biomass into carbon for electrode material includes (i) versatility and (ii) *in-situ* pores. Since a wide range of biomass is converted into carbon similar synthesis procedures can be carried out like carbonization, activation, and purification. Various chemicals used for the activation process and introduced metallic compounds during conversion may drastically improve its electrochemical performance [15–19].

Next, the biomolecule precursors have their selector morphology which almost remains unchanged after the carbonization process and could inherently offer well interconnected conductive substrate with pores and voids. Finally compared to the synthesis procedure of other carbon like CNT, Graphene Oxide, and biomass-derived (BDC) offers harsh free chemicals with low-pressure synthetic procedures making them an energy-efficient method [20–23].

An example of BDC based flexible supercapacitor with its electrochemical performance was shown below in Figure 18.1.

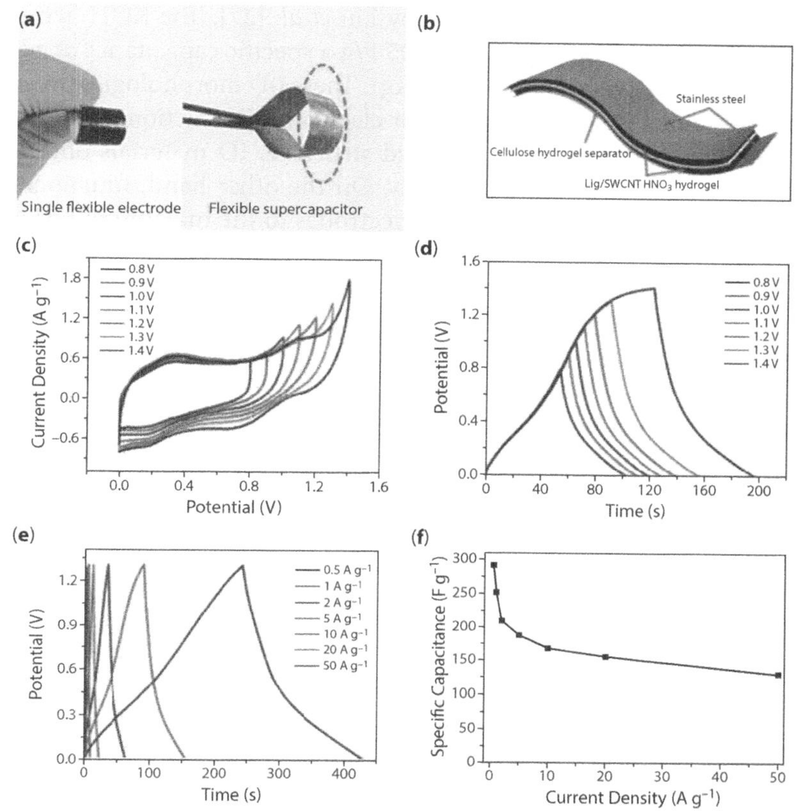

Figure 18.1 (a) Picture of single Lig/SWCNT HNO_3 electrode and (b) the scheme. (c) Cyclic voltammetry at 5 mV s^{-1} (d) GCD curves at 1 A g^{-1} (e) GCD curves at different current density and (f) the specific capacitance of FSC *(reproduced with permission [21])*.

18.2 Tuneable Physiochemical Properties

Different sources of biomass have been used as a precursor in the preparation process. The physio chemical property of the prepared sample has a great impact on the electrochemical properties of the BDCs. For instance physical dimensions like fiber, as its own higher accessible surface area, in comparison to hierarchical three dimensions [24]. Also only, a range of pore size distribution is encouraged became a wide pore may result in structural collapse and a narrow pore constrains widespread accessibility of electrolytes [25]. In some cases, based, on the source presence of self-dopant in carbon is observed [26]. This has to be observed and optimized carefully because not every depend enhances the electrochemical behavior of BDCs.

18.2.1 Effect of Morphology

Based on inheritance and evolution, biomass can be of different dimensions. The aim is to prepare the material with a high surface area, and a better ionic pathway which intern promotes the efficiency of the material. Zero dimensional materials like carbon spheres were synthesized from high porous glucose by Liu *et al.* [27]. The KOH-activated carbon as active material in the electrode provides 405 F/g a specific capacitance at a 0.5 A/g with better cyclability and energy density (53 Wh/Kg). These 0D morphologies not only provide rich ionic channels and a large surface area for electrolyte interaction but also provide low resistance pathway due to well-interconnected structure. ID materials offer direct channels thus achieving exceptional rate capability. On the other hand, functional groups on the surface improve the wetting behavior of electrodes to the maximum. For instance, the N-doped carbon nanofiber (CNF) prepared via electrospinning of cellulose was investigated as an active electrode for the device it showed capacitance of 235F/g and enhanced metal oxide was composited [28]. Compared to 0D and 1D materials 2D has a great interest due to (i) strong in-plane covalent band and (ii) Optimized layer thickness. Through direct pyrolysis of a leaf, a porous nanosheet was prepared whose energy density was 14.8Wh/L with good cycling stability [29].

As an ideal building block, 3D material offers high contact between the active site and electrolyte. Rhys typhina derived interconnected carbon by pyrolysis. After the procedure, the biomolecular morphology was retained to its maximum, facilitating optimal porous distribution. According to the report, from the point of benefit due to dimension biomass is considered a suitable material for electrode application. However interfacial issues should be paid attention to for better electrochemical performance [30].

18.2.2 Effect of the Activation Process

By varying the activating agents, reaction temperature, timing, and rate of heating, the effective size of pores can be attired. For example, Dhinesh *et al.* [31] prepared activated carbon derived from *Datura stramonium* and analyzed the effect of activating agents. KOH as an activating agent delivered a surface area of 1390m^2/g with a specific capacitance of 665F/g at

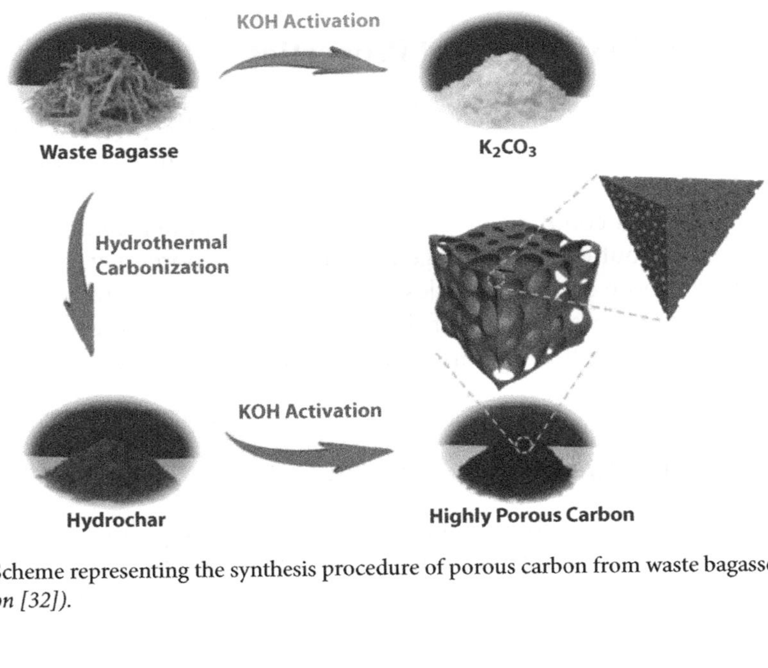

Figure 18.2 Scheme representing the synthesis procedure of porous carbon from waste bagasse *(reproduced with permission [32])*.

5/mv/s. the cycling stability remains 65% after consecutive charging and discharging whose energy density range 29.25Wh/Kg at 2700 W/Kg.

Peifeng Yu *et al.* [32] prepared porous carbon from waste bagasse through a facile method. The obtained materials display a hierarchically porous structure with a honeycomb-like texture. Gaining from the combination of hierarchical pore structure and excellent porosity, it yields a high capacitance of 413 F g^{-1} at 1 A g^{-1} and good cycling stability of 93.4% capacitance retention after 10000 cycles. The scheme of the synthesis process was shown above (Figure 18.2).

18.2.3 Effect of Doping

The incorporation of heteroatoms and self-doped biomass was considered a promising way to enhance the active sites and improve electronic conductivity and wettability.

For example, Zhang *et al.* [33] synthesized carbon aerogel with nitrogen dopant through the sol-gel technique using KOH as activating agent. Additional nitrogen was provided by urea used in the synthesis process. However, the prepared sample shows an efficiency of 400 F/g at 0.5 A/g. In addition to nitrogen, oxygen and sulphur can be also used as doping heteroatoms. Dual or multiple atom doping provides strong electronegativity, variable electron-donor properties, and a large operating window.

Priyadharshini *et al.* [34] have prepared a carbon dot and composited it with NiP_2O_7, the electrochemical analysis revealed that nanocomposite shows higher specific capacitance than pure pyrophosphate. The capacitance of the CD composited materials has been substantially increased with the specific capacitance of about 247 from 72 Fg^{-1} of pure $Ni_2P_2O_7$. This transformation of the capacitance behavior correlated to the optimization of the charge transfer and active area of the nanohybrid compared with the $Ni_2P_2O_7$ sample. Furthermore, the $Ni_2P_2O_7$ with CQDs holds 92% of its capacitance after 5000 cycles. This work thus signifies improvement of specific capacitance of the metal pyrophosphate is possible with the addition of a carbon quantum dot. The schematic representation of the work is shown in Figure 18.3.

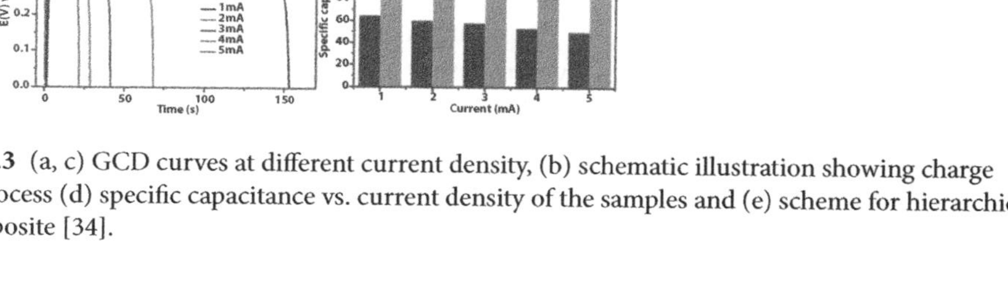

Figure 18.3 (a, c) GCD curves at different current density, (b) schematic illustration showing charge storage process (d) specific capacitance vs. current density of the samples and (e) scheme for hierarchical nanocomposite [34].

18.3 Synthesis Procedure

A suitable synthesis procedure has to be chosen to obtain a large amount of yield without compromising the quality of the sample. Several parameters including temperature, time and so on play an important part in deciding the electrochemical characteristics of carbon [35].

18.3.1 Pyrolysis

It is the thermal decomposition process under an inert atmosphere whose temperature ranges up to 900 degrees Celsius. The obtained quantity of carbon relies on the feedstock's nature and operating parameters [36].

18.3.2 Hydrothermal Carbonization

Biomass conversion to carbon through this method consumes less energy whose temperature ranges from 182-250 degrees Celsius. By varying solvent medium desired functional groups can be functionalized on the surface [37].

18.3.3 Torre Faction

The disadvantage of high temperature in pyrolysis could be overcome by a slow pyrolysis technique whose temperature ranges between 250 to 300°C. Inert-controlled O_2 surroundings can be utilized for the process [38].

18.3.4 Gasification

The gasification of this technique involves the synthesis process at elevated temperatures. The steam gasification of the region enforces force ranging in Micron size and has high mineral content because of partial oxidation of biomass [39].

18.4 Main Categories of Biomass

Major contributions of biomass include cellulose hemicellulose and Lignin and protecting a small ratio of wax and inorganic elements may also be observed. Hemicellulose whose molecule weighs less than cellulose contains xyloglucans, xylenes, galactomannans, and manna's. In addition, several functional groups include cinnamic acetyl and methyl groups. Although the Biomass were collected from different sources, major categories were planted, marine organisms and animal waste [40].

18.4.1 Plant-Based Biomass

Lignocellulose was a major component of plant-based Biomass. The ratio of each component varies from plant to plant however mostly the components remain the same and their quantitative ratios were given in Table 18.1. Different parts including the stem leaf flower seed and fruit can be utilized to prepare BDC. Due to the exchange of fluids through the stem, the ion content in this region is high. Jiang *et al.* [41] synthesized BDC from cornstalk in an open atmosphere. In the activation step the *in situ* salts themselves act as

Table 18.1 The chemical constituents of various plant-based biomass (in dry weight % basis) [50].

Biomass	Lignin (%)	Cellulose (%)	Hemicellulose (%)	Extractives (%)
Coconut coir	46.48	21.46	12.36	8.77
Coconut sheath	29.7	31.05	19.22	1.74
Bagasse	22.56	39.45	26.97	4.33
Banana leaf	24.84	25.65	17.04	9.84
Sunflower seed hull	20.0	18.5	14.7	43.4
Apricot pit shell	31.91	34.31	-	-
Willow leaves	28.7	31.3	25.2	-

activating reagents. The presence of salt in the molten state not only avoids burning but also facilitates a facile etching through Cl-ion. The obtained thin sheets possess a higher surface area of 588m^2/g. The electrochemical analysis states that it has an efficiency of 407 F/g with a cycling efficiency of 92.6% capacity retention over 20000 consecutive cycles.

Mostly mesophyll tissues with the protein are mainly present in the leaf due to photosynthesis. Sulphur and nitrogen self-doped BDC were prepared from ginkgo leaf. During the synthesis process, Ammonium Nitrate was used for activation which gives a wide range of Mesoporous and microporous morphology using this BDC. An asymmetric device was constructed that exhibits a specific capacitance of 330F/g at 0.5 A/g and 85.8% of capacitance after 10000 cycles [42]. Generally, petals consist of saccharides, vitamins, and proteins with calcium and magnesium also found in some flowers. Presence of calcium and magnesium act as templates during the calcination process. For instance, bougainvillea flower-based graphene showed an efficiency of 458F/g with high energy and power density ranging 64Wh/kg and 273Wh/kg respectively [43].

The ideal precursor of plant-based BDC is considered a seed. In nature, we could find the spread of seeds to distant areas by using wind like catkins. The hierarchical structure composed of hollow thin-walled tubes was prepared from Dandelion [44], benefiting from the synergistic property the electrochemical capability Ranges 12Wh/kg and 700W/kg energy and power density respectively.

18.4.2 Microorganism-Based Biomass

Apart from the plant, BDCs were also prepared from several microorganisms like cellulose of bacteria, fungus, etc. Microorganism possesses hyphae which form 3D interconnects while conversion of these tissues transforms into honeycomb morphology through which several hierarchical structures were designed. Carbohydrates, proteins, fibers, and ash are the constituents of microorganism-based biomass and few quantitative values were reported in Table 18.2. The addition of a small number of bacteria could be a cheap and promising method of doping nitrogen in active material that could likely improve the electrochemical properties [45]. Fungus as a source of biomass yields a cost-effective eco-friendly carbon

Table 18.2 The chemical constituents of various microorganism (in dry weight % basis) [50].

Microorganisms	Carbohydrates	Crude fiber	Crude protein	Crude fat	Ash
Agaricus bisporus	42.56	13.21	33.85	2.41	7.97
B. aereus	34.0	17.0	26.9	2.1	8.5
Lactarius deliciosus	25.0	36.3	20.2	2.5	7.5
R. virescens	13.4	32.8	28.3	1.5	11.9
T. matsutake	36.7	29.1	14.3	5.0	8.9
Tricholoma portentosum	34.6	30.1	19.6	5.8	9.9

for a large-scale application. Long *et al.* [16] 2014 prepared fungus-based carbon nanosheet by hydrothermal followed carbonization process. The addition of KOH during synthesis acts as both template and activating agent. As a template it facilitates an agglomeration-free environment and as an activating agent yields porous morphology. Overall, the prepared active carbon exhibits good specific capacitance of 360 F cm^3 and better cycle stability of 99% retention over 10000 cycles). Zhen-Yu Wu *et al.* [46] prepared P-doped, N, P-co doped, or B, P-co doped CNF aerogels from bacterial cellulose and utilized them as the electrode for both symmetric and asymmetric supercapacitors. In this work, the amount of ammonia dihydrogen phosphate was optimized and efficiency reached a higher capacitance of about ~204 F/g with a low transfer resistance of R_{ct} 0.367Ω and a higher power density of 186 kW/kg. On the other hand, asymmetric devices were fabricated. The schematic figure

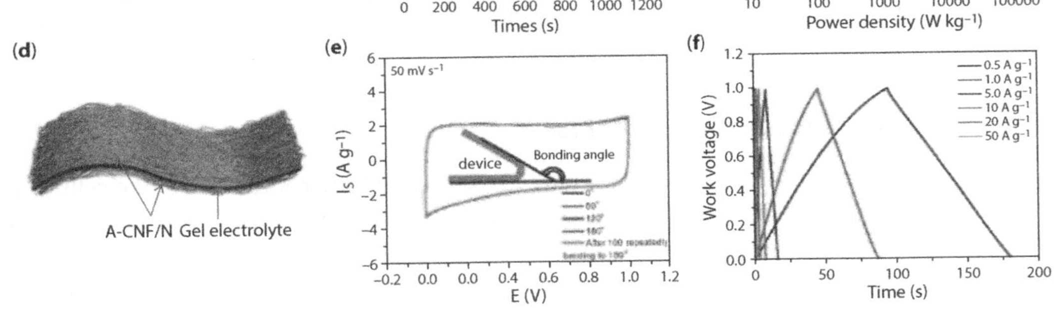

Figure 18.4 (a) A schematic diagram of CNF/MnO.$_2$//CNF/N-ASC device. (b) Galvanostatic charge–discharge curves under various voltages. (c) Ragone plots. (d) Scheme of all-solid-state supercapacitor. (e) CV curves of A-CNF/N under different bending angles. (f) GCD plot of the A-CNF/N at various current densities *(reproduced with permission* [46]*).*

of the fabricated device along with their electrochemical performance was shown in the Figure 18.4. Followed by him Kwang *et al.* [47], prepared bacteria-loaded polyaniline@ multi walled carbon nanotubes (MWNT) through *in situ* polymerization of polyaniline in MWNT's dispersion. Here, the bacteria are attached to the composites through electrostatic interaction and yielded porous morphology with doping after the carbonization process. Then, bacteria-loaded materials exhibited a maximum specific capacitance value of 57.7 F/g at 0.2A/g within a potential window of -1 to 0 V. The efficiency retained up to 70% after 1000 cycles. Hence, after loading with bacteria, the electrochemical performance is further enhanced compared with that of PANI@MWNI'S. It suggests that bacteria-assisted electrode materials are promising materials for high performance for supercapacitor electrodes.

18.4.3 Animal-Based Biomass

Animal-based Biomass major constituents of this type of biomass include chitin which is nothing but Keratin. The quantitative amount of chitin in the various organism was given in Table 18.3. The presence of a large amount of Nitrogen in chitin makes it suitable for SC application. Different from cellulose, both in terms of thermal and chemical chitin is more stable because of its high degree of intermolecular bonding and crosslinking networks. In a review article by Arabi *et al.*, the author blended various studies relating to insect molluces and so on which acted as sources for the chitin extraction. Before its direct application, the Titan undergoes several synthetic procedures including mechanical grinding deprotonation and demineralization. On the other hand, keratin from different parts of the chemical also processes excellent chemical and thermal stability due to intermolecular bonding. Hong *et al.*, [48] prepared n-doped porous carbon from endothelium carenum gigeria, Galli, by carbonization at 450 degrees Celsius followed by KOH activation. Compared to all prepared carbon ECGG prepared at 900 degrees Celsius shows better electrochemical behavior of 198 F/g current density of 1 A/g.

In cyclic analysis, the prepared sample exhibits cycling retention of 91.6% after a consecutive 3000 charge-discharge cycle. Delvina Japhet Tarimo *et al.* [49] synthesized activated carbon from cooked chicken bone waste (CCBW). By varying the ratio of activating agent (KOH) the efficiency in electrochemical performance was analyzed. On comparing various

Table 18.3 The quantitative amount of chitin in various organism (in dry weight % basis) [50].

Organism	Chitin (%)
Cancer (crab)	72.1
Crangon and Pandalus (shrimp)	17–40
Nephro (lobster)	69.8
Lepas (goose barnacle)	58.3
Pieris pupae (butterfly)	64.0
Squid pen	41.0

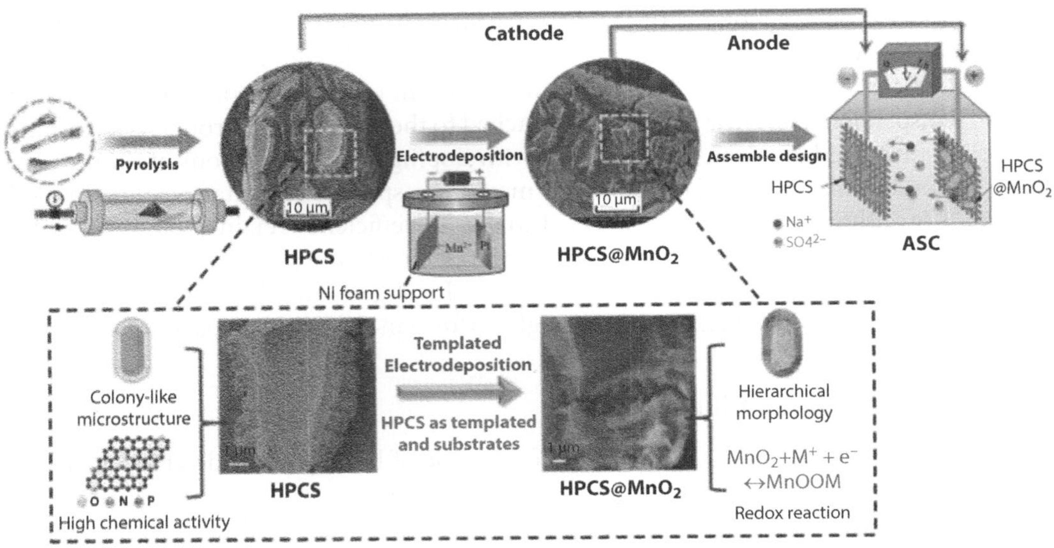

Figure 18.5 Schematic representation of preparation process of HPCS and its MnO_2 composite also the microstructural images of prepared nanocomposite were shown *(reproduced with permission [49])*.

ratios 1: 1 offered better electrochemical properties. The fabricated symmetric device with CCBW as shown in Figure 18.5 shows good energy and power density of 17.1 Wh/kg and 425 W/kg respectively which may be ascribed to a high surface area of 2235.8 m^2/g. The device also demonstrates enhanced capacitance after the device was exposed to a floating time of 300 hours. Xiaomin Yuan *et al.* [18] prepared a hierarchical porous carbon scaffold (HPCS) from animal bone composited with MnO_2. The prepared hierarchical porous carbon scaffold acts as a conductive matrix and offered a good gravimetric capacitance of ~232 F/g with negligible loss in cycling stability of 5% over 10,000 constant charging and discharging cycles. Due to better performance, the device was assembled using HPCS||HPCS@MnO_2 as cathode respectively. The device exhibited an energy density of ~61 W/kg. The author concludes that instead of micropores, the presence of mesopores could efficiently improve capacitance in both surface and diffusion-controlled mechanisms. Overall, morphology and functional groups on the surface could effectively modify the electrochemical performance of the electrode.

18.5 Conclusion and Future Perspective

In conclusion, different sources of novel biomass for energy storage applications have been discussed. Mostly the power and energy density rely on the dimension of the bio precursor chosen for the active material preparation. Apart from that, several parameters like calcination temperature, activating agent, time, and dopant may vary the efficiency significantly [45]. Also, little literature highlighting the major components of different biomass with its application in hybrid devices was elaborately explained. From the view of the above points, the below conclusions were given:

(1) The alteration of physicochemical properties by numerous methods not only urges for new material but also forces us to study existing material in detail or hybrids were made with suitable material.
(2) Among several synthesis procedures hydrothermal method developed carbonaceous active material for supercapacitor electrode which seems to be more efficient, but possess high temperatures and pressure involvement raised the cost. In future, this discrepancy can be removed well by introducing a catalyst that operates at a lower temperature.
(3) For the most part, biomass-derived carbon exhibits high efficiency in terms of power density and cycling stability.
(4) The achieved gravimetric capacitance was low compared to commercially available material which could be significantly improved by adding heteroatom doping or compositing with metallic materials. Thus proper optimization has to be considered for its wide range of applications.
(5) The carbonization methods had depicted high suitability in the preparation of carbon materials with optimum porosity; as well as, a combination of methods may be proposed for advanced improvement of the materials and devices.

References

1. Pazhanivel, T., Dhinesh, S., Priyadharshini, M., R.G., Aerogels utilizations in batteries, in: *Aerogels II: Preparation, Properties and Applications*, pp. 99–120, Materials Research Forum LLC, Millersville PA, USA, 2021.
2. Thangavelu, P., Matheswaran, P., Duraisamy, N., *Polymer-based binary nanocomposites*, Boca Raton, Florida, 2019.
3. Priyadharshini, M., Sandhiya, M., Sathish, M., Pazhanivel, T., Mani, G., Alothman, A.A., Alqahtani, K.N., Surfactant-dependant self-organisation of nickel pyrophosphate for electrochemical supercapacitors. *J. Mater. Sci. Mater. Electron.*, 305, 9269–9276, 2021.
4. Matheswaran, P., Karuppiah, P., Thangavelu, P., Contribution of different charge storage mechanisms in cobalt pyrophosphate-based supercapattery. *Ionics (Kiel)*, 292, 1769–1780, 2021.
5. Matheswaran, P., Karuppiah, P., Chen, S.M., Thangavelu, P., A binder-free Ni2P2O7/Co2P2O7 nanograss array as an efficient cathode for supercapacitors. *New J. Chem.*, 44, 30, 13131–13140, 2020.
6. Conway, B.E., Electrochemical supercapacitors scientific fundamentals and technological applications, in: *Adv. Lithium-Ion Battery*, pp. 481–505, 1999.
7. Simon, P., Gogotsi, Y., Dunn, B., Where do batteries end and supercapacitors begin? *Sci. (80-.)*, 343, 6176, 1210–1211, 2014.
8. Matheswaran, P., Thangavelu, P., Ganapathi, B., Controlled approach on spinel lithium manganese oxide as possible cathode for high-performance Li-ion supercapacitors. *Mater. Technol.*, 34, 689–696, 2019.
9. Matheswaran, P., Karuppiah, P., Chen, S.M., Thangavelu, P., Ganapathi, B., Fabrication of g-C3N4 nanomesh-anchored amorphous NiCoP2O7: Tuned cycling life and the dynamic behavior of a hybrid capacitor. *ACS Omega*, 3, 12, 18694–18704, 2018.
10. Yu, A., Davies, A., Chen, Z., Electrochemical supercapacitors. *Electrochem. Technol. Energy Storage Convers.*, 1, 317–382, 2012.

11. Priyadharshini, M., Pazhanivel, T., Maiyalagan, T., Albaqami, M.D., Ganesh, P.S., Electrochemical investigation on hierarchical sea urchin shaped zinc nickel selenide for efficient supercapacitor. *Ceram. Int.*, 48, 29011–29017, 2022.
12. Dhinesh, S., Priyadharshini, M., Pazhanivel, T., Gobi, R., Biomass-derived N, S self-doped activated carbon embedded MnO2 as cathode for supercapacitor. *Mater. Technol.*, 37, 1837–1845, 2021.
13. Gogotsi, Y. and Simon, P., Materials for electrochemical capacitors. *Nat. Mater.*, 7, 11, 845–854, 2018.
14. Matheswaran, P., Thangavelu, P., Palanivel, B., Carbon dot sensitized integrative g-C3N4/AgCl hybrids: An synergetic interaction for enhanced visible light driven photocatalytic process. *Adv. Powder Technol.*, 30, 8, 1715–1723, 2019.
15. Paulraj Gundupalli, M., Cheng, Y.S., Chuetor, S., Bhattacharyya, D., Sriariyanun, M., Effect of dewaxing on saccharification and ethanol production from different lignocellulosic biomass. *Bioresour. Technol.*, 339, 125596, 2021.
16. Long, C., Chen, X., Jiang, L., Zhi, L., Fan, Z., Porous layer-stacking carbon derived from in-built template in biomass for high volumetric performance supercapacitors. *Nano Energy*, 12, 141–151, 2015.
17. Rodiahwati, W. and Sriariyanun, M., Lignocellulosic biomass to biofuel production: Integration of chemical and extrusion (screw press) pretreatment. *King Mongkut's Univ. Technol. North Bangk. Inter. J. Appl. Sci. Technol.*, 9, 4, 289–298, 2016.
18. Yuan, X., Zhang, Y., Yan, Y., Wei, B., Qiao, K., Zhu, B., Cai, X., Chou, T.W., Tunable synthesis of biomass-based hierarchical porous carbon scaffold@MnO2 nanohybrids for asymmetric supercapacitor. *Chem. Eng. J.*, 393, 121214, 2020.
19. Gupta, G.K. and Shukla, P., Lignocellulosic biomass for the synthesis of nanocellulose and its eco-friendly advanced applications. *Front. Chem.*, 8, 601256, 2020.
20. Asim, N., Badiei, M., Alghoul, M.A., Mohammad, M., Fudholi, A., Akhtaruzzaman, M., Amin, N., Sopian, K., Biomass and industrial wastes as resource materials for aerogel preparation: Opportunities, challenges, and research directions. *Ind. Eng. Chem. Res.*, 58, 38, 17621–17645, 2019.
21. Peng, Z., Zou, Y., Xu, S., Zhong, W., Yang, W., High-performance biomass-based flexible solid-state supercapacitor constructed of pressure-sensitive lignin-based and cellulose hydrogels. *ACS Appl. Mater. Interfaces*, 10, 26, 22190–22200, 2018.
22. Jiang, C., Yakaboylu, G.A., Yumak, T., Zondlo, J.W., Sabolsky, E.M., Wang, J., Activated carbons prepared by indirect and direct CO2 activation of lignocellulosic biomass for supercapacitor electrodes. *Renew. Energy*, 155, 38–52, 2020.
23. Rico-García, D., Ruiz-Rubio, L., Pérez-Alvarez, L., Hernández-Olmos, S.L., Guerrero-Ramírez, G.L., Vilas-Vilela, J.L., Lignin-based hydrogels: Synthesis and applications. *Polym. (Basel)*, 12, 1, 1–23, 2020.
24. Zhou, X., Wang, P., Zhang, Y., Wang, L., Zhang, L., Zhang, L., Xu, L., Liu, L., Biomass based nitrogen-doped structure-tunable versatile porous carbon materials. *J. Mater. Chem. A*, 5, 25, 12958–12968, 2017.
25. Li, Q., Jiang, R., Dou, Y., Wu, Z., Huang, T., Feng, D., Yang, J., Yu, A., Zhao, D., Synthesis of mesoporous carbon spheres with a hierarchical pore structure for the electrochemical double-layer capacitor. *Carbon N. Y.*, 49, 4, 1248–1257, 2011.
26. Liu, Y., Zhu, C., Gao, Y., Yang, L., Xu, J., Zhang, X., Lu, C., Wang, Y., Zhu, Y., Biomass-derived nitrogen self-doped carbon dots via a simple one-pot method: Physicochemical, structural, and luminescence properties. *Appl. Surf. Sci.*, 510, 145437, 2020.

27. Liu, S., Cai, Y., Zhao, X., Liang, Y., Zheng, M., Hu, H., Dong, H., Jiang, S., Liu, Y., Xiao, Y., Sulfur-doped nanoporous carbon spheres with ultrahigh specific surface area and high electrochemical activity for supercapacitor. *J. Power Sources*, 360, 373–382, 2017.
28. Cai, J., Niu, H., Li, Z., Du, Y., Cizek, P., Xie, Z., Xiong, H., Lin, T., High-performance supercapacitor electrode materials from cellulose-derived carbon nanofibers. *ACS Appl. Mater. Interfaces*, 7, 27, 14946–14953, 2015.
29. Liu, B., Liu, Y., Chen, H., Yang, M., Li, H., Oxygen and nitrogen co-doped porous carbon nanosheets derived from Perilla frutescens for high volumetric performance supercapacitors. *J. Power Sources*, 341, 309–317, 2017.
30. Wei, X., Wei, J.S., Li, Y., Zou, H., Robust hierarchically interconnected porous carbons derived from discarded Rhus typhina fruits for ultrahigh capacitive performance supercapacitors. *J. Power Sources*, 414, 13–23, 2019.
31. Subramanian, D., Raju, G., Palanivel, B., Al-Zaqri, N., Hossain, M.S., Exploration of O-N-S heteroatom self-doped mesoporous activated carbon derived from Datura stramonium seed pods as a potential electrode for supercapacitor application. *Ionics (Kiel)*, 28, 5, 2363–2375, 2022.
32. Yu, P., Liang, Y., Dong, H., Hu, H., Liu, S., Peng, L., Zheng, M., Xiao, Y., Liu, Y., Rational synthesis of highly porous carbon from waste bagasse for advanced supercapacitor application. *ACS Sustain. Chem. Eng.*, 6, 11, 15325–15332, 2018.
33. Zhang, F., Liu, T., Zhang, J., Cui, E., Yue, L., Jiang, R., Hou, G., The potassium hydroxide-urea synergy in improving the capacitive energy-storage performance of agar-derived carbon aerogels. *Carbon N. Y.*, 147, 451–459, 2019.
34. Priyadharshini, M., Pazhanivel, T., Bharathi, G., Carbon quantum dot incorporated nickel pyrophosphate as alternate cathode for supercapacitors. *ChemistrySelect*, 5, 8, 2643–2652, 2020.
35. Rawat, S., Mishra, R.K., Bhaskar, T., Biomass derived functional carbon materials for supercapacitor applications. *Chemosphere*, 286, 131961, 2022.
36. Wei, J., Tu, C., Yuan, G., Liu, Y., Bi, D., Xiao, L., Lu, J., Theng, B.K.G., Wang, H., Zhang, L., Zhang, X., Assessing the effect of pyrolysis temperature on the molecular properties and copper sorption capacity of a halophyte biochar. *Environ. Pollut.*, 251, 56–65, 2019.
37. Funke, A. and Ziegler, F., Hydrothermal carbonization of biomass: A summary and discussion of chemical mechanisms for process engineering. *Biofuels Bioprod. Biorefin.*, 4, 2, 160–177, 2010.
38. Nguila Inari, G., Petrissans, M., Gerardin, P., Chemical reactivity of heat-treated wood. *Wood Sci. Technol.*, 41, 2, 157–168, 2007.
39. Prabakar, D., Manimudi, V.T., Suvetha K, S., Sampath, S., Mahapatra, D.M., Rajendran, K., Pugazhendhi, A., Advanced biohydrogen production using pretreated industrial waste: Outlook and prospects. *Renew. Sustain. Energy Rev.*, 96, 306–324, 2018.
40. Laurichesse, S. and Avérous, L., Chemical modification of lignins: Towards biobased polymers. *Prog. Polym. Sci.*, 39, 7, 1266–1290, 2014.
41. Wang, C., Wu, D., Wang, H., Gao, Z., Xu, F., Jiang, K., A green and scalable route to yield porous carbon sheets from biomass for supercapacitors with high capacity. *J. Mater. Chem. A*, 6, 3, 1244–1254, 2018.
42. Zheng, L., Wang, S., Yang, Y., Fu, X., Jiang, T., Yang, J., Ammonium nitrate-assisted synthesis of nitrogen/sulfur-codoped hierarchically porous carbons derived from ginkgo leaf for supercapacitors. *ACS Omega*, 4, 3, 5904–5914, 2019.
43. Panmand, R.P., Patil, P., Sethi, Y., Kadam, S.R., Kulkarni, M.V., Gosavi, S.W., Munirathnam, N.R., Kale, B.B., Unique perforated graphene derived from Bougainvillea flowers for high-power supercapacitors: A green approach. *Nanoscale*, 9, 14, 4801–4809, 2017.
44. Zhao, J., Li, Y., Wang, G., Wei, T., Liu, Z., Cheng, K., Ye, K., Zhu, K., Cao, D., Fan, Z., Enabling high-volumetric-energy-density supercapacitors: Designing open, low-tortuosity

heteroatom-doped porous carbon-tube bundle electrodes. *J. Mater. Chem. A*, 5, 44, 23085–23093, 2017.

45. Bi, Z., Kong, Q., Cao, Y., Sun, G., Su, F., Wei, X., Li, X., Ahmad, A., Xie, L., Chen, C.M., Biomass-derived porous carbon materials with different dimensions for supercapacitor electrodes: A review. *J. Mater. Chem. A*, 7, 27, 16028–16045, 2019.
46. Wu, Z.Y., Liang, H.W., Chen, L.F., Hu, B.C., Yu, S.H., Bacterial cellulose: A robust platform for design of three dimensional carbon-based functional nanomaterials. *Acc. Chem. Res.*, 49, 1, 96–105, 2016.
47. Lee, K.S., Park, C.W., Phiri, I., Ko, J.M., New design for Polyaniline@multiwalled carbon nanotubes composites with bacteria doping for supercapacitor electrodes. *Polym. (Guildf)*, 210, 123014, 2020.
48. Hong, X., Hui, K.S., Zeng, Z., Hui, K.N., Zhang, L., Mo, M., Li, M., Hierarchical nitrogen-doped porous carbon with high surface area derived from endothelium corneum gigeriae galli for high-performance supercapacitor. *Electrochim. Acta*, 130, 464–469, 2014.
49. Tarimo, D.J., Oyedotun, K.O., Sylla, N.F., Mirghni, A.A., Ndiaye, N.M., Manyala, N., Waste chicken bone-derived porous carbon materials as high performance electrode for supercapacitor applications. *J. Energy Storage*, 51, 104378, 2022.
50. Liu, Y., Chen, J., Cui, B., Yin, P., Zhang, C., Design and preparation of biomass-derived carbon materials for supercapacitors: A review. *C*, 4, 4, 53, 2018.

Index

Printed and bound by CPI Group (UK) Ltd, Croydon, CR0 4YY
17/09/2026
14964281-0002